How to Build Android Apps with Kotlin

Second Edition

A practical guide to developing, testing, and publishing your first Android apps

Alex Forrester
Eran Boudjnah
Alexandru Dumbravan
Jomar Tigcal

BIRMINGHAM—MUMBAI

How to Build Android Apps with Kotlin
Second Edition

Copyright © 2023 Packt Publishing

All rights reserved. No part of this book may be reproduced, stored in a retrieval system, or transmitted in any form or by any means, without the prior written permission of the publisher, except in the case of brief quotations embedded in critical articles or reviews.

Every effort has been made in the preparation of this book to ensure the accuracy of the information presented. However, the information contained in this book is sold without warranty, either express or implied. Neither the authors, nor Packt Publishing or its dealers and distributors, will be held liable for any damages caused or alleged to have been caused directly or indirectly by this book.

Packt Publishing has endeavored to provide trademark information about all of the companies and products mentioned in this book by the appropriate use of capitals. However, Packt Publishing cannot guarantee the accuracy of this information.

Group Product Manager: Rohit Rajkumar
Publishing Product Manager: Nitin Nainani
Content Development Editor: Abhishek Jadhav
Technical Editor: Simran Ali
Copy Editor: Safis Editing
Project Coordinator: Aishwarya Mohan
Proofreader: Safis Editing
Indexer: Manju Arasan
Production Designer: Joshua Misquitta
Marketing Coordinator: Nivedita Pandey

First published: February 2021
Second edition: May 2023
Production reference: 1210423

Published by Packt Publishing Ltd.
Livery Place
35 Livery Street
Birmingham
B3 2PB, UK.

ISBN 978-1-83763-493-4

www.packtpub.com

Dedicated to my wife Angela and daughter Catherine for all their love and support.

– Alex Forrester

To my endlessly supportive wife, Lea, for always being there for support. To my parents, Amos and Tirtsa, for spending some of their best years teaching and encouraging me. I could not have asked for better people in my life, so thank you all.

– Eran Boudjnah

Dedicated to Niki for her constant support.

– Alexandru Dumbravan

To my loving wife, Celine, for her support and encouragement. To my parents for all their sacrifices and for raising me well.

– Jomar Tigcal

Contributors

About the authors

Alex Forrester is an experienced software developer with more than 20 years of experience in mobile and web development and content management systems. He has worked with Android since 2010, creating flagship apps for blue-chip companies across a broad range of industries at Sky, The Automobile Association, HSBC, Discovery Channel, and O2. Alex lives in Hertfordshire with his wife and daughter. When he's not developing, he likes rugby and running in the Chiltern hills.

Eran Boudjnah is a developer with over 20 years of experience in developing desktop applications, websites, interactive attractions, and mobile applications. He has worked with Android since 2011, developing apps and leading mobile teams for a wide range of clients, from start-ups (JustEat and Plume Design) to large-scale companies (Sky and HSBC) and conglomerates. He is passionate about board games (with a modest collection of a few hundred games) and has a Transformers collection he's quite proud of. Eran lives in Brentwood, England, with Lea, his wife.

Alexandru Dumbravan is an Android developer with more than 10 years of experience building Android apps, focusing on fintech applications since 2016 when he moved to London. In his spare time, Alex enjoys video games, movies, and the occasional gym visit.

Jomar Tigcal is an Android developer with over 14 years of experience in mobile and software development. He has worked on various stages of Android app development for small start-ups and large companies since 2012. Jomar has also given talks and conducted training and workshops on Android. In his free time, he likes running and reading. He lives in Vancouver, BC, Canada, with his wife Celine.

About the reviewers

Ed Holloway-George is an Android developer and Google Developer Expert originally from Oxford, England, but currently living in Nottingham. An Android developer for just over 10 years, Ed now works for ASOS as a lead developer having previously worked on well-known applications such as National Trust, My Oxfam, Snoop, Carling Tap, and more.

In his spare time, Ed can be found speaking at conferences, writing blog posts, and sharing pictures of his dog.

Guruprasad Bagade is a senior developer who has led teams and has over a decade of experience in mobile and software development. He has witnessed changes in Android development from Java to Kotlin with the most recent framework libraries. He primarily worked in the banking domain for Barclays and JP Morgan clients. He has hired everyone from freshers to experienced developers for organizations and helped set up teams while also publishing knowledge articles on Android on the internal portals of the various organizations where he has worked.

He has published technical research papers at the **Institute of Electrical and Electronics Engineers (IEEE)** and international and national conferences. He also contributes to open source projects. In his spare time, he keeps himself up to date with the latest technologies.

Table of Contents

Preface — xv

Part 1: Android Foundation

1

Creating Your First App — 3

Technical requirements	4
Creating an Android project with Android Studio	4
Exercise 1.01 – creating an Android Studio project for your app	4
Setting up a virtual device and running your app	8
Exercise 1.02 – setting up a virtual device and running your app on it	9
The Android manifest	16
Exercise 1.03 – configuring the Android manifest internet permission	18
Using Gradle to build, configure, and manage app dependencies	22
The project-level build.gradle file	22
The app-level build.gradle file	23
Exercise 1.04 – exploring how Material Design is used to theme an app	27
Android application structure	30
Exercise 1.05 – adding interactive UI elements to display a bespoke greeting to the user	38
Accessing Views in layout files	46
Further input validation	46
Activity 1.01 – producing an app to create RGB colors	47
Summary	49

2

Building User Screen Flows — 51

Technical requirements	51
The Activity lifecycle	52
Exercise 2.01 – logging the Activity Callbacks	55

Saving and restoring the Activity state	63	Exercise 2.05 – retrieving a result from an Activity	84
Exercise 2.02 – saving and restoring the state in layouts	63	**Intents, Tasks, and Launch Modes**	95
Exercise 2.03 – saving and restoring the state with Callbacks	71	Exercise 2.06 – setting the Launch Mode of an Activity	96
Activity interaction with Intents	77	Activity 2.01 – creating a login form	101
Exercise 2.04 – an introduction to Intents	78	**Summary**	102

3

Developing the UI with Fragments — 103

Technical requirements	104	Exercise 3.02 – adding fragments statically to an activity	117
The fragment lifecycle	104	**Static fragments and dual-pane layouts**	127
onAttach	105	Exercise 3.03 – dual-pane layouts with static fragments	128
onCreate	106		
onCreateView	106	**Dynamic fragments**	145
onViewCreated	106	Exercise 3.04 – adding fragments dynamically to an activity	146
onActivityCreated	106		
onStart	106	**Jetpack Navigation**	150
onResume	106	Exercise 3.05 – adding a Jetpack navigation graph	151
onPause	107		
onStop	107	Activity 3.01 – creating a quiz on the planets	156
onDestroyView	107		
onDestroy	107	**Summary**	160
onDetach	107		
Exercise 3.01 – adding a basic fragment and the fragment lifecycle	108		

4

Building App Navigation — 161

Technical requirements	162	Exercise 4.01 – creating an App with a navigation drawer	164
Navigation overview	162		
Navigation drawer	162	**Bottom navigation**	181

Exercise 4.02 – adding bottom navigation to your app	181	Activity 4.01 – building primary and secondary app navigation	198
Tabbed navigation	191	Summary	199
Exercise 4.03 – using tabs for app navigation	191		

Part 2: Displaying Network Calls

5

Essential Libraries: Retrofit, Moshi, and Glide 203

Technical requirements	203	Loading images from a remote URL	218
Introducing REST, API, JSON, and XML	203	Exercise 5.03 – loading the image from the obtained URL	221
Fetching data from a network endpoint	205	Activity 5.01 – displaying the current weather	228
Exercise 5.01 – reading data from an API	208	Summary	229
Parsing a JSON response	213		
Exercise 5.02 – extracting the image URL from the API response	215		

6

Adding and Interacting with RecyclerView 231

Technical requirements	232	Exercise 6.03 – responding to clicks	253
Adding RecyclerView to our layout	232	Supporting different Item types	256
Exercise 6.01 – adding an empty RecyclerView to your main activity	233	Exercise 6.04 – adding titles to RecyclerView	260
Populating RecyclerView	235	Swiping to remove Items	266
Exercise 6.02 – populating your RecyclerView	243	Exercise 6.05 – adding swipe to delete functionality	269
Responding to clicks in RecyclerView	251	Adding items interactively	272
		Exercise 6.06 – implementing an Add A Cat button	274

Activity 6.01 – managing a list of Items 277   Summary 279

7

Android Permissions and Google Maps 281

Technical requirements	282	Map clicks and custom markers	299
Requesting permission from the user	282	Exercise 7.03 – adding a custom marker where the map was clicked	303
Exercise 7.01 – requesting the location permission	286	Activity 7.01 – creating an app to find the location of a parked car	308
Showing a map of the user's location	292	Summary	309
Exercise 7.02 – obtaining the user's current location	295		

8

Services, WorkManager, and Notifications 311

Technical requirements	312	Exercise 8.02 – tracking your SCA's work with a Foreground Service	326
Starting a background task using WorkManager	312	Activity 8.01 – reminder to drink water	334
Exercise 8.01 – executing background work with the WorkManager class	316	Summary	335
Background operations noticeable to the user – using a Foreground Service	321		

9

Building User Interfaces Using Jetpack Compose 337

Technical requirements	337	Handling user actions	345
What is Jetpack Compose?	337	Exercise 9.02 – handling user inputs	347
Exercise 9.01 – first Compose screen	342	Theming in Compose	350

Part 3: Testing and Code Structure

10

Unit Tests and Integration Tests with JUnit, Mockito, and Espresso — 371

Technical requirements	372	Exercise 10.02 – double integration	403
Types of testing	372	UI tests	412
JUnit	374	Testing in Jetpack Compose	417
Android Studio testing tips	382	Exercise 10.03 – random waiting times	420
Mockito	386	TDD	431
Exercise 10.01 – testing the sum of numbers	392	Exercise 10.04 – using TDD to calculate the sum of numbers	433
Integration tests	396	Activity 10.01 – developing with TDD	436
Robolectric	396	Summary	438
Espresso	401		

11

Android Architecture Components — 441

Technical requirements	442	Room	463
Android components background	442	Entities	464
		DAO	467
ViewModel	443	Setting up the database	469
Exercise 11.01 – shared ViewModel	445	Third-party frameworks	473
Data streams	453	Exercise 11.03 – making a little room	475
LiveData	454	Activity 11.01 – a shopping notes app	480
Additional data streams	461	Summary	482

12

Persisting Data 483

Technical requirements	483	FileProvider	499
Preferences and DataStore	484	The Storage Access Framework (SAF)	500
SharedPreferences	484	Asset files	500
Exercise 12.01 – wrapping SharedPreferences	485	Exercise 12.03 – copying files	501
DataStore	489	Scoped storage	508
Exercise 12.02 – Preference DataStore	491	Camera and media storage	509
Files	495	Exercise 12.04 – taking photos	511
Internal storage	496	Activity 12.01 – dog downloader	518
External storage	498	Summary	521

13

Dependency Injection with Dagger, Hilt, and Koin 523

Technical requirements	523	Scopes	536
The necessity of dependency injection	524	Subcomponents	537
		Exercise 13.02 – Dagger injection	538
Manual DI	524	Hilt	543
Exercise 13.01 – manual injection	527	Exercise 13.03 – Hilt injection	547
Dagger 2	531	Koin	550
Consumers	532	Exercise 13.04 – Koin injection	553
Providers	533	Activity 13.01 – injected repositories	556
Connectors	534	Summary	557
Qualifiers	535		

Part 4: Polishing and Publishing an App

14

Coroutines and Flow   561

Technical requirements	562	Using Flow on Android	573
Using Coroutines on Android	562	Collecting Flows on Android	574
Creating coroutines	563	Creating Flows with Flow Builders	576
Adding coroutines to your project	564	Using operators with Flows	577
Exercise 14.01 – using coroutines in an Android app	565	Exercise 14.03 – using Flow in an Android application	578
		Activity 14.01 – creating a TV Guide app	581
Transforming LiveData	570	Summary	584
Exercise 14.02 – LiveData transformations	571		

15

Architecture Patterns   585

Technical requirements	586	Exercise 15.02 – using Repository with Room in an Android project	595
Getting started with MVVM	586		
Binding data on Android with data binding	588	Using WorkManager	598
Exercise 15.01– using data binding in an Android project	590	Exercise 15.03 – adding WorkManager to an Android Project	598
		Activity 15.01 – revisiting the TV Guide app	601
Using Retrofit and Moshi	593	Summary	603
Implementing the Repository pattern	593		

16

Animations and Transitions with CoordinatorLayout and MotionLayout   605

Technical requirements	606	Adding activity transitions through XML	606
Activity transitions	606	Adding activity transitions through code	607

Starting an activity with an activity transition	608	Adding MotionLayout	616
Exercise 16.01 – creating activity transitions in an app	608	Creating animations with MotionLayout	617
Adding a shared element transition	610	Exercise 16.03 – adding animations with MotionLayout	618
Starting an activity with the shared element transition	611	The Motion Editor	620
Exercise 16.02 – creating the shared element transition	612	Debugging MotionLayout	625
		Modifying the MotionLayout path	627
Animations with CoordinatorLayout	**614**	Exercise 16.04 – modifying the animation path with keyframes	629
Animations with MotionLayout	**616**	Activity 16.01 – Password Generator	636
		Summary	**638**

17

Launching Your App on Google Play — 639

Preparing your apps for release	**640**	**Uploading an app to Google Play**	**653**
Versioning apps	640	Creating a store listing	653
Creating a keystore	641	Preparing the release	654
Exercise 17.01 – creating a keystore in Android Studio	641	Rolling out a release	656
Storing the keystore and passwords	644	**Managing app releases**	**657**
Signing your apps for release	646	Release tracks	657
Exercise 17.02 – creating a signed APK	646	Staged rollouts	658
Android app bundle	649	Managed publishing	660
Exercise 17.03 – creating a signed app bundle	650	Activity 17.01 – publishing an app	662
App signing by Google Play	652	**Summary**	**663**
Creating a developer account	**652**		

Index — 665

Other Books You May Enjoy — 674

Preface

Android has ruled the app market for the past decade, and developers are increasingly looking to start building their own Android apps. *How to Build Android Apps with Kotlin* starts with the building blocks of Android development, teaching you how to use Android Studio, the **integrated development environment** (IDE) for Android, with the Kotlin programming language for app development.

Then, you'll learn how to create apps and run them on virtual devices using guided exercises. You'll cover the fundamentals of Android development, from structuring an app to building out the UI with activities, fragments, and various navigation patterns. Progressing through the chapters, you'll delve into Android's RecyclerView to make the most of displaying lists of data and become comfortable with fetching data from a web service and handling images.

You'll then learn about mapping, location services, and the permissions model before working with notifications and how to persist data. Next, you'll build user interfaces using Jetpack Compose. Moving on, you'll get to grips with testing, covering the full spectrum of the test pyramid. You'll also learn how **Android Architecture Components** (**AAC**) is used to cleanly structure your code and explore various architecture patterns and the benefits of dependency injection.

Coroutines and the Flow API are covered for asynchronous programming. The focus then returns to the UI, demonstrating how to add motion and transitions when users interact with your apps. Toward the end, you'll build an interesting app to retrieve and display popular movies from a movie database, and then see how to publish your apps on Google Play.

By the end of this book, you'll have the skills and confidence needed to build fully-fledged Android apps using Kotlin.

Who this book is for

If you want to build your own Android apps using Kotlin but are unsure of how to begin, then this book is for you. A basic understanding of the Kotlin programming language will help you grasp the topics covered in this book more quickly.

What this book covers

Chapter 1, *Creating Your First App*, shows how to use Android Studio to build your first Android app. Here, you will create an Android Studio project, understand what it's made up of, and explore the tools necessary for building and deploying an app on a virtual device. You will also learn about the structure of an Android app.

Chapter 2, *Building User Screen Flows*, dives into the Android ecosystem and the building blocks of an Android application. Concepts such as activities and their lifecycle, intents, and tasks will be introduced, as well as restoring the state and passing data between screens or activities.

Chapter 3, *Developing the UI with Fragments*, teaches you the fundamentals of using fragments for the user interface of an Android application. You will learn how to use fragments in multiple ways to build application layouts for phones and tablets, including using the Jetpack Navigation component.

Chapter 4, *Building App Navigation*, goes through the different types of navigation in an application. You will learn about navigation drawers with sliding layouts, bottom navigation, and tabbed navigation.

Chapter 5, *Essential Libraries: Retrofit, Moshi, and Glide*, gives you an insight into how to build apps that fetch data from a remote data source with the use of the Retrofit library and the Moshi library to convert data into Kotlin objects. You will also learn about the Glide library, which loads remote images into your app.

Chapter 6, *Adding and Interacting with RecyclerView*, introduces the concept of building lists and displaying them with the help of the RecyclerView widget.

Chapter 7, *Android Permissions and Google Maps*, presents the concept of permissions and how to request them from the user in order for your app to execute specific tasks, as well as introducing you to the Maps API.

Chapter 8, *Services, WorkManager, and Notifications*, details the concept of background work in an Android app and how you can have your app execute certain tasks in a way that is invisible to the user, as well as covering how to show a notification of this work.

Chapter 9, *Building User Interfaces Using Jetpack Compose*, shows how Jetpack Compose works, how to apply styles and themes, and how to use Jetpack Compose in projects started with layout files.

Chapter 10, *Unit Tests and Integration Tests with JUnit, Mockito, and Espresso*, teaches you about the different types of tests for an Android application, what frameworks are used for each type of test, and the concept of test-driven development.

Chapter 11, *Android Architecture Components*, provides an insight into components from the Android Jetpack libraries, such as ViewModel, which will help separate the business logic from the user interface code. We will then look at how we can use observable data streams such as LiveData to deliver data to the user interface. Finally, we will look at the Room library to analyze how we can persist data.

Chapter 12, *Persisting Data*, shows you the various ways to store data on a device, from SharedPreferences to files. The Repository concept will also be introduced, giving you an idea of how to structure your app in different layers.

Chapter 13, *Dependency Injection with Dagger, Hilt, and Koin*, explains the concept of dependency injection and the benefits it provides to an application. Frameworks such as Dagger, Hilt, and Koin are introduced to help you manage your dependencies.

Chapter 14, *Coroutines and Flow*, introduces you to doing background operations and data manipulations with coroutines and Flow. You'll also learn about manipulating and displaying data using Flow operators and LiveData transformation.

Chapter 15, *Architecture Patterns*, explains the architecture patterns you can use to structure your Android projects to separate them into different components with distinct functionality. These make it easier for you to develop, test, and maintain your code.

Chapter 16, *Animations and Transitions with CoordinatorLayout and MotionLayout*, discusses how to enhance your apps with animations and transitions with `CoordinatorLayout` and `MotionLayout`.

Chapter 17, *Launching Your App on Google Play*, concludes this book by showing you how to publish your apps on Google Play: from preparing a release to creating a Google Play Developer account, and finally launching your app.

To get the most out of this book

Each great journey begins with a humble step. Before we can do awesome things in Android, we need to be prepared with a productive environment. In this section, we will see how to do that.

Minimum hardware requirements

For an optimal learning experience, we recommend the following hardware configuration:

- **Processor**: Intel Core i5 or equivalent or higher
- **Memory**: 8 GB RAM or more
- **Storage**: 8 GB available space minimum

Software requirements

You'll also need the following software installed in advance:

- **OS**: 64-bit Windows 8/10/11, macOS, or 64-bit Linux
- Android Studio Electric Eel or higher

Installation and setup

Before you start this book, you will need to install Android Studio Electric Eel (or higher), which is the software you will be using throughout the chapters. You can download Android Studio from `https://developer.android.com/studio`.

On macOS, launch the DMG file and drag and drop Android Studio into the `Applications` folder. Once this is done, open Android Studio. On Windows, launch the EXE file. If you're using Linux, unpack the ZIP file into your preferred location. Open your Terminal and navigate to the `android-studio/bin/` directory and execute `studio.sh`.

Next, the **Data Sharing** dialog will pop up; click either the **Send usage statistics to Google** button or the **Don't send** button to disable sending anonymous usage data to Google:

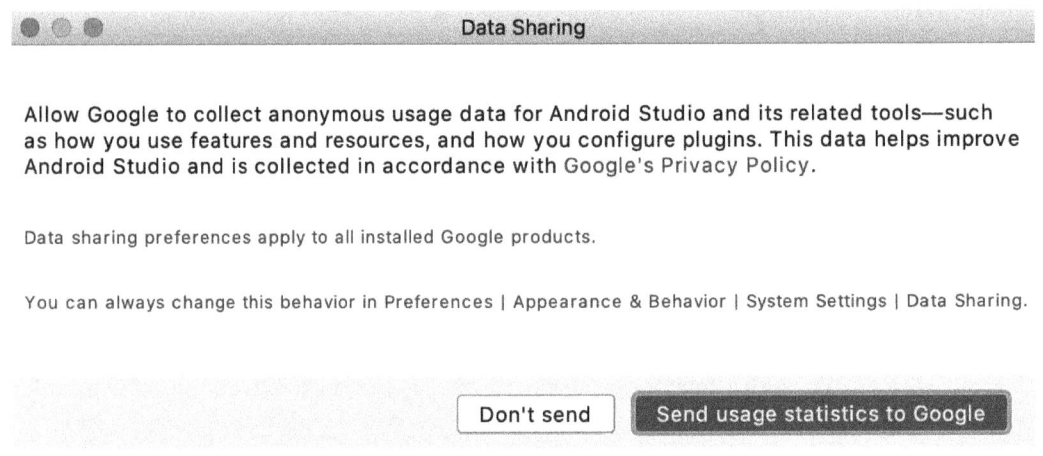

The Data Sharing dialog

In the **Welcome** dialog, click the **Next** button to start the setup:

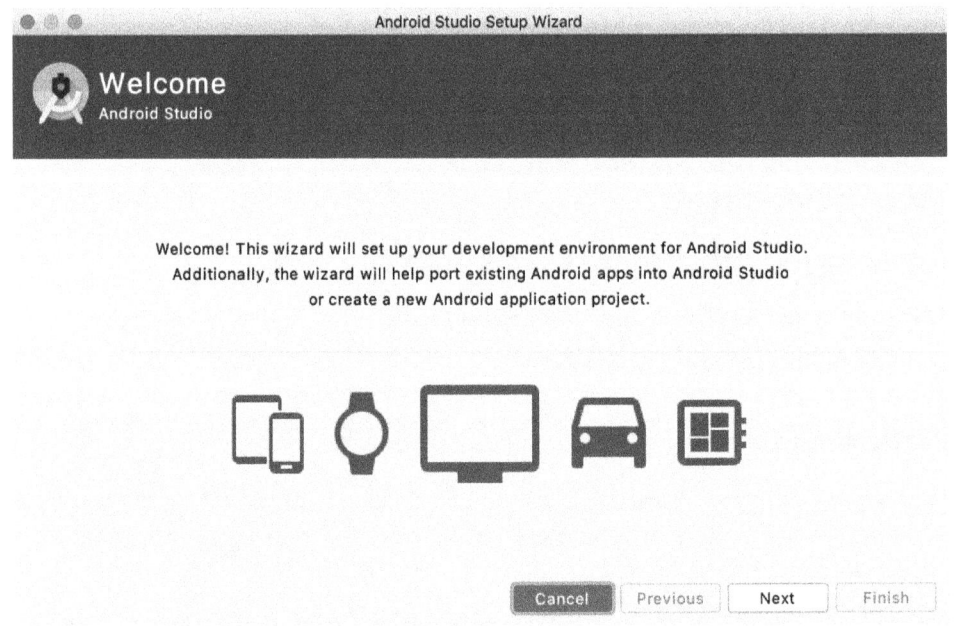

The Welcome dialog

In the **Install Type** dialog, select **Standard** to install the recommended settings. Then, click the **Next** button:

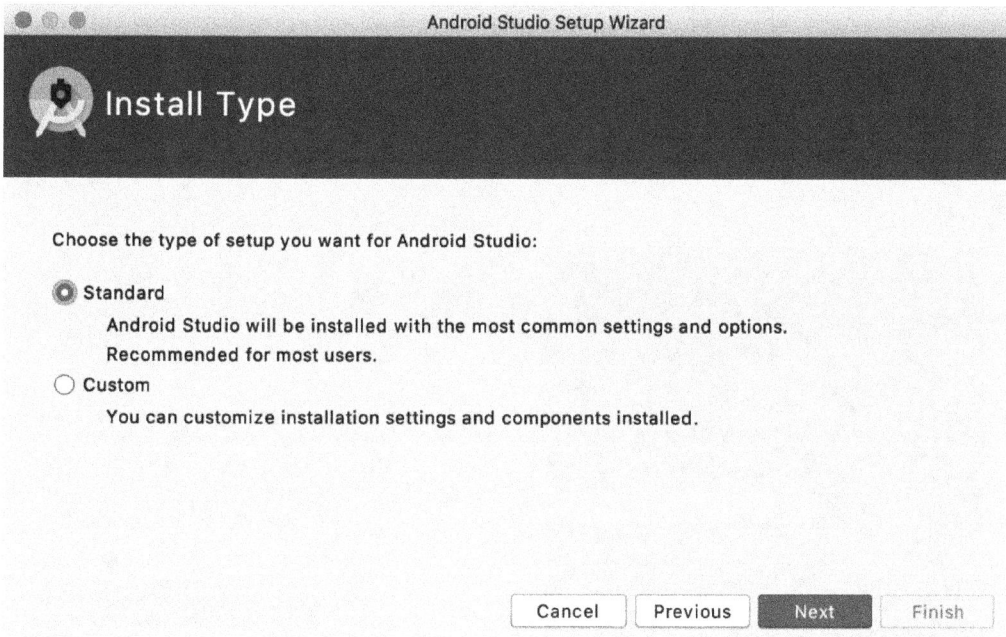

The Install Type dialog

In the **Select UI Theme** dialog, choose your preferred IDE theme—either **Light** or **Darcula** (dark theme)—then click the **Next** button:

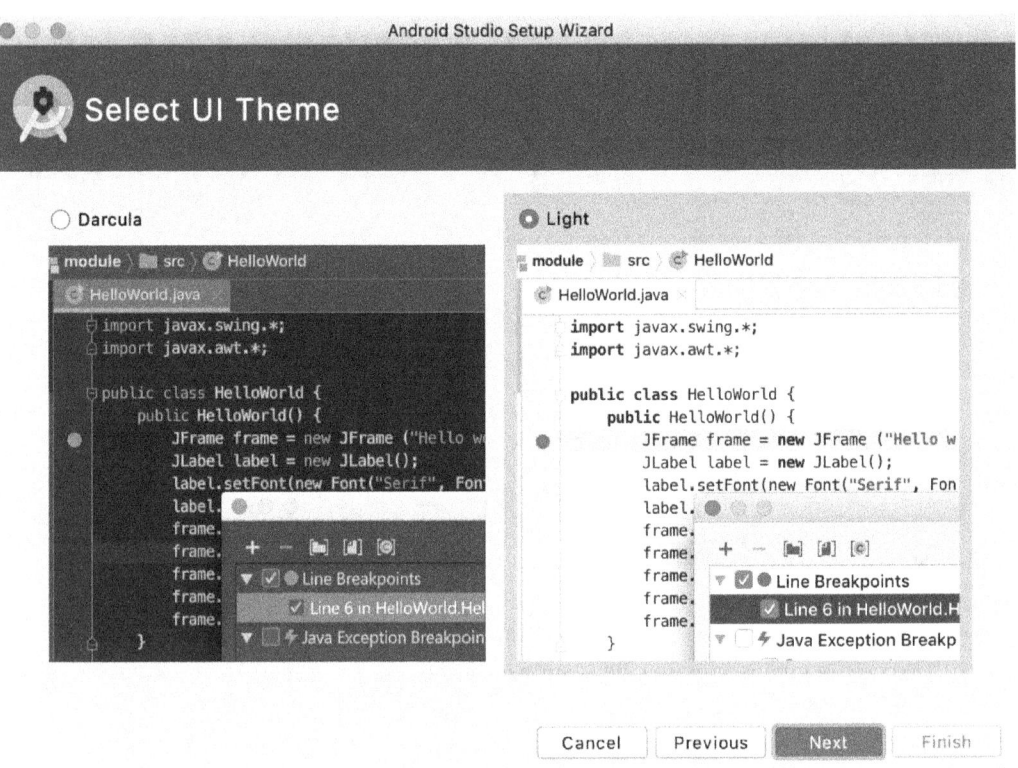

The Select UI Theme dialog

In the **Verify Settings** dialog, review your settings and then click the **Finish** button. The setup wizard downloads and installs additional components, including the Android SDK:

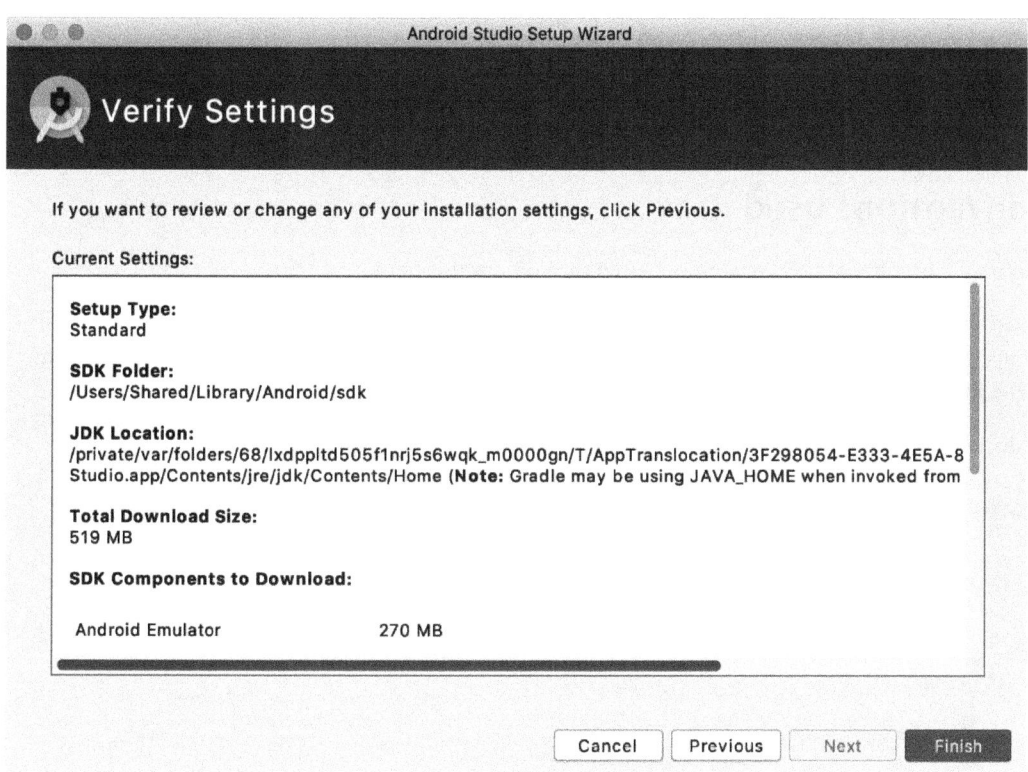

The Verify Settings dialog

Once the download finishes, you can click the **Finish** button. You are now ready to create your Android project.

If you are using the digital version of this book, we advise you to type the code yourself or access the code from the book's GitHub repository (a link is available in the next section). Doing so will help you avoid any potential errors related to the copying and pasting of code.

Download the example code files

You can download the example code files for this book from GitHub at `https://github.com/PacktPublishing/How-to-Build-Android-Apps-with-Kotlin-Second-Edition`. If there's an update to the code, it will be updated in the GitHub repository.

We also have other code bundles from our rich catalog of books and videos available at `https://github.com/PacktPublishing/`. Check them out!

Download the color images

We also provide a PDF file that has color images of the screenshots and diagrams used in this book. You can download it here: https://packt.link/vnOCn.

Conventions used

There are a number of text conventions used throughout this book.

`Code in text`: Indicates code words in text, database table names, folder names, filenames, file extensions, pathnames, dummy URLs, user input, and Twitter handles. Here is an example: "You can find it in the main project window under `MyApplication` | `app` | `src` | `main`."

A block of code is set as follows:

```
<resources>
    <string name="app_name">My Application</string>
</resources>
```

When we wish to draw your attention to a particular part of a code block, the relevant lines or items are set in bold:

```
<?xml version="1.0" encoding="utf-8"?>
<resources>
    <string name="app_name">My Application</string>
    <string name="first_name_text">First name:</string>
    <string name="last_name_text">Last name:</string>
</resources>
```

Bold: Indicates a new term, an important word, or words that you see onscreen. For instance, words in menus or dialog boxes appear in **bold**. Here is an example: "Click **Finish** and your virtual device will be created."

> **Tips or important notes**
> Appear like this.

Get in touch

Feedback from our readers is always welcome.

General feedback: If you have questions about any aspect of this book, email us at `customercare@packtpub.com` and mention the book title in the subject of your message.

Errata: Although we have taken every care to ensure the accuracy of our content, mistakes do happen. If you have found a mistake in this book, we would be grateful if you would report this to us. Please visit `www.packtpub.com/support/errata` and fill in the form.

Piracy: If you come across any illegal copies of our works in any form on the internet, we would be grateful if you would provide us with the location address or website name. Please contact us at `copyright@packt.com` with a link to the material.

If you are interested in becoming an author: If there is a topic that you have expertise in and you are interested in either writing or contributing to a book, please visit `authors.packtpub.com`.

Share your thoughts

Once you've read *How to Build Android Apps with Kotlin, Second Edition*, we'd love to hear your thoughts! Scan the QR code below to go straight to the Amazon review page for this book and share your feedback.

`https://www.amazon.in/review/create-review/error?asin=1837634939`

Your review is important to us and the tech community and will help us make sure we're delivering excellent quality content.

Download a free PDF copy of this book

Thanks for purchasing this book!

Do you like to read on the go but are unable to carry your print books everywhere?

Is your eBook purchase not compatible with the device of your choice?

Don't worry, now with every Packt book you get a DRM-free PDF version of that book at no cost.

Read anywhere, any place, on any device. Search, copy, and paste code from your favorite technical books directly into your application.

The perks don't stop there, you can get exclusive access to discounts, newsletters, and great free content in your inbox daily

Follow these simple steps to get the benefits:

1. Scan the QR code or visit the link below

```
https://packt.link/free-ebook/9781837634934
```

2. Submit your proof of purchase
3. That's it! We'll send your free PDF and other benefits to your email directly

Part 1: Android Foundation

This first part introduces the user to Android Studio, the **integrated development environment (IDE)** used for Android development, and then guides them through the building blocks of Android development. It's a comprehensive overview of the Android framework, working through guided exercises that reinforce the learning objectives so this knowledge can be retained.

We will cover the following chapters in this section:

- *Chapter 1, Creating Your First App*
- *Chapter 2, Building User Screen Flows*
- *Chapter 3, Developing the UI with Fragments*
- *Chapter 4, Building App Navigation*

1
Creating Your First App

This chapter is an introduction to Android, where you will set up your environment and focus on the fundamentals of Android development. By the end of this chapter, you will have gained the knowledge required to create an Android app from scratch and install it on a virtual or physical Android device.

You will be able to analyze and understand the importance of the `AndroidManifest.xml` file and use the Gradle build tool to configure your app and implement **user interface** (**UI**) elements from Material Design.

Android is the most widely used mobile phone operating system in the world, with over three billion active devices. This presents great opportunities to contribute and make an impact by learning Android and building apps that have a global reach. However, for a developer who is new to Android, there are many issues you must contend with in order to get started learning and becoming productive.

This book will address these issues. After learning the tooling and development environment, you will explore fundamental practices to build Android apps. We will cover a wide range of real-world development challenges faced by developers and explore various techniques to overcome them.

In this chapter, you will learn how to create a basic Android project and add features to it. You will be introduced to the comprehensive development environment of Android Studio and learn about the core areas of the software to enable you to work productively.

Android Studio provides all the tooling for application development but not the knowledge. This first chapter will guide you through using the software effectively to build an app and configure the most common areas of an Android project.

We will cover the following topics in the chapter:

- Creating an Android project with Android Studio
- Setting up a virtual device and running your app
- The Android Manifest
- Using Gradle to build, configure, and manage app dependencies
- Android application structure

Technical requirements

The complete code for all the exercises and the activity in this chapter is available on GitHub at https://packt.link/9611D

Creating an Android project with Android Studio

In order to be productive in terms of building Android apps, it is essential to become confident with how to use **Android Studio**. This is the official **integrated development environment** (**IDE**) for Android development, built on JetBrains' **IntelliJ IDEA IDE** and developed by the Android Studio team at Google. You will use it throughout this course to create apps and progressively add more advanced features.

The development of Android Studio has followed the development of the IntelliJ IDEA IDE. The fundamental features of an IDE are, of course, present, enabling you to optimize your code with suggestions, shortcuts, and standard refactoring. The programming language you will use throughout this course to create Android apps is Kotlin. Previously the standard language to create Android apps was Java.

Since Google I/O 2017 (the annual Google developer conference), this has been Google's preferred language for Android app development. What really sets Android Studio apart from other Android development environments is that **Kotlin** was created by JetBrains, the company that created IntelliJ IDEA, the software Android Studio is built on. Therefore, you can benefit from established and evolving first-class support for Kotlin.

Kotlin was created to address some of the shortcomings of Java in terms of verbosity, handling null types, and adding more functional programming techniques, amongst many other issues. As Kotlin has been the preferred language for Android development since 2017, taking over from Java, you will use it in this book.

Getting to grips and familiarizing yourself with Android Studio will enable you to feel confident working on and building Android apps. So, let's get started creating your first project.

> **Note**
> The installation and setup of Android Studio are covered in the *Preface*. Please ensure you have completed those steps before you continue.

Exercise 1.01 – creating an Android Studio project for your app

This is the starting point for creating a project structure your app will be built upon. The template-driven approach will enable you to create a basic project in a short timeframe while setting up the building blocks you can use to develop your app.

Creating an Android project with Android Studio   5

To complete this exercise, perform the following steps:

1. Upon opening Android Studio, you will see a window asking whether you want to create a new project or open an existing one. Select **Create New Project**.

2. Now, you'll enter a simple wizard-driven flow, which greatly simplifies the creation of your first Android project. The next screen you will see has a large number of options for the initial setup you'd like your app to have:

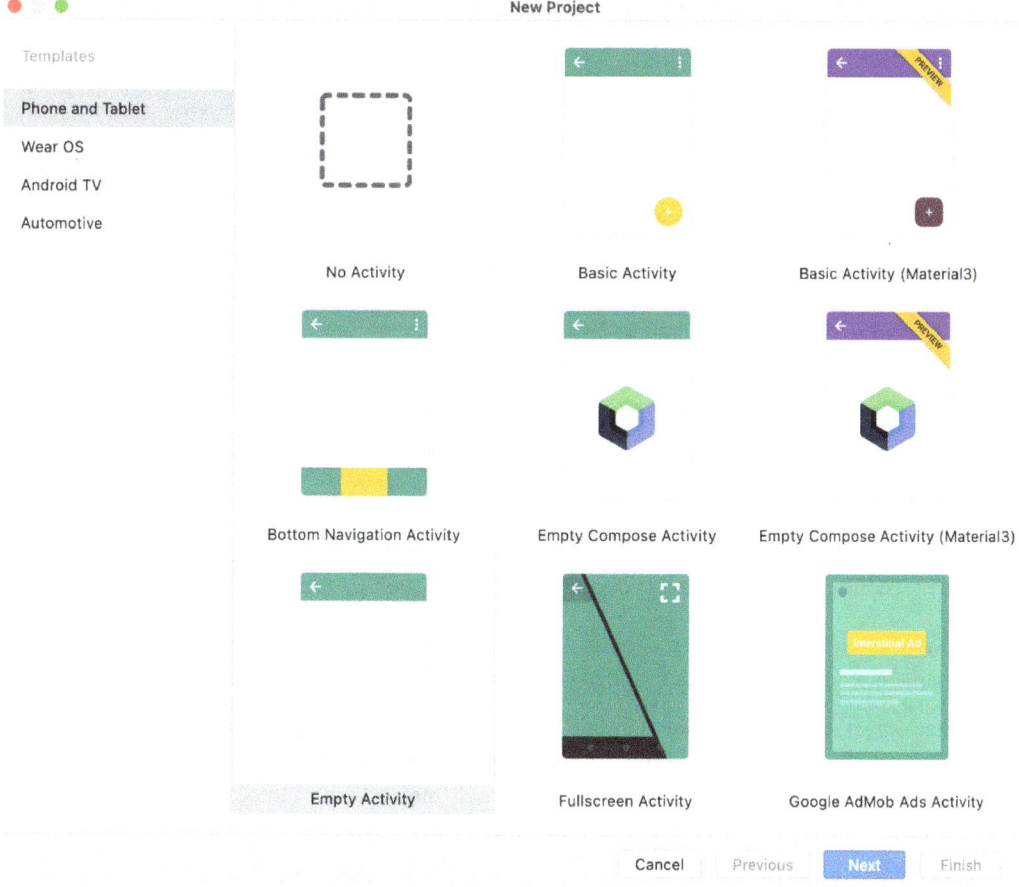

Figure 1.1 – Starting a project template for your app

3. Welcome to your first introduction to the Android development ecosystem. The word displayed in most of the project types is *Activity*. In Android, an Activity is a page or screen. The options you can choose from all create this initial screen differently.

The descriptions describe how the first screen of the app will look. These are templates to build your app with. Select **Empty Activity** from the template and click on **Next**.

The project configuration screen is as follows:

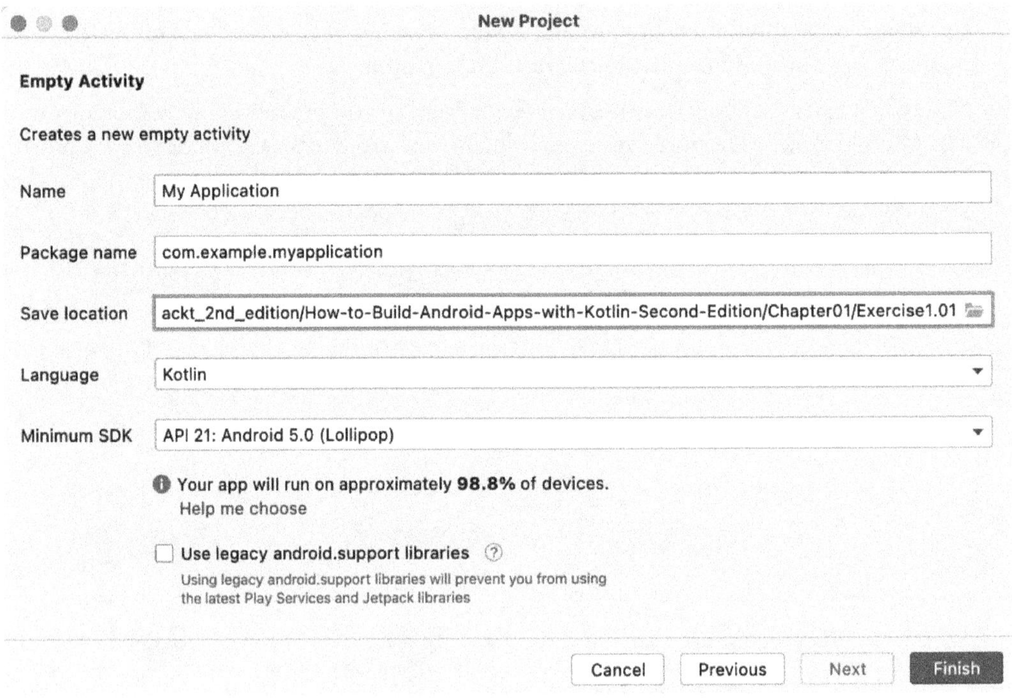

Figure 1.2 – Project configuration

4. The preceding screen configures your app. Let's go through all the options:

   - **Name**: Similar to the name of your Android project, this name will appear as the default name of your app when it's installed on a phone and visible on Google Play.

   - **Package name**: This uses the standard reverse domain name pattern to create a name. It will be used as an address identifier for source code and assets in your app. It is best to make this name as clear and descriptive and as closely aligned with the purpose of your app as possible. Therefore, it's probably best to change this to use one or more sub-domains (such as com.sample.shop.myshop). As shown in *Figure 1.2*, the **Name** value of the app (in lowercase with spaces removed) is appended to the domain.

   - **Save location**: This is the local folder on your machine where the app will initially be stored. This can be changed in the future, so you can probably keep the default or edit it to something different (such as Users/MyUser/android/projects). The default location will vary with the operating system you are using. By default, the project will be saved into a new folder with the name of the application with spaces removed. This results in a MyApplication project folder being created. Please change this to the Exercise or Activity that you are working on, so for this project, name the folder Exercise1.01.

- **Language**: **Kotlin** is Google's preferred language for Android app development.
- **Minimum SDK**: Depending on which version of Android Studio you download, the default might be the same as shown in *Figure 1.2* or a different version. Keep this the same. Most of Android's new features are made backward compatible, so your app will run fine on the vast majority of older devices. However, if you do want to target newer devices, you should consider raising the minimum API level. There is a **Help Me Choose** link to a dialog that explains the feature set that you have access to with a view to development on different versions of Android and the current percentage of devices worldwide running each Android version.
- **Use legacy android.support libraries**: Leave this unchecked. You will be using AndroidX libraries, which are the replacement for the support libraries that were designed to make features on newer versions of Android backward compatible with older versions, but it provides much more than this. It also contains new Android components called **Jetpack**, which, as the name suggests, *boosts* your Android development and provide a host of rich features you will want to use in your app, thereby simplifying common operations.

Once you have filled in all these details, select **Finish**. Your project will be built, and you will then be presented with the following screen or similar. You can immediately see the activity that has been created (MainActivity) in one tab and the layout used for the screen in the other tab (activity_main.xml). The application structure folders are in the left panel:

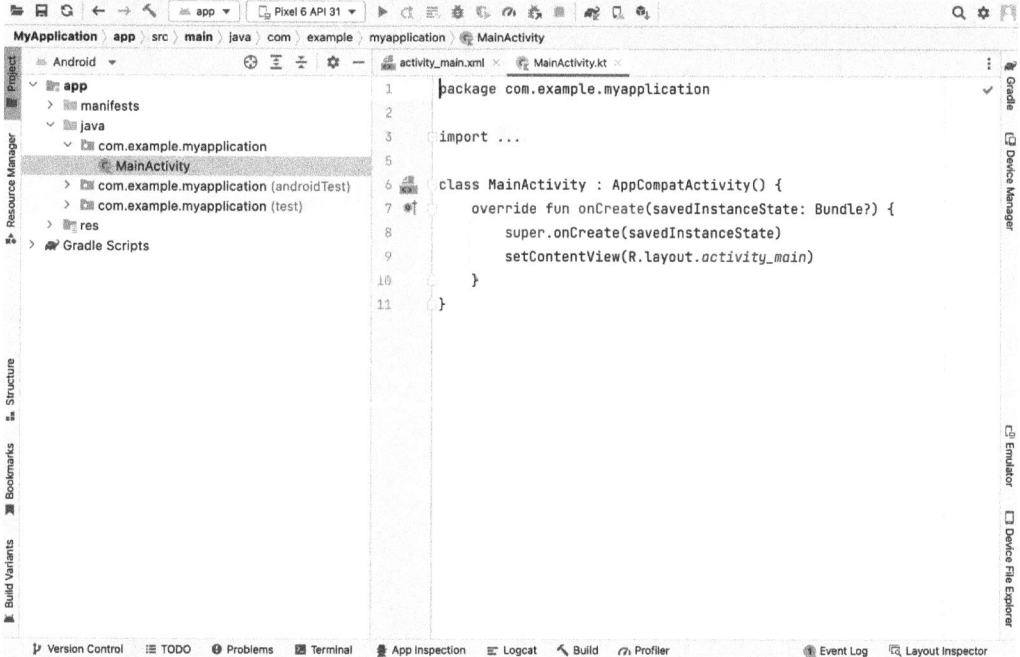

Figure 1.3 – Android Studio default project

In this exercise, you have gone through the steps to create your first Android app using Android Studio. This template-driven approach has shown you the core options you need to configure for your app.

In the next section, you will set up a virtual device and see your app run for the first time.

Setting up a virtual device and running your app

As a part of installing Android Studio, you downloaded and installed the latest Android **software development kit** (**SDK**) components. These included a base emulator, which you will configure to create a virtual device to run Android apps on. An emulator mimics the hardware and software features and configuration of a real device. The benefit is that you can make changes and quickly see them on your desktop while developing your app. Although virtual devices do not have all the features of a real device, the feedback cycle is often quicker than going through the steps of connecting a real device.

Also, although you should ensure your app runs as expected on different devices, you can standardize it by targeting a specific device by downloading a device profile, even if you don't have a real device if this is a requirement of your project.

The screen you will have seen (or something similar) when installing Android Studio is as follows:

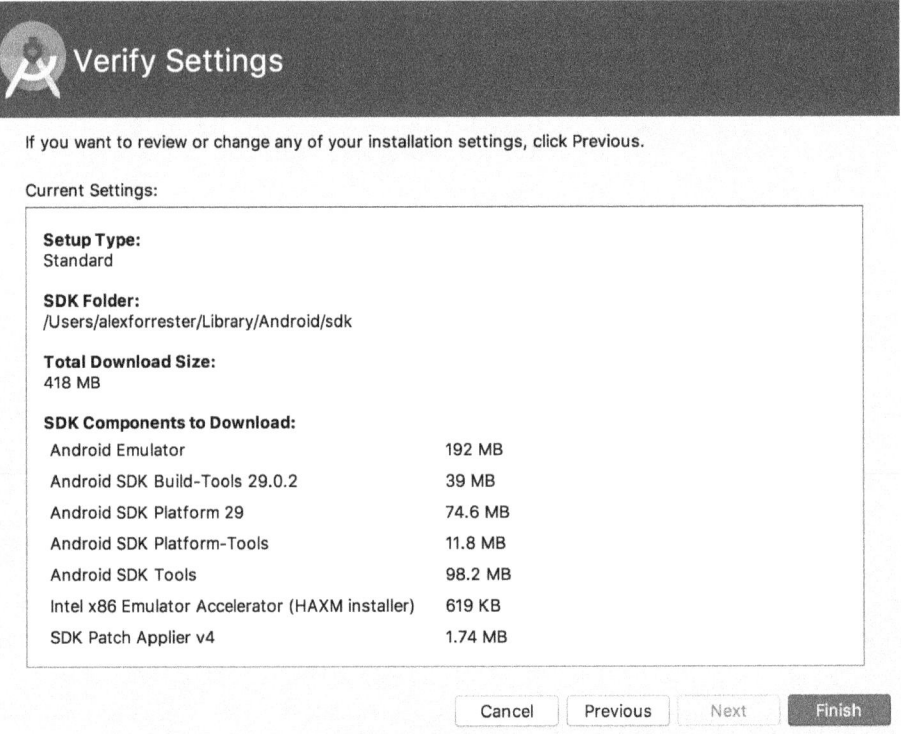

Figure 1.4 – SDK components

Let's take a look at the SDK components that are installed and how the virtual device fits in:

- **Android Emulator**: This is the base emulator, which we will configure to create virtual devices of different Android makes and models.
- **Android SDK Build-Tools**: Android Studio uses the build tools to build your app. This process involves compiling, linking, and packaging your app to prepare it for installation on a device.
- **Android SDK Platform**: This is the version of the Android platform that you will use to develop your app. The platform refers to the API level.
- **Android SDK Platform-Tools**: These are tools you can use, ordinarily, from the command line, to interact with and debug your app.
- **Android SDK Tools**: In contrast to the platform tools, these are tools that you use predominantly from within Android Studio in order to accomplish certain tasks, such as the virtual device for running apps and the SDK manager to download and install platforms and other components of the SDK.
- **Intel x86 Emulator Accelerator (HAXM installer)**: If your OS provides it, this is a feature at the hardware level of your computer you will be prompted to enable, which allows your emulator to run more quickly.
- **SDK Patch Applier v4**: As newer versions of Android Studio become available, this enables patches to be applied to update the version you are running.

With this knowledge, let's start with the next exercise of this chapter.

Exercise 1.02 – setting up a virtual device and running your app on it

We set up an Android Studio project to create our app in *Exercise 1.01, Creating an Android Studio project for your app*, and we are now going to run it on a virtual device. You can also run your app on a real device, but you will use a virtual device in this exercise. This process is a continuous cycle while working on your app. Once you have implemented a feature, you can verify its look and behavior as you require.

For this exercise, you will create a single virtual device, but you should ensure you run your app on multiple devices to verify that its look and behavior are consistent. Perform the following steps:

1. In the toolbar in Android Studio, you will see two drop-down boxes next to each other with **app** and **No devices** pre-selected:

10 | Creating Your First App

Figure 1.5 – The Android Studio toolbar

app is the configuration of the app that we will run. As we haven't set up a virtual device yet, it says **No devices**.

2. In order to create a virtual device, click on **Device Manager**, as shown in *Figure 1.5*, to open the virtual devices window/screen. The option to do this can also be accessed from the **Tools** menu:

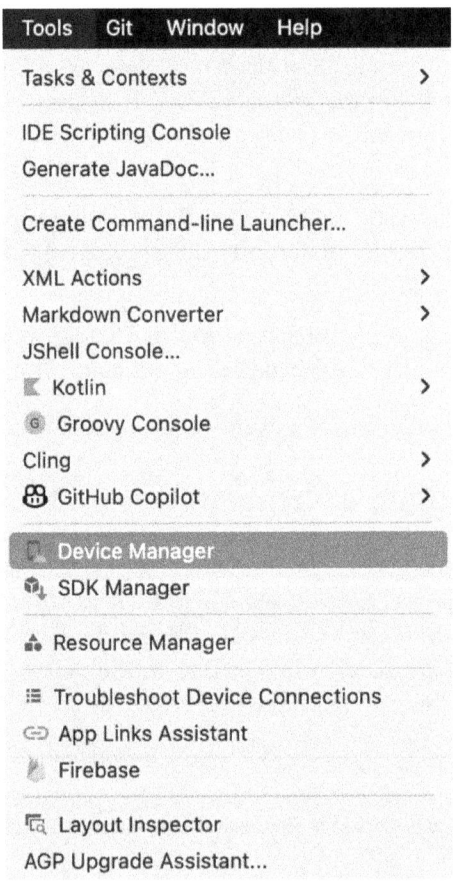

Figure 1.6 – Device Manager in the Tools menu

3. Click the button or toolbar option to open the **Device Manager** window and click the **Create device** button, as shown in *Figure 1.7*:

Figure 1.7 – The Device Manager window

You will then be presented with a screen, as shown in *Figure 1.8*:

Figure 1.8 – Device definition creation

4. We are going to choose the **Pixel 6** device. The real (non-virtual device) Pixel range of devices is developed by Google and has access to the most up-to-date versions of the Android platform. Once selected, click the **Next** button:

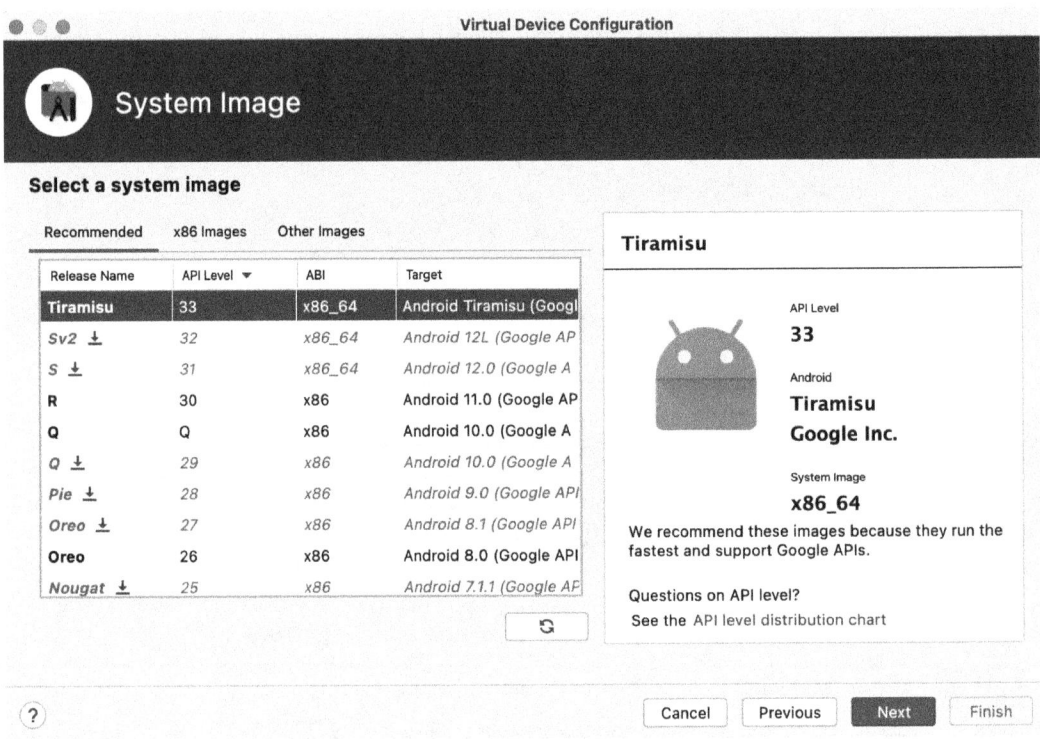

Figure 1.9 – System Image

The **Tirimasu** name displayed here is the initial code/release name for Android 13. Select the latest system image available. The **Target** column might also show **(Google Play)** or **(Google APIs)** in the name. Google APIs mean that the system image comes pre-installed with Google Play Services.

This is a rich feature set of Google APIs and Google apps that your app can use and interact with. On first running the app, you will see apps such as Maps and Chrome instead of a plain emulator image. A Google Play system image means that, in addition to the Google APIs, the Google Play app will also be installed.

5. You should develop your app with the latest version of the Android platform to benefit from the latest features. On first creating a virtual device, you will have to download the system image. If a **Download** link is displayed next to **Release Name**, click on it, and wait for the download to complete. Select the **Next** button to see the virtual device you have set up:

Setting up a virtual device and running your app 13

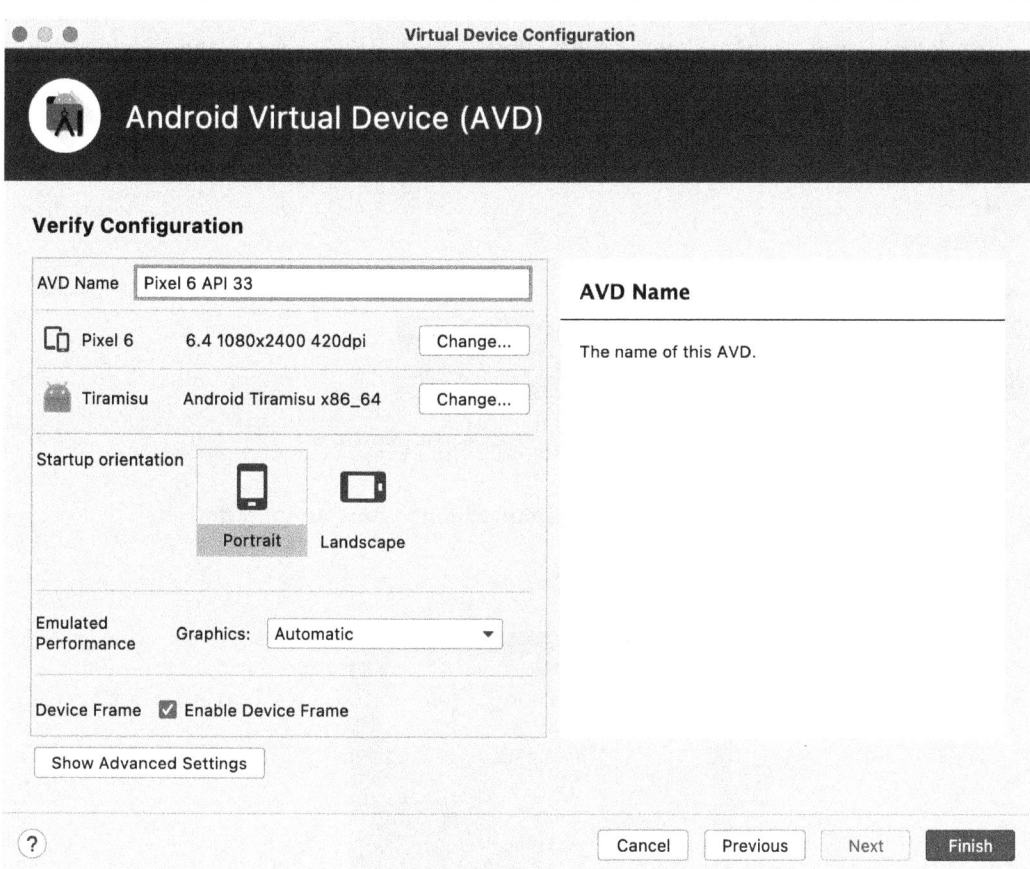

Figure 1.10 – Virtual device configuration

6. Click **Finish**, and your virtual device will be created. You will then see your device highlighted:

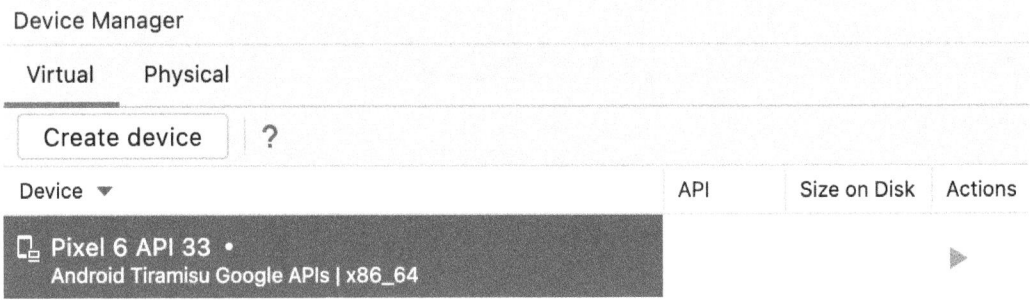

Figure 1.11 – Virtual devices listed

7. Press the play arrow button under the **Actions** column to run the virtual device:

Figure 1.12 – Virtual device launched

You will then see the virtual device running within Android Studio in the **Emulator** tool window. Now that you've created the virtual device and it's running, you can go back into Android Studio to run your app.

8. The virtual device you have set up and started will be selected. Press the green triangle/play button to launch your app:

Figure 1.13 – App launch configuration

This will load the app into the emulator as shown in *Figure 1.14*.

Figure 1.14 – The app running on a virtual device

In this exercise, you have gone through the steps to create a virtual device and run the app you created on it. The Android Virtual Device Manager, which you have used to do this, enables you to create the device (or range of devices) you would like to target your app for. Running your app on the virtual device allows a quick feedback cycle to verify how a new feature development behaves and that it displays the way you expect it to.

Next, you will explore the `AndroidManifest.xml` file of your project, which contains the information and configuration of your app.

The Android manifest

The app you have just created, although simple, encompasses the core building blocks that you will use in all of the projects you create. The app is driven from the `AndroidManifest.xml` file, a manifest file that details the contents of your app. It is located at app | manifests | AndroidManifest.xml:

```xml
<?xml version="1.0" encoding="utf-8"?>
<manifest xmlns:android=
    "http://schemas.android.com/apk/res/android"
    xmlns:tools="http://schemas.android.com/tools">
    <application
        android:allowBackup="true"
        android:dataExtractionRules="@xml/data_extraction_
            rules"
        android:fullBackupContent="@xml/backup_rules"
        android:icon="@mipmap/ic_launcher"
        android:label="@string/app_name"
        android:roundIcon="@mipmap/ic_launcher_round"
        android:supportsRtl="true"
        android:theme="@style/Theme.MyApplication"
        tools:targetApi="31">
        <activity
            android:name=".MainActivity"
            android:exported="true">
            <intent-filter>
                <action android:name="android.intent.action.
                    MAIN" />
                <category android:name="android.intent.
                    category.LAUNCHER" />
            </intent-filter>
```

```
        </activity>
    </application>
</manifest>
```

A typical manifest file, in general terms, is a top-level file that describes the enclosed files or other data and associated metadata that forms a group or unit. The Android manifest applies this concept to your Android app as an XML file.

Every Android app has an application class that allows you to configure the app. After the `<application>` element opens, you define your app's components. As we have just created our app, it only contains the first screen shown in the following code:

```
<activity android:name=".MainActivity">
```

The next child XML node specified is as follows:

```
<intent-filter>
```

Android uses intents as a mechanism for interacting with apps and system components. Intents get sent, and the intent filter registers your app's capability to react to these intents. `<android.intent.action.MAIN>` is the main entry point into your app, which, as it appears in the enclosing XML of `.MainActivity`, specifies that this screen will be started when the app is launched. `Android.intent.category.LAUNCHER` states that your app will appear in the launcher of your user's device.

As you have created your app from a template, it has a basic manifest that will launch the app and display an initial screen at startup through an `Activity` component. Depending on which other features you want to add to your app, you may need to add permissions in the Android manifest file.

Permissions are grouped into three different categories: normal, signature, and dangerous:

- **Normal**: These permissions include accessing the network state, Wi-Fi, the internet, and Bluetooth. These are usually permitted without asking for the user's consent at runtime.
- **Signature**: These permissions are shared by the same group of apps that must be signed with the same certificate. This means these apps can share data freely, but other apps can't get access.
- **Dangerous**: These permissions are centered around the user and their privacy, such as sending SMS, access to accounts and location, and reading and writing to the filesystem and contacts.

These permissions have to be listed in the manifest, and in the case of dangerous permissions, from Android Marshmallow API 23 (Android 6 Marshmallow) onward, you must also ask the user to grant the permissions at runtime.

In the next exercise, we will configure the Android Manifest. Detailed documentation on this file can be found at `https://developer.android.com/guide/topics/manifest/manifest-intro`.

Exercise 1.03 – configuring the Android manifest internet permission

The key permission that most apps require is access to the internet. This is not added by default. In this exercise, we will fix that and, in the process, load a `WebView`, which enables the app to show web pages. This use case is very common in Android app development as most commercial apps will display a privacy policy, terms and conditions, and so on. As these documents are likely common to all platforms, the usual way to display them is to load a web page. To do this, perform the following steps:

1. Create a new Android Studio project as you did in *Exercise 1.01, Creating an Android Studio project for your app*.

2. Switch tabs to the `MainActivity` class. From the main project window, it's located at `app | java | com | example | myapplication`.

   You can change what the project window displays by opening up the **Tool** window by selecting **View | Tool Windows | Project** – this will select **Project** view. The drop-down options on the top of the **Project** window allow you to change the way you view your project, with the most commonly used displays being **Project** and **Android**:

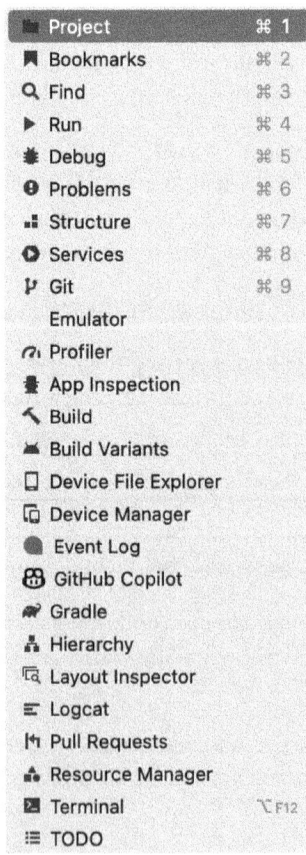

Figure 1.15 – The Tool Windows drop-down menu

On opening the `MainActivity` class, you'll see that it has the following content or similar:

```
package com.example.myapplication

import androidx.appcompat.app.AppCompatActivity
import android.os.Bundle

class MainActivity : AppCompatActivity() {
    override fun onCreate(savedInstanceState: Bundle?)
    {
        super.onCreate(savedInstanceState)
        setContentView(R.layout.activity_main)
    }
}
```

You'll examine the contents of this file in more detail in the next section of this chapter, but for now, you just need to be aware that the `setContentView(R.layout.activity_main)` statement sets the layout of the UI you saw when you first ran the app in the virtual device.

3. Use the following code to change this to the following:

```
package com.example.myapplication

import androidx.appcompat.app.AppCompatActivity
import android.os.Bundle
import android.webkit.WebView

class MainActivity : AppCompatActivity() {
    override fun onCreate(savedInstanceState: Bundle?) {
        super.onCreate(savedInstanceState)
        val webView = WebView(this)
        webView.settings.javaScriptEnabled = true
        setContentView(webView)
        webView.loadUrl("https://www.google.com")
    }
}
```

So, you are replacing the layout file with `WebView`. The `val` keyword is a read-only property reference, which can't be changed once it has been set. JavaScript needs to be enabled in `WebView` to execute JavaScript.

> **Note**
> We are not setting the type, but Kotlin has type inference, so it will infer the type if possible. So, specifying the type explicitly with `val webView: WebView = WebView(this)` is not necessary. Depending on which programming languages you have used in the past, the order of defining the parameter name and type may or may not be familiar. Kotlin follows Pascal notation, that is, name followed by type.

4. Now, run the app up, and the text will appear as shown in the screenshot here:

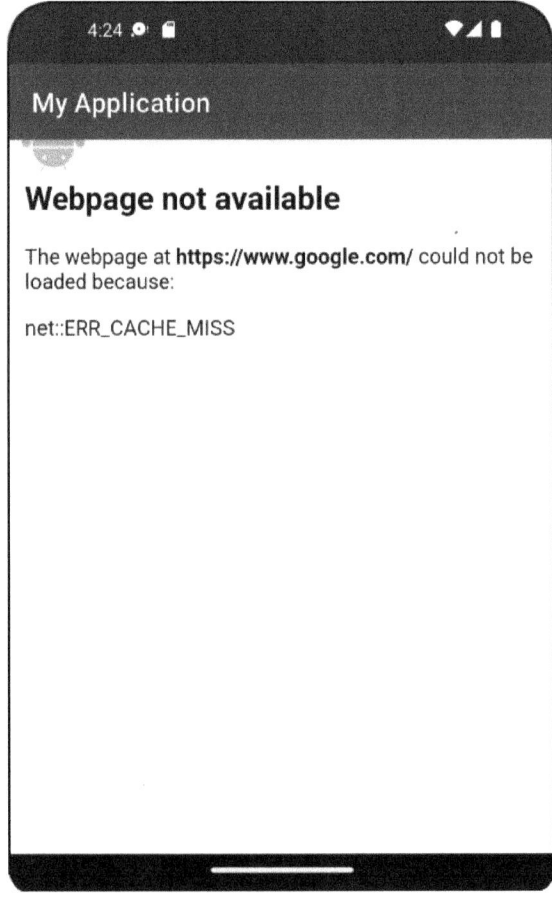

Figure 1.16 – No internet permission error message

5. This error occurs because there is no `INTERNET` permission added to your `AndroidManifest.xml` file. (If you get the `net::ERR_CLEARTEXT_NOT_PERMITTED` error, this is because the URL you are loading into `WebView` is not HTTPS, and non-HTTPS traffic is disabled from API level 28, Android 9.0 Pie and above).

6. Let's fix that by adding the `INTERNET` permission to the manifest. Open up the Android manifest and add the following above the `<application>` tag:

   ```
   <uses-permission android:name="android.permission.
   INTERNET" />
   ```

   You can find the full Android manifest file with the permission added here: https://packt.link/smzpl

   Uninstall the app from the virtual device before running up the app again. You need to do this, as app permissions can sometimes get cached.

   Do this by long-pressing on the app icon and selecting the **App Info** option that appears and then pressing the Bin icon with the **Uninstall** text below it. Alternatively, long press the app icon and then drag it to the Bin icon with the **Uninstall** text beside it in the top-right corner of the screen.

7. Install the app again and see the web page appear in `WebView`:

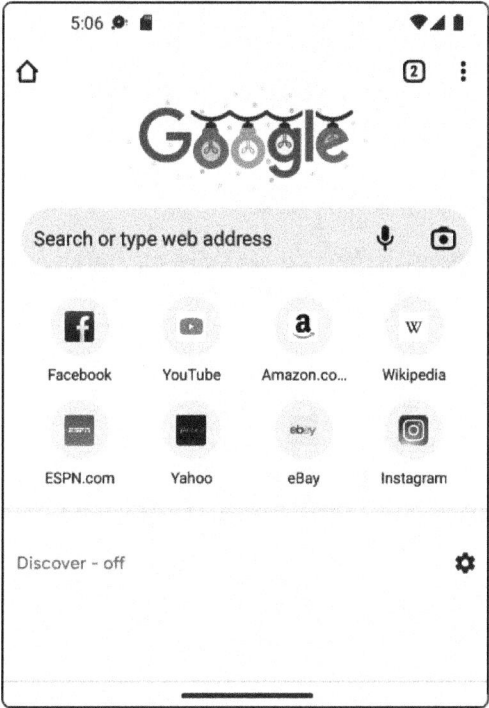

Figure 1.17 – App displaying WebView

In this example, you learned how to add a permission to the manifest. The Android Manifest can be thought of as a table of contents of your app. It lists all the components and permissions your app uses. As you have seen from starting the app from the launcher, it also provides the entry points into your app.

In the next section, you will explore the Android build system, which uses the Gradle build tool to get your app up and running.

Using Gradle to build, configure, and manage app dependencies

In the course of creating this project, you have principally used the Android platform SDK. The necessary Android libraries were downloaded when you installed Android Studio. However, these are not the only libraries that are used to create your app. To configure and build your Android project or app, a build tool called **Gradle** is used.

Gradle is a multi-purpose build tool that Android Studio uses to build your app. By default, Android Studio uses Groovy, a dynamically typed **Java virtual machine** (**JVM**) language, to configure the build process and allows easy dependency management so you can add libraries to your project and specify the versions.

Android Studio can also be configured to use Kotlin to configure builds, but as the default language is Groovy, you will be using this. The files that this build and configuration information is stored in are named `build.gradle`.

When you first create your app, there are two `build.gradle` files, one at the root/top level of the project and one specific to your app in the app `module` folder.

The project-level build.gradle file

Let's now have a look at the project-level `build.gradle` file. This is where you set up all the root project settings, which can be applied to sub-modules/projects:

```
plugins {
    id 'com.android.application' version '7.4.2' apply
    false
    id 'com.android.library' version '7.4.2' apply false
    id 'org.jetbrains.kotlin.android' version '1.8.0'
    apply false
}
```

Gradle works on a plugin system, so you can write your own plugin that does a task or series of tasks and plug it into your build pipeline. The three plugins listed previously do the following:

- `com.android.application`: This adds support to create an Android application
- `com.android.library`: This enables sub-projects/modules to be Android libraries
- `org.jetbrains.kotlin.android`: This provides integration and language support for Kotlin in the project

The `apply false` statement enables these plugins only to sub-projects/modules, and not the project's root level. The `version '7.3.1'` specifies the plugin version, which is applied to all sub-projects/modules.

The app-level build.gradle file

The `build.gradle` app is specific to your project configuration:

```
plugins {
    id 'com.android.application'
    id 'org.jetbrains.kotlin.android'
}
android {
    namespace 'com.example.myapplication'
    compileSdk 33
    defaultConfig {
        applicationId "com.example.myapplication"
        minSdk 24
        targetSdk 33
        versionCode 1
        versionName "1.0"
        testInstrumentationRunner
            "androidx.test.runner.AndroidJUnitRunner" }
    buildTypes {
        release {
            minifyEnabled false
            proguardFiles getDefaultProguardFile('proguard-
                android-optimize.txt'), 'proguard-rules.pro'
        }
    }
```

```
        compileOptions {
            sourceCompatibility JavaVersion.VERSION_1_8
            targetCompatibility JavaVersion.VERSION_1_8
        }
        kotlinOptions {
            jvmTarget = '1.8'
        }
    }
    dependencies {…}
```

The plugins for Android and Kotlin, detailed in the root `build.gradle` file, are applied to your project here by ID in the `plugins` lines.

The `android` block, provided by the `com.android.application` plugin, is where you configure your Android-specific configuration settings:

- `namespace`: This is set from the package name you specified when creating the project. It will be used for generating build and resource identifiers.
- `compileSdk`: This is used to define the API level the app has been compiled with, and the app can use the features of this API and lower.
- `defaultConfig`: This is the base configuration of your app.
- `applicationId`: This is set to your app's package and is the app identifier that is used on Google Play to uniquely identify your app. It can be changed to be different from the package name if required.
- `minSdk`: This is the minimum API level your app supports. This will filter out your app from being displayed in Google Play for devices that are lower than this.
- `targetSdk`: This is the API level you are targeting. This is the API level your built app is intended to work and has been tested with.
- `versionCode`: This specifies the version code of your app. Every time an update needs to be made to the app, the version code needs to be increased by one or more.
- `versionName`: A user-friendly version name that usually follows semantic versioning of *X.Y.Z*, where *X* is the major version, *Y* is the minor version, and *Z* is the patch version, for example, *1.0.3*.
- `testInstrumentationRunner`: This is the test runner to use for your UI tests.

- buildTypes: Under buildTypes, a release is added that configures your app to create a release build. The minifyEnabled value, if set to true, will shrink your app size by removing any unused code, as well as obfuscating your app. This obfuscation step changes the name of the source code references to values such as a.b.c(). This makes your code less prone to reverse engineering and further reduces the size of the built app.
- compileOptions: This is the language level of the Java source code (sourceCompatibility) and byte code (targetCompatibility).
- kotlinOptions: This is the jvm library the kotlin gradle plugin should use.

The dependencies block specifies the libraries your app uses on top of the Android platform SDK, as shown here (with added comments):

```
dependencies {
// Kotlin extensions, jetpack component with Android
    Kotlin language features
implementation 'androidx.core:core-ktx:1.7.0'
// Provides backwards compatible support libraries and
    jetpack components
implementation 'androidx.appcompat:appcompat:1.6.1'
// Material design components to theme and style your
    app
implementation
    'com.google.android.material:material:1.8.0'
// The ConstraintLayout ViewGroup updated separately
    from main Android sources
implementation
    'androidx.constraintlayout:constraintlayout:2.1.4'
// Standard Test library for unit tests
testImplementation 'junit:junit:4.13.2'
// UI Test runner
androidTestImplementation
    'androidx.test.ext:junit:1.1.5'
// Library for creating Android UI tests
androidTestImplementation
    'androidx.test.espresso:espresso-core:3.5.1'
}
```

The dependencies follow the Maven **Project Object Model** (**POM**) convention of `groupId`, `artifactId`, and `versionId` separated by `:`. So, as an example, the compatible support library specified earlier is shown as:

```
'androidx.appcompat:appcompat:1.6.1'
```

The `groupId` is `android.appcompat`, `artifactId` is `appcompat`, and `versionId` is `1.5.1`. The build system locates and downloads these dependencies to build the app from the `repositories` block detailed in the `settings.gradle` file explained in the following section.

> **Note**
> The dependency versions specified in the previous code section and in the following sections of this and other chapters are subject to change and are updated over time, so they are likely to be higher when you create these projects.

The `implementation` notation for adding these libraries means that their internal dependencies will not be exposed to your app, making compilation faster.

Here, the `androidx` components are added as dependencies rather than in the Android platform source. This is so that they can be updated independently from Android versions. `androidx` contains the suite of Android Jetpack libraries and the repackaged support library.

The next Gradle file to examine is `settings.gradle`, which initially looks like this:

```
pluginManagement {
    repositories {
        google()
        mavenCentral()
        gradlePluginPortal()
    }
}
dependencyResolutionManagement {
repositoriesMode.set(RepositoriesMode.FAIL_ON_PROJECT_REPOS)
    repositories {
        google()
        mavenCentral()
    }
}
rootProject.name = "My Application"
include ':app'
```

On first creating a project with Android Studio, there will only be one module, app, but when you add more features, you can add new modules that are dedicated to containing the source of a feature rather than packaging it in the main app module.

These are called **feature modules**, and you can supplement them with other types of modules, such as shared modules, which are used by all other modules, like a networking module. This file also contains the repositories of the plugins and dependencies to download from in separate blocks for plugins and dependencies.

Setting the value of RepositoriesMode.FAIL_ON_PROJECT_REPOS ensures all dependencies repositories are defined here; otherwise, a build error will be triggered.

Exercise 1.04 – exploring how Material Design is used to theme an app

In this exercise, you will learn about Google's new design language, **Material Design**, and use it to load a Material Design-themed app. Material Design is a design language created by Google that adds enriched UI elements based on real-world effects such as lighting, depth, shadows, and animations. Perform the following steps to complete the exercise:

1. Create a new Android Studio project as you did in *Exercise 1.01, Creating an Android Studio project for your app*.

2. First, look at the dependencies block and find the Material Design dependency:

    ```
    implementation
    'com.google.android.material:material:1.8.0'
    ```

3. Next, open the themes.xml file located at app | src | main | res | values | themes.xml: There is also a themes.xml file in the values-night folder used for a dark mode, which we will explore later:

    ```
    <resources xmlns:tools="http://schemas.android.com/tools">
        <!-- Base application theme. -->
        <style name="Theme.MyApplication" parent="Theme.
            MaterialComponents.DayNight.DarkActionBar">
            <!-- Primary brand color. -->
            <item name="colorPrimary">@color/purple_500
                </item>
            <item name="colorPrimaryVariant">@color/
                purple_700</item>
            <item name="colorOnPrimary">@color/white</item>
            <!-- Secondary brand color. -->
    ```

```xml
            <item name="colorSecondary">@color/teal_200
                </item>
            <item name="colorSecondaryVariant">@color/
                teal_700</item>
            <item name="colorOnSecondary">@color/black</item>
            <!-- Status bar color. -->
            <item name="android:statusBarColor">?attr/
                colorPrimaryVariant</item>
            <!-- Customize your theme here. -->
        </style>
    </resources>
```

Notice that the parent of `Theme.MyApplication` is `Theme.MaterialComponents.DayNight.DarkActionBar`.

The Material Design dependency added in the `dependencies` block is being used here to apply the theme of the app. One of the key differences that **Material Design Components** (**MDC**) offer over the `AppCompat` themes that preceded them is the ability to provide variations to the primary and secondary colors of your app.

For example, `colorPrimaryVariant` enables you to add a tint to the primary color, which can be either lighter or darker than the `colorPrimary` color. In addition, you can style view element colors in the foreground of your app with `colorOnPrimary`.

Together these bring cohesive branding to theme your app. To see this in effect, make the following changes to invert the primary and secondary colors:

```xml
    <resources xmlns:tools="http://schemas.android.com/
        tools">
        <!-- Base application theme. -->
        <style name="Theme.MyApplication" parent="Theme.
            MaterialComponents.DayNight.DarkActionBar">
            <!-- Primary brand color. -->
            <item name="colorPrimary">@color/teal_200</item>
            <item name="colorPrimaryVariant">@color/
                teal_700</item>
            <item name="colorOnPrimary">@color/white</item>
            <!-- Secondary brand color. -->
            <item name="colorSecondary">@color/purple_200
                </item>
            <item name="colorSecondaryVariant">@color/
```

```xml
            purple_700</item>
        <item name="colorOnSecondary">@color/black</item>
        <!-- Status bar color. -->
        <item name="android:statusBarColor">?attr/
            colorPrimaryVariant</item>
        <!-- Customize your theme here. -->
    </style>
</resources>
```

4. Run the app now, and you will see the app themed differently. The action bar and status bar have changed background color in contrast to the default Material themed app, as shown in *Figure 1.18*:

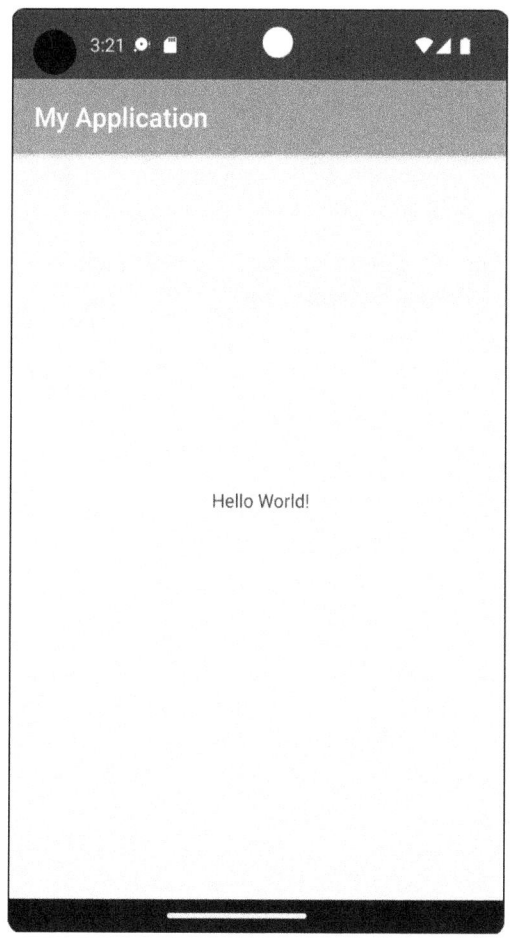

Figure 1.18 – App with primary and secondary colors inverted

In this exercise, you've learned how Material Design can be used to theme an app. As you are currently only displaying `TextView` on the screen, it is not clear what benefits material design provides, but this will change when you start using Material UI design widgets more.

Now that you've learned how the project is built and configured, in the next section, you'll explore the project structure in detail, learn how it has been created, and gain familiarity with the core areas of the development environment.

Android application structure

Now that we have covered how the Gradle build tool works, we'll explore the rest of the project. The simplest way to do this is to examine the folder structure of the app. There is a tool window at the top left of Android Studio called **Project**, which allows you to browse the contents of your app.

By default, it is set to **open**/**selected** when your Android project is first created. When you select it, you will see a view similar to the screenshot in *Figure 1.19*. If you can't see any window bars on the left-hand side of the screen, then go to the top toolbar and select **View** | **Appearance** | **Tool Window Bars** and make sure it is ticked.

There are many different options for how to browse your project, but **Android** will be pre-selected. This view neatly groups the app folder structure, so let's take a look at it.

Here is an overview of these files with more detail about the most important ones. On opening it, you will see that it consists of the following folder structure:

Android application structure

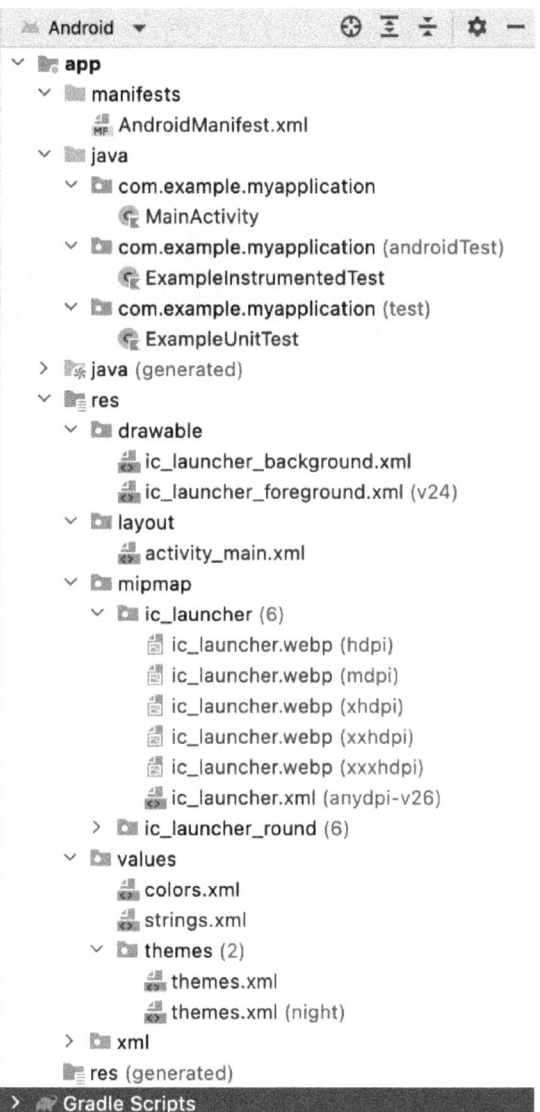

Figure 1.19 – Overview of the files and folder structure in the app

The Kotlin file (`MainActivity`), which you've specified as running when the app starts, is as follows:

```
package com.example.myapplication

import androidx.appcompat.app.AppCompatActivity
import android.os.Bundle
```

```
class MainActivity : AppCompatActivity() {
    override fun onCreate(savedInstanceState: Bundle?) {
        super.onCreate(savedInstanceState)
        setContentView(R.layout.activity_main)
    }
}
```

The `import` statements include the libraries and the source of what this activity uses. The `class MainActivity : AppCompatActivity()` class header creates a class that extends `AppCompatActivity`. In Kotlin, the `:` character is used for both deriving from a class (also known as inheritance) and implementing an interface.

`MainActivity` derives from `androidx.appcompat.app.AppCompatActivity`, which is the backward-compatible activity designed to make your app work on older devices.

Android activities have many callback functions you can override at different points of the activity's life. This is known as the **activity lifecycle**. For this activity, as you want to display a screen with a layout, you override the `onCreate` function as shown here:

```
override fun onCreate(savedInstanceState: Bundle?)
```

The `override` keyword in Kotlin specifies that you are providing a specific implementation for a function defined in the parent class. The `fun` keyword (as you may have guessed) stands for *function*. The `savedInstanceState: Bundle?` parameter is Android's mechanism for restoring previously saved state. For this simple activity, you haven't stored any state, so this value will be `null`. The question mark, `?`, that follows the type declares that this type can be `null`.

The `super.onCreate(savedInstanceState)` line calls through to the overridden method of the base class, and finally, `setContentView(R.layout.activity_main)` loads the layout we want to display in the activity; otherwise, it would be displayed as a blank screen as no layout has been defined.

Let's have a look at some other files (*Figure 1.19*) present in the folder structure:

- `ExampleInstrumentedTest`: This is an example UI test. You can check and verify the flow and structure of your app by running tests on the UI when the app is running.

- `ExampleUnitTest`: This is an example unit test. An essential part of creating an Android app is writing unit tests to verify that the source code works as expected.

- `ic_launcher_background.xml` and `ic_launcher_foreground.xml`: These two files together make up the launcher icon of your app in vector format, which will be used by the `ic_launcher.xml` launcher icon file in Android API 26 (Oreo) and above.

- `activity_main.xml`: This is the layout file that was created by Android Studio when we created the project. It is used by `MainActivity` to draw the initial screen content, which appears when the app runs:

  ```xml
  <?xml version="1.0" encoding="utf-8"?>
  <androidx.constraintlayout.widget.ConstraintLayout
  xmlns:android=
      "http://schemas.android.com/apk/res/android"
      xmlns:app="http://schemas.android.com/apk/res-auto"
      xmlns:tools="http://schemas.android.com/tools"
      android:layout_width="match_parent"
      android:layout_height="match_parent"
      tools:context=".MainActivity">
      <TextView
          android:layout_width="wrap_content"
          android:layout_height="wrap_content"
          android:text="Hello World!"
          app:layout_constraintBottom_toBottomOf="parent"
          app:layout_constraintEnd_toEndOf="parent"
          app:layout_constraintStart_toStartOf="parent"
          app:layout_constraintTop_toTopOf="parent" />
  </androidx.constraintlayout.widget.ConstraintLayout>
  ```

Screen displays in Android can be created using XML or Jetpack Compose, which uses a declarative API to dynamically build your UI. You will learn Jetpack Compose in *Chapter 9*. For XML, the documents start with an XML header followed by a top-level `ViewGroup` (which here is `ConstraintLayout`) and then one or more nested `Views` and `ViewGroups`.

The `ConstraintLayout` `ViewGroup` allows very precise positioning of views on a screen, constraining views with parent and sibling views, guidelines, and barriers. Detailed documentation on `ConstraintLayout` can be found at `https://developer.android.com/reference/androidx/constraintlayout/widget/ConstraintLayout`.

`TextView`, currently the only child view of `ConstraintLayout`, displays text on the screen through the `android:text` attribute. The horizontal positioning of the view is done by constraining the view to both the start and end of the parent, which centers the view horizontally as both constraints are applied.

From start to end, left-to-right languages (`ltr`) are read left to right, while `non ltr` languages are read right to left. The view is positioned vertically in the center by constraining the view to both the top and the bottom of its parent. The result of applying all four constraints centers `TextView` both horizontally and vertically within `ConstraintLayout`.

There are three XML namespaces in the `ConstraintLayout` tag:

- `xmlns:android`: This refers to the Android-specific namespace and it is used for all attributes and values within the main Android SDK.
- `xmlns:app`: This namespace is for anything not in the Android SDK. So, in this case, `ConstraintLayout` is not part of the main Android SDK but is added as a library.
- `xmnls:tools`: This refers to a namespace used for adding metadata to the XML, which indicates where the layout is used (`tools:context=".MainActivity"`). It is also used to show sample text visible in previews.

The two most important attributes of an Android XML layout file are `android:layout_width` and `android:layout_height`.

These can be set to absolute values, usually of density-independent pixels (known as `dip` or `dp`) that scale pixel sizes to be roughly equivalent on different density devices. More commonly, however, these attributes have the `wrap_content` or `match_parent` values set for them. `wrap_content` will be as big as required to only enclose its contents. `match_parent` will be sized according to its parent.

There are other `ViewGroups` you can use to create layouts. For example, `LinearLayout` lays out views vertically or horizontally, `FrameLayout` is usually used to display a single child view, and `RelativeLayout` is a simpler version of `ConstraintLayout`, which lays out views positioned relative to the parent and sibling views.

The `ic_launcher.webp` files are the `.webp` launcher icons that have an icon for every different density of devices. This image format was created by Google and has greater compression compared to the `.png` images. As the minimum version of Android we are using is API 21: Android 5.0 (Jelly Bean), these `.webp` images are included, as support for the launcher vector format was not introduced until Android API 26 (Oreo).

The `ic_launcher.xml` file uses the vector files (`ic_launcher_background.xml` and `ic_launcher_foreground.xml`) to scale to different density devices in Android API 26 (Oreo) and above.

> **Note**
> To target different density devices on the Android platform, besides each one of the `ic_launcher.png` icons, you will see in brackets the density it targets. As devices vary widely in their pixel densities, Google created density buckets so that the correct image would be selected to be displayed depending on how many dots per inch the device has.

The different density qualifiers and their details are as follows:

- `nodpi`: Density-independent resources
- `ldpi`: Low-density screens of 120 dpi
- `mdpi`: Medium-density screens of 160 dpi (the baseline)
- `hdpi`: High-density screens of 240 dpi
- `xhdpi`: Extra-high-density screens of 320 dpi
- `xxhdpi`: Extra-extra-high-density screens of 480 dpi
- `xxxhdpi`: Extra-extra-extra-high-density screens of 640 dpi
- `tvdpi`: Resources for televisions (approx 213 dpi)

The baseline density bucket was created at `160` dots per inch for medium-density devices and is called **mdpi**. This represents a device where an inch of the screen is `160` dots/pixels, and the largest display bucket is `xxxhdpi`, which has `640` dots per inch. Android determines the appropriate image to display based on the individual device.

So, the Pixel 6 emulator has a density of approximately `411dpi`, so it uses resources from the extra-extra-high-density bucket (`xxhdpi`), which is the closest match. Android has a preference for scaling down resources to best match density buckets, so a device with `400dpi`, which is halfway between the `xhdpi` and `xxhdpi` buckets, is likely to display the `480dpi` asset from the `xxhdpi` bucket.

To create alternative bitmap drawables for different densities, you should follow the `3:4:6:8:12:16` scaling ratio between the six primary densities. For example, if you have a bitmap drawable that's `48x48` pixels for medium-density screens, all the different sizes should be as follows:

- `36x36` (`0.75x`) for low density (`ldpi`)
- `48x48` (`1.0x` baseline) for medium density (`mdpi`)
- `72x72` (`1.5x`) for high density (`hdpi`)
- `96x96` (`2.0x`) for extra-high density (`xhdpi`)
- `144x144` (`3.0x`) for extra-extra-high density (`xxhdpi`)
- `192x192` (`4.0x`) for extra-extra-extra-high density (`xxxhdpi`)

For a comparison of these physical launcher icons per density bucket, refer to the following table:

mdpi	hdpi	xhdpi	xxhdpi	xxxhdpi

Figure 1.20 – Comparison of principal density bucket launcher image sizes

> **Note**
> Launcher icons are made slightly larger than normal images within your app as they will be used by the device's launcher. As some launchers can scale up the image, this ensures there is no pixelation and blurring of the image.

Now you are going to look at some of the resources the app uses. These are referenced in XML files and keep the display and formatting of your app consistent.

In the `colors.xml` file, you define the colors you want to use in your app in hexadecimal format:

```xml
<?xml version="1.0" encoding="utf-8"?>
<resources>
    <color name="purple_200">#FFBB86FC</color>
    <color name="purple_500">#FF6200EE</color>
    <color name="purple_700">#FF3700B3</color>
    <color name="teal_200">#FF03DAC5</color>
    <color name="teal_700">#FF018786</color>
    <color name="black">#FF000000</color>
    <color name="white">#FFFFFFFF</color>
</resources>
```

The format is based on the **ARGB** color space, so the first two characters are for **A**lpha (transparency), the next two for **R**ed, the next two for **G**reen, and the last two for **B**lue. For Alpha, #00 is completely transparent through to #FF, which is completely opaque. For the colors, #00 means none of the color is added to make up the composite color, and #FF means all of the color is added.

If no transparency is required, you can omit the first two characters. So, to create fully blue and 50% transparent blue colors, here's the format:

```xml
<color name="colorBlue">#0000FF</color>
<color name=
    "colorBlue50PercentTransparent">#770000FF</color>
```

The `strings.xml` file displays all the text displayed in the app:

```xml
<resources>
    <string name="app_name">My Application</string>
</resources>
```

You can use hardcoded strings in your app, but this leads to duplication and also means you cannot customize the text if you want to make the app multilingual. By adding strings as resources, you can also update the string in one place if it is used in different places in the app.

Common styles you would like to use throughout your app are added to the `themes.xml` file:

```xml
<resources xmlns:tools="http://schemas.android.com/tools">
    <!-- Base application theme. -->
    <style name="Theme.MyApplication" parent=
        "Theme.MaterialComponents.DayNight.DarkActionBar">
        <!-- Primary brand color. -->
        <item name="colorPrimary">@color/purple_500</item>
        <item name="colorPrimaryVariant">@color/purple_700
            </item>
        <item name="colorOnPrimary">@color/white</item>
        <!-- Secondary brand color. -->
        <item name="colorSecondary">@color/teal_200</item>
        <item name="colorSecondaryVariant">@color/teal_700
            </item>
        <item name="colorOnSecondary">@color/black</item>
        <!-- Status bar color. -->
        <item name="android:statusBarColor"
            tools:targetApi="l">?attr/colorPrimaryVariant
            </item>
        <!-- Customize your theme here. -->
    </style></resources>
```

It is possible to apply style information directly to views by setting `android:textStyle="bold"` as an attribute on `TextView`. However, you would have to repeat this in multiple places for every `TextView` you wanted to display in bold. Furthermore, when you start to have multiple style attributes added to individual views, it adds a lot of duplication and can lead to errors when you want to make a change to all similar views and miss changing a style attribute in one view.

If you define a style, you only have to change the style, and it will update all the views that have that style applied to them. A top-level theme was applied to the application tag in the `AndroidManifest.xml` file when you created the project and is referred to as a theme that styles all views contained within the app.

The colors you have defined in the `colors.xml` file are used here. In effect, if you change one of the colors defined in the `colors.xml` file, it will now propagate to style the app as well.

You've now explored the core areas of the app. You have added the `TextView` views to display labels, headings, and blocks of text. In the next exercise, you will be introduced to UI elements allowing the user to interact with your app.

Exercise 1.05 – adding interactive UI elements to display a bespoke greeting to the user

The goal of this exercise is to add the capability of users to add and edit text and then submit this information to display a bespoke greeting with the entered data. You will need to add editable text views to achieve this. The `EditText` view is typically how this is done and can be added in an XML layout file like this:

```
<EditText
    android:id="@+id/full_name"
    style="@style/TextAppearance.AppCompat.Title"
    android:layout_width="wrap_content"
    android:layout_height="wrap_content"
    android:hint="@string/first_name" />
```

This uses an Android `TextAppearance.AppCompat.Title` style to display a title, as shown in *Figure 1.21*:

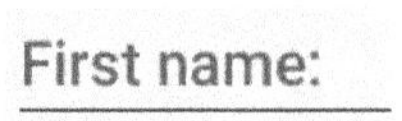

First name:

Figure 1.21 – EditText with a hint

Although this is perfectly fine to enable the user to add/edit text, the TextInputEditText material and its wrapper TextInputLayout view give some polish to the EditText display. Here's how EditText can be updated:

```
<com.google.android.material.textfield.TextInputLayout
    android:id="@+id/first_name_wrapper"
    style="@style/text_input_greeting"
    android:layout_width="match_parent"
    android:layout_height="wrap_content"
    android:hint="@string/first_name_text">
    <com.google.android.material.textfield
        .TextInputEditText
        android:id="@+id/first_name"
        android:layout_width="match_parent"
        android:layout_height="wrap_content" />
</com.google.android.material.textfield.TextInputLayout>
```

The output is as follows:

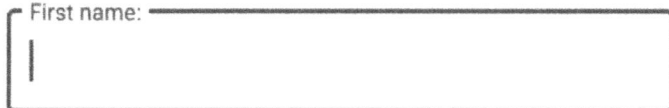

Figure 1.22 – The TextInputLayout/TextInputEditText material with a hint

TextInputLayout allows us to create a label for the TextInputEditText view and does a nice animation when the TextInputEditText view is focused (moving to the top of the field) while still displaying the label. The label is specified with android:hint.

You will change the Hello World text in your app so a user can enter their first and last name and further display a greeting by pressing a button. To do this, perform the following steps:

1. Create a new Android Studio project as you did in *Exercise 1.01, Creating an Android Studio project for your app*, called My Application.
2. Create the labels and text you are going to use in your app by adding these entries to app | src | main | res | values | strings.xml:

```
<string name="first_name_text">First name:</string>
<string name="last_name_text">Last name:</string>
<string name="enter_button_text">Enter</string>
```

```xml
<string name="welcome_to_the_app">Welcome to the app</string>
<string name="please_enter_a_name">Please enter a full name!</string>
```

3. Next, we will update our styles to use in the layout by adding the following styles to the app | src | main | res | values | themes.xml theme:

```xml
<style name="text_input_greeting" parent="Widget.MaterialComponents.TextInputLayout.OutlinedBox">
    <item name="android:layout_margin">8dp</item>
</style>
<style name="button_greeting">
    <item name="android:layout_margin">8dp</item>
    <item name="android:gravity">center</item>
</style>
<style name="greeting_display" parent="@style/TextAppearance.MaterialComponents.Body1">
    <item name="android:layout_margin">8dp</item>
    <item name="android:gravity">center</item>
    <item name="android:layout_height">40dp</item>
</style>
<style name="screen_layout_margin">
    <item name="android:layout_margin">12dp</item>
</style>
```

> **Note**
> The parents of some of the styles refer to Material styles, so these styles will be applied directly to the views and the styles specified.

4. Now that we have added the styles we want to apply to views in the layout and the text, we can update the layout in `activity_main.xml` in the app | src | main | res | layout folder:

```xml
<?xml version="1.0" encoding="utf-8"?>
<androidx.constraintlayout.widget.ConstraintLayout
    xmlns:android=
       "http://schemas.android.com/apk/res/android"
    xmlns:app="http://schemas.android.com/apk/res-auto"
    xmlns:tools="http://schemas.android.com/tools"
```

```xml
        android:layout_width="match_parent"
        android:layout_height="match_parent"
        style="@style/screen_layout_margin"
        tools:context=".MainActivity">
        <com.google.android.material.textfield.TextInputLayout
            android:id="@+id/first_name_wrapper"
            style="@style/text_input_greeting"
            android:layout_width="match_parent"
            android:layout_height="wrap_content"
            android:hint="@string/first_name_text"
            app:layout_constraintTop_toTopOf="parent"
            app:layout_constraintStart_toStartOf="parent">
            <com.google.android.material.textfield.
                TextInputEditText android:id="@+id/first_name"
                android:layout_width="match_parent" android:layout_
                height="wrap_content" />
        </com.google.android.material.textfield.TextInputLayout>
        <com.google.android.material.textfield.TextInputLayout
            android:id="@+id/last_name_wrapper"
            style="@style/text_input_greeting"
            android:layout_width="match_parent"
            android:layout_height="wrap_content"
            android:hint="@string/last_name_text"
            app:layout_constraintTop_toBottomOf="@id/first_name_
                wrapper"
            app:layout_constraintStart_toStartOf="parent">
            <com.google.android.material.textfield.
                TextInputEditText android:id="@+id/last_name"
                android:layout_width="match_parent" android:layout_
                height="wrap_content" />
        </com.google.android.material.textfield.TextInputLayout>
        <com.google.android.material.button.MaterialButton
            android:layout_width="match_parent"
            android:layout_height="wrap_content"
            style="@style/button_greeting"
            android:id="@+id/enter_button"
```

```
                android:text="@string/enter_button_text"
                app:layout_constraintTop_toBottomOf="@id/last_name_
                    wrapper"
                app:layout_constraintStart_toStartOf="parent"/>
        <TextView
                android:id="@+id/greeting_display"
                android:layout_width="match_parent"
                style="@style/greeting_display"
                app:layout_constraintTop_toBottomOf="@id/enter_
                    button"
                app:layout_constraintStart_toStartOf="parent" />
    </androidx.constraintlayout.widget.ConstraintLayout>
```

5. Run the app and see the look and feel. You have added IDs for all the views so they can be constrained against their siblings and also provide a way in the activity to get the values of the `TextInputEditText` views. The `style="@style.."` notation applies the style from the `themes.xml` file.

If you select one of the `TextInputEditText` views, you'll see the label animated and move to the top of the view:

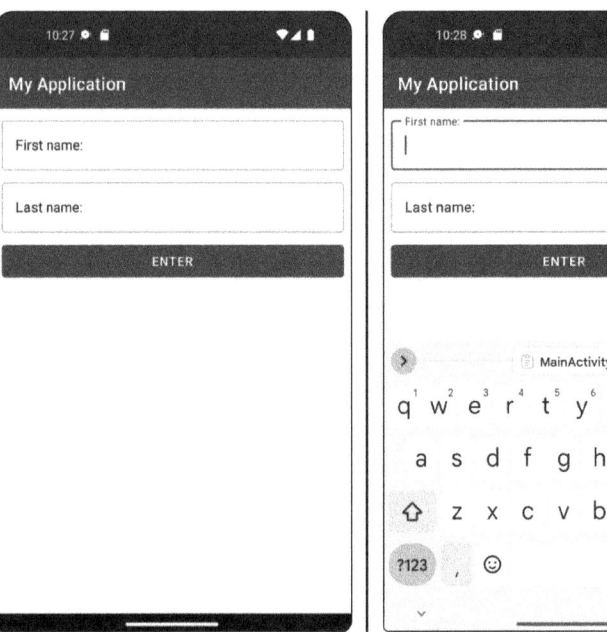

Figure 1.23 – The TextInputEditText fields with label states with no focus and with focus

6. Now, we must add the interaction with the view in our activity. The layout by itself doesn't do anything other than allow the user to enter text into the `EditText` fields. Clicking the button at this stage will not do anything. You will accomplish this by capturing the entered text by using the IDs of the form fields when the button is pressed and then using the text to populate a `TextView` message.
7. Open `MainActivity` and complete the next steps to process the entered text and use this data to display a greeting and handle any form input errors.
8. In the `onCreate` function, set a `ClickListener` on the button so we can respond to the button click and retrieve the form data by updating `MainActivity` to what is displayed in the following code block:

```kotlin
package com.example.myapplication

import androidx.appcompat.app.AppCompatActivity
import android.os.Bundle
import android.view.Gravity
import android.widget.Button
import android.widget.TextView
import android.widget.Toast
import com.example.myapplication.R
import com.google.android.material.textfield.TextInputEditText

class MainActivity : AppCompatActivity() {
    override fun onCreate(savedInstanceState: Bundle?) {
        super.onCreate(savedInstanceState)
        setContentView(R.layout.activity_main)
        findViewById<Button>(R.id.enter_button)?.
        setOnClickListener {
            //Get the greeting display text
            val greetingDisplay =
                findViewById<TextView>(R.id.greeting_
                display)
            //Get the first name TextInputEditText value
            val firstName =
                findViewById<TextInputEditText>(R.
                id.first_name)
                ?.text.toString().trim()
```

```
                //Get the last name TextInputEditText value
                val lastName =
                    findViewById<TextInputEditText>(R.
                    id.last_name)
                    ?.text.toString().trim()
                //Add code below this line in step 9 to Check
                    names are not empty here:


            }
        }
    }
```

9. Then, check that the trimmed names are not empty and format the name using Kotlin's string templates:

```
if (firstName.isNotEmpty() && lastName.isNotEmpty()) {
    val nameToDisplay = firstName.plus(" ")
        .plus(lastName)
    //Use Kotlin's string templates feature to display
        the name
    greetingDisplay?.text = " ${getString(R.string.
        welcome_to_the_app)} ${nameToDisplay}!"
}
```

10. Finally, show a message if the form fields have not been filled in correctly:

```
else {
    Toast.makeText(this, getString(R.string.please_
        enter_a_name), Toast.LENGTH_LONG)
        .apply {
            setGravity(Gravity.CENTER, 0, 0)
            show()
        }
}
```

The `Toast` specified is a small text dialog that appears above the main layout for a short time to display a message to the user before disappearing.

11. Run the app and enter text into the fields and verify that a greeting message is shown when both text fields are filled in, and a pop-up message appears with why the greeting hasn't been set if both fields are not filled in. You should see the following display for each one of these cases:

 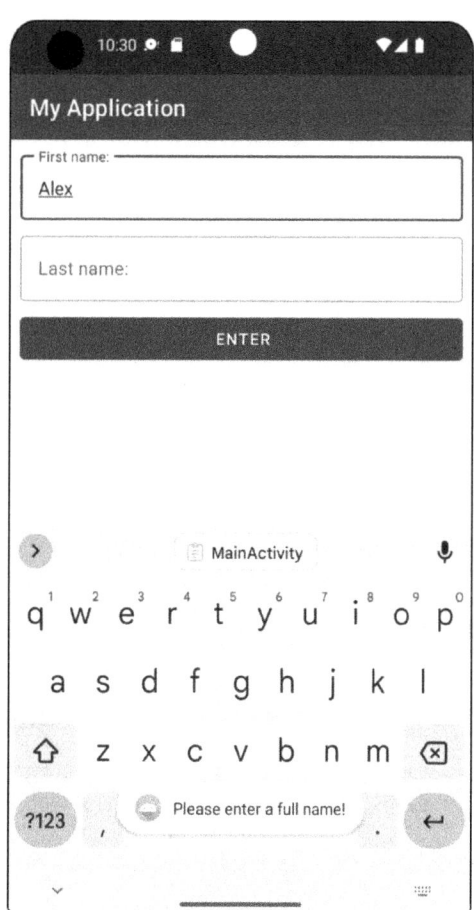

Figure 1.24 – The app with the name filled in correctly and with an error

The full exercise code can be viewed at https://packt.link/UxbOu.

The preceding exercise introduced you to adding interactivity to your app with the `EditText` fields that a user can fill in, adding a click listener to respond to button events, and performing some validation.

Accessing Views in layout files

The established way to access Views in layout files is to use `findViewById` with the name of the View's ID. So the `enter_button` button is retrieved by the `findViewById<Button>(R.id.enter_button)` syntax after the layout has been set in `setContentView(R.layout.activity_main)` in the Activity.

You will use this technique in this course. Google has also introduced **ViewBinding** to replace `findViewById`, which creates a binding class to access Views and has the advantage of null and type safety. You can read about this at `https://developer.android.com/topic/libraries/view-binding`.

Further input validation

Validating user input is a key concept in processing user data, and you must have seen it in action many times when you've not filled in a required field in a form. This is what the previous exercise validated when it checked that the user had entered values into both the first name and last name fields.

There are other validation options that are available directly within XML view elements. Let's say, for instance, you wanted to validate an IP address entered into a field. You know that an IP address can be four numbers separated by periods/dots where the maximum length of a number is three.

So, the maximum number of characters that can be entered into the field is 15, and only numbers and periods/dots can be entered. Two XML attributes can help us with the validation:

- `android:digits="0123456789."`: This restricts the characters that can be entered into the field by listing all the permitted individual characters
- `android:maxLength="15"`: This restricts the user from entering more than the maximum number of characters an IP address will consist of

So, this is how you can display this in a form field:

```xml
<com.google.android.material.textfield.TextInputLayout style="@style/Widget.MaterialComponents.TextInputLayout.OutlinedBox"
    android:layout_width="match_parent"
    android:layout_height="wrap_content">
    <com.google.android.material.textfield.TextInputEditText  android:id="@+id/ip_address"
    android:digits="0123456789."
    android:layout_width="match_parent"
    android:layout_height="wrap_content"
    android:maxLength="15" />
</com.google.android.material.textfield.TextInputLayout>
```

This validation restricts the characters that can be input and the maximum length. Additional validation would be required on the sequence of characters and whether they are periods/dots or numbers, as per the IP address format, but it is the first step to assist the user in entering the correct characters. There is also an `android:inputType` XML attribute, which can be used to specify permitted characters and configure the input options, `android:inputType="textPassword"`, for example, ensures that the characters entered are hidden. `android:inputType="Phone"` is the input method for a phone number.

With the knowledge gained from the chapter, let's start with the following activity.

Activity 1.01 – producing an app to create RGB colors

In this activity, we will look into a scenario that uses validation. Suppose you have been tasked with creating an app that shows how the RGB channels of red, green, and blue are added together in the RGB color space to create a color.

Each RGB channel should be added as two hexadecimal characters, where each character can be a value of 0–9 or A–F. The values will then be combined to produce a six-character hexadecimal string that is displayed as a color within the app.

This activity aims to produce a form with editable fields in which the user can add two hexadecimal values for each color. After filling in all three fields, the user should click a button that takes the three values and concatenates them to create a valid hexadecimal color string. This should then be converted to a color and displayed in the UI of the app.

The following steps will help you to complete the activity:

1. Create a new Android Studio project as you did in *Exercise 1.01, Creating an Android Studio project for your app*.
2. Add a `Title` constrained to the top of the layout.
3. Add a brief description to the user on how to complete the form.
4. Add three material `TextInputLayout` fields wrapping three `TextInputEditText` fields that appear under `Title`. These should be constrained so that each view is above the other (rather than to the side). Name the `TextInputEditText` fields `Red Channel`, `Green Channel`, and `Blue Channel`, respectively, and add a restriction to each field to allow entry only of two characters and add hexadecimal characters.
5. Add a button that takes the inputs from the three color fields.
6. Add a view that displays the produced color in the layout.
7. Finally, display the RGB color created from the three channels in the layout when the button is pressed and all input is valid.

The final output should look like this (the color will vary depending on the inputs):

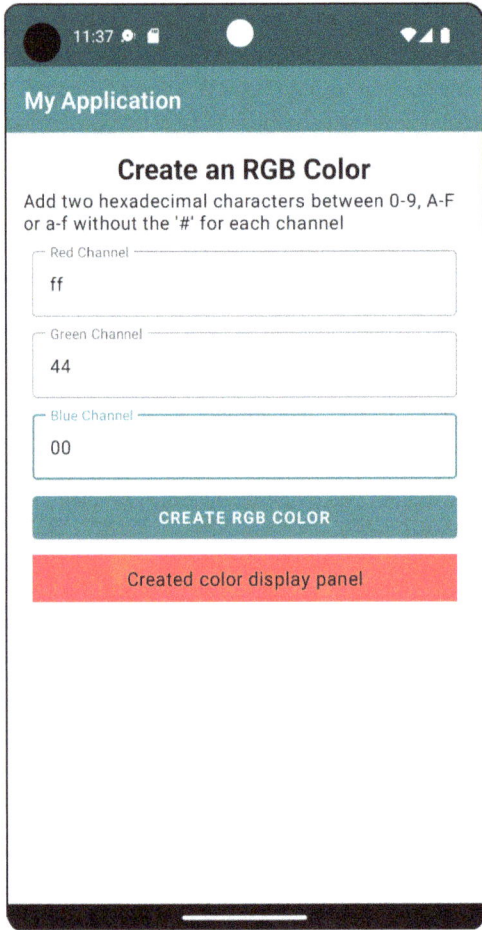

Figure 1.25 – Output when the color is displayed

Note

The solution to this activity can be found at `https://packt.link/By7eE`.

> **Note**
> When loading all completed projects from the GitHub repository for this course into Android Studio for the first time, do *not* open the project using **File | Open** from the top menu. Always use **File | New | Import Project**. This ensures the app builds correctly. When opening projects after the initial import, you can use **File | Open** or **File | Open Recent**.

Summary

This chapter has covered a lot about the foundations of Android development. You started with how to create Android projects using Android Studio and then created and ran apps on a virtual device.

The chapter then progressed by exploring the `AndroidManifest` file, which details the contents of your app and the permission model, followed by an introduction to Gradle and the process of adding dependencies and building your app.

This was then followed by going into the details of an Android application and the files and folder structure. Layouts and views were introduced, and exercises iterated to illustrate how to construct UIs with an introduction to Google's Material Design.

The next chapter will build on this knowledge by learning about the activity lifecycle, activity tasks, and launch modes, persisting and sharing data between screens, and how to create robust user journeys through your apps.

2
Building User Screen Flows

This chapter covers the Android activity lifecycle and explains how the Android system interacts with your app. By the end of this chapter, you'll have learned how to build user journeys through different screens. You'll also be able to use activity tasks and launch modes, save and restore the state of your activity, use logs to report on your application, and share data between screens.

The previous chapter introduced you to the core elements of Android development, from configuring your app using the `AndroidManifest.xml` file, working with simple activities, and the Android resource structure to building an app with `gradle` and running an app on a virtual device.

In this chapter, you'll go further and learn how the Android system interacts with your app through the Android lifecycle, how you are notified of changes to your app's state, and how you can use the Android lifecycle to respond to these changes.

You'll then progress to learning how to create user journeys through your app and how to share data between screens. You'll be introduced to different techniques to achieve these goals so that you'll be able to use them in your own apps and recognize them when you see them used in other apps.

We will cover the following topics in the chapter:

- The Activity lifecycle
- Saving and restoring the Activity state
- Activity interaction with Intents
- Intents, Tasks, and Launch Modes

Technical requirements

The complete code for all the exercises and the activity in this chapter is available on GitHub at `https://packt.link/PmKJ6`

The Activity lifecycle

In the previous chapter, we used the `onCreate(saveInstanceState: Bundle?)` method to display a layout in the UI of our screen. Now, we'll explore in more detail how the Android system interacts with your application to make this happen. As soon as an Activity is launched, it goes through a series of steps to take it through initialization, from preparing to be displayed to being partially displayed and then fully displayed.

There are also steps that correspond with your application being hidden, backgrounded, and then destroyed. This process is called the **Activity lifecycle**. For every one of these steps, there is a **callback** that your Activity can use to perform actions such as creating and changing the display and saving data when your app has been put into the background and then restoring that data after your app comes back into the foreground.

These callbacks are made on your Activity's parent, and it's up to you to decide whether you need to implement them in your own Activity to take any corresponding action. Each of these callback functions has the `override` keyword. The `override` keyword in Kotlin means that either this function is providing an implementation of an interface or an abstract method, or, in the case of your Activity here, which is a subclass, it is providing the implementation that will override its parent.

Now that you know how the Activity lifecycle works in general, let's go into more detail about the principal callbacks you will work with in order, from creating an Activity to the Activity being destroyed:

- `override fun onCreate(savedInstanceState: Bundle?)`: This is the callback that you will use the most for activities that draw a full-sized screen. It's here where you prepare your Activity layout to be displayed. At this stage, after the method has completed, it is still not displayed to the user, although it will appear that way if you don't implement any other callbacks. You usually set up the UI of your Activity here by calling the `setContentView(R.layout.activity_main)` method and carrying out any initialization that is required.

    This method is only called once in its lifecycle unless the Activity is created again. This happens by default for some actions (such as rotating the phone from portrait to landscape orientation). The `savedInstanceState` parameter of the `Bundle?` type (? means the type can be null) in its simplest form is a map of key-value pairs optimized to save and restore data.

    It will be null if this is the first time that the Activity has been run after the app has started, if the Activity is being created for the first time, or if the Activity is being recreated without any states being saved.

- `override fun onRestart()`: When the Activity restarts, this is called immediately before `onStart()`. It is important to be clear about the difference between restarting an Activity and recreating an activity. When the Activity is backgrounded by pressing the home button, when it comes back into the foreground again `onRestart()` will be called. Recreating an Activity is what happens when a configuration change happens, such as the device being rotated. The Activity is finished and then created again, in which case `onRestart()` will not be called.
- `override fun onStart()`: This is the first callback made when the Activity is brought from the background to the foreground.
- `override fun onRestoreInstanceState(savedInstanceState: Bundle?)`: If the state has been saved using `onSaveInstanceState(outState: Bundle?)`, this is the method that the system calls after `onStart()` where you can retrieve the `Bundle` state instead of restoring the state using `onCreate(savedInstanceState: Bundle?)`.
- `override fun onResume()`: This callback is run as the final stage of creating an Activity for the first time, and also when the app has been backgrounded and then is brought into the foreground. Upon the completion of this callback, the screen/activity is ready to be used, receive user events, and be responsive.
- `override fun onSaveInstanceState(outState: Bundle?)`: If you want to save the state of the activity, this function can do so. You add key-value pairs using one of the convenience functions depending on the data type. The data will then be available if your Activity is recreated in `onCreate(saveInstanceState: Bundle?)` and `onRestoreInstanceState(savedInstanceState: Bundle?)`.
- `override fun onPause()`: This function is called when the Activity starts to be backgrounded or another dialog or Activity comes into the foreground.
- `override fun onStop()`: This function is called when the Activity is hidden, either because it is being backgrounded or another Activity is being launched on top of it.
- `override fun onDestroy()`: This is called by the system to kill the Activity when system resources are low, when `finish()` is called explicitly on the Activity, or, more commonly, when the Activity is killed by the user closing the app from the recents/overview button.

The flow of callbacks/events is illustrated in the following diagram:

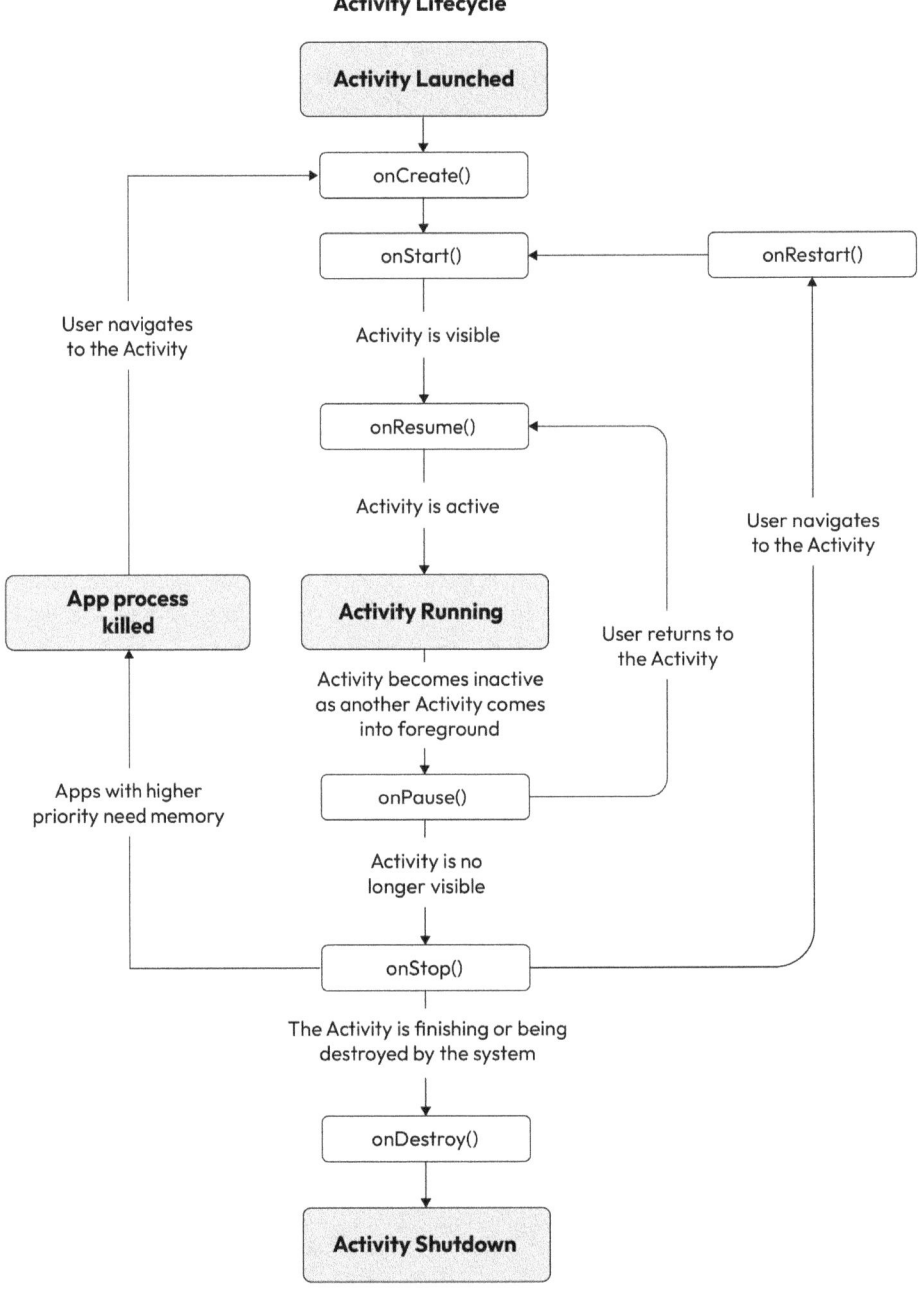

Figure 2.1 – Activity lifecycle

Now that you understand what these common lifecycle callbacks do, let's implement them to see when they are called.

Exercise 2.01 – logging the Activity Callbacks

Create an application called `Activity Callbacks` with an empty Activity. The aim of this exercise is to log the Activity callbacks and the order that they occur for common operations:

1. In order to verify the order of the callbacks, let's add a log statement at the end of each callback. Open up `MainActivity` and prepare the Activity for logging by adding `import android.util.Log` to the `import` statements. Then, add a constant to the class to identify your Activity. Constants in Kotlin are identified by the `const` keyword and can be declared at the top level (outside the class) or in an object within the class.

   Top-level constants are generally used if they are required to be public. For private constants, Kotlin provides a convenient way to add static functionality to classes by declaring a companion object. Add the following at the bottom of the class below `onCreate(savedInstanceState: Bundle?)`:

   ```
   companion object {
       private const val TAG = "MainActivity"
   }
   ```

   Then, add a log statement at the end of `onCreate(savedInstanceState: Bundle?)`:

   ```
   Log.d(TAG, "onCreate")
   ```

   Our Activity should now have the following code:

   ```
   package com.example.activitycallbacks

   import android.os.Bundle
   import android.util.Log
   import androidx.appcompat.app.AppCompatActivity

   class MainActivity : AppCompatActivity() {

       override fun onCreate(savedInstanceState: Bundle?) {
           super.onCreate(savedInstanceState)
           setContentView(R.layout.activity_main)
           Log.d(TAG, "onCreate")
       }
   ```

```
        companion object {
            private const val TAG = "MainActivity"
        }
    }
```

d in the preceding log statement refers to *debug*. There are six different log levels that can be used to output message information from the least to most important – v for *verbose*, d for *debug*, i for *info*, w for *warn*, e for *error*, and wtf for *what a terrible failure* (this last log level highlights an exception that should never occur):

```
Log.v(TAG, "verbose message")
Log.d(TAG, "debug message")
Log.i(TAG, "info message")
Log.w(TAG, "warning message")
Log.e(TAG, "error message")
Log.wtf(TAG, "what a terrible failure message")
```

2. Now, let's see how the logs are displayed in Android Studio. Open the **Logcat** window. It can be accessed by clicking on the **Logcat** tab at the bottom of the screen and also from the toolbar by going to **View** | **Tool Windows** | **Logcat**.

3. Run the app on the virtual device and examine the **Logcat** window output. You should see the log statement you have added formatted like the following line in *Figure 2.2*. If the **Logcat** window looks different, you might have to enable the newest version of **Logcat** by going to **Android Studio** | **Settings** | **Experimental** and checking the box that says **Enable New Logcat Tool Window**.

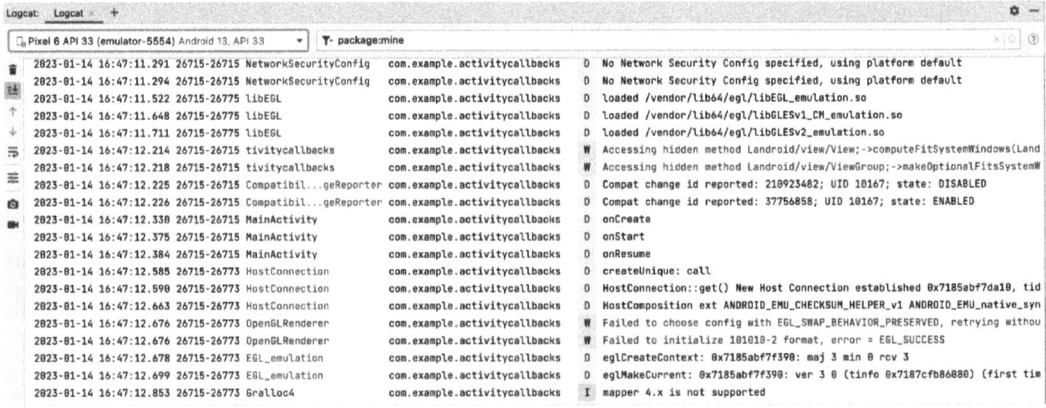

Figure 2.2 – Log output in Logcat

4. Log statements can be quite difficult to interpret at first glance, so let's break down the following statement into its separate parts:

```
2023-01-14 16:47:12.330 26715-26715/com.example.
activitycallbacks/D/onCreate
```

Let's examine the elements of the log statement in detail:

Fields	Values
Date	`2023-01-14`
Time	`16:47:12.330`
Process identifier and thread identifier (your app process ID and current thread ID)	`26715-26715`
Class name	`MainActivity`
Package name	`com.example.activitycallbacks`
Log level	`D (for Debug)`
Log message	`onCreate`

Figure 2.3 – Table explaining a log statement

By default, in the log filter (the text box above the log window), it says `package:mine`, which is your app logs. You can examine the output of the different log levels of all the processes on the device by changing the log filter from `level:debug` to other options in the drop-down menu. If you select `level:verbose`, as the name implies, you will see a lot of output.

5. What's great about the `tag` option of the log statement is that it enables you to filter the log statements that are reported in the **Logcat** window of Android Studio by typing in `tag` followed by the text of the tag, `tag:MainActivity`, as shown in *Figure 2.4*:

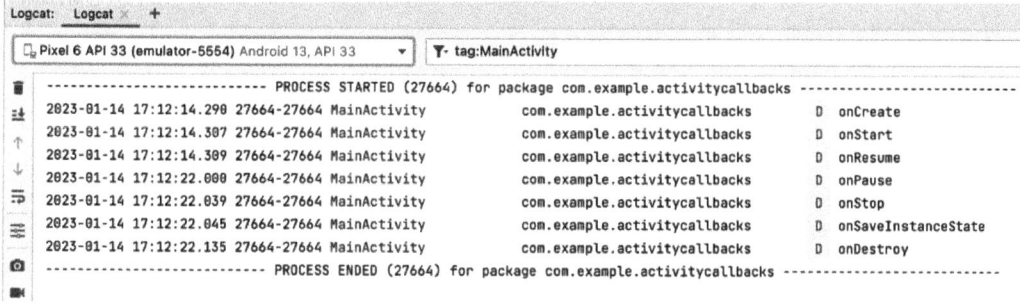

Figure 2.4 – Filtering log statements by the tag name

So, if you are debugging an issue in your Activity, you can type in the tag name and add logs to your Activity to see the sequence of log statements. This is what you are going to do next by implementing the principal Activity callbacks and adding a log statement to each one to see when they are run.

6. Place your cursor on a new line after the closing brace of the onCreate(savedInstanceState: Bundle?) function and then add the onRestart() callback with a log statement. Make sure you call through to super.onRestart() so that the existing functionality of the Activity callback works as expected:

```
override fun onRestart() {
    super.onRestart()
    Log.d(TAG, "onRestart")
}
```

> **Note**
> In Android Studio, you can start typing the name of a function and autocomplete options will pop up with suggestions for functions to override. Alternatively, if you go to the top menu and then **Code | Generate | Override methods**, you can select the methods to override.

Do this for all of the following callback functions:

```
onCreate(savedInstanceState: Bundle?)
onRestart()
onStart()
onRestoreInstanceState(savedInstanceState: Bundle)
onResume()
onPause()
onStop()
onSaveInstanceState(outState: Bundle)
onDestroy()
```

7. The completed activity will now override the callbacks with your implementation, which adds a log message. The following truncated code snippet shows a log statement in onCreate(savedInstanceState: Bundle?). The complete class is available at https://packt.link/Lj2GT:

```
package com.example.activitycallbacks
import android.os.Bundle
import android.util.Log
import androidx.appcompat.app.AppCompatActivity
```

```kotlin
class MainActivity : AppCompatActivity() {
    override fun onCreate(savedInstanceState: Bundle?) {
        super.onCreate(savedInstanceState)
        setContentView(R.layout.activity_main)
        Log.d(TAG, "onCreate")
    }
    companion object {
        private const val TAG = "MainActivity"
    }
}
```

8. Run the app, and then once it has loaded, as in *Figure 2.5*, look at the **Logcat** output; you should see the following log statements (this is a shortened version):

   ```
   D/MainActivity: onCreate
   D/MainActivity: onStart
   D/MainActivity: onResume
   ```

   The Activity has been created, started, and then prepared for the user to interact with:

Figure 2.5 – The app loaded and displaying MainActivity

9. Press the round home button in the center of the navigation controls in the emulator window above the virtual device and background the app. Not all devices use the same three-button navigation of *back* (triangle icon), *home* (circle icon), and *recents/overview* (square icon). Gesture navigation can also be enabled so all these actions can be achieved by swiping and optionally holding. You should now see the following **Logcat** output:

   ```
   D/MainActivity: onPause
   D/MainActivity: onStop
   D/MainActivity: onSaveInstanceState
   ```

   For apps that target versions below Android Pie (API 28), `onSaveInstanceState(outState: Bundle)` may also be called before `onPause()` or `onStop()`.

10. Now, bring the app back into the foreground by pressing the recents/overview square button in the emulator controls and selecting the app. You should now see the following:

    ```
    D/MainActivity: onRestart
    D/MainActivity: onStart
    D/MainActivity: onResume
    ```

    The Activity has been restarted. You might have noticed that the `onRestoreInstanceState(savedInstanceState: Bundle)` function was not called. This is because the Activity was not destroyed and recreated.

11. Press the recents/overview square button again and then swipe the app image upward to kill the activity. This is the output:

    ```
    D/MainActivity: onPause
    D/MainActivity: onStop
    D/MainActivity: onDestroy
    ```

12. Launch your app again and then rotate the phone. You might find that the phone does not rotate, and the display is sideways. If this happens, drag down the status bar at the very top of the virtual device, look for a button with a rectangular icon with arrows called **Auto-rotate**, and select it.

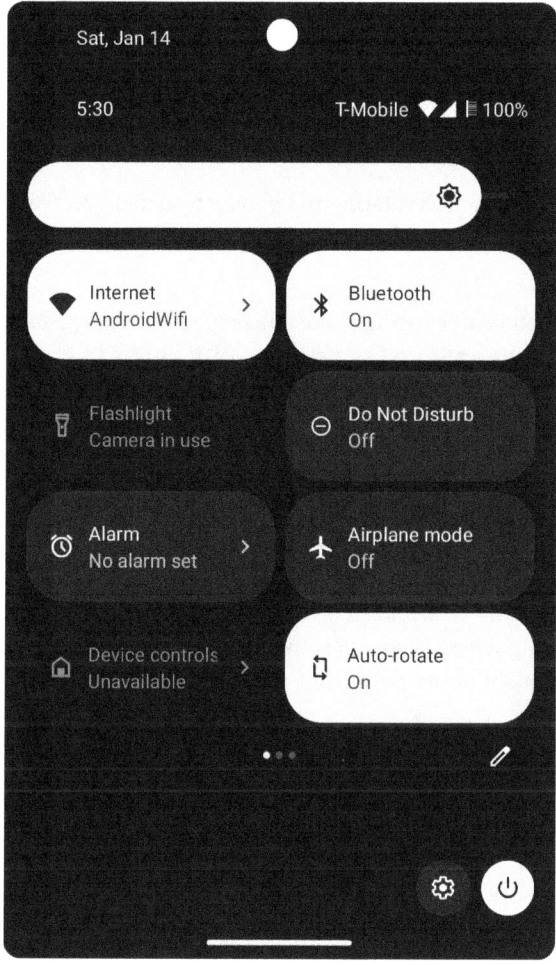

Figure 2.6 – Quick settings bar with Wi-Fi and Auto-rotate button selected

You should see the following callbacks:

```
D/MainActivity: onCreate
D/MainActivity: onStart
D/MainActivity: onResume
D/MainActivity: onPause
D/MainActivity: onStop
D/MainActivity: onSaveInstanceState
D/MainActivity: onDestroy
D/MainActivity: onCreate
D/MainActivity: onStart
```

```
D/MainActivity: onRestoreInstanceState
D/MainActivity: onResume
```

Please note that as stated in step 9, the order of the `onSaveInstanceState(outState: Bundle)` callback may vary.

Configuration changes, such as rotating the phone, by default recreate the activity. You can choose not to handle certain configuration changes in the app, which will then not recreate the activity.

13. To not recreate the activity for rotation, add `android:configChanges= "orientation|screenSize|screenLayout"` to `MainActivity` in the `AndroidManifest.xml` file. Launch the app and then rotate the phone, and these are the only callbacks that you have added to `MainActivity` that you will see:

    ```
    D/MainActivity: onCreate
    D/MainActivity: onStart
    D/MainActivity: onResume
    ```

    The `orientation` and `screenSize` values have the same function for different Android API levels for detecting screen orientation changes. The `screenLayout` value detects other layout changes that might occur on foldable phones.

    These are some of the config changes you can choose to handle yourself (another common one is `keyboardHidden` to react to changes in accessing the keyboard). The app will still be notified by the system of these changes through the following callback:

    ```
    override fun onConfigurationChanged(newConfig:
    Configuration) {
        super.onConfigurationChanged(newConfig)
        Log.d(TAG, "onConfigurationChanged")
    }
    ```

    If you add this callback function to `MainActivity`, and you have added `android: configChanges="orientation|screenSize|screenLayout"` to `Main Activity` in the manifest, you will see it called on rotation.

This approach of not restarting the activity is not recommended as the system will not apply alternative resources automatically. So, rotating a device from portrait to landscape won't apply a suitable landscape layout.

In this exercise, you have learned about the principal Activity callbacks and how they run when a user carries out common operations with your app through the system's interaction with `MainActivity`. In the next section, we will cover saving the state and restoring it, as well as see more examples of how the Activity lifecycle works.

Saving and restoring the Activity state

In this section, you'll explore how your Activity saves and restores the state. As you've learned in the previous section, configuration changes, such as rotating the phone, cause the Activity to be recreated. This can also happen if the system has to kill your app in order to free up memory.

In these scenarios, it is important to preserve the state of the Activity and then restore it. In the next two exercises, you'll work through an example ensuring that the user's data is restored when `TextView` is created and populated from a user's data after filling in a form.

Exercise 2.02 – saving and restoring the state in layouts

In this exercise, firstly create an application called `Save and Restore` with an empty activity. The app you are going to create will have a simple form that offers a discount code for a user's favorite restaurant if they enter some personal details (no actual information will be sent anywhere, so your data is safe):

1. Open up the `strings.xml` file (located in app | src | main | res | values | strings.xml) and add the following strings that you'll need for your app:

   ```
   <string name="header_text">Enter your name and email for
   a discount code at Your Favorite Restaurant!</string>
       <string name="first_name_label">First Name:</string>
       <string name="email_label">Email:</string>
       <string name="last_name_label">Last Name:</string>
       <string name="discount_code_button">GET DISCOUNT
           </string>
       <string name="discount_code_confirmation">Hey   %s!
           Here is your discount code</string>
       <string name="add_text_validation">Please fill in all
           form fields</string>
   ```

2. In `R.layout.activity_main`, replace the contents with the following XML that creates a containing layout file and adds a `TextView` header with the text `Enter your name and email for a discount code at Your Favorite Restaurant!` This is done by adding the `android:text` attribute with the `@string/header_text` value:

   ```
   <?xml version="1.0" encoding="utf-8"?>
   <androidx.constraintlayout.widget.ConstraintLayout
       xmlns:android=
           "http://schemas.android.com/apk/res/android"
       xmlns:app="http://schemas.android.com/apk/res-auto"
   ```

```xml
    xmlns:tools="http://schemas.android.com/tools"
    android:layout_width="match_parent"
    android:layout_height="match_parent"
    android:padding="4dp"
    android:layout_marginTop="4dp"
    tools:context=".MainActivity">
    <TextView
        android:id="@+id/header_text"
        android:gravity="center"
        android:textSize="20sp"
        android:paddingStart="8dp"
        android:paddingEnd="8dp"
        android:layout_width="wrap_content"
        android:layout_height="wrap_content"
        android:text="@string/header_text"
        app:layout_constraintTop_toTopOf="parent"
        app:layout_constraintEnd_toEndOf="parent"
        app:layout_constraintStart_toStartOf="parent" />
</androidx.constraintlayout.widget.ConstraintLayout>
```

You will see here that `android:textSize` is specified in `sp` which stands for Scale-independent pixels. This unit type represents the same values as density-independent pixels, which define the size measurement according to the density of the device that your app is being run on, and also change the text size according to the user's preference, defined in **Settings | Display | Font style** (this might be **Font size and style** or something similar, depending on the exact device you are using).

Other attributes in the layout affect positioning. The most common ones are padding and margin. Padding is applied on the inside of Views and is the space between the text and the border. Margins are specified on the outside of Views and are the space from the outer edges of Views. For example, `android:padding` sets the padding for the View with the specified value on all sides. Alternatively, you can specify the padding for one of the four sides of a View with `android:paddingTop`, `android:paddingBottom`, `android:paddingStart`, and `android:paddingEnd`. This pattern also exists to specify margins, so `android:layout_margin` specifies the margin value for all four sides of a View, and `android:layout_marginTop`, `android:layout_marginBottom`, `android:layout_marginStart`, and `android:layout_marginEnd` allow setting the margin for individual sides.

In order to have consistency and uniformity throughout the app with these positioning values, you can define the margin and padding values as dimensions contained within a `dimens.xml` file so they can be used in multiple layouts. A dimension value of `<dimen name="grid_4">4dp</`

dimen> can then be used as a View attribute like this: `android:paddingStart="@dimen/grid_4"`. To position the content within a View, you can specify `android:gravity`. The `center` value constrains the content both vertically and horizontally within the View.

3. Next, add three `EditText` views below `header_text` for the user to add their first name, last name, and email:

   ```
   <EditText
       android:id="@+id/first_name"
       android:textSize="20sp"
       android:layout_marginStart="24dp"
       android:layout_marginEnd="16dp"
       android:layout_width="wrap_content"
       android:layout_height="wrap_content"
       android:hint="@string/first_name_label"
       android:inputType="text"
       app:layout_constraintTop_toBottomOf="@id/header_text"
       app:layout_constraintStart_toStartOf="parent"
       />
   <EditText
       android:textSize="20sp"
       android:layout_marginEnd="24dp"
       android:layout_width="wrap_content"
       android:layout_height="wrap_content"
       android:hint="@string/last_name_label"
       android:inputType="text"
       app:layout_constraintTop_toBottomOf="@id/header_text"
       app:layout_constraintStart_toEndOf="@id/first_name"
       app:layout_constraintEnd_toEndOf="parent" />
   <!-- android:inputType="textEmailAddress" is not
   enforced, but is a hint to the IME (Input Method
   Editor) usually a keyboard to configure the
   display for an email -typically by showing the '@'
   symbol -->
   <EditText
       android:id="@+id/email"
       android:textSize="20sp"
       android:layout_marginStart="24dp"
   ```

```
            android:layout_marginEnd="32dp"
            android:layout_width="match_parent"
            android:layout_height="wrap_content"
            android:hint="@string/email_label"
            android:inputType="textEmailAddress"
            app:layout_constraintTop_toBottomOf="@id/first_name"
            app:layout_constraintEnd_toEndOf="parent"
            app:layout_constraintStart_toStartOf="parent" />
```

The `EditText` fields have an `inputType` attribute to specify the type of input that can be entered into the form field. Some values, such as `number` on `EditText`, restrict the input that can be entered into the field, and on selecting the field, suggest how the keyboard is displayed. Others, such as `android:inputType="textEmailAddress"`, will not enforce an @ symbol being added to the form field, but will give a hint to the keyboard to display it.

4. Finally, add a button for the user to press to generate a discount code, a `TextView` to display the discount code, and a `TextView` for the confirmation message:

```
<Button
        android:id="@+id/discount_button"
        android:textSize="20sp"
        android:layout_marginTop="12dp"
        android:gravity="center"
        android:layout_width="wrap_content"
        android:layout_height="wrap_content"
        android:text="@string/discount_code_button"
        app:layout_constraintTop_toBottomOf="@id/email"
        app:layout_constraintEnd_toEndOf="parent"
        app:layout_constraintStart_toStartOf="parent"/>
<TextView
        android:id="@+id/discount_code_confirmation"
        android:gravity="center"
        android:textSize="20sp"
        android:paddingStart="16dp"
        android:paddingEnd="16dp"
        android:layout_marginTop="8dp"
        android:layout_width="match_parent"
```

```
            android:layout_height="wrap_content"
            app:layout_constraintTop_toBottomOf="@id/discount_
                button"
            app:layout_constraintEnd_toEndOf="parent"
            app:layout_constraintStart_toStartOf="parent"
            tools:text="Hey John Smith! Here is your discount
                code" />
    <TextView
            android:id="@+id/discount_code"
            android:gravity="center"
            android:textSize="20sp"
            android:textStyle="bold"
            android:layout_marginTop="8dp"
            android:layout_width="match_parent"
            android:layout_height="wrap_content"
            app:layout_constraintTop_toBottomOf="@id/discount_
                code_confirmation"
            app:layout_constraintEnd_toEndOf="parent"
            app:layout_constraintStart_toStartOf="parent"
            tools:text="XHFG6H9O" />
```

There are also some attributes that you haven't seen before. The `xmlns:tools="http://schemas.android.com/tools"` tools namespace, which was specified at the top of the XML layout file, enables certain features that can be used when creating your app to assist with configuration and design.

The attributes are removed when you build your app, so they don't contribute to the overall size of the app. You are using the `tools:text` attribute to show the text that will typically be displayed in the form fields. This helps when you switch to the `Design` view from viewing the XML in the `Code` view in Android Studio as you can see an approximation of how your layout displays on a device.

5. Run the app and you should see the output displayed in *Figure 2.7*. The **GET DISCOUNT** button has not been enabled and so currently will not do anything.

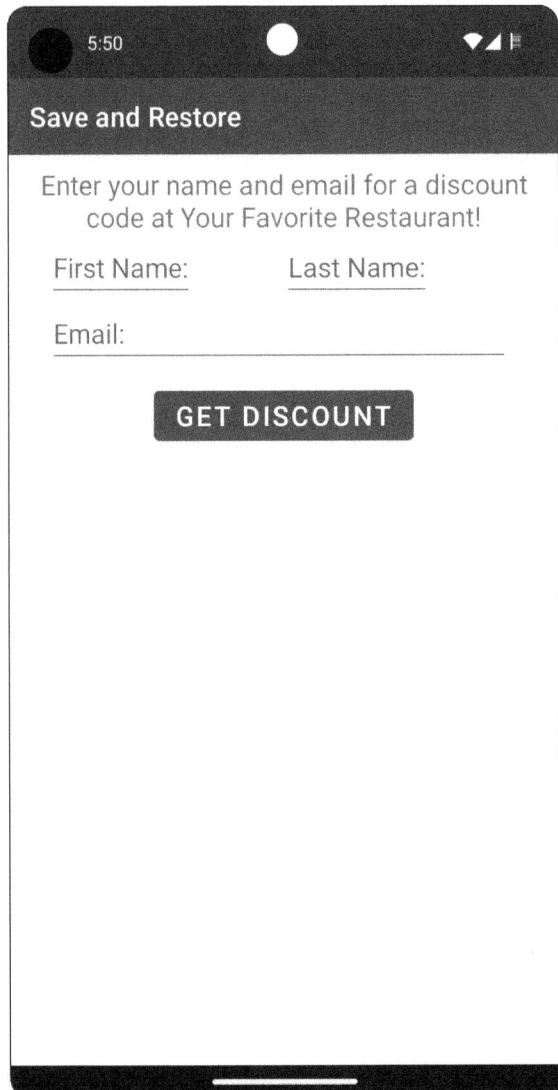

Figure 2.7 – The Activity screen on the first launch

6. Enter some text into each of the form fields:

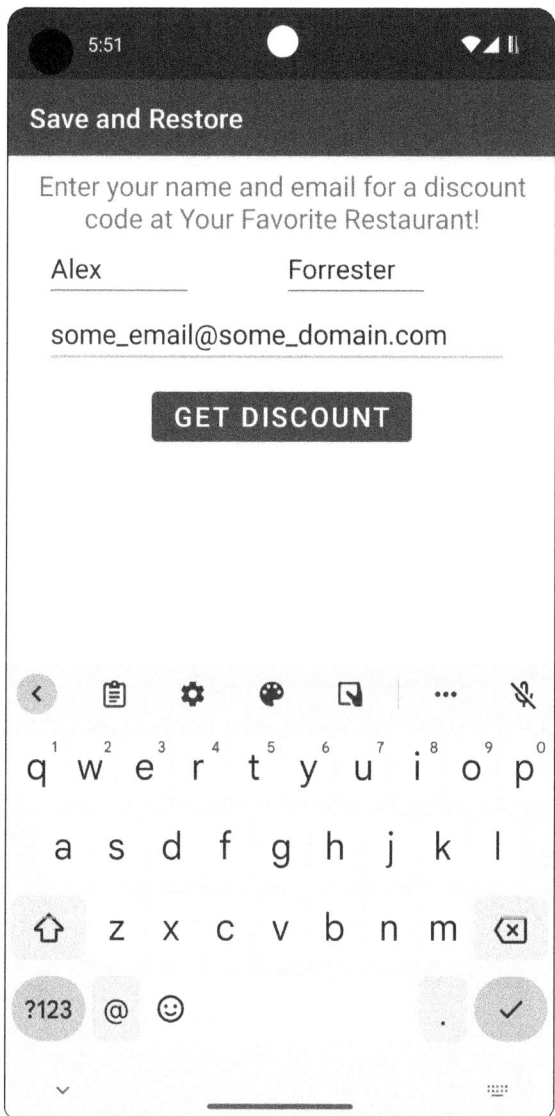

Figure 2.8 – The EditText fields filled in

7. Now, use the second rotate button in the virtual device controls (⟲) to rotate the phone 90 degrees to the right:

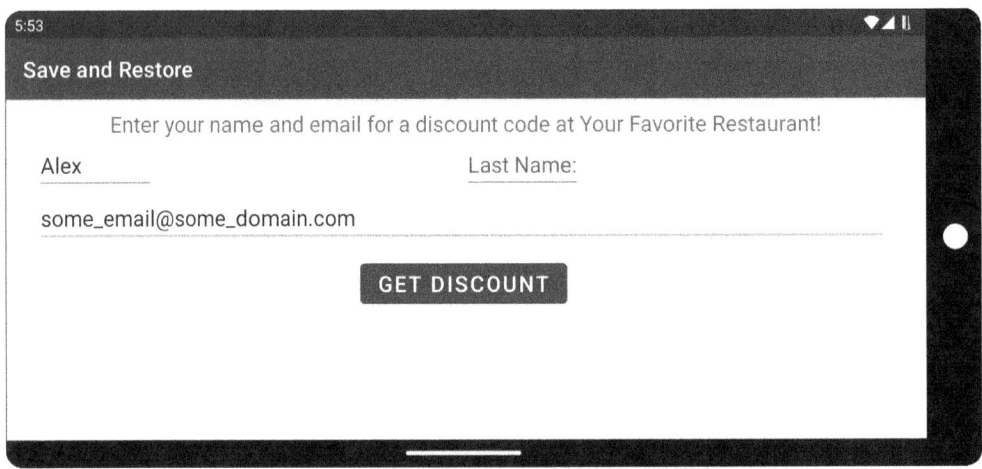

Figure 2.9 – The virtual device turned to landscape orientation

Can you spot what has happened? The **Last Name** field value is no longer set. It has been lost in the process of recreating the activity. Why is this? Well, in the case of the `EditText` fields, the Android framework will preserve the state of the fields if they have an ID set on them.

8. Go back to the `activity_main.xml` layout file and add an ID to the `Last Name` `EditText` which appears below the `First Name EditText`:

```
<EditText
    android:id="@+id/first_name"

<EditText
    android:id="@+id/last_name"
    ...
```

When you run up the app again and rotate the device, it will preserve the value you have entered. You've now seen that you need to set an ID on the `EditText` fields to preserve the state. For the `EditText` fields, it's common to retain the state on a configuration change when the user is entering details into a form so that it is the default behavior if the field has an ID.

Obviously, you want to get the details of the `EditText` field once the user has entered some text, which is why you set an ID, but setting an ID for other field types, such as `TextView`, does not retain the state if you update them and you need to save the state yourself. Setting IDs for Views that enable scrolling, such as `RecyclerView`, is also important as it enables the scroll position to be maintained when the Activity is recreated.

Now, you have defined the layout for the screen, but you have not added any logic for creating and displaying the discount code. In the next exercise, we will work through this.

The layout created in this exercise is available at https://packt.link/ZJleK.

You can find the code for the entire exercise at https://packt.link/Kh0kR.

Exercise 2.03 – saving and restoring the state with Callbacks

The aim of this exercise is to bring all the UI elements in the layout together to generate a discount code after the user has entered their data. In order to do this, you will have to add logic to the button to retrieve all the `EditText` fields and then display a confirmation to the user, as well as generate a discount code:

1. Open up `MainActivity.kt` and replace the contents with the following:

    ```
    package com.example.saveandrestore

    import android.content.Context
    import android.os.Bundle
    import android.util.Log
    import android.view.inputmethod.InputMethodManager
    import android.widget.Button
    import android.widget.EditText
    import android.widget.TextView
    import android.widget.Toast
    import androidx.appcompat.app.AppCompatActivity
    import java.util.*

    class MainActivity : AppCompatActivity() {

        private val discountButton: Button
            get() = findViewById(R.id.discount_button)
        private val firstName: EditText
            get() = findViewById(R.id.first_name)
        private val lastName: EditText
    ```

```kotlin
        get() = findViewById(R.id.last_name)
    private val email: EditText
        get() = findViewById(R.id.email)
    private val discountCodeConfirmation: TextView
        get() = findViewById(R.id.discount_code_
            confirmation)
    private val discountCode: TextView
        get() = findViewById(R.id.discount_code)

    override fun onCreate(savedInstanceState: Bundle?) {
        super.onCreate(savedInstanceState)
        setContentView(R.layout.activity_main)
        Log.d(TAG, "onCreate")
        // here we handle the Button onClick event
        discountButton.setOnClickListener {
            val firstName = firstName.text.toString().
                trim()
            val lastName = lastName.text.toString().
                trim()
            val email = email.text.toString()
            if (firstName.isEmpty() || lastName.isEmpty()
            || email.isEmpty()) {
                Toast.makeText(this, getString(R.string.
                    add_text_validation),
                    Toast.LENGTH_LONG)
                    .show()
            } else {
                val fullName = firstName.plus(" ")
                    .plus(lastName)
                discountCodeConfirmation.text =
                    getString(R.string.discount_code_
                    confirmation, fullName)
                // Generates discount code
                discountCode.text = UUID.randomUUID().
                    toString().take(8).uppercase()
                hideKeyboard()
```

```
                    }
                }
            }
            private fun hideKeyboard() {
                if (currentFocus != null) {
                    val imm = getSystemService(Context.INPUT_
                        METHOD_SERVICE) as InputMethodManager
                    imm.hideSoftInputFromWindow(currentFocus?.
                        windowToken, 0)
                }
            }
            companion object {
                private const val TAG = "MainActivity"
            }
        }
```

`get() = ...` is a custom accessor for a property.

Upon clicking the discount button, you retrieve the values from the `first_name` and `last_name` fields, concatenate them with a space, and then use a string resource to format the discount code confirmation text. The string you reference in the `strings.xml` file is as follows:

```
<string name="discount_code_confirmation">Hey %s! Here
is your discount code</string>
```

The `%s` value specifies a string value to be replaced when the string resource is retrieved. This is done by passing in the full name when getting the string:

```
getString(R.string.discount_code_confirmation,
fullName)
```

The code is generated by using the **Universally Unique Identifier** (**UUID**) library from the `java.util` package. This creates a unique ID, and then the `take()` Kotlin function is used to get the first eight characters before setting these to uppercase. Finally, `discountCode` is set in the view, the keyboard is hidden, and all the form fields are set back to their initial values.

2. Run the app and enter some text into the name and email fields, and then click on GET DISCOUNT:

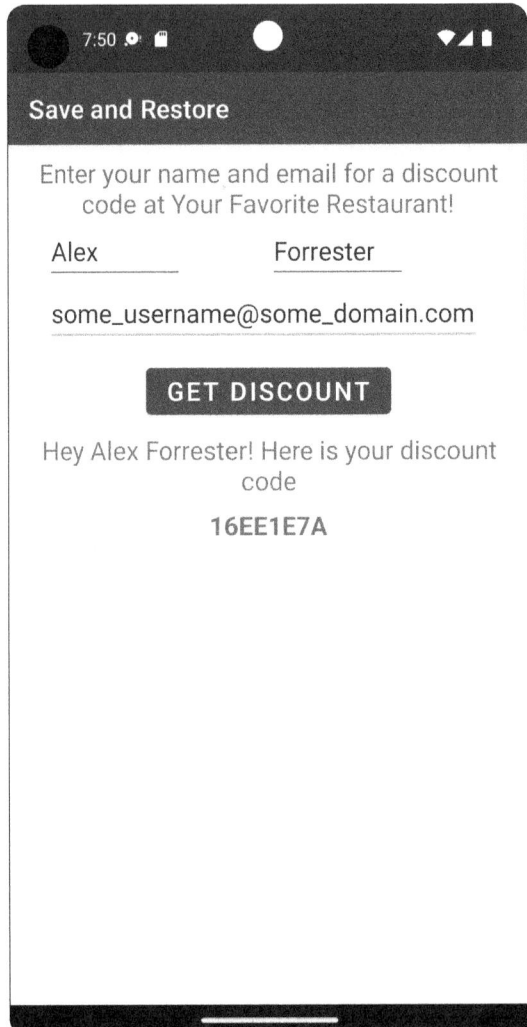

Figure 2.10 – Screen displayed after the user has generated a discount code

The app behaves as expected, showing the confirmation.

3. Now, rotate the phone by pressing the second rotate button ( ) in the emulator controls and observe the result:

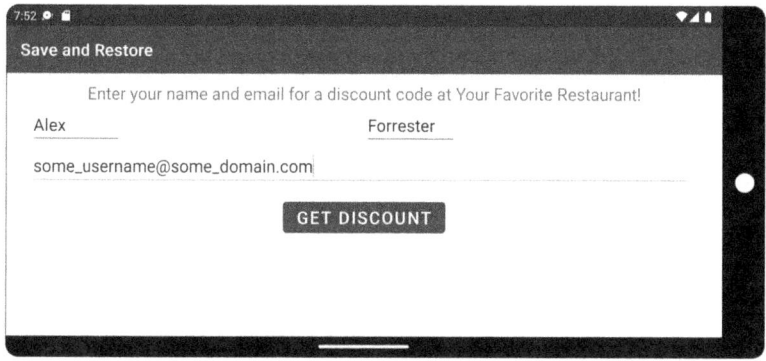

Figure 2.11 – Discount code no longer displaying on the screen

Oh, no! The discount code has gone. The `TextView` fields do not retain the state, so you will have to save the state yourself.

4. Go back into `MainActivity.kt` and add the following Activity callbacks:

   ```
   override fun onRestoreInstanceState(savedInstanceState: 
   Bundle) {
       super.onRestoreInstanceState(savedInstanceState)
       Log.d(TAG, "onRestoreInstanceState")
   }
   override fun onSaveInstanceState(outState: Bundle) {
       super.onSaveInstanceState(outState)
       Log.d(TAG, "onSaveInstanceState")
   }
   ```

   These callbacks, as the names declare, enable you to save and restore the instance state. `onSaveInstanceState(outState: Bundle)` allows you to add key-value pairs from your Activity when it is being backgrounded or destroyed, which you can retrieve in either `onCreate(savedInstanceState: Bundle?)` or `onRestoreInstanceState(savedInstanceState: Bundle)`.

   So, you have two callbacks to retrieve the state once it has been set. If you are doing a lot of initialization in `onCreate(savedInstanceState: Bundle)`, it might be better to use `onRestoreInstanceState(savedInstanceState: Bundle)` to retrieve this instance state when your Activity is being recreated. In this way, it's clear which state is being recreated. However, you might prefer to use `onCreate(savedInstanceState: Bundle)` if there is minimal setup required.

   Whichever of the two callbacks you decide to use, you will have to get the state you set in the `onSaveInstanceState(outState: Bundle)` call. For the next step in the exercise, you will use `onRestoreInstanceState(savedInstanceState: Bundle)`.

5. Add two constants to the `MainActivity` companion object, which is at the bottom of `MainActivity`:

   ```
   private const val DISCOUNT_CONFIRMATION_MESSAGE =
   "DISCOUNT_CONFIRMATION_MESSAGE"
   private const val DISCOUNT_CODE = "DISCOUNT_CODE"
   ```

6. Now, add these constants as keys for the values you want to save and retrieve and make the following changes to the Activity in the onSaveInstanceState(outState: Bundle) and onRestoreInstanceState(savedInstanceState: Bundle) functions.

   ```
   override fun onRestoreInstanceState(savedInstanceState:
   Bundle) {
       super.onRestoreInstanceState(savedInstanceState)
       Log.d(TAG, "onRestoreInstanceState")
       // Get the discount code or an empty string if it
          hasn't been set
       discountCode.text = savedInstanceState.
          getString(DISCOUNT_CODE,"")
       // Get the discount confirmation message or an empty
          string if it hasn't been set
       discountCodeConfirmation.text = savedInstanceState.
          getString(DISCOUNT_CONFIRMATION_MESSAGE,"")
   }
   override fun onSaveInstanceState(outState: Bundle) {
       super.onSaveInstanceState(outState)
       Log.d(TAG, "onSaveInstanceState")
       outState.putString(DISCOUNT_CODE, discountCode.text.
       toString())
       outState.putString(DISCOUNT_CONFIRMATION_MESSAGE,
       discountCodeConfirmation.text.toString())
   }
   ```

7. Run the app, enter the values into the `EditText` fields, and then generate a discount code. Then, rotate the device and you will see that the discount code is restored in *Figure 2.12*:

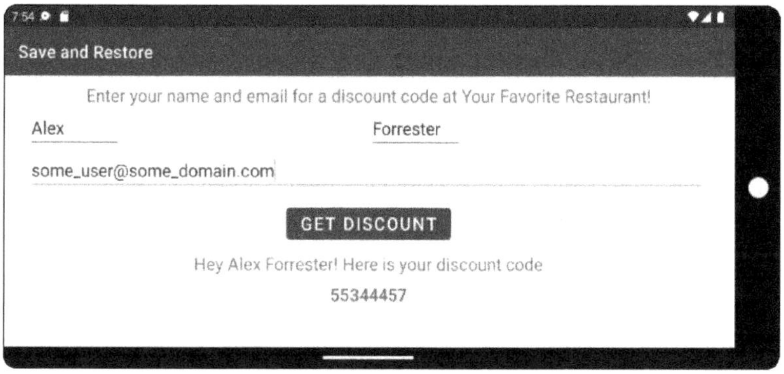

Figure 2.12 – Discount code continues to be displayed on the screen

In this exercise, you first saw how the state of the `EditText` fields is maintained on configuration changes. You also saved and restored the instance state using the Activity lifecycle `onSaveInstanceState(outState: Bundle)` and `onCreate(savedInstanceState: Bundle?)/onRestoreInstanceState(savedInstanceState: Bundle)` functions. These functions provide a way to save and restore simple data. The Android framework also provides `ViewModel`, an Android architecture component that is lifecycle-aware. The mechanisms of how to save and restore this state (with `ViewModel`) are managed by the framework, so you don't have to explicitly manage it as you have done in the preceding example. You will learn how to use this component in *Chapter 11*, *Android Architecture Components*.

The complete code for this exercise can be found here: `https://packt.link/zsGW3`.

In the next section, you will add another Activity to an app and navigate between the activities.

Activity interaction with Intents

An intent in Android is a communication mechanism between components. Within your own app, a lot of the time, you will want another specific Activity to start when some action happens in the current activity. Specifying exactly which Activity will start is called an **explicit intent**. On other occasions, you will want to get access to a system component, such as the camera. As you can't access these components directly, you will have to send an intent, which the system resolves in order to open the camera. These are called **implicit intents**. An intent filter has to be set up in order to register to respond to these events. Go to the `AndroidManifest.xml` file from the previous exercise and you will see an example of two intent filters set within the `<intent-filter>` XML element of the `MainActivity`:

```
<intent-filter>
    <action android:name="android.intent.action.MAIN"/>
```

```xml
        <category android:name="android.intent.category. LAUNCHER"/>
    </intent-filter>
```

The one specified with `<action android:name="android.intent.action.MAIN" />` means that this is the main entry point into the app. Depending on which category is set, it governs which Activity first starts when the app is started. The other intent filter that is specified is `<category android:name="android.intent.category.LAUNCHER" />`, which defines that the app should appear in the launcher. When combined, the two intent filters define that when the app is started from the launcher, `MainActivity` should be started. Removing the `<action android:name="android.intent.action.MAIN" />` intent filter results in the "Error running 'app': Default Activity not found" message. As the app has not got a main entry point, it can't be launched. If you remove `<category android:name="android.intent.category.LAUNCHER" />` then there is nowhere that it can be launched from.

For the next exercise, you will see how intents work to navigate around your app.

Exercise 2.04 – an introduction to Intents

The goal of this exercise is to create a simple app that uses intents to display text to the user based on their input:

1. Create a new project in Android Studio called `Intents Introduction` and select an empty Activity. Once you have set up the project, go to the toolbar and select File | New | Activity | Empty Activity. Call it `WelcomeActivity` and leave all the other defaults as they are. It will be added to the `AndroidManifest.xml` file, ready to use. The issue you have now that you've added `WelcomeActivity` is knowing how to do anything with it. `MainActivity` starts when you launch the app, but you need a way to launch `WelcomeActivity` and then, optionally, pass data to it, which is when you use intents.

2. In order to work through this example, add the following strings to the `strings.xml` file.

```xml
        <string name="header_text">Please enter your name and
            then we\'ll get started!</string>
        <string name="welcome_text">Hello %s, we hope you
            enjoy using the app!</string>
        <string name="full_name_label">Enter your full
            name:</string>
        <string name="submit_button_text">SUBMIT</string>
```

3. Next, change the `MainActivity` layout in `activity_main.xml` and replace the content with the following code to add an `EditText` and a `Button`:

```xml
<?xml version="1.0" encoding="utf-8"?>
<androidx.constraintlayout.widget.ConstraintLayout
    xmlns:android=
        "http://schemas.android.com/apk/res/android"
    xmlns:app="http://schemas.android.com/apk/res-auto"
    xmlns:tools="http://schemas.android.com/tools"
    android:layout_width="match_parent"
    android:padding="28dp"
    android:layout_height="match_parent"
    tools:context=".MainActivity">
    <EditText
        android:id="@+id/full_name"
        android:layout_width="wrap_content"
        android:layout_height="wrap_content"
        android:textSize="28sp"
        android:hint="@string/full_name_label"
        android:layout_marginBottom="24dp"
        app:layout_constraintTop_toTopOf="parent"
        app:layout_constraintStart_toStartOf="parent"
        app:layout_constraintEnd_toEndOf="parent"/>
    <Button
        android:id="@+id/submit_button"
        android:textSize="24sp"
        android:layout_width="wrap_content"
        android:layout_height="wrap_content"
        android:text="@string/submit_button_text"
        app:layout_constraintTop_toBottomOf="@id/full_
            name"
        app:layout_constraintEnd_toEndOf="parent"
        app:layout_constraintStart_toStartOf="parent"/>
</androidx.constraintlayout.widget.ConstraintLayout>
```

The app, when run, looks as in *Figure 2.13*:

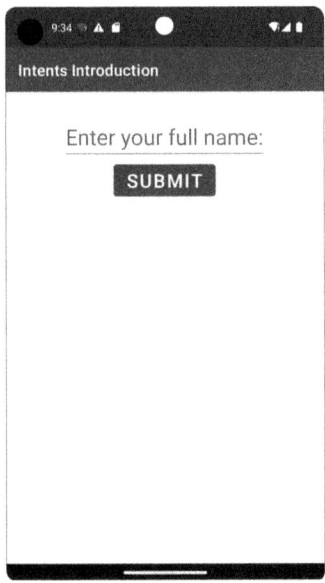

Figure 2.13 – The app display after adding the EditText full name field and SUBMIT button

You now need to configure the button so that when it's clicked, it retrieves the user's full name from the `EditText` field and then sends it in an intent, which starts `WelcomeActivity`.

4. Update the `activity_welcome.xml` layout file to prepare to do this:

   ```
   <?xml version="1.0" encoding="utf-8"?>
   <androidx.constraintlayout.widget.ConstraintLayout
       xmlns:android=
         "http://schemas.android.com/apk/res/android"
       xmlns:app="http://schemas.android.com/apk/res-auto"
       xmlns:tools="http://schemas.android.com/tools"
       android:layout_width="match_parent"
       android:layout_height="match_parent"
       tools:context=".WelcomeActivity">
       <TextView
           android:id="@+id/welcome_text"
           android:textSize="24sp"
           android:padding="24sp"
           android:layout_width="wrap_content"
   ```

```
                android:layout_height="wrap_content"
                app:layout_constraintTop_toTopOf="parent"
                app:layout_constraintEnd_toEndOf="parent"
                app:layout_constraintStart_toStartOf="parent"
                app:layout_constraintBottom_toBottomOf="parent"
                tools:text="Welcome John Smith we hope you enjoy
                    using the app!"/>
</androidx.constraintlayout.widget.ConstraintLayout>
```

You are adding a `TextView` field to display the full name of the user with a welcome message. The logic to create the full name and welcome message will be shown in the next step.

5. Now, open `MainActivity` and add a constant value above the class header and update the imports:

   ```
   package com.example.intentsintroduction

   import android.content.Intent
   import android.os.Bundle
   import android.widget.Button
   import android.widget.EditText
   import android.widget.Toast
   import androidx.appcompat.app.AppCompatActivity
   const val FULL_NAME_KEY = "FULL_NAME_KEY"

   class MainActivity : AppCompatActivity() {
       ...
   }
   ```

   You will use the constant to set the key to hold the full name of the user by setting it in the intent.

6. Then, add the following code to the bottom of `onCreate(savedInstanceState: Bundle?)`:

   ```
   findViewById<Button>(R.id.submit_button).
   setOnClickListener {
       val fullName = findViewById<EditText>(R.id.full_
           name).text.toString()

       if (fullName.isNotEmpty()) {
           // Set the name of the Activity to launch
   ```

```
            val welcomeIntent = Intent(this,
                WelcomeActivity::class.java)
            welcomeIntent.putExtra(FULL_NAME_KEY,
                fullName)
            startActivity(welcomeIntent)
        } else {
            Toast.makeText(this, getString(
                R.string.full_name_label),
                Toast.LENGTH_LONG).show()
        }
    }
```

There is logic to retrieve the value of the full name and verify that the user has filled this in; otherwise, a pop-up toast message will be shown if it's blank. The main logic, however, takes the `fullName` value of the `EditText` field and creates an explicit intent to start `WelcomeActivity`, and then puts an `Extra` key with a value into the Intent. The last step is to use the intent to start `WelcomeActivity`.

7. Now, run the app, enter your name, and press **SUBMIT**, as shown in *Figure 2.14*:

Figure 2.14 – The default screen displayed when the intent extras data is not processed

Well, that's not very impressive. You've added the logic to send the user's name, but not to display it.

8. To enable this, please open `WelcomeActivity` and add the import `import android.widget.TextView` to the imports list and add the following to the bottom of `onCreate(savedInstanceState: Bundle?)`:

```
if (intent != null) {
    val fullName = intent.getStringExtra(FULL_NAME_KEY)
    findViewById<TextView>(R.id.welcome_text).text =
        getString(R.string.welcome_text, fullName)
}
```

We check that the intent that started the Activity is not null and then retrieve the string value that was passed from the `MainActivity` intent by getting the string `FULL_NAME_KEY` extra key. We then format the `<string name="welcome_text">Hello %s, we hope you enjoy using the app!</string>` resource string by getting the string from the resources and passing in the `fullname` value retrieved from the intent. Finally, this is set as the text of `TextView`.

9. Run the app again, and a simple greeting will be displayed, as in *Figure 2.15*:

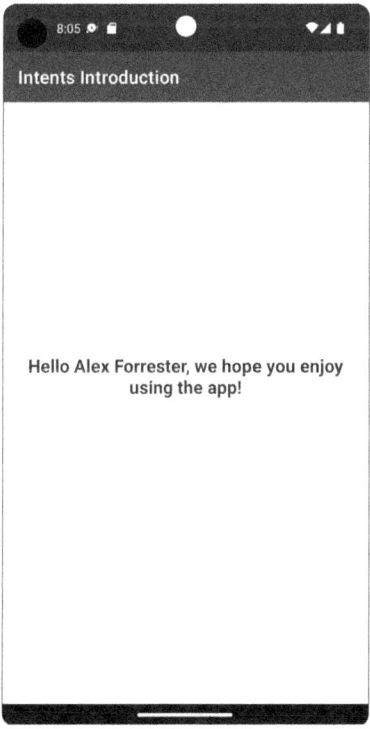

Figure 2.15 – User welcome message displayed

This exercise, although very simple in terms of layouts and user interaction, allows the demonstration of some core principles of intents. You will use them to add navigation and create user flows from one section of your app to another. In the next section, you will see how you can use intents to launch an Activity and receive a result back from it.

Exercise 2.05 – retrieving a result from an Activity

For some user flows, you will only launch an Activity for the sole purpose of retrieving a result back from it. This pattern is often used to ask permission to use a particular feature, popping up a dialog with a question about whether the user gives their permission to access contacts, the calendar, and so on, and then reporting the result of yes or no back to the calling Activity. In this exercise, you will ask the user to pick their favorite color of the rainbow, and then once that is chosen, display the result in the calling activity:

1. Create a new project named `Activity Results` with an empty activity and add the following strings to the `strings.xml` file:

   ```xml
   <string name="header_text_main">Please click the button below to choose your favorite color of the rainbow!</string>
   <string name="header_text_picker">Rainbow Colors</string>
   <string name="footer_text_picker">Click the button above which is your favorite color of the rainbow.</string>
   <string name="color_chosen_message">%s is your favorite color!</string>
   <string name="submit_button_text">CHOOSE COLOR</string>
   <string name="red">RED</string>
   <string name="orange">ORANGE</string>
   <string name="yellow">YELLOW</string>
   <string name="green">GREEN</string>
   <string name="blue">BLUE</string>
   <string name="indigo">INDIGO</string>
   <string name="violet">VIOLET</string>
   <string name="unexpected_color">Unexpected color</string>
   ```

2. Add the following colors to `colors.xml`:

   ```xml
   <!--Colors of the Rainbow -->
   <color name="red">#FF0000</color>
   <color name="orange">#FF7F00</color>
   <color name="yellow">#FFFF00</color>
   <color name="green">#00FF00</color>
   ```

```xml
        <color name="blue">#0000FF</color>
        <color name="indigo">#4B0082</color>
        <color name="violet">#9400D3</color>
```

3. Now, you have to set up the Activity that will set the result you receive in `MainActivity`. Go to **File** | **New** | **Activity** | **EmptyActivity** and create an Activity called `RainbowColorPickerActivity`.

4. Update the `activity_main.xml` layout file to display a header, a button, and then a hidden `android:visibility="gone"` View, which will be made visible and set with the user's favorite color of the rainbow when the result is reported:

```xml
<?xml version="1.0" encoding="utf-8"?>
<androidx.constraintlayout.widget.ConstraintLayout
    xmlns:android=
       "http://schemas.android.com/apk/res/android"
    xmlns:app="http://schemas.android.com/apk/res-auto"
    xmlns:tools="http://schemas.android.com/tools"
    android:layout_width="match_parent"
    android:layout_height="match_parent"
    tools:context=".MainActivity">
    <TextView
        android:id="@+id/header_text"
        android:textSize="20sp"
        android:padding="10dp"
        android:gravity="center"
        android:layout_width="wrap_content"
        android:layout_height="wrap_content"
        android:text="@string/header_text_main"
        app:layout_constraintTop_toTopOf="parent"
        app:layout_constraintEnd_toEndOf="parent"
        app:layout_constraintStart_toStartOf="parent"/>
    <Button
        android:id="@+id/submit_button"
        android:textSize="18sp"
        android:layout_width="wrap_content"
        android:layout_height="wrap_content"
        android:text="@string/submit_button_text"
```

```xml
            app:layout_constraintTop_toBottomOf="@id/header_
                text"
            app:layout_constraintEnd_toEndOf="parent"
            app:layout_constraintStart_toStartOf="parent"/>
        <TextView
            android:id="@+id/rainbow_color"
            android:layout_width="320dp"
            android:layout_height="50dp"
            android:layout_margin="12dp"
            android:textSize="22sp"
            android:textColor="@android:color/white"
            android:gravity="center"
            android:visibility="gone"
            tools:visibility="visible"
            app:layout_constraintTop_toBottomOf="@id/submit_
                button"
            app:layout_constraintStart_toStartOf="parent"
            app:layout_constraintEnd_toEndOf="parent"
            tools:text="BLUE is your favorite color"
            tools:background="@color/blue"/>
    </androidx.constraintlayout.widget.ConstraintLayout>
```

5. You'll be using the `registerForActivityResult(ActivityResultContracts.StartActivityForResult())` function to get a result back from the Activity you launch. Add two constant keys for the values we want to use in the intent, as well as a default color constant above the class header in `MainActivity`, and update the imports so it is displayed as follows with the package name and imports:

    ```
    package com.example.activityresults

    import android.content.Intent
    import android.graphics.Color
    import android.os.Bundle
    import android.widget.Button
    import android.widget.TextView
    import androidx.activity.result.contract.
    ActivityResultContracts
    import androidx.appcompat.app.AppCompatActivity
    ```

```
import androidx.core.content.ContextCompat
import androidx.core.view.isVisible

const val RAINBOW_COLOR_NAME = "RAINBOW_COLOR_NAME" //
Key to return rainbow color name in intent
const val RAINBOW_COLOR = "RAINBOW_COLOR" // Key to
return rainbow color in intent
const val DEFAULT_COLOR = "#FFFFFF" // White

class MainActivity : AppCompatActivity()…
```

6. Then, create a property below the class header that is used to both launch the new activity and return a result from it:

```
private val startForResult =
    registerForActivityResult(ActivityResultContracts.
    StartActivityForResult()) { activityResult ->
        val data = activityResult.data
        val backgroundColor = data?.getIntExtra(RAINBOW_
            COLOR, Color.parseColor(DEFAULT_COLOR))
            ?: Color.parseColor(DEFAULT_COLOR)
        val colorName = data?.getStringExtra(RAINBOW_
            COLOR_NAME) ?: ""
        val colorMessage = getString(R.string.color_
            chosen_message, colorName)

        val rainbowColor = findViewById<TextView>(R.
            id.rainbow_color)
        rainbowColor.setBackgroundColor(ContextCompat.
        getColor(this, backgroundColor))
        rainbowColor.text = colorMessage
        rainbowColor.isVisible = true
    }
```

Once the result is returned, you can proceed to query the intent data for the values you are expecting. For this exercise, we want to get the background color name (`colorName`) and the hexadecimal value of the color (`backgroundColor`) so that we can display it. The ? operator checks whether the value is null (that is, not set in the intent), and if so, the Elvis operator (?:) sets the default value. The color message uses string formatting to set a message, replacing the placeholder in the resource value with the color name. Now that you've got the colors, you can make the `rainbow_color` TextView field visible and set the background color of the View to `backgroundColor` and add text displaying the name of the user's favorite color of the rainbow.

7. First, add the logic to launch the Activity from the property defined previously by adding the following to the bottom of `onCreate(savedInstanceState: Bundle?)`::

    ```
    findViewById<Button>(R.id.submit_button)
    .setOnClickListener {
        startForResult.launch(Intent(this,
        RainbowColorPickerActivity::class.java)
        )
    }
    ```

    This creates an Intent that is launched for its result: `Intent(this, RainbowColorPickerActivity::class.java)`.

8. For the layout of the `RainbowColorPickerActivity` activity, you are going to display a button with a background color and color name for each of the seven colors of the rainbow: RED, ORANGE, YELLOW, GREEN, BLUE, INDIGO, and VIOLET. These will be displayed in a `LinearLayout` vertical list. For most of the layout files in the book, you will be using `ConstraintLayout`, as it provides fine-grained positioning of individual Views. For situations where you need to display a vertical or horizontal list of a small number of items, `LinearLayout` is also a good choice. If you need to display a large number of items, then `RecyclerView` is a better option as it can cache layouts for individual rows and recycle views that are no longer displayed on the screen. You will learn about `RecyclerView` in *Chapter 6, RecyclerView*.

9. The first thing you need to do in `RainbowColorPickerActivity` is create the layout. This will be where you present the user with the option to choose their favorite color of the rainbow.

10. Open `activity_rainbow_color_picker.xml` and replace the layout, inserting the following:

    ```
    <?xml version="1.0" encoding="utf-8"?>
    <ScrollView xmlns:android=
        "http://schemas.android.com/apk/res/android"
        xmlns:app="http://schemas.android.com/apk/res-auto"
        xmlns:tools="http://schemas.android.com/tools"
        android:layout_width="match_parent"
        android:layout_height="wrap_content">
    </ScrollView>
    ```

We are adding `ScrollView` to allow the contents to scroll if the screen height cannot display all of the items. `ScrollView` can only take one child View, which is the layout to scroll.

11. Next, add `LinearLayout` within `ScrollView` to display the contained views in the order that they are added with a header and a footer. The first child View is a header with the title of the page and the last View that is added is a footer with instructions for the user to pick their favorite color:

```xml
<LinearLayout
    android:layout_width="match_parent"
    android:layout_height="wrap_content"
    android:gravity="center_horizontal"
    android:orientation="vertical"
    tools:context=".RainbowColorPickerActivity">
    <TextView
        android:id="@+id/header_text"
        android:layout_width="wrap_content"
        android:layout_height="wrap_content"
        android:layout_marginTop="10dp"
        android:padding="10dp"
        android:text="@string/header_text_picker"
        android:textAllCaps="true"
        android:textSize="24sp"
        android:textStyle="bold"
        app:layout_constraintEnd_toEndOf="parent"
        app:layout_constraintStart_toStartOf="parent"
        app:layout_constraintTop_toTopOf="parent" />
    <TextView
        android:layout_width="380dp"
        android:layout_height="wrap_content"
        android:gravity="center"
        android:padding="10dp"
        android:text="@string/footer_text_picker"
        android:textSize="20sp"
        app:layout_constraintEnd_toEndOf="parent"
        app:layout_constraintStart_toStartOf="parent"
        app:layout_constraintTop_toTopOf="parent" />
</LinearLayout>
```

The layout should now look as in *Figure 2.16* in the app:

Figure 2.16 – Rainbow colors screen with a header and footer

12. Now, finally, add the button views between the header and the footer to select a color of the rainbow, and then run the app (the following code only displays the first button). The full layout is available at https://packt.link/ZgdHX:

```
<Button
    android:id="@+id/red_button"
    android:layout_width="120dp"
    android:layout_height="wrap_content"
    android:backgroundTint="@color/red"
    android:text="@string/red" />
```

These buttons are displayed in the order of the colors of the rainbow with the color text and background. The XML id attribute is what you will use in the Activity to prepare the result of what is returned to the calling activity.

13. Now, open `RainbowColorPickerActivity` and replace the content with the following:

    ```
    package com.example.activityresults

    import android.app.Activity
    import android.content.Intent
    import androidx.appcompat.app.AppCompatActivity
    import android.os.Bundle
    import android.view.View
    import android.widget.Toast

    class RainbowColorPickerActivity : AppCompatActivity() {
        override fun onCreate(savedInstanceState: Bundle?) {
            super.onCreate(savedInstanceState)
            setContentView(R.layout.activity_rainbow_color_
                picker)
        }
        private fun setRainbowColor(colorName: String, color:
        Int) {
            Intent().let { pickedColorIntent ->
                pickedColorIntent.putExtra(RAINBOW_COLOR_
                    NAME, colorName)
                pickedColorIntent.putExtra(RAINBOW_COLOR,
                    color)
                setResult(Activity.RESULT_OK,
                    pickedColorIntent)
                finish()
            }
        }
    }
    ```

    The `setRainbowColor` function creates an intent and adds the rainbow color name and the rainbow color hex value as String extras. The result is then returned to the calling Activity, and as you have no further use for this Activity, you call `finish()` so that the calling Activity is displayed. The way that you retrieve the rainbow color that the user has chosen is by adding a listener for all the buttons in the layout.

14. Now, add the following to the bottom of `onCreate(savedInstanceState: Bundle?)`:

    ```
    val colorPickerClickListener = View.OnClickListener {
    view ->
        when (view.id) {
            R.id.red_button -> setRainbowColor(getString(R.
                string.red), R.color.red)
            R.id.orange_button -> setRainbowColor(getString(R.
                string.orange), R.color.orange)
            R.id.yellow_button -> setRainbowColor(getString(R.
                string.yellow), R.color.yellow)
            R.id.green_button -> setRainbowColor(getString(R.
                string.green), R.color.green)
            R.id.blue_button -> setRainbowColor(getString(R.
                string.blue), R.color.blue)
            R.id.indigo_button -> setRainbowColor(getString(R.
                string.indigo), R.color.indigo)
            R.id.violet_button -> setRainbowColor(getString(R.
                string.violet), R.color.violet)
            else -> {
                Toast.makeText(this, getString(R.string.
                    unexpected_color), Toast.LENGTH_LONG)
                    .show()
            }
        }
    }
    ```

    The `colorPickerClickListener` added in the preceding code determines which colors to set for the `setRainbowColor(colorName: String, color: Int)` function by using a when statement. The when statement is the equivalent of the `switch` statement in Java and languages based on C. It allows multiple conditions to be satisfied with one branch and is more concise. In the preceding example, `view.id` is matched against the IDs of the rainbow layout buttons and, when found, executes the branch, setting the color name and hex value from the string resources to pass into `setRainbowColor(colorName: String, color: Int)`.

15. Now, add this click listener to the buttons from the layout below the preceding code:

    ```
    findViewById<View>(R.id.red_button).
    setOnClickListener(colorPickerClickListener)
    ```

```
findViewById<View>(R.id.orange_button).
setOnClickListener(colorPickerClickListener)
findViewById<View>(R.id.yellow_button).
setOnClickListener(colorPickerClickListener)
findViewById<View>(R.id.green_button).
setOnClickListener(colorPickerClickListener)
findViewById<View>(R.id.blue_button).
setOnClickListener(colorPickerClickListener)
findViewById<View>(R.id.indigo_button).
setOnClickListener(colorPickerClickListener)
findViewById<View>(R.id.violet_button).
setOnClickListener(colorPickerClickListener)
```

Every button has a `ClickListener` interface attached, and as the operation is the same, they have the same `ClickListener` interface attached. Then, when the button is pressed, it sets the result of the color that the user has chosen and returns it to the calling activity.

16. Now, run the app and press the CHOOSE COLOR button, as shown in *Figure 2.17*:

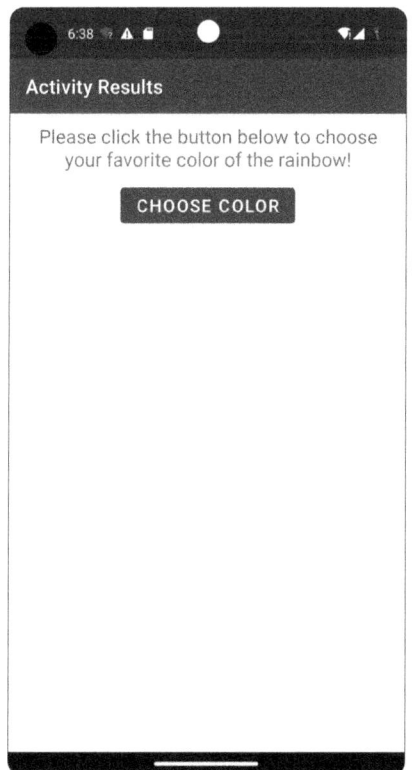

Figure 2.17 – The rainbow colors app start screen

17. Now, select your favorite color of the rainbow:

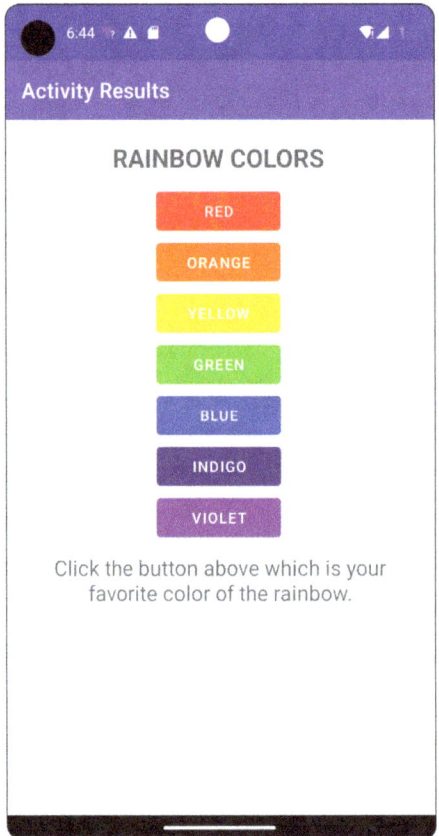

Figure 2.18 – The rainbow colors selection screen

18. Once you've chosen your favorite color, a screen with your favorite color will be displayed, as shown in *Figure 2.19*:

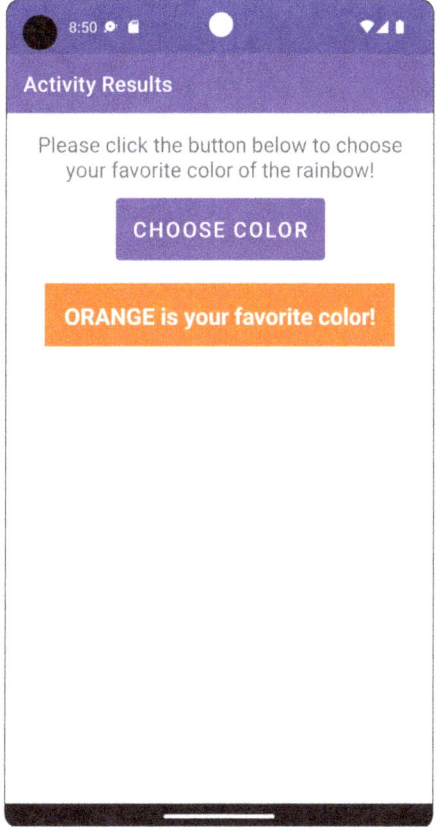

Figure 2.19 – The app displaying the selected color

As you can see, the app displays the color that you've selected as your favorite color in *Figure 2.19*.

This exercise introduced you to another way of creating user flows using `registerFor ActivityResult`. This can be very useful for carrying out a dedicated Task where you need a result before proceeding with the user's flow through the app. Next, you will explore launch modes and how they impact the flow of user journeys when building apps.

Intents, Tasks, and Launch Modes

Up until now, you have been using the standard behavior for creating Activities and moving from one Activity to the next. When you open the app from the launcher with the default behavior, it creates its own Task, and each Activity you create is added to a back stack, so when you open three Activities one after the other as part of your user's journey, pressing the back button three times will move the user back through the previous screens/Activities and then go back to the device's home screen, while still keeping the app open.

The launch mode for this type of Activity is called `Standard`; it is the default and doesn't need specifying in the Activity element of `AndroidManifest.xml`. Even if you launch the same Activity three times, one after the other, there will be three instances of the same activity that exhibit the behavior described previously.

For some apps, you may want to change this behavior so the same instance is used. The launch mode that can help here is called `singleTop`. If a `singleTop` Activity is the most recently added, when the same `singleTop` Activity is launched again, then instead of creating a new Activity, it uses the same Activity and runs the `onNewIntent` callback. In this callback, you receive an intent, and you can then process this intent as you have done previously in `onCreate`.

There are three other launch modes to be aware of called `singleTask`, `singleInstance` and `singleInstancePerTask`. These are not for general use and are only used for special scenarios. Detailed documentation of all launch modes can be viewed here: https://developer.android.com/guide/topics/manifest/activity-element#lmode.

You'll explore the differences in behavior of the `Standard` and `singleTop` launch modes in the next exercise.

Exercise 2.06 – setting the Launch Mode of an Activity

This exercise has many different layout files and Activities to illustrate the two most commonly used launch modes. Please download the code from https://packt.link/DQrGI:

1. Open up the `activity_main.xml` file and examine it.

   This illustrates a new concept when using layout files. If you have a layout file and you would like to include it in another layout, you can use the `<include>` XML element (have a look at the following snippet of the layout file):

   ```
   <include layout="@layout/letters"
       android:id="@+id/letters_layout"
       android:layout_width="wrap_content"
       android:layout_height="wrap_content"
       app:layout_constraintLeft_toLeftOf="parent"
       app:layout_constraintRight_toRightOf="parent"
       app:layout_constraintTop_toBottomOf="@id/launch_mode_
           standard"/>
   <include layout="@layout/numbers"
       android:layout_width="wrap_content"
       android:layout_height="wrap_content"
   ```

```
            app:layout_constraintLeft_toLeftOf="parent"
            app:layout_constraintRight_toRightOf="parent"
            app:layout_constraintTop_toBottomOf="@id/launch_mode_
                single_top"/>
```

The preceding layout uses the `include` XML element to include the two layout files: `letters.xml` and `numbers.xml`.

2. Open up and inspect the `letters.xml` and `numbers.xml` files found in the res | layout folder. These are very similar and are only differentiated from the buttons they contain by the ID of the buttons themselves and the text label they display.

3. Run the app and you will see the following screen:

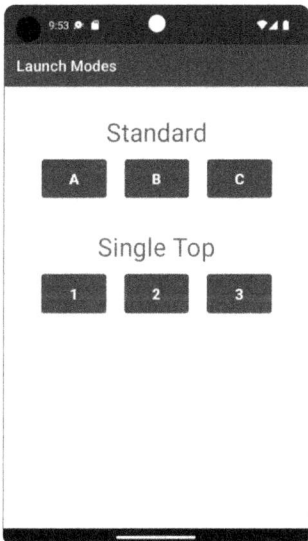

Figure 2.20 – App displaying both the standard and single top modes

In order to demonstrate/illustrate the difference between the `standard` and `singleTop` activity launch modes, you have to launch two or three activities one after the other.

4. Open up `MainActivity` and examine the contents of the code block (truncated) in `onCreate(savedInstanceState: Bundle?)`:

```
val buttonClickListener = View.OnClickListener { view ->
    when (view.id) {
        R.id.letterA -> startActivity(Intent(this,
            ActivityA::class.java))
```

```
                // Other letters and numbers follow the same 
                   pattern/flow
                else -> {
                    Toast.makeText(
                        this, getString(
                            R.string.unexpected_button_pressed
                        ),
                        Toast.LENGTH_LONG
                    )
                        .show()
                }
            }
        }
        findViewById<View>(R.id.letterA).
        setOnClickListener(buttonClickListener)
        // The buttonClickListener is set on all the number and 
        letter views
```

The logic contained in the main Activity and the other activities is basically the same. It displays an Activity and allows the user to press a button to launch another Activity using the same logic of creating a `ClickListener` and setting it on the button you saw in *Exercise 2.05, Retrieving a result from an Activity*.

5. Open the `AndroidManifest.xml` file and you will see the following activities displayed:

   ```
   <activity android:name=".ActivityA"
   android:launchMode="standard"/>
   <activity android:name=".ActivityB"
   android:launchMode="standard"/>
   <activity android:name=".ActivityC"
   android:launchMode="standard"/>
   <activity android:name=".ActivityOne"
   android:launchMode="singleTop"/>
   <activity android:name=".ActivityTwo"
   android:launchMode="singleTop"/>
   <activity android:name=".ActivityThree"
   android:launchMode="singleTop"/>
   ```

   You launch an Activity based on a button pressed on the main screen, but the letter and number activities have a different launch mode, which you can see specified in the `AndroidManifest.xml` file.

The `standard` launch mode is specified here to illustrate the difference between `standard` and `singleTop`, but `standard` is the default and would be how the Activity is launched if the `android:launchMode` XML attribute was not present.

6. Press one of the letters under the `Standard` heading and you will see the following screen (with **A**, **B**, or **C**):

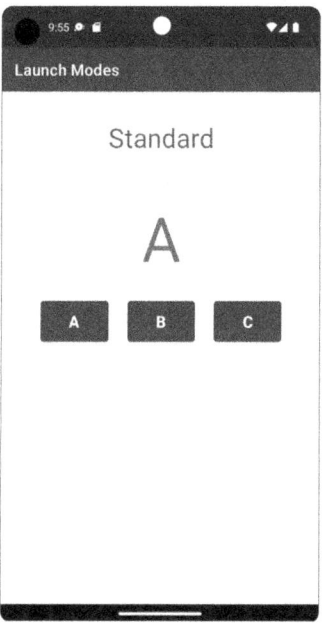

Figure 2.21 – The app displaying standard activity

7. Keep on pressing any of the letter buttons, which will launch another Activity. Logs have been added to show this sequence of launching activities. Here is the log after pressing 10 letter Activities randomly:

```
MainActivity  com.example.launchmodes  onCreate
Activity A    com.example.launchmodes  onCreate
Activity A    com.example.launchmodes  onCreate
Activity B    com.example.launchmodes  onCreate
Activity B    com.example.launchmodes  onCreate
Activity C    com.example.launchmodes  onCreate
Activity B    com.example.launchmodes  onCreate
Activity B    com.example.launchmodes  onCreate
Activity A    com.example.launchmodes  onCreate
Activity C    com.example.launchmodes  onCreate
```

```
Activity B    com.example.launchmodes onCreate
Activity C    com.example.launchmodes onCreate
```

If you observe the preceding log, every time the user presses a character button in launch mode, a new instance of the character Activity is launched and added to the back stack.

8. Close the app, making sure it is not backgrounded (or in the recents/overview menu) but is actually closed, and then open the app again and press one of the number buttons under the **Single Top** heading:

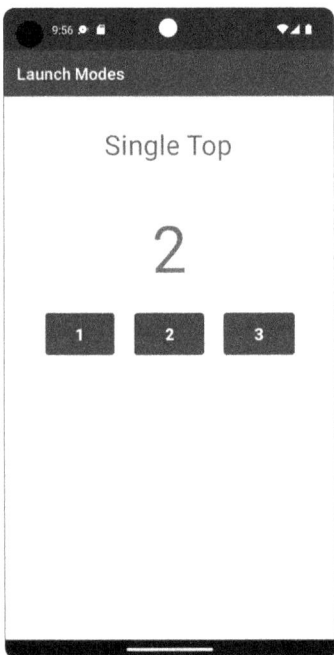

Figure 2.22 – The app displaying the Single Top activity

9. Press the number buttons 10 times, but make sure you press the same number button at least twice sequentially before pressing another number button.

The logs you should see in the **Logcat** window (**View** | **Tool Windows** | **Logcat**) should be similar to the following:

```
MainActivity com.example.launchmodes onCreate
Activity 1   com.example.launchmodes onCreate
Activity 1   com.example.launchmodes onNewIntent
Activity 2   com.example.launchmodes onCreate
Activity 2   com.example.launchmodes onNewIntent
```

```
Activity 3    com.example.launchmodes    onCreate
Activity 2    com.example.launchmodes    onCreate
Activity 3    com.example.launchmodes    onCreate
Activity 3    com.example.launchmodes    onNewIntent
Activity 1    com.example.launchmodes    onCreate
Activity 1    com.example.launchmodes    onNewIntent
```

You'll notice that instead of calling `onCreate` when you pressed the same button again at least twice sequentially, the Activity is not created, but a call is made to `onNewIntent`. If you press the back button, you'll notice that it will take you less than 10 clicks to back out of the app and return to the home screen, reflecting the fact that 10 activities have not been created.

Activity 2.01 – creating a login form

The aim of this activity is to create a login form with username and password fields. Once the values in these fields are submitted, check these entered values against the hardcoded values and display a welcome message if they match, or an error message if they don't, and return the user to the login form. The steps needed to achieve this are the following:

1. Create a form with username and password `EditText` Views and a `LOGIN` button.
2. Add a `ClickListener` interface to the button to react to a button press event.
3. Validate that the form fields are filled in.
4. Check the submitted username and password fields against the hardcoded values.
5. Display a welcome message with the username if successful and hide the form.
6. Display an error message if not successful and redirect the user back to the form.

There are a few possible ways that you could go about trying to complete this activity. Here are three ideas for approaches you could adopt:

- Use a `singleTop` Activity and send an intent to route to the same Activity to validate the credentials
- Use a `standard` Activity to pass a username and password to another Activity and validate the credentials
- Use `registerForActivityResult` to carry out the validation in another Activity and then return the result

The completed app, upon its first loading, should look as in *Figure 2.23*:

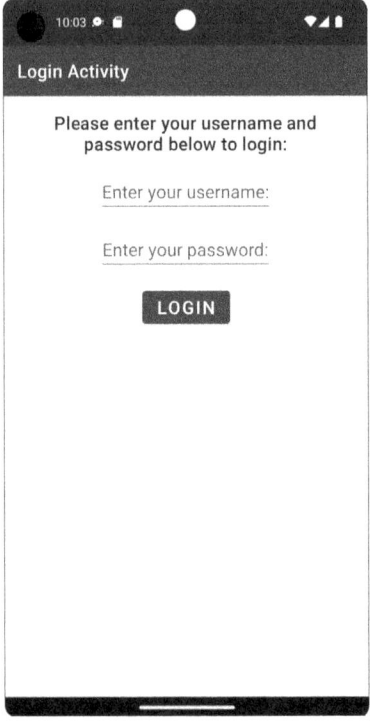

Figure 2.23 – The app display when first loaded

Note

The solution to this activity can be found at https://packt.link/PmKJ6.

Summary

In this chapter, you have covered a lot of the groundwork of how your application interacts with the Android framework, from the Activity lifecycle callbacks to retaining the state in your activities, navigating from one screen to another, and how intents and launch modes make this happen. These are core concepts that you need to understand in order to move on to more advanced topics.

In the next chapter, you will be introduced to fragments and how they fit into the architecture of your application, as well as exploring more of the Android resources framework.

3
Developing the UI with Fragments

This chapter covers fragments and the fragment lifecycle. It demonstrates how to use them to build efficient and dynamic layouts that respond to different screen sizes and configurations and allow you to divide your UI into different sections. By the end of this chapter, you will be able to create static and dynamic fragments, pass data to and from fragments and activities, and use the Jetpack `Navigation` component to detail how fragments fit together.

In the previous chapter, we explored the Android activity lifecycle and looked into how it is used in apps to navigate between screens. We also analyzed various types of launch modes, which defined how transitioning between screens happened. In this chapter, you'll explore fragments. A fragment is a section, portion, or, as the name implies, fragment of an Android activity.

Throughout the chapter, you'll learn how to use fragments, see how they can exist in more than one activity, and discover how multiple fragments can be used in one activity. You'll start by adding simple fragments to an activity and then progress to learning about the difference between static and dynamic fragments.

Fragments can be used to simplify creating layouts for Android tablets that have larger form factors using dual-pane layouts. For example, if you have an average-sized phone screen and you want to include a list of news stories, you might only have enough space to display the list.

If you viewed the same list of stories on a tablet, you'd have more space available so you could display the same list and also a story itself to the right of the list. Each of these different areas of the screen can use a fragment. You can then use the same fragment on both the phone and the tablet. You benefit from reusing and simplifying the layouts and don't have to repeat creating similar functionality.

Once you've explored how fragments are created and used, you'll then learn how to organize your user journeys with fragments. You'll apply some established practices for using fragments in this way. Finally, you'll learn how to simplify fragment use by creating a navigation graph with the Android Jetpack `Navigation` component, which allows you to specify linking fragments together with destinations.

We will cover the following topics in this chapter:

- The fragment lifecycle
- Static fragments and dual-pane layouts
- Dynamic fragments
- Jetpack Navigation

Technical requirements

The complete code for all the exercises and the activity in this chapter is available on GitHub at `https://packt.link/KmdBZ`

The fragment lifecycle

A fragment is a component with its own lifecycle. Understanding the fragment lifecycle is critical as it provides callbacks at certain stages of fragment creation, the running state, and destruction that configure the initialization, display, and cleanup. Fragments run in an activity, and a fragment's lifecycle is bound to the activity's lifecycle.

In many ways, the fragment lifecycle is very similar to the activity lifecycle, and at first glance, it appears that the former replicates the latter. There are as many callbacks that are the same or similar in the fragment lifecycle as there are in the activity lifecycle, such as `onCreate(savedInstanceState: Bundle?)`.

The fragment lifecycle is tied to the activity lifecycle, so wherever fragments are used, the fragment callbacks are interleaved with the activity callbacks.

The same steps are gone through to initialize the fragment and prepare for it to be displayed to the user before being available for the user to interact with. The same teardown steps that an activity goes through happen to the fragment as well when the app is backgrounded, hidden, and exited. Fragments, like activities, have to extend/derive from a parent `Fragment` class, and you can choose which callbacks to override depending on your use case. The lifecycle is displayed in the following diagram, followed by more detail about each function.

The fragment lifecycle

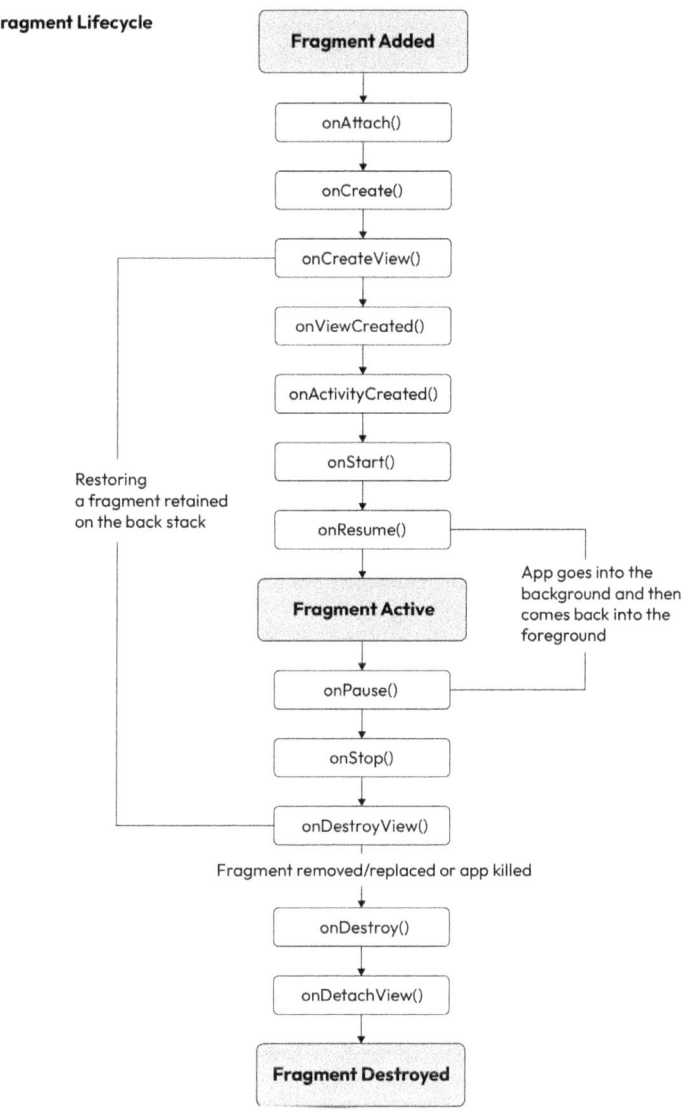

Figure 3.1 – Fragment lifecycle

Let's now explore these callbacks, the order they appear in, and what they do.

onAttach

`override fun onAttach(context: Context)`: This is the point where your fragment becomes linked to the activity it is used in. It allows you to reference the activity, although at this stage neither the fragment nor the activity has been fully created.

onCreate

`override fun onCreate(savedInstanceState: Bundle?)`: This is where you do any initialization of your fragment. This is not where you set the layout of your fragment, as at this stage, there is no UI available to display and no `setContentView` available as there is in an activity. As is the same in the activity's `onCreate()` function, you can use the `savedInstanceState` parameter to restore the state of the fragment when it is being re-created.

onCreateView

`override fun onCreateView(inflater: LayoutInflater, container: ViewGroup?, savedInstanceState: Bundle?): View?`: Now, this is where you get to create the layout of your fragment. The most important thing to remember here is that instead of setting the layout (as is the case with an activity), the fragment will actually return the layout `View?` from this function.

The views you have in your layout are available to refer to here, but there are a few caveats. You need to create the layout before you can reference the views contained within it, which is why it's preferred to do view manipulation in `onViewCreated`.

onViewCreated

`override fun onViewCreated(view View, savedInstanceState: Bundle?)`: This callback is the one in between your fragment being fully created and being visible to the user. It's where you'll typically set up your views and add any functionality and interactivity to these views. This might be adding an `OnClickListener` to a button and then calling a function when it's clicked.

onActivityCreated

`override fun onActivityCreated(context: Context)`: This is called immediately after the activity's `onCreate` has been run. Most of the initialization of the view state of the fragment will have been done, and this is the place to do the final setup if required.

onStart

`override fun onStart()`: This is called when the fragment is about to become visible to the user but is not yet available for the user to interact with.

onResume

`override fun onResume()`: At the end of this call, your fragment is available for the user to interact with. Normally, there is minimal setup or functionality defined in this callback, as when the app goes into the background and then comes back into the foreground, this callback will always be called.

onPause

`override fun onPause()`: Like its counterpart, `onPause()` in an activity signals that your app is going into the background or has been partially covered by something else on the screen. Use this to save any changes to the fragment's state.

onStop

`override fun onStop()`: The fragment is no longer visible at the end of this call and goes into the background.

onDestroyView

`override fun onDestroyView()`: This is usually called for doing a final cleanup before the fragment is destroyed. You should use this callback if it is necessary to clean up any resources. If the fragment is pushed to the back stack and retained, then it can also be called without destroying the fragment. On completion of this callback, the fragment's layout view is removed.

onDestroy

`override fun onDestroy()`: The fragment is being destroyed. This can occur because the app is being killed or because this fragment is being replaced by another fragment.

onDetach

`override fun onDetach()`: This is called when the fragment has been detached from its activity.

There are more fragment callbacks, but these are the ones you will use for the majority of cases. Typically, you'll only use a subset of these callbacks: `onAttach()` to associate an activity with the fragment, `onCreate` to initialize the fragment, `onCreateView` to set the layout, and then `onViewCreated`/`onActivityCreated` to do further initialization, and perhaps `onPause()` to do some cleanup.

> **Note**
> Further details of these callbacks can be found in the official documentation at `https://developer.android.com/guide/fragments`.

Now that we've gone through some of the theory of the fragment lifecycle and how it is affected by the host activity's lifecycle, let's see those callbacks being run in action.

Exercise 3.01 – adding a basic fragment and the fragment lifecycle

In this exercise, we will create and add a basic fragment to an app. The aim of this exercise is to gain familiarity with how fragments are added to an activity and the layout they display. To do this, you will create a new blank fragment with a layout in Android Studio. You will then add the fragment to the activity and verify the fragment has been added by the display of the fragment layout. Perform the following steps:

1. Create an application in Android Studio with an empty activity called `Fragment Lifecycle`.
2. Once the application has been built, create a new fragment by going to **File** | **New** | **Fragment** **(Blank)**. You just want to create a plain vanilla fragment at this stage, so you use the `Fragment (Blank)` option. When you've selected this option, you will be presented with the screen shown in *Figure 3.2*:

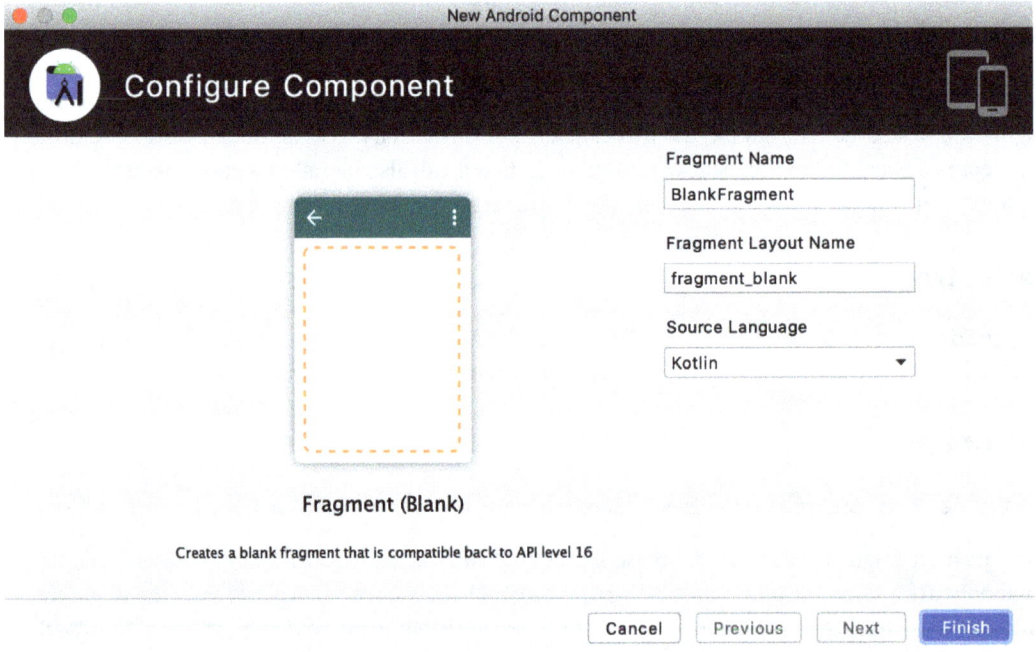

Figure 3.2 – Creating a new fragment

3. Rename the fragment `MainFragment` and the layout `fragment_main`. Then, press **Finish** and the `Fragment` class will be created and opened. There are two functions that have been added: `onCreate`, which initializes the fragment, and `onCreateView` (displayed in the following code snippet), which inflates the layout file used for the fragment:

```
override fun onCreateView(
    inflater: LayoutInflater, container:
```

```
            ViewGroup?,
        savedInstanceState: Bundle?
): View? {
    // Inflate the layout for this fragment
    return inflater.inflate(
        R.layout.fragment_main, container, false)
}
```

4. When you open up the `fragment_main.xml` layout file, you'll see the following code:

```
<?xml version="1.0" encoding="utf-8"?>
<FrameLayout
xmlns:android=
  "http://schemas.android.com/apk/res/android"
    xmlns:tools="http://schemas.android.com/tools"
    android:layout_width="match_parent"
    android:layout_height="match_parent"
    tools:context=".MainFragment">
    <!-- TODO: Update blank fragment layout -->
    <TextView
        android:layout_width="match_parent"
        android:layout_height="match_parent"
        android:text="@string/hello_blank_fragment" />
</FrameLayout>
```

A simple layout has been added with a `TextView` and some example text using `@string/hello_blank_fragment`. This string resource has the text hello blank fragment. As `layout_width` and `layout_height` are specified as `match_parent`, `TextView` will occupy the whole of the screen. The text itself, however, will be added at the top left of the view with the default position.

5. Add the `android:gravity="center"` attribute and value to the `TextView` so that the text appears in the center of the screen:

```
<TextView
    android:layout_width="match_parent"
    android:layout_height="match_parent"
    android:gravity="center"
    android:text="@string/hello_blank_fragment" />
```

If you run up the UI now, you'll see the **Hello World!** display in *Figure 3.3*:

Figure 3.3 – Initial app layout display without a fragment added

Well, you can see some **Hello World!** text, but not the `hello blank fragment` text you might have been expecting. The fragment and its layout do not automatically get added to an activity when you create it. This is a manual process.

6. Open the `activity_main.xml` file and replace the contents with the following:

```
<?xml version="1.0" encoding="utf-8"?>
<androidx.constraintlayout.widget.ConstraintLayout
    xmlns:android=
      "http://schemas.android.com/apk/res/android"
    xmlns:tools="http://schemas.android.com/tools"
    android:layout_width="match_parent"
    android:layout_height="match_parent"
    tools:context="com.example.fragmentlifecycle
        .MainActivity">
```

```xml
    <fragment
        android:id="@+id/main_fragment"
        android:name="com.example.fragmentlifecycle
            .MainFragment"
        android:layout_width="match_parent"
        android:layout_height="match_parent" />
</androidx.constraintlayout.widget.ConstraintLayout>
```

Just as there are view declarations you can add to layouts in XML, there is also a `fragment` element. You've added the fragment to `ConstraintLayout` with constraints of `match_parent` for `layout_width` and `layout_height` so it will occupy the whole of the screen.

The most important `xml` attribute to examine here is `android:name`. It's here where you specify the fully qualified name of the package and the `Fragment` class that you are going to add to the layout with `com.example.fragmentlifecycle.MainFragment`. Now run the app, and you will see the output shown in *Figure 3.4*:

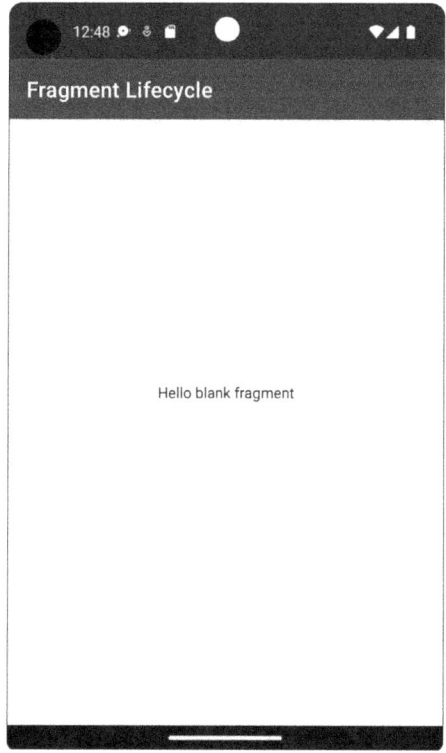

Figure 3.4 – App layout display with a fragment added

This proves that your fragment with the text `Hello blank fragment` has been added to the activity and the layout you defined is being displayed. Next, you'll examine the callback methods between the activity and the fragment and how this happened.

7. Open up the `MainFragment` class and add a `TAG` constant to the companion object with the value `"MainFragment"` to identify the class.

   ```
   private const val TAG = "MainFragment"
   ```

   Then, add/update the functions with appropriate log statements.

   You will need to add the imports for the `Log` statement and `context` to the imports at the top of the class. The following code snippet is truncated. Follow the link shown to see the full code block you need to use:

MainFragment.kt

```
import android.content.Context
import android.util.Log

    override fun onAttach(context: Context) {
        super.onAttach(context)
        Log.d(TAG, "onAttach: ")
    }
    override fun onCreate(savedInstanceState: Bundle?)
    {
        super.onCreate(savedInstanceState)
        Log.d(TAG,"onCreate: ")
        arguments?.let {
            param1 = it.getString(ARG_PARAM1)
            param2 = it.getString(ARG_PARAM2)
        }
    }
    override fun onCreateView(
        inflater: LayoutInflater, container:
            ViewGroup?, savedInstanceState: Bundle?
    ): View? {
        Log.d(TAG,"onCreateView: ")
        // Inflate the layout for this fragment
        return inflater.inflate(
```

```
                    R.layout.fragment_main, container, false)
        }
```

You can find the complete code for this step at https://packt.link/XcOJ4.

You will need to add Log.d(TAG, "onCreateView") to the onCreateView callback and Log.d(TAG, "onCreate") to the onCreate callback that already exist, and then override the onAttach function, adding Log.d(TAG, "onAttach"), and onViewCreated, adding Log.d(TAG, "onViewCreated").

8. Next, open the MainActivity class and add a companion object with a TAG constant with the value "MainActivity". Then add the Log import to the top of the class and then the common onStart and onResume callback methods, as shown in the following code snippet.

```
import android.util.Log
override fun onCreate(savedInstanceState: Bundle?)
{
    super.onCreate(savedInstanceState)
    setContentView(R.layout.activity_main)
    Log.d(TAG, "onCreate")
}
override fun onStart() {
    super.onStart()
    Log.d(TAG, "onStart")
}
override fun onResume() {
    super.onResume()
    Log.d(TAG, "onResume")
}
companion object {
    private const val TAG = "MainActivity"
}
```

You'll see that you also have to add the onCreate log statement, Log.d(TAG, "onCreate"), as this callback was already there when you added the activity to the project.

You learned in *Chapter 2, Building User Screen Flows*, how to view log statements, and you are going to open the **Logcat** window in Android Studio to examine the logs and the order they are called in when you run the app.

In *Chapter 2, Building User Screen Flows*, you viewed logs from a single activity so you could see the order they were called in. Now you'll examine the order in which the MainActivity and MainFragment callbacks happen.

9. Open up the **Logcat** window. (Just to remind you, it can be accessed by clicking the **Logcat** tab at the bottom of the screen and also from the toolbar with **View | Tool Windows | Logcat**.) As both MainActivity and MainFragment start with Main, you can type tag:Main in the search box to filter the logs to only show statements with this text. Run the app, and you should see the following:

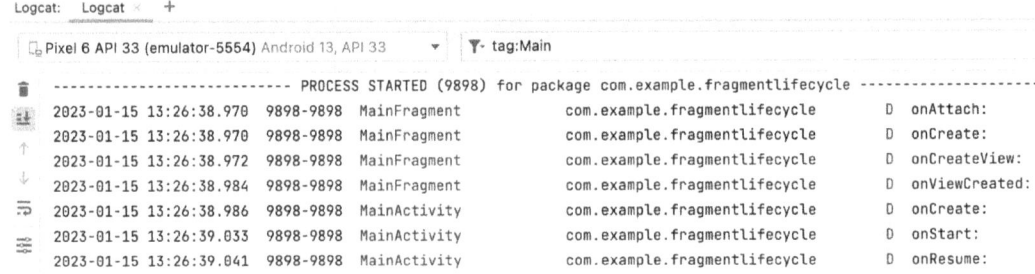

Figure 3.5 – Logcat statements shown when starting the app

What's interesting here is that the first few callbacks are from the fragment. It is linked to the activity it has been placed in with the onAttach callback. The fragment is initialized, and its view is displayed in onCreate and onCreateView, before another callback, onViewCreated, is called, confirming that the fragment UI is ready to be displayed.

This is before the activity's onCreate method is called. This makes sense as the activity creates its UI based on what it contains. As this is a fragment that defines its own layout, the activity needs to know how to measure, lay out, and draw the fragment as it does in the onCreate method.

Then, the fragment receives confirmation that this has been done with the onActivityCreated callback before both the fragment and activity start to display the UI in onStart, before preparing the user to interact with it after their respective onResume callbacks have finished.

> **Note**
> The interaction between the activity and fragment lifecycles detailed previously is for cases when static fragments, which are those defined in the layout of an activity, are created. For dynamic fragments, which can be added when the activity is already running, the interaction can differ.

So, now that the fragment and the containing activity are shown, what happens when the app is backgrounded or closed? The callbacks are still interleaved when the fragment and activity are paused, stopped, and finished.

10. Add the following callbacks to the `MainFragment` class:

    ```
    override fun onPause() {
        super.onPause()
        Log.d(TAG, "onPause")
    }
    override fun onStop() {
        super.onStop()
        Log.d(TAG, "onStop")
    }
    override fun onDestroyView() {
        super.onDestroyView()
        Log.d(TAG, "onDestroyView")
    }
    override fun onDestroy() {
        super.onDestroy()
        Log.d(TAG, "onDestroy")
    }
    override fun onDetach() {
        super.onDetach()
        Log.d(TAG, "onDetach")
    }
    ```

11. Then, add these callbacks to `MainActivity`:

    ```
    override fun onPause() {
        super.onPause()
        Log.d(TAG, "onPause")
    }
    override fun onStop() {
        super.onStop()
        Log.d(TAG, "onStop")
    }
    override fun onDestroy() {
        super.onDestroy()
        Log.d(TAG, "onDestroy")
    }
    ```

12. Build the app, and once it is running, you'll see the callbacks from before starting both the fragment and activity. You can use the dustbin icon at the top left of the `Logcat` window to clear the statements. Then, rotate the app and review the output log statements:

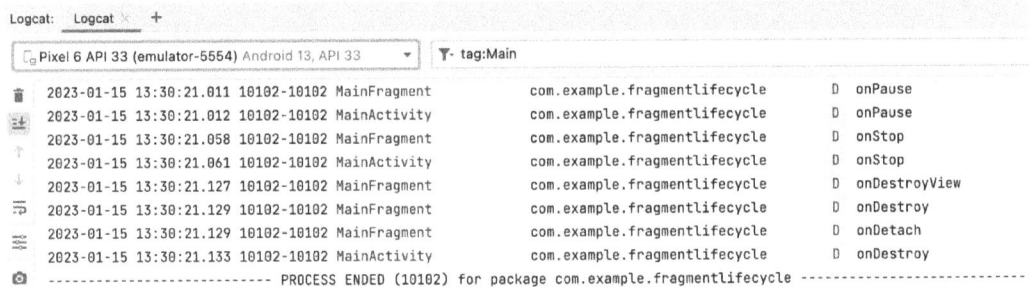

Figure 3.6 – Logcat statements shown when closing the app

The `onPause` and `onStop` statements are as you might expect in that the fragment gets notified of these callbacks first as it is contained within the activity. You can think of this as being inward to outward in that the child elements are notified before the containing parent, so the parent knows how to respond.

The fragment is then torn down, removed from the activity, and then destroyed with the `onDestroyView`, `onDestroy`, and `onDetach` functions before the activity itself is destroyed after any final cleanup is done in `onDestroy`. It doesn't make sense for the activity to finish until all the component parts that make up the activity are themselves removed.

The full fragment lifecycle callbacks and how they relate to the activity callbacks are complicated in Android because which callbacks are applied in which situation can differ quite substantially. To view a more detailed overview, see the official documentation at https://developer.android.com/guide/fragments.

For the majority of situations, you will only use the preceding fragment callbacks. This example demonstrates both how self-contained fragments are in their creation, display, and destruction, and also their interdependence with the containing activity.

Now that we've gone through a basic example of adding a fragment to an activity and examining the interaction between the fragment and the activity, let's see a more detailed example of how you add two fragments to an activity.

Exercise 3.02 – adding fragments statically to an activity

This exercise will demonstrate how to add two fragments to an activity with their own UI and separate functionality. You'll create a simple counter class that increments and decrements a number and a color class that changes the color applied programmatically to some `Hello World` text. Perform the following steps:

1. Create an application in Android Studio with an empty activity called `Fragment Intro`. Then, add the following strings required for the exercise in the res | values | strings.xml file:

   ```
   <string name="hello_world">Hello World</string>
   <string name="red_text">Red</string>
   <string name="green_text">Green</string>
   <string name="blue_text">Blue</string>
   <string name="zero">0</string>
   <string name="plus">+</string>
   <string name="minus">-</string>
   <string name="counter_text">Counter</string>
   ```

   These strings are used in both the counter fragment as well as the color fragment, which you will create next.

2. Add a new blank fragment by going to **File** | **New** | **Fragment (Blank)** called `CounterFragment` with the `fragment_counter` layout name.

3. Now, make changes to the `fragment_counter.xml` file. The following code is truncated for space. Follow the link shown for the full code you need to use:

fragment_counter.xml

```
<TextView
    android:id="@+id/counter_text"
    android:layout_width="wrap_content"
    android:layout_height="wrap_content"
    android:text="@string/counter_text"
    android:paddingTop="10dp"
    android:textSize="44sp"
    app:layout_constraintEnd_toEndOf="parent"
    app:layout_constraintStart_toStartOf="parent"
    app:layout_constraintTop_toTopOf="parent"/>
<TextView
    android:id="@+id/counter_value"
```

```
            android:layout_width="wrap_content"
            android:layout_height="wrap_content"
            android:text="@string/zero"
            android:textSize="54sp"
            android:textStyle="bold"
            app:layout_constraintEnd_toEndOf="parent"
            app:layout_constraintStart_toStartOf="parent"
            app:layout_constraintTop_toBottomOf=
                "@id/counter_text"
            app:layout_constraintBottom_toTopOf="@id/plus"/>
```

You can find the complete code for this step at https://packt.link/ca4EK.

We are using a simple `ConstraintLayout` file that has a `TextView` set up for the `@+id/counter_text` header and a `TextView` for the value of the counter, `android:id="@+id/counter_value"` (with a default of `@string/zero`), which will be changed by the `android:id="@+id/plus"` and `android:id="@+id/minus"` buttons.

> **Note**
> For a simple example such as this, you are not going to set individual styles on the views with `style="@some_style"` notation, which would be best practice to avoid repeating these values on each view.

4. Now open `CounterFragment` and add a property to be the counter below the class header (it is a `var` so it is mutable and can be changed):

   ```
   var counter = 0
   ```

5. Now open and add the following `onViewCreated` function. You will also need to add the following imports to the top of the class:

   ```
   import android.widget.Button
   import android.widget.TextView

   override fun onViewCreated(view: View,
   savedInstanceState: Bundle?) {
       super.onViewCreated(view, savedInstanceState)
   ```

```
        val counterValue =
            view.findViewById<TextView>(R.id.counter_value)

        view.findViewById<Button>(R.id.plus)
        .setOnClickListener {
            counter++
            counterValue.text = counter.toString()
        }
        view.findViewById<Button>(R.id.minus)
        .setOnClickListener {
            if (counter > 0) {
                counter--
                counterValue.text = counter.toString()
            }
        }
    }
```

We've added `onViewCreated`, which is the callback run when the layout has been applied to your fragment. The `onCreateView` callback, which creates the view, was run when the fragment itself was created.

The buttons you've specified in the preceding fragment have `OnClickListener` set up on them to increment and decrement the value of the `counter`.

6. Firstly, within the plus button `OnClickListener`, you are incrementing the counter and setting this value on the view:

    ```
    counter++
    counterValue.text = counter.toString()
    ```

7. Then, within the minus button `OnClickListener`, it decrements the value by one but only if the value is greater than one, so no negative numbers are set:

    ```
    if (counter > 0) {
        counter--
        counterValue.text = counter.toString()
    }
    ```

8. You have not added the fragment to the `MainActivity` layout. To do this, go into `activity_main.xml` and replace the contents with the following code:

```xml
<?xml version="1.0" encoding="utf-8"?>
<LinearLayout
xmlns:android=
    "http://schemas.android.com/apk/res/android"
    xmlns:tools="http://schemas.android.com/tools"
    android:layout_width="match_parent"
    android:layout_height="match_parent"
    android:orientation="vertical"
    tools:context=".MainActivity">
    <fragment
        android:id="@+id/counter_fragment"
        android:name="com.example.fragmentintro
            .CounterFragment"
        android:layout_width="match_parent"
        android:layout_height="match_parent"/>
</LinearLayout>
```

You are going to change the layout from `ConstraintLayout` to `LinearLayout` for simplicity as you can easily add one fragment above the other when you come to the next stage. You specify the fragment to be used within the `fragment` XML element with the name attribute with the fully qualified package name used for the class: `android:name="com.example.fragmentintro.CounterFragment`.

If you used a different package name when you created the app, then this must refer to the `CounterFragment` you created. The important thing to grasp here is that you have added a fragment to your main activity layout and the fragment also has a layout.

This shows some of the power of using fragments as you can encapsulate the functionality of one feature of your app, complete with a layout file and fragment class, and add it to multiple activities.

Once you've done this, run the fragment in the virtual device as in *Figure 3.7*:

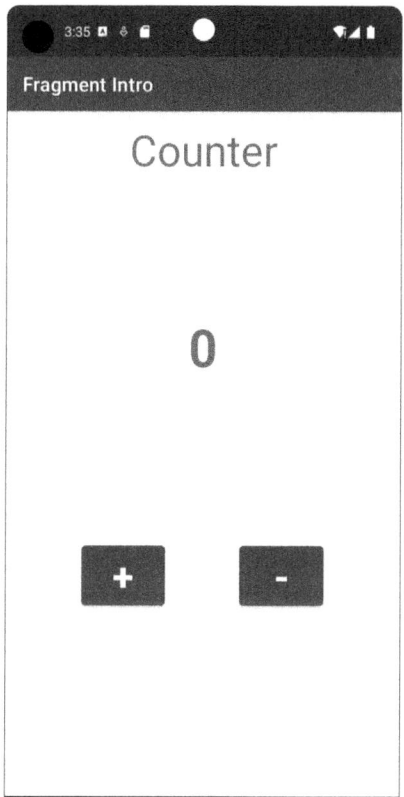

Figure 3.7 – App displaying the counter fragment

You have created a simple counter. The basic functionality works as expected, incrementing and decrementing a counter value.

9. In the next step, you are going to add another fragment to the bottom half of the screen. This demonstrates the versatility of fragments. You can have encapsulated pieces of the UI with functionality and features in different areas of the screen.

10. Now, create a new fragment using the earlier steps for creating the CounterFragment called ColorFragment with the fragment_color layout name.

11. Next, open up the fragment_color.xml file that has been created and replace the contents with the code at the following link. The following snippet is truncated – see the link for the full code:

```
<TextView
        android:id="@+id/hello_world"
```

```
            android:layout_width="wrap_content"
            android:layout_height="0dp"
            android:textSize="34sp"
            android:paddingBottom="12dp"
            android:text="@string/hello_world"
            app:layout_constraintEnd_toEndOf="parent"
            app:layout_constraintStart_toStartOf="parent"
            app:layout_constraintTop_toTopOf="parent" />

    <Button
        android:id="@+id/red_button"
        android:layout_width="wrap_content"
        android:layout_height="0dp"
        android:textSize="24sp"
        android:text="@string/red_text"
app:layout_constraintEnd_toStartOf="@+id/green_button"
app:layout_constraintStart_toStartOf="parent"
app:layout_constraintTop_toBottomOf="@id/hello_world"
/>
```

You can find the complete code for this step at https://packt.link/GCYDR.

The layout adds a `TextView` with three buttons. The `TextView` text and the text for all the buttons are set as string resources (`@string`).

12. Next, go into the `activity_main.xml` file and add ColorFragment below CounterFragment inside LinearLayout:

```
<?xml version="1.0" encoding="utf-8"?>
<LinearLayout
xmlns:android=
  "http://schemas.android.com/apk/res/android"
    xmlns:tools="http://schemas.android.com/tools"
    android:layout_width="match_parent"
    android:layout_height="match_parent"
    android:orientation="vertical"
    tools:context=".MainActivity">
    <fragment
```

```
            android:id="@+id/counter_fragment"
            android:name="com.example.fragmentintro
                .CounterFragment"
            android:layout_width="match_parent"
            android:layout_height="match_parent"/>
    <fragment
            android:id="@+id/color_fragment"
            android:name="com.example.fragmentintro
                .ColorFragment"
            android:layout_width="match_parent"
            android:layout_height="match_parent"/>
</LinearLayout>
```

When you run the app, you will see that `ColorFragment` is not visible, as in *Figure 3.8*:

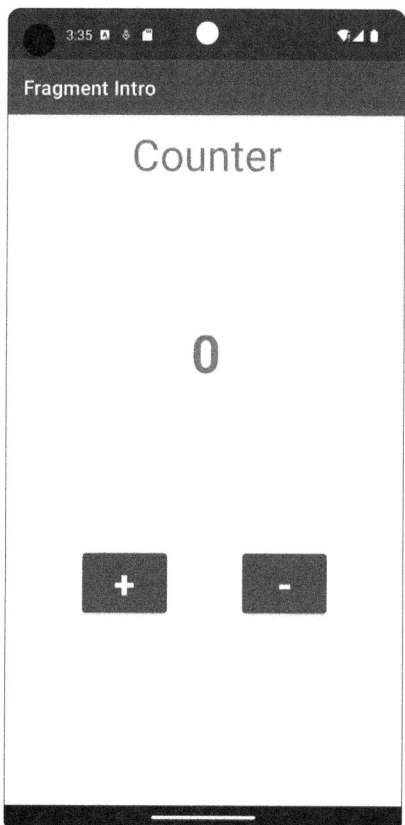

Figure 3.8 – App shown without ColorFragment displayed

You've included `ColorFragment` in the layout, but because `CounterFragment` has its width and height set to match its parent (`android:layout_width="match_parent" android:layout_height="match_parent"`) and it is the first view in the layout, it takes up all the space.

What you need is some way to specify the proportion of the height that each fragment should occupy. The `LinearLayout` orientation is set to vertical so the fragments will appear one on top of the other when `layout_height` is not set to `match_parent`.

In order to define the proportion of this height, you need to add another attribute, `layout_weight`, to each fragment in the `activity_main.xml` layout file. When you use `layout_weight` to determine the proportional height the fragments should occupy, you set the `layout_height` value of the fragments to `0dp`.

13. Update the `activity_main.xml` layout with the following changes, setting `layout_height` for both fragments to `0dp` and adding the `layout_weight` attributes with the following values:

```xml
<?xml version="1.0" encoding="utf-8"?>
<LinearLayout
xmlns:android=
    "http://schemas.android.com/apk/res/android"
    xmlns:tools="http://schemas.android.com/tools"
    android:layout_width="match_parent"
    android:layout_height="match_parent"
    android:orientation="vertical"
    tools:context=".MainActivity">
    <fragment
        android:id="@+id/counter_fragment"
        android:name="com.example.fragmentintro
            .CounterFragment"
        android:layout_width="match_parent"
        android:layout_height="0dp"
        android:layout_weight="2"/>
    <fragment
        android:id="@+id/color_fragment"
        android:name="com.example.fragmentintro
            .ColorFragment"
        android:layout_width="match_parent"
        android:layout_height="0dp"
```

```
            android:layout_weight="1"/>
    </LinearLayout>
```

These changes make `CounterFragment` occupy twice the height of `ColorFragment`, as shown in *Figure 3.9*:

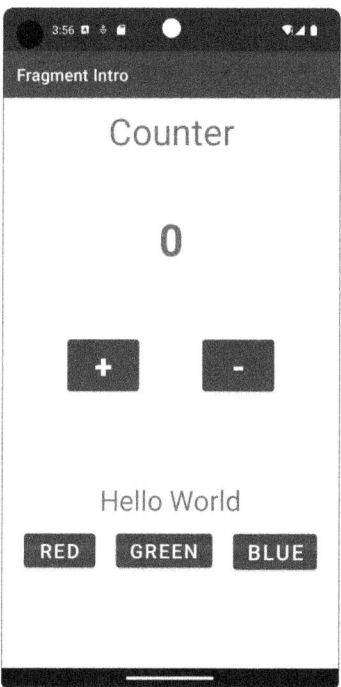

Figure 3.9 – CounterFragment with twice the amount of vertical space allocated

You can experiment by changing the weight values to see the differences you can make to the display of the layout.

14. At this point, pressing the **RED**, **GREEN**, and **BLUE** color buttons will have no effect on the **Hello World** text. The button actions have not been specified. The next step involves adding interactivity to the buttons to make changes to the style of the **Hello World** text.

15. Add the following `onViewCreated` function in `ColorFragment`, which overrides its parent to add behavior to the fragment after the layout view has been set up. You will also need to add the `TextView`, `Color`, and `Button` imports to change the color of the text:

```
import android.widget.Button
import android.widget.TextView
import android.graphics.Color
```

```kotlin
override fun onViewCreated(view: View,
savedInstanceState: Bundle?) {
    super.onViewCreated(view, savedInstanceState)
    val redButton =
        view.findViewById<Button>(R.id.red_button)
    val greenButton =
        view.findViewById<Button>(R.id.green_button)
    val blueButton =
        view.findViewById<Button>(R.id.blue_button)
    val helloWorldTextView =
        view.findViewById<TextView>(R.id.hello_world)
    redButton.setOnClickListener {
        helloWorldTextView.setTextColor(Color.RED)
    }
    greenButton.setOnClickListener {
        helloWorldTextView.setTextColor(Color.GREEN)
    }
    blueButton.setOnClickListener {
        helloWorldTextView.setTextColor(Color.BLUE)
    }
}
```

Here, you are adding an `OnClickListener` to each button defined in the layout and setting the **Hello World** text to the desired color.

16. Run the app and press the **RED**, **GREEN**, and **BLUE** buttons. You should see a similar display to the one in *Figure 3.10*:

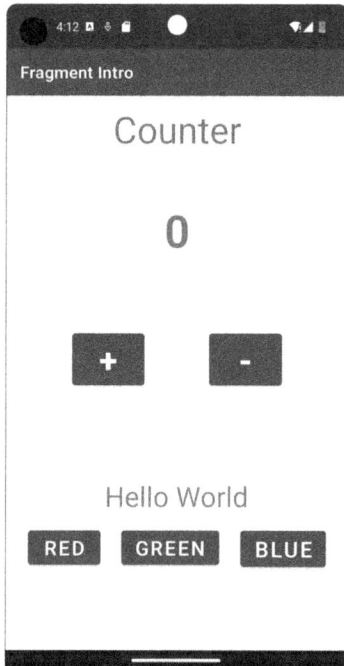

Figure 3.10 – ColorFragment setting text color to red, green, and blue

This exercise, although simple, has demonstrated some fundamental concepts of using fragments. The features of your app that the user can interact with can be developed independently and not rely on bundling two or more features into one layout and activity. This makes fragments reusable and means you can focus your attention when developing your app on adding a well-defined UI, logic, and features into a single fragment.

Static fragments and dual-pane layouts

The previous exercise introduced you to static fragments, those that can be defined in the activity XML layout file. You can also create different layouts and resources for different screen sizes. This is used for deciding which resources to display depending on whether the device is a phone or a tablet.

The space for laying out UI elements can increase substantially with a larger size tablet. Android allows specifying different resources depending on many different form factors. The qualifier frequently used to define a tablet in the `res` (resources) folder is `sw600dp`.

This states that if the shortest width (`sw`) of the device is over 600 dp, then use these resources. This qualifier is used for 7" tablets and larger. Tablets facilitate what is known as dual-pane layouts. A pane represents a self-contained part of the user interface. If the screen is large enough, then two panes (dual-pane layouts) can be supported. This also provides the opportunity for one pane to interact with another to update content.

Exercise 3.03 – dual-pane layouts with static fragments

In this exercise, you are going to create a simple app that displays a list of star signs and specific information about each star sign. It will use different displays for phones and tablets.

The phone will display a list and then open the selected list item's content on another screen while the tablet will display the same list in one pane and open the list item's content in another pane on the same screen in a dual-pane layout.

In order to do this, you have to create another layout that will only be used for 7" tablets and above. Perform the following steps:

1. Firstly, create a new Android Studio project with an `Empty Activity` called `Dual Pane Layouts`.

2. Then, making sure the Android view is selected in the project view on the left-hand side, go to **File** | **New** | **Android Resource File** and fill in the following fields of the dialog (you need to select **Smallest Screen Width** in the dialog's left-hand pane – once you have selected it, the option changes to **Screen Width**):

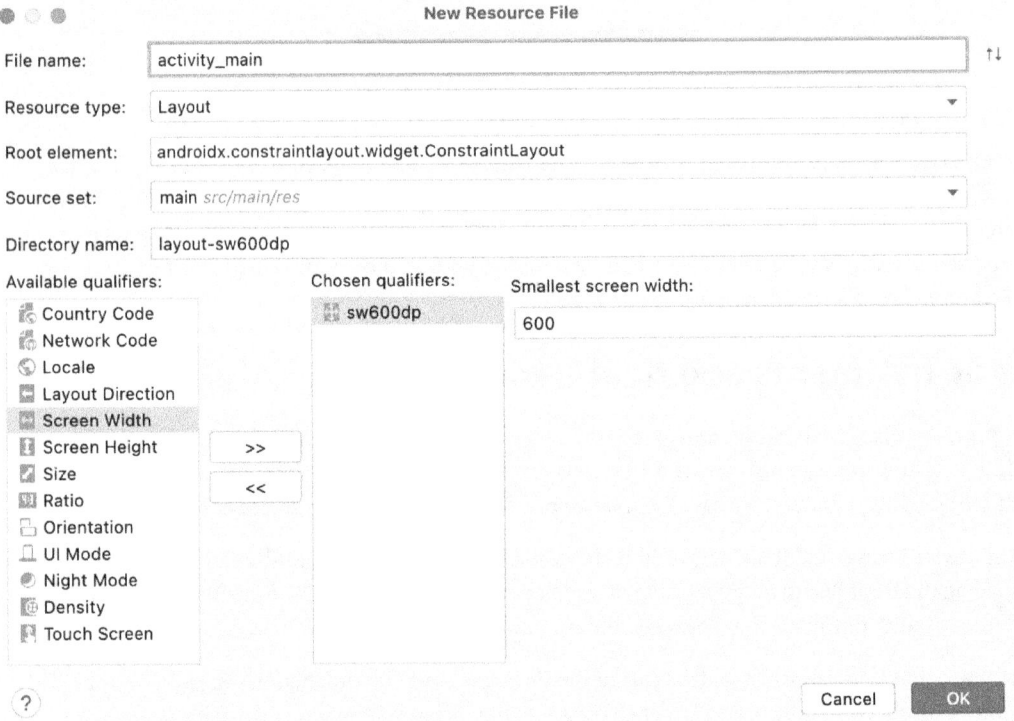

Figure 3.11 – Device Variations View

Static fragments and dual-pane layouts

3. This creates a new folder in the `main | res` folder named `'layout-sw600dp'` with the `activity_main.xml` layout file added.

   At the moment, it is a duplicate of the `activity_main.xml` file added when you created the app, but you will change it to customize the screen display for tablets.

   In order to demonstrate the use of dual-pane layouts, you are going to create a list of star signs so that, when a list item is selected, it will display some basic information about the star sign.

4. Go to the top toolbar and select **File | New | Fragment | Fragment (Blank)**. Call it `ListFragment`.

   For this exercise, you need to update the `strings.xml` and `themes.xml` files, adding the following entries:

strings.xml

```xml
<string name="star_signs">Star Signs</string>
<string name="symbol">Symbol: %s</string>
<string name="date_range">Date Range: %s</string>
<string name="aquarius">Aquarius</string>
<string name="pisces">Pisces</string>
<string name="aries">Aries</string>
<string name="taurus">Taurus</string>
<string name="gemini">Gemini</string>
<string name="cancer">Cancer</string>
<string name="leo">Leo</string>
<string name="virgo">Virgo</string>
<string name="libra">Libra</string>
<string name="scorpio">Scorpio</string>
<string name="sagittarius">Sagittarius</string>
<string name="capricorn">Capricorn</string>
<string name="unknown_star_sign">Unknown Star Sign
</string>
```

themes.xml

```xml
<style name="StarSignTextView"
parent="Base.TextAppearance.AppCompat.Large" >
    <item name="android:padding">18dp</item>
</style>
<style name="StarSignTextViewHeader"
```

```
            parent="Base.TextAppearance.AppCompat.Display1" >
                <item name="android:padding">18dp</item>
            </style>
```

5. Open the main | res | layout | fragment_list.xml file and replace the default contents with the following:

```
<?xml version="1.0" encoding="utf-8"?>
<ScrollView
xmlns:android=
    "http://schemas.android.com/apk/res/android"
    xmlns:tools="http://schemas.android.com/tools"
    android:layout_width="match_parent"
    android:layout_height="wrap_content"
    tools:context=".ListFragment">
    <LinearLayout
        android:layout_width="match_parent"
        android:layout_height="wrap_content"
        android:orientation="vertical">
        <TextView
            android:layout_width="match_parent"
            android:layout_height="wrap_content"
            android:gravity="center"
            android:textSize="24sp"
            android:textStyle="bold"
            style="@style/StarSignTextView"
            android:text="@string/star_signs" />
        <View
            android:layout_width="match_parent"
            android:layout_height="1dp"
            android:background="?android:attr/
                dividerVertical" />
        <TextView
            android:id="@+id/aquarius"
            style="@style/StarSignTextView"
            android:layout_width="match_parent"
            android:layout_height="wrap_content"
```

```
                android:text="@string/aquarius" />
        </LinearLayout>
</ScrollView>
```

You will see that the first xml element is a `ScrollView`. A `ScrollView` is a `ViewGroup` that allows the contents to scroll, and as you will be adding 12 star signs into the `LinearLayout` it contains, this is likely to occupy more vertical space than is available on the screen.

Adding `ScrollView` prevents the contents from being cut off vertically when there is no more room to display them and scrolls the layout. A `ScrollView` can only contain one child view. Here, it's a `LinearLayout`, and as the contents will be displayed vertically, the orientation is set to vertical (`android:orientation="vertical"`). Below the first title `TextView`, you have added a divider `View` and a `TextView` for the first star sign, Aquarius.

6. Add the other 11 star signs with the same format, adding the divider first and then the `TextView`. The name of the string resource and the `id` value should be the same for each `TextView`. The names of the star signs you will create a view from are specified in the `strings.xml` file.

> **Note**
> The technique used to lay out a list is fine for an example, but in a real-world app, you would create a `RecyclerView` dedicated to displaying lists that can scroll, with the data bound to the list by an adapter. You will cover this in a later chapter.

7. Next, create `StarSignListener` above the `MainActivity` class header and make `MainActivity` implement it by adding the following:

```
interface StarSignListener {
    fun onSelected(id: Int)
}
class MainActivity : AppCompatActivity(),
StarSignListener {
    ...
    override fun onSelected(id: Int) {
        TODO("not implemented yet")
    }
}
```

This is how the fragments will communicate back to the activity when a user selects a star sign from `ListFragment` and logic will be added depending on whether a dual pane is available or not.

8. Once you've created the layout file, go into the `ListFragment` class and update it with the following contents, keeping `onCreateView()` in place. You can see in the fragment in the `onAttach()` callback you are stating that the activity implements the `StarSignListener` interface so it can be notified when the user clicks an item in the list. Add the import for the `Context` required for `onAttach` with the other imports at the top of the file:

```kotlin
import android.content.Context
class ListFragment : Fragment(), View.OnClickListener
{
    // TODO: Rename and change types of parameters
    private var param1: String? = null
    private var param2: String? = null

    private lateinit var starSignListener:
        StarSignListener
    override fun onAttach(context: Context) {
        super.onAttach(context)
        if (context is StarSignListener) {
            starSignListener = context
        } else {
            throw RuntimeException("Must implement
                StarSignListener")
        }
    }
    override fun onCreateView(...)
    override fun onViewCreated(view: View,
    savedInstanceState:Bundle?) {
      super.onViewCreated(view, savedInstanceState)
      val starSigns = listOf<View>(
          view.findViewById(R.id.aquarius),
          view.findViewById(R.id.pisces),
          view.findViewById(R.id.aries),
          view.findViewById(R.id.taurus),
          view.findViewById(R.id.gemini),
          view.findViewById(R.id.cancer),
          view.findViewById(R.id.leo),
          view.findViewById(R.id.virgo),
```

```
                view.findViewById(R.id.libra),
                view.findViewById(R.id.scorpio),
                view.findViewById(R.id.sagittarius),
                view.findViewById(R.id.capricorn)
            )
            starSigns.forEach {
                it.setOnClickListener(this)
            }
        }
        override fun onClick(v: View?) {
            v?.let { starSign ->
                starSignListener.onSelected(starSign.id)
            }
        }
    }
```

The remaining callbacks are similar to what you have seen in the previous exercises. You create the fragment view with `onCreateView`. You set up the buttons with an `OnClickListener` in `onViewCreated` and then you handle clicks in `onClick`.

The `listOf` syntax in `onViewCreated` is a way of creating a `readonly` list with the specified elements, which in this case are your star sign `TextView` instances. Then, in the following code, you loop over these `TextViews`, setting the `OnClickListener` for each of the individual `TextViews` by iterating over the `TextView` list with the `forEach` statement. The `it` syntax here refers to the element of the list that is being operated on, which will be one of the 12 star sign `TextViews`.

9. Finally, the `onClick` statement communicates back to the activity through `StarSignListener` when one of the star signs in the list has been clicked:

    ```
    v?.let { starSign ->
        starSignListener.onSelected(starSign.id)
    }
    ```

    You check whether the view specified as v is null with ? and then only operate upon it with the `let` scope function if it isn't, before passing the `id` value of the star sign to `Activity/StarSignListener`.

> **Note**
>
> Listeners are a common way to react to changes. By specifying a `Listener` interface, you are specifying a contract to be fulfilled. The implementing class is then notified of the results of the listener operation.

10. Next, create `DetailFragment`, which will display the star sign details. Create a blank fragment as you have done before and call it `DetailFragment`. Replace the `fragment_detail` layout file contents with the following XML file:

```xml
<?xml version="1.0" encoding="utf-8"?>
<LinearLayout
xmlns:android=
   "http://schemas.android.com/apk/res/android"
    xmlns:tools="http://schemas.android.com/tools"
    android:layout_width="match_parent"
    android:layout_height="match_parent"
    android:orientation="vertical"
    tools:context=".DetailFragment">
    <TextView
        android:id="@+id/star_sign"
        style="@style/StarSignTextViewHeader"
        android:textStyle="bold"
        android:gravity="center"
        android:layout_width="match_parent"
        android:layout_height="wrap_content"
        tools:text="Aquarius"/>
    <TextView
        android:id="@+id/symbol"
        style="@style/StarSignTextView"
        android:layout_width="match_parent"
        android:layout_height="wrap_content"
        tools:text="Water Carrier"/>
    <TextView
        android:id="@+id/date_range"
        style="@style/StarSignTextView"
        android:layout_width="match_parent"
        android:layout_height="wrap_content"
        tools:text="Date Range:
            January 20 - February 18" />
</LinearLayout>
```

Here, you create a simple `LinearLayout`, which will display the star sign name, the symbol of the star sign, and the date range. You'll set these values in `DetailFragment`.

11. Open `DetailFragment` and update the contents with the following text and also add `TextView` and `Toast` imports to the imports list:

```kotlin
import android.widget.TextView
import android.widget.Toast

class DetailFragment : Fragment() {
    // TODO: Rename and change types of parameters
    private var param1: String? = null
    private var param2: String? = null

    private val starSign: TextView?
        get() = view?.findViewById(R.id.star_sign)
    private val symbol: TextView?
        get() = view?.findViewById(R.id.symbol)
    private val dateRange: TextView?
        get() = view?.findViewById(R.id.date_range)

    override fun onCreate(...)
    override fun onCreateView(...)

    fun setStarSignData(starSignId: Int) {
        when (starSignId) {
            R.id.aquarius -> {
                starSign?.text =
                    getString(R.string.aquarius)
                symbol?.text =
                    getString(R.string.symbol,
                        "Water Carrier")
                dateRange?.text =
                    getString(R.string.date_range,
                        "January 20 - February 18")
            }
        }
    }
}
```

The onCreateView inflates the layout as normal. The setStarSignData() function is what populates the data from the passed-in starSignId. The when expression is used to determine the ID of the star sign and set the appropriate contents.

The setStarSignData function here formats text passed with the getString function – getString(R.string.symbol, "Water Carrier"), for example, passes the text Water Carrier into the symbol string, <string name="symbol">Symbol: %s</string>, and replaces %s with the passed-in value. You can see what other string formatting options there are in the official docs: https://developer.android.com/guide/topics/resources/string-resource.

Following the pattern introduced by aquarius, add the other 11 star signs below the aquarius block from the completed file here: https://packt.link/C9sWZ.

Right now, you have added both ListFragment and DetailFragment. Currently, however, they have not been added to the activity layout and/or synced together, so selecting the star sign item in ListFragment will not load contents into DetailFragment. Let's look at how you can change that.

12. Firstly, you need to change the layout of activity_main.xml in both the layout folder and layout-sw600dp.

13. Open up res | layout | activity_main.xml if in the Project view. In the default Android view, open up res | layout | activity_main.xml and select the top activity_main.xml file without (sw600dp). Replace the contents with the following:

```xml
<?xml version="1.0" encoding="utf-8"?>
<androidx.constraintlayout.widget.ConstraintLayout
xmlns:android=
  "http://schemas.android.com/apk/res/android"
    xmlns:tools="http://schemas.android.com/tools"
    android:layout_width="match_parent"
    android:layout_height="match_parent"
    tools:context=".MainActivity">
    <fragment
        android:id="@+id/star_sign_list"
        android:name="com.example.dualpanelayouts
            .ListFragment"
        android:layout_height="match_parent"
        android:layout_width="match_parent"/>
</androidx.constraintlayout.widget.ConstraintLayout>
```

If you run the app and select one of the star signs now, you will get a `NotImplementedError` as we need to replace the `TODO` item with this functionality.

14. Then, open up res | layout-sw600dp | activity_main.xml if in the Project view. In the default Android view, open up res | layout | activity_main.xml (sw600dp). Replace the contents with the following:

    ```xml
    <?xml version="1.0" encoding="utf-8"?>
    <LinearLayout
    xmlns:android=
       "http://schemas.android.com/apk/res/android"
        xmlns:tools="http://schemas.android.com/tools"
        android:layout_width="match_parent"
        android:layout_height="match_parent"
        android:orientation="horizontal"
        tools:context=".MainActivity">
        <fragment
            android:id="@+id/star_sign_list"
            android:name="com.example.dualpanelayouts
                .ListFragment"
            android:layout_height="match_parent"
            android:layout_width="0dp"
            android:layout_weight="1"/>
        <View
            android:layout_width="1dp"
            android:layout_height="match_parent"
            android:background="?android:attr/
                dividerVertical" />
        <fragment
            android:id="@+id/star_sign_detail"
            android:name="com.example.dualpanelayouts
                .DetailFragment"
            android:layout_height="match_parent"
            android:layout_width="0dp"
            android:layout_weight="2"/>
    </LinearLayout>
    ```

    You are adding `LinearLayout`, which will by default lay out its content horizontally.

You add `ListFragment`, a divider, and then `DetailFragment` and assign the fragments appropriate IDs. Notice also that you are using the concept of weights to assign the space available for each fragment. When you do this, you specify `android:layout_width="0dp"`. `layout_weight` then sets the proportion of the width available by the weight measurements as `LinearLayout` is set to lay out the fragments horizontally.

`ListFragment` is specified as `android:layout_weight="1"` and `DetailFragment` is specified as `android:layout_weight="2"`, which tells the system to assign `DetailFragment` twice the width of `ListFragment`. In this case, where there are three views including the divider, which is a fixed dp width, this will result roughly in `ListFragment` occupying one-third of the width and `DetailFragment` occupying two-thirds.

15. To see the app, create a new virtual device as shown in *Chapter 1, Creating Your First App*, and select **Category** | **Tablet** | **Nexus 7**.
16. This will create a 7" tablet. Then, launch the virtual device and run the app. This is the initial view you will see when you launch the tablet in portrait mode:

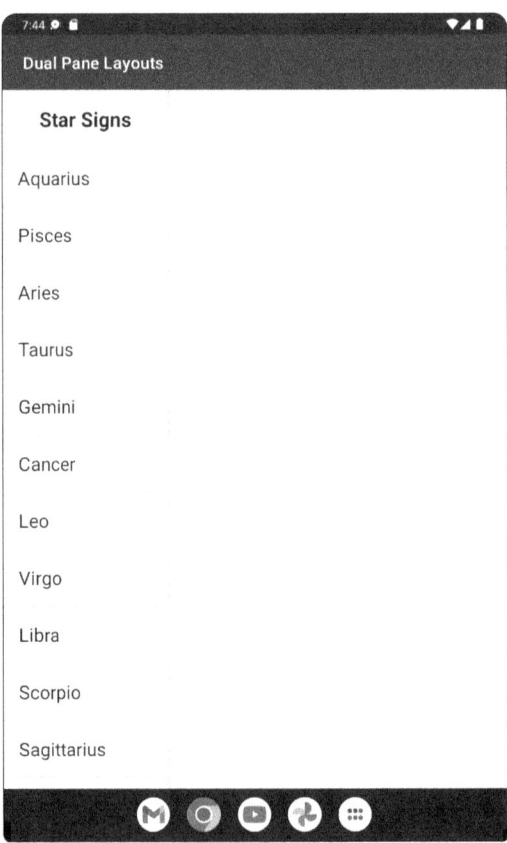

Figure 3.12 – Initial star sign app UI display

You can see that the list takes up about a third of the screen and the blank space two-thirds of the screen.

17. Click the bottom rotate button on the virtual device to turn the virtual device 90 degrees clockwise.

18. Once you've done that, the virtual device will go into landscape mode. It will not, however, change the screen orientation to landscape.

19. In order to do this, click on the rotate button in the bottom-left corner of the virtual device. You can also select the status bar at the top of the virtual device and hold and drag down to display the quick settings bar where you can turn on auto-rotation by selecting the rotate button. (You might have to swipe left/right within the quick settings bar to show the **Auto-rotate** option.)

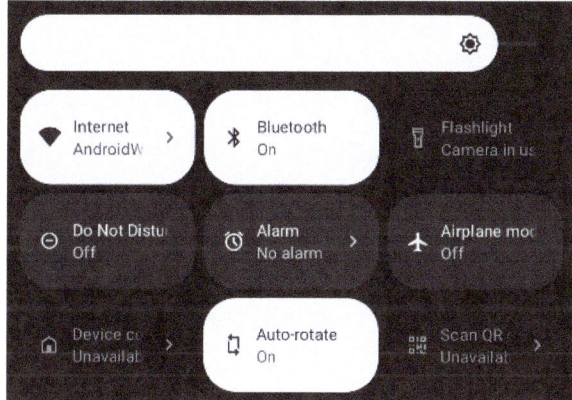

Figure 3.13 – Quick settings bar with Auto-rotate selected

20. This will then change the tablet layout to landscape:

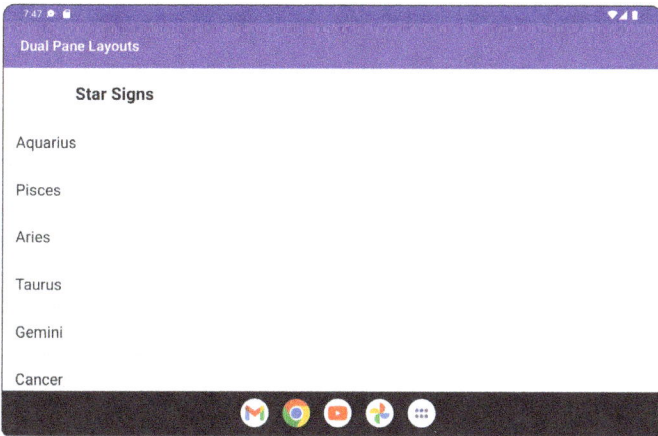

Figure 3.14 – Initial star sign app UI display in landscape on a tablet

21. The next thing to do is enable selecting a list item to load contents into the Detail pane of the screen. For that, we need to make changes in MainActivity. Update MainActivity with the following code to retrieve fragments by their ID in the pattern of retrieving views by their IDs (there will be some unused imports that will be required when the phone implementation is done):

```kotlin
package com.example.dualpanelayouts

import android.content.Intent
import android.os.Bundle
import android.view.View
import androidx.appcompat.app.AppCompatActivity
const val STAR_SIGN_ID = "STAR_SIGN_ID"
interface StarSignListener {
    fun onSelected(id: Int)
}
class MainActivity : AppCompatActivity(),
StarSignListener {
    var isDualPane: Boolean = false
    override fun onCreate(savedInstanceState: Bundle?)
    {
        super.onCreate(savedInstanceState)
        setContentView(R.layout.activity_main)
        isDualPane =
            findViewById<View>(R.id.star_sign_detail)
                != null
    }
    override fun onSelected(id: Int) {
        if (isDualPane) {
            val detailFragment =
                supportFragmentManager
                .findFragmentById(
                R.id.star_sign_detail) as
                DetailFragment
            detailFragment.setStarSignData(id)
        }
    }
}
```

> **Note**
>
> Fragments were introduced in API 11 in 2011 with a `FragmentManager` class to manage their interactions with activities. `SupportFragmentManager` was introduced to support using fragments in Android versions before API 11 in the Android Support Library. `SupportFragmentManager` has been further developed as the base for the Jetpack Fragment library, which adds improvements for managing fragments.

This example and those that follow use `supportFragmentManager.findFragmentById` to retrieve fragments. You can also, however, retrieve fragments with `Tag` if you add a tag to the fragment XML by using `android:tag="MyFragmentTag"`.

22. You can then retrieve the fragment by using `supportFragmentManager.findFragmentByTag("MyFragmentTag")`.

23. In order to retrieve data from the fragment, the activity needs to implement `StarSignListener`. This completes the contract set in the fragment to pass back details to the implementing class. The `onCreate` function sets the layout and then checks whether `DetailFragment` is in the activity's inflated layout by checking whether the `R.id.star_sign_detail` ID exists.

24. From the Project view, the res | layout | activity_main.xml file only contains ListFragment, but you've added the code in the res | layout-sw600dp | activity_main.xml file to contain `DetailFragment` with `android:id="@+id/star_sign_detail"`.

25. This will be used for the layout of the Nexus 7 tablet. In the default Android view, open up res | layout | activity_main.xml and select the top activity_main.xml file without (sw600dp) and then select activity_main.xml (sw600dp) to see these differences.

26. So now we can retrieve the star sign ID passed from `ListFragment` back to `MainActivity` by `StarSignListener` and pass it into `DetailFragment`. This is achieved by checking the `isDualPane` Boolean, and if that evaluates to `true`, you know you can pass the star sign ID to `DetailFragment` with this code:

    ```
    val detailFragment = supportFragmentManager
    .findFragmentById (R.id.star_sign_detail) as
    DetailFragment
    detailFragment.setStarSignData(id)
    ```

27. You cast the fragment from `id` to `DetailFragment` and call the following:

    ```
    detailFragment.setStarSignData(id)
    ```

28. As you've implemented this function in the fragment and are checking by the `id` value which contents to display, the UI is updated:

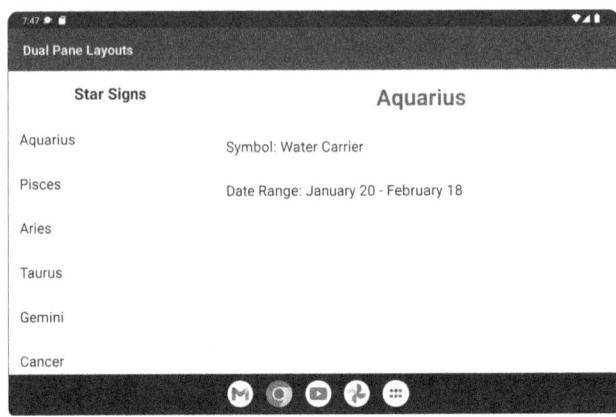

Figure 3.15 – Star sign app dual-pane display in landscape on a tablet

29. Now clicking on a list item works as intended, showing the dual-pane layout with the contents set correctly.
30. If the device is not a tablet, however, even when a list item is clicked, nothing will happen as there is not an `else` branch condition to do anything if the device is not a tablet, which is defined by the `isDualPane` Boolean. The display will be as in *Figure 3.16* and won't change when items are selected:

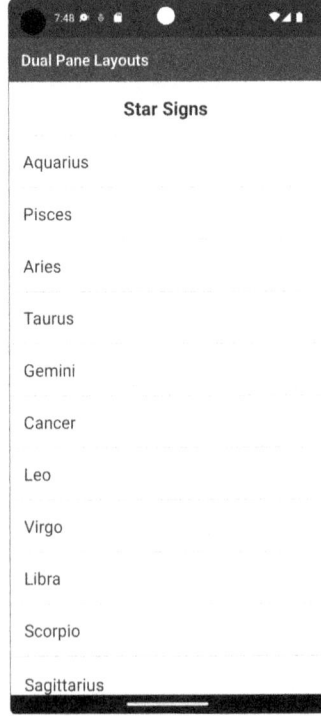

Figure 3.16 – Initial star sign app UI display on a phone

Static fragments and dual-pane layouts   143

31. You are going to display the star sign detail in another activity. Create a new `DetailActivity` by going to **File** | **New** | **Activity** | **Empty Activity**. Once created, update `activity_detail.xml` with this layout:

```xml
<?xml version="1.0" encoding="utf-8"?>
<androidx.constraintlayout.widget.ConstraintLayout
xmlns:android=
   "http://schemas.android.com/apk/res/android"
    xmlns:app="http://schemas.android.com/apk/res-auto"
    xmlns:tools="http://schemas.android.com/tools"
    android:layout_width="match_parent"
    android:layout_height="match_parent"
    tools:context=".DetailActivity">
    <fragment
        android:id="@+id/star_sign_detail"
        android:name="com.example.dualpanelayouts
            .DetailFragment"
        android:layout_height="match_parent"
        android:layout_width="match_parent"/>
</androidx.constraintlayout.widget.ConstraintLayout>
```

32. This adds `DetailFragment` as the only fragment in the layout. Now, update the `onCreate` function of `DetailActivity` with the following contents:

```kotlin
override fun onCreate(savedInstanceState: Bundle?) {
    super.onCreate(savedInstanceState)
    setContentView(R.layout.activity_detail)
    val starSignId =
        intent.extras?.getInt(STAR_SIGN_ID, 0) ?: 0
    val detailFragment =
        supportFragmentManager.findFragmentById(
        R.id.star_sign_detail) as DetailFragment
    detailFragment.setStarSignData(starSignId)
}
```

33. The star sign `id` is expected to be passed from another activity to this one by setting a key in the intent's extras (also called a bundle). We covered intents in *Chapter 2, Building User Screen Flows*, but as a reminder, they enable communication between different components and also can send data.

34. In this case, the intent that opened this activity has set a star sign ID. It will use `id` to set the star sign ID in `DetailFragment`. Next, you need to implement the `else` branch of the `isDualPane` check to launch `DetailActivity`, passing through the star sign ID in the intent.

35. Update `MainActivity` as follows to do this:

    ```
    override fun onSelected(id: Int) {
        if (isDualPane) {
            val detailFragment = supportFragmentManager
                .findFragmentById(R.id.star_sign_detail)
                    as DetailFragment
            detailFragment.setStarSignData(id)
        } else {
            val detailIntent = Intent(this,
                DetailActivity::class.java)
                    .apply {
                        putExtra(STAR_SIGN_ID, id)
                    }
            startActivity(detailIntent)
        }
    }
    ```

36. Once you click on one of the star sign names on a phone display, it shows the contents in `DetailActivity` occupying the whole of the screen without the list:

Figure 3.17 – Single-pane star sign detail screen on a phone

This exercise has demonstrated the flexibility of fragments. They can encapsulate the logic and display different features of your app, which can be integrated in different ways depending on the form factor of the device. They can be arranged onscreen in a variety of ways, which are constrained by the layout they are included in; therefore, they can feature as a part of dual-pane layouts or all or part of a single-pane layout.

This exercise showed fragments being laid out side by side on a tablet, but they can also be laid out on top of each other and in a variety of other ways. The next topic illustrates how the configuration of fragments used in your app doesn't have to be specified statically in XML but can also be done dynamically.

Dynamic fragments

So far, you've only seen fragments added in XML at compile time. Although this can satisfy many use cases, you might want to add fragments dynamically at runtime to respond to the user's actions. This can be achieved by adding `ViewGroup` as a container for fragments and then adding, replacing, and removing fragments from `ViewGroup`.

This technique is more flexible as the fragments can be active until they are no longer needed and then removed, instead of always being inflated in XML layouts as you have seen with static fragments. If three or four more fragments are required to fulfill separate user journeys in one activity, then the preferred option is to react to the user's interaction in the UI by adding/replacing fragments dynamically.

Using static fragments works better when the user's interaction with the UI is fixed at compile time and you know in advance how many fragments you need. For example, this would be the case for selecting items from a list to display the contents.

Exercise 3.04 – adding fragments dynamically to an activity

In this exercise, we will build the same star sign app as before but will demonstrate how the list and detail fragments can be added to screen layouts dynamically and not directly within an XML layout. You can also pass arguments into a fragment. For simplicity, you are going to create the same configuration for both phones and tablets. Perform the following steps:

1. Create a new project with an `Empty Activity` called `Dynamic Fragments`.

2. Once you have done that, add the following dependency – you need to use `FragmentContainerView`, an optimized ViewGroup for handling fragment transactions with `app/build.gradle` within the `dependences{ }` block:

    ```
    implementation 'androidx.fragment:fragment-ktx:1.5.6'
    ```

3. Copy the contents of the following XML resource files from *Exercise 3.03 – dual-pane layouts with static fragments*, and add them to the corresponding files in this exercise: `strings.xml` (changing the `app_name` string from `Dual Pane Layouts` to `Dynamic Fragments`), `fragment_detail.xml`, and `fragment_list.xml`. All of these files exist in the project created in the previous exercise and you simply add the contents to this new project.

4. Then, copy `DetailFragment` and `ListFragment` to the new project. You will have to change the package name from `package com.example.dualpanelayouts` to `package com.example.dynamicfragments` in these two files. Finally, add the styles defined below the base application style in `themes.xml` from the last exercise to `themes.xml` in this project.

5. You now have the same fragments set up as in the previous exercise. Now, open the `activity_main.xml` layout and replace the contents with this:

    ```
    <?xml version="1.0" encoding="utf-8"?>
    <androidx.fragment.app.FragmentContainerView
    xmlns:android=
      "http://schemas.android.com/apk/res/android"
        android:id="@+id/fragment_container"
    ```

```
            android:layout_width="match_parent"
            android:layout_height="match_parent" />
```

This is the `FragmentContainerView` you will add the fragments to. You'll notice that there are no fragments added in the layout XML as these will be added dynamically.

6. Go into `MainActivity` and replace the content with the following:

```
package com.example.dynamicfragments

import androidx.appcompat.app.AppCompatActivity
import android.os.Bundle
import androidx.fragment.app.FragmentContainerView
const val STAR_SIGN_ID = "STAR_SIGN_ID"
interface StarSignListener {
    fun onSelected(id: Int)
}

class MainActivity : AppCompatActivity() {
    override fun onCreate(savedInstanceState: Bundle?)
    {
        super.onCreate(savedInstanceState)
        setContentView(R.layout.activity_main)
        if (savedInstanceState == null) {
            findViewById<FragmentContainerView>(
            R.id.fragment_container)?.let {
            frameLayout ->
              val listFragment = ListFragment()
              supportFragmentManager.beginTransaction()
                  .add(frameLayout.id,
                  listFragment).commit()
            }
        }
    }
}
```

You are getting a reference to the `FragmentContainerView` specified in `activity_main.xml`, creating a new `ListFragment`, and then adding this fragment to the `FragmentContainerView` with the ID of the `fragment_container`.

The fragment transaction specified is add as you are adding a fragment to FrameLayout for the first time. You call commit() to execute the transaction immediately. There is a null check with savedInstanceState to only add this ListFragment if there is no state to restore, which there would be if a fragment had been previously added.

7. Next, make MainActivity implement StarSignListener by adding the following:

```
class MainActivity : AppCompatActivity(),
StarSignListener {
    ...
    override fun onSelected(id: Int) {
    }
}
```

8. Now if you run the app, you will see the star sign list being displayed on mobile and tablet.

The problem you now come to is how to pass the star sign ID to DetailFragment now that it's not in an XML layout.

One option would be to use the same technique as in the last example by creating a new activity and passing the star sign ID in an intent, but you shouldn't have to create a new activity to add a new fragment; otherwise, you might as well dispense with fragments and just use activities.

You are going to replace ListFragment in FragmentContainerView with DetailFragment, but first, you need to find a way to pass the star sign ID into DetailFragment. You do this by passing this id value as an argument when you create the fragment. The standard way to do this is by using a Factory method in a fragment.

9. Add the following code to the bottom of DetailFragment (a sample factory method will have been added when you created the fragment using the template/wizard, which you can update here):

```
companion object {
    private const val STAR_SIGN_ID = "STAR_SIGN_ID"
    fun newInstance(starSignId: Int) =
    DetailFragment().apply {
        arguments = Bundle().apply {
            putInt(STAR_SIGN_ID, starSignId)
        }
    }
}
```

A companion object allows you to add Java's equivalent of static members into your class. Here, you are instantiating a new DetailFragment and setting arguments passed into the fragment. The arguments of the fragment are stored in a `Bundle()`, so in the same way as an activity's intent extras (which is also a bundle), you add the values as key pairs. In this case, you are adding the `STAR_SIGN_ID` key with the value `starSignId`.

10. The next thing to do is override one of the `DetailFragment` lifecycle functions to use the passed-in argument:

    ```
    override fun onViewCreated(view: View,
    savedInstanceState: Bundle?) {
        val starSignId = arguments?.getInt(STAR_SIGN_ID,
            0) ?: 0
        setStarSignData(starSignId)
    }
    ```

11. You do this in `onViewCreated` as at this stage, the layout of the fragment has been set up and you can access the view hierarchy (whereas if you accessed the arguments in `onCreate`, the fragment layout would not be available as this is done in `onCreateView`):

    ```
    val starSignId = arguments?.getInt(STAR_SIGN_ID, 0) ?: 0
    ```

12. This line gets the star sign ID from the passed-in fragment arguments, setting a default of 0 if the `STAR_SIGN_ID` key cannot be found. Then, you call `setStarSignData(starSignId)` to display the star sign contents.

13. Now you just need to implement the `StarSignListener` interface in `MainActivity` to retrieve the star sign ID from `ListFragment`:

    ```
    class MainActivity : AppCompatActivity(),
    StarSignListener {
        override fun onCreate(savedInstanceState: Bundle?)
        {
            super.onCreate(savedInstanceState)
            setContentView(R.layout.activity_main)
            if (savedInstanceState == null) {
                findViewById<FragmentContainerView>(
                R.id.fragment_container)?.let {
                frameLayout ->
                    val listFragment = ListFragment()
                    supportFragmentManager.beginTransaction()
                        .add(frameLayout.id,
                        listFragment).commit()
    ```

```
                    }
                }
            }
            override fun onSelected(starSignId: Int) {
                findViewById<FragmentContainerView>(
                    R.id.fragment_container)?.let { frameLayout ->
                    val detailFragment =
                        DetailFragment.newInstance(starSignId)
                    supportFragmentManager.beginTransaction()
                        .replace(frameLayout.id,
                            detailFragment)
                        .addToBackStack(null)
                        .commit()
                }
            }
        }
```

You create `DetailFragment` as explained earlier with the `factory` method passing in the star sign ID: `DetailFragment.newInstance(starSignId)`.

At this stage, `ListFragment` is still the fragment that has been added to the activity `FrameLayout`. You need to replace it with `DetailFragment`, which requires another transaction. This time, however, you use the `replace` function to replace `ListFragment` with `DetailFragment`.

Before you commit the transaction, you call `.addToBackStack(null)` so the app does not exit when the back button is pressed but instead goes back to `ListFragment` by popping `DetailFragment` off the fragment stack.

This exercise has introduced adding fragments dynamically to your activity. The next topic introduces a more well-defined structure for creating fragments, called a navigation graph.

Jetpack Navigation

Using dynamic and static fragments, although very flexible, introduces a lot of boilerplate code into your app and can become quite complicated when user journeys require adding, removing, and replacing multiple fragments while managing the back stack.

Google introduced Jetpack components, as you learned in *Chapter 1, Creating Your First App*, to use established best practices in your code. The `Navigation` component within the suite of Jetpack components enables you to reduce boilerplate code and simplify navigation within your app. We are going to use it now to update the Star Sign app.

Exercise 3.05 – adding a Jetpack navigation graph

In this exercise, we are going to reuse most of the classes and resources from the last exercise. We will first create an empty project and copy the resources. Next, we will add the dependencies and create a navigation graph.

Using a step-by-step approach, we will configure the navigation graph and add destinations to navigate between fragments. Perform the following steps:

1. Create a new project with an `Empty Activity` called `Jetpack Fragments`.
2. Copy `strings.xml`, `fragment_detail.xml`, `fragment_list.xml`, `DetailFragment`, and `ListFragment` from the previous exercise, remembering to change the `app_name` string in `strings.xml` and the package name for the fragment classes. You will need to change the following line for the resources from `import com.example.dynamicfragments.R` to `import com.example.jetpacknavigation.R`.
3. Finally, add the styles defined below the base application style in `themes.xml` from the last exercise to `themes.xml` in this project. You will also need to add the following above the class header in `MainActivity`:

   ```
   const val STAR_SIGN_ID = "STAR_SIGN_ID"
   interface StarSignListener {
       fun onSelected(id: Int)
   }
   ```

4. Once you have done that, add the following dependencies – you need to use the `Navigation` component in `app/build.gradle` within the `dependences{ }` block:

   ```
   implementation
   "androidx.navigation:navigation-fragment-ktx:2.5.3"
   implementation
   "androidx.navigation:navigation-ui-ktx:2.5.3"
   ```

5. It will prompt you to **Sync now** in the top right-hand corner of the screen to update the dependencies. Click the button, and after they've updated, make sure the app module is selected and go to **File | New | Android Resource**.

6. Once this dialog appears, change **Resource type** to **Navigation** and then name the file nav_graph:

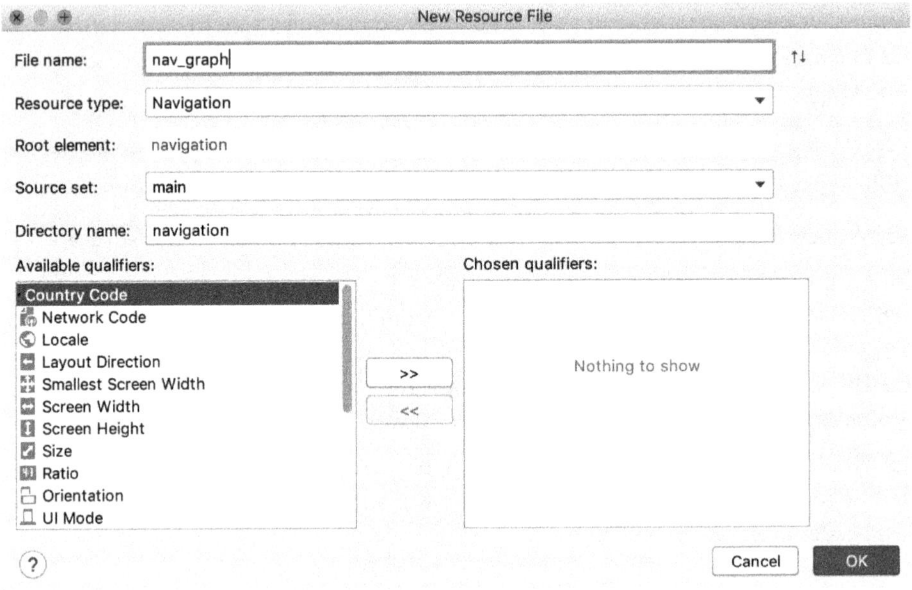

Figure 3.18 – New Resource File dialog

Click **OK** to proceed. This creates a new folder in the res folder called Navigation with nav_graph.xml inside it.

7. Update the nav_graph.xml navigation file with the following code:

```
<?xml version="1.0" encoding="utf-8"?>
<navigation
xmlns:android=
  "http://schemas.android.com/apk/res/android"
    xmlns:app="http://schemas.android.com/apk/res-auto"
    xmlns:tools="http://schemas.android.com/tools"
    android:id="@+id/nav_graph"
    app:startDestination="@id/starSignList">
    <fragment
        android:id="@+id/starSignList"
        android:name="com.example.jetpackfragments
            .ListFragment"
        android:label="List"
        tools:layout="@layout/fragment_list">
        <action
```

```
                android:id="@+id/star_sign_id_action"
                app:destination="@id/starSign">
            </action>
        </fragment>
        <fragment
            android:id="@+id/starSign"
            android:name="com.example.jetpackfragments
                .DetailFragment"
            android:label="Detail"
            tools:layout="@layout/fragment_detail" />
</navigation>
```

The preceding file is a working navigation graph. Although the syntax is unfamiliar, it is quite straightforward to understand:

A. `ListFragment` and `DetailFragment` are present as they would be if you were adding static fragments.

B. There is an `id` value to identify the graph at the root `<navigation>` element and IDs on the fragments themselves. Navigation graphs introduce the concept of destinations, so at the root `navigation` level, there is `app:startDestination`, which has the ID of `starSignList`, which is `ListFragment`, then within the `<fragment>` tag, there is the `<action>` element.

C. Actions are what link the destinations within the navigation graph together. The destination action here has an ID, so you can refer to it in code, and has a destination, which, when used, it will direct to.

Now that you've added the navigation graph, you need to use it to link the activity and fragments together.

8. Open up `activity_main.xml` and replace `TextView` inside `ConstraintLayout` with the following `FragmentContainerView`:

```
<?xml version="1.0" encoding="utf-8"?>
<androidx.fragment.app.FragmentContainerView
xmlns:android=
    "http://schemas.android.com/apk/res/android"
    xmlns:app="http://schemas.android.com/apk/res-auto"
    android:id="@+id/nav_host_fragment"
    android:name="androidx.navigation.fragment
        .NavHostFragment"
```

```
            android:layout_width="match_parent"
            android:layout_height="match_parent"
            app:defaultNavHost="true"
            app:navGraph="@navigation/nav_graph" />
```

FragmentContainerView has been added with the name android:name="androidx.navigation.fragment.NavHostFragment". It will host the fragments from the app:navGraph="@navigation/nav_graph" that you have just created.

app:defaultNavHost states that it is the app's default navigation graph. It also controls the back navigation when one fragment replaces another. You can have more than one NavHostFragment in a layout for controlling two or more areas of the screen that manage their own fragments, which you might use for dual-pane layouts in tablets, but there can only be one default.

There are a few changes you need to make to make the app work as expected in the ListFragment.

9. Firstly, remove the class file header and references to StarSignListener. So, the following will be replaced:

```
interface StarSignListener {
    fun onSelected(starSignId: Int)
}
class ListFragment : Fragment(), View.OnClickListener
{
    private lateinit var starSignListener:
        StarSignListener
    override fun onAttach(context: Context) {
        super.onAttach(context)
        if (context is StarSignListener) {
            starSignListener = context
        } else {
            throw RuntimeException("Must implement
            StarSignListener")
        }
    }
}
```

And it will be replaced with the following line of code:

```
class ListFragment : Fragment() {
```

10. Next, at the bottom of the class, remove the `onClick` overridden method as you are not implementing `View.OnClicklistener`:

    ```
    override fun onClick(v: View?) {
        v?.let { starSign ->
            starSignListener.onSelected(starSign.id)
        }
    }
    ```

11. In the `onViewCreated` method, replace the `forEach` statement that loops over the star sign views:

    ```
    starSigns.forEach {
        it.setOnClickListener(this)
    }
    ```

    Replace it with the following code and add the `Navigation` import to the imports list:

    ```
    import androidx.navigation.Navigation
    starSigns.forEach { starSign ->
        val fragmentBundle = Bundle()
        fragmentBundle.putInt(STAR_SIGN_ID, starSign.id)
        starSign.setOnClickListener(
            Navigation.createNavigateOnClickListener(
                R.id.star_sign_id_action, fragmentBundle)
        )
    }
    ```

    Here, you are creating a bundle to pass STAR_SIGN_ID with the view ID of the selected star sign to `NavigationClickListener`. It uses the ID of the `R.id.star_sign_id_action` action to load `DetailFragment` when clicked as that is the destination for the action. `DetailFragment` does not need any changes and uses the passed-in `fragment` argument to load the details of the selected star sign ID.

12. Run up the app, and you'll see that the app behaves as it did before.

Now you've been able to remove a significant amount of boilerplate code and document the navigation within the app in the navigation graph. In addition, you have offloaded more of the management of the fragment lifecycle to the Android framework, saving more time to work on features.

Jetpack Navigation is a powerful `androidx` component and enables you to map your whole app and the relationships between fragments, activities, and so on. You can also use it selectively to manage different areas of your app that have a defined user flow, such as the startup of your app and guiding the user through a series of welcome screens, or some wizard layout user journey, for example.

156 Developing the UI with Fragments

With this knowledge, let's try completing an activity using the techniques learned from all these exercises.

Activity 3.01 – creating a quiz on the planets

For this activity, you will create a quiz where users have to answer one of three questions on the planets of the Solar System. The number of fragments you choose to use is up to you. However, considering this chapter's content, which is separating the UI and logic into separate fragment components, it is likely you will use two fragments or more to achieve this.

The screenshots that follow show one way this could be done, but there are multiple ways to create this app. You can use one of the approaches detailed in this chapter, such as static fragments, dynamic fragments, or the Jetpack `Navigation` component, or something custom that uses a combination of these and other approaches.

The content of the quiz is as follows. In the UI, you need to ask the user these three questions:

- What is the largest planet?
- Which planet has the most moons?
- Which planet spins on its side?

Then, you need to provide a list of planets so the user can choose the planet that they believe is the answer to the question:

- MERCURY
- VENUS
- EARTH
- MARS
- JUPITER
- SATURN
- URANUS
- NEPTUNE

Once they have given their answer, you need to show them whether they are correct or wrong. The correct answer should be accompanied by some text that gives more detail about the question's answer:

```
Jupiter is the largest planet and is 2.5 times the mass of all
the other planets put together.
Saturn has the most moons and has 82 moons.
Uranus spins on its side with its axis at nearly a right angle
to the Sun.
```

The following are some screenshots of how the UI might look to achieve the requirements of the app you need to build:

Questions screen:

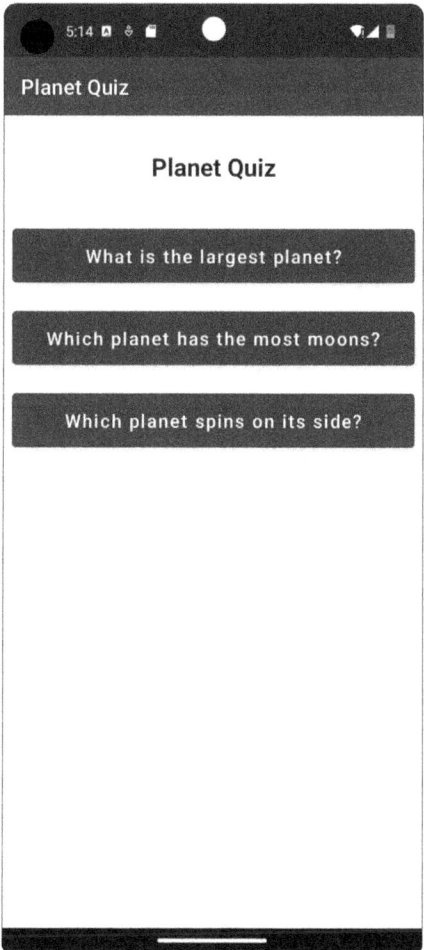

Figure 3.19 – Planet Quiz questions screen

Answer options screen:

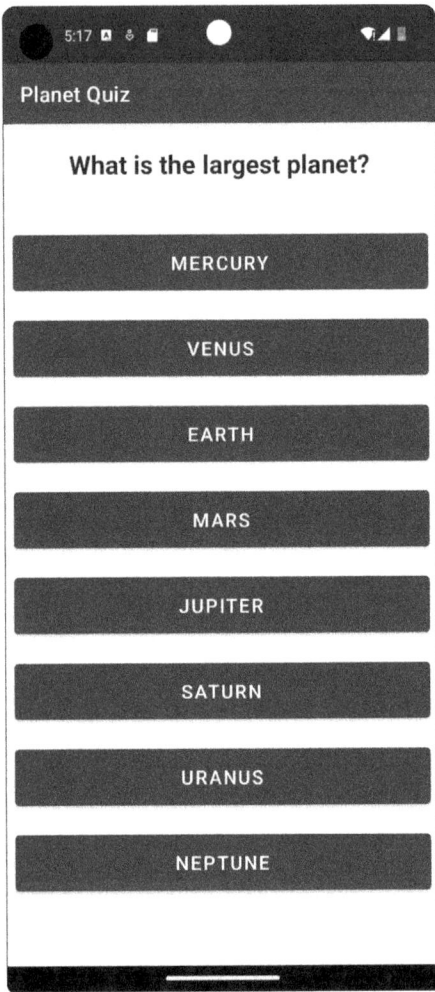

Figure 3.20 – Planet Quiz multiple-choice answer screen

Answer screen:

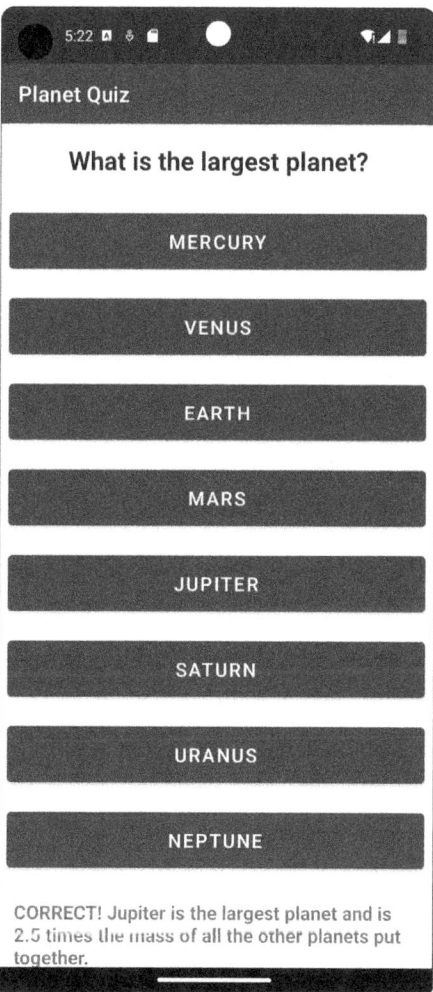

Figure 3.21 – Planet Quiz answer screen with detailed answer

The following steps will help to complete the activity:

1. Create an Android project with an `Empty Activity`.
2. Update the `strings.xml` file with the entries you need for the project.
3. Amend the `themes.xml` file with styles for the project.
4. Create a `QuestionsFragment`, update the layout with the questions, and add interaction with buttons and `OnClickListener`(s).

5. Optionally, create a multiple-choice fragment and add answer options and button-click handling (this can also be done by adding the possible answer options to `QuestionsFragment`).

6. Create an `AnswersFragment` that displays the relevant question's answer and also displays more details about the answer itself.

> **Note**
> The solution to this activity can be found at https://packt.link/By7eE.

Summary

This chapter has covered fragments in depth, starting with learning about the fragment lifecycle and the key functions to override in your own fragments. We then moved on to adding simple fragments statically to an app in XML and demonstrating how the UI display and logic can be self-contained in individual fragments. Other options for how to add fragments to an app using a `FragmentContainerView` and dynamically adding and replacing fragments were then covered. We then finished with how this can be simplified by using the Jetpack `Navigation` component.

Fragments are one of the fundamental building blocks of Android development. The concepts you have learned about here will allow you to build upon them and progress to create increasingly more advanced apps. Fragments are at the core of building effective navigation into your apps in order to bind features and functionality that are simple and easy to use.

The next chapter will explore this area in detail by using established UI patterns to build clear and consistent navigation and illustrate how fragments are used to enable this.

4
Building App Navigation

In this chapter, you will build user-friendly app navigation through three primary patterns: bottom navigation, the navigation drawer, and tabbed navigation. Through guided theory and practice, you will learn how each of these patterns works so that users can easily access your app's content. This chapter will also focus on making the user aware of where they are in the app and which level of your app's hierarchy they can navigate to.

By the end of this chapter, you will know how to use these three primary navigation patterns and understand how they work with the app bar to support navigation.

In the previous chapter, you explored fragments and the **fragment lifecycle** and employed Jetpack navigation to simplify their use in your apps. In this chapter, you will learn how to add different types of navigation to your app while continuing to use Jetpack navigation.

You will start off by learning about the navigation drawer, the earliest widely adopted navigational pattern used in Android apps, before exploring bottom navigation and tab navigation. You'll learn about the Android navigation user flow, how it is built around destinations, and how they govern navigation within the app.

The difference between primary and secondary destinations will be explained, as well as which one of the three primary navigation patterns is more suitable, depending on your app's use case.

We will cover the following topics in this chapter:

- Navigation overview
- Navigation drawer
- Bottom navigation
- Tabbed navigation

Technical requirements

The complete code for all the exercises and the activity in this chapter is available on GitHub at https://packt.link/B2rz6.

Navigation overview

The Android navigation user flow is built around **destinations** within your app. There are primary destinations available at the top level of your app and, subsequently, are always displayed in the main app navigation and secondary destinations. A guiding principle of each of the three navigation patterns is to contextually provide information about the main section of the app the user is in at any point in time.

This can take the form of a label in the top app bar of the destination the user is in, optionally displaying an arrow hint that the user is not at the top level, and/or providing highlighted text and icons in the **user interface** (**UI**) that indicate the section the user is in. Navigation in your app should be fluid and natural, intuitively guiding the user while also providing some context of where they are at any given point in time.

Each of the three navigation patterns you will explore accomplishes this goal in varying ways. Some of these navigational patterns are more suitable for use with a higher number of top-level primary destinations to display, and others are suitable for less.

Navigation drawer

The **navigation drawer** is one of the most common navigation patterns used in Android apps and was certainly the first pattern to be widely adopted. The following is a screenshot of the culmination of the next exercise, which shows a simple navigation drawer in its closed state:

Figure 4.1 – App with the navigation drawer closed

The navigation drawer is accessed through what has become commonly known as the hamburger menu, which is the icon with three horizontal lines at the top left of *Figure 4.1*. The navigation options are not visible on the screen, but contextual information about the screen you are on is displayed in the top app bar.

An overflow menu can also accompany this on the right-hand side of the screen, through which other contextually relevant navigation options can be accessed. The following screenshot is of a navigation drawer in the open state, showing all the navigation options:

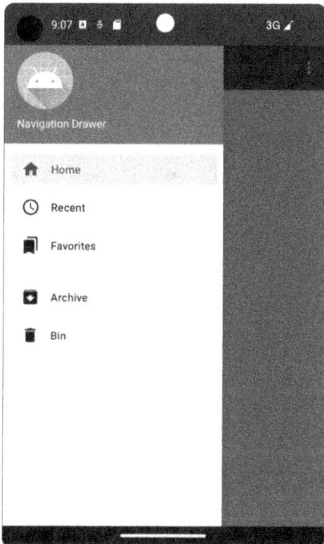

Figure 4.2 – App with the navigation drawer open

Upon selecting the hamburger menu, the navigation drawer slides out from the left with the current section highlighted. This can be displayed with or without an icon. Due to the nature of the navigation occupying the height of the screen, it is best suited to five or more top-level destinations.

The destinations can also be grouped together to indicate multiple hierarchies of primary destinations (shown by the dividing line in the preceding screenshot), and these hierarchies can also have labels. In addition, the drawer content is also scrollable. In summary, the navigation drawer is a very convenient way to provide quick access to many different app destinations.

A weakness of the navigation drawer is that it requires the user to select the hamburger menu for the destinations to become visible. Tabs and bottom navigation (with fixed tabs), in contrast, are always visible. However, this is conversely also a strength of the navigation drawer as more screen space can be used for the app's content.

Let's get started with the first exercise of this chapter and create a navigation drawer so that we can access all the sections of an app.

Exercise 4.01 – creating an App with a navigation drawer

In this exercise, you will create a new app in Android Studio named `Navigation Drawer` using the `Empty Activity` project template while leaving all the other defaults as they are. There are wizard options where you can create a new project with all the navigation patterns you are going to produce in the exercises within this chapter, but we will build the apps incrementally to guide you through the steps.

You will build an app that often uses a navigation drawer, such as a news or mail app. The sections we will be adding are **Home**, **Favorites**, **Recents**, **Archive**, **Bin**, and **Settings**.

Perform the following steps to complete this exercise:

1. Create a new project with an `Empty Activity` called `Navigation Drawer`. Do not use the **Navigation Drawer Activity** project template, as we are going to use incremental steps to build the app.

2. Add the Gradle dependencies you will require to `app/build.gradle`:

   ```
   implementation
   'androidx.navigation:navigation-fragment-ktx:2.5.3'
   implementation
   'androidx.navigation:navigation-ui-ktx:2.5.3'
   ```

3. Update `strings.xml` and `themes.xml` in the `res/values` folder with the following content:

strings.xml

```
        <string name="nav_header_desc">Navigation header</string>
<string name="home">Home</string>
    <string name="settings">Settings</string>
    <string name="content">Content</string>
    <string name="archive">Archive</string>
    <string name="recent">Recent</string>
    <string name="favorites">Favorites</string>
    <string name="bin">Bin</string>
    <string name="home_fragment">Home Fragment</string>
    <string name="settings_fragment">Settings Fragment</string>
    <string name="content_fragment">Content Fragment</string>
    <string name="archive_fragment">Archive Fragment</
```

```
    string>
        <string name="recent_fragment">Recent Fragment</
string>
        <string name="favorites_fragment">Favorites
Fragment</string>
        <string name="bin_fragment">Bin Fragment</string>
        <string name="link_to_content_button">Link to Content
Button</string>
```

themes.xml

```
        <style name="Theme.NavigationDrawer.NoActionBar">
            <item name="windowActionBar">false</item>
            <item name="windowNoTitle">true</item>
        </style>

        <style name="Theme.NavigationDrawer.AppBarOverlay"
parent="ThemeOverlay.AppCompat.Dark.ActionBar" />

        <style name="Theme.NavigationDrawer.PopupOverlay"
parent="ThemeOverlay.AppCompat.Light" />
```

4. Next, update the activity element with name `MainActivity` in the `AndroidManifest.xml` to NOT use an Action Bar. This will be provided by the Navigation Drawer layout. Go to app | manifests | AndroidManifest.xml and add the `android:theme` attribute with the `NoActionBar` style as in the snippet of code shown here:

```
<activity
    android:name=".MainActivity"
    android:exported="true"
    android:theme=
        "@style/Theme.NavigationDrawer.NoActionBar">
```

5. Create the following fragments (**File** | **New** | **Fragment** | **Fragment (Blank)**) making sure you have **app** selected in the Project window:

   - HomeFragment
   - FavoritesFragment
   - RecentFragment
   - ArchiveFragment

- `SettingsFragment`
- `BinFragment`
- `ContentFragment`

6. Change each of these fragment layouts to use the following content:

   ```xml
   <?xml version="1.0" encoding="utf-8"?>
   <androidx.constraintlayout.widget.ConstraintLayout
   xmlns:android=
      "http://schemas.android.com/apk/res/android"
     xmlns:app="http://schemas.android.com/apk/res-auto"
     android:layout_width="match_parent"
     android:layout_height="match_parent">
     <TextView
         android:layout_width="match_parent"
         android:layout_height="wrap_content"
         android:layout_margin="8dp"
         android:text="@string/home_fragment"
         android:textAlignment="center"
         android:layout_gravity="center_horizontal"
         android:textSize="20sp"
         app:layout_constraintBottom_toBottomOf="parent"
         app:layout_constraintEnd_toEndOf="parent"
         app:layout_constraintStart_toStartOf="parent"
         app:layout_constraintTop_toTopOf="parent" />
   </androidx.constraintlayout.widget.ConstraintLayout>
   ```

   The only difference is the `android:text` attribute, which will have the corresponding string from the `strings.xml` file. So, create these fragments with the correct string, indicating which fragment the user is viewing.

   This may seem a bit repetitive, and one single fragment could be updated with this text, but it demonstrates how you would separate different sections in a real-world app.

7. Update the `fragment_home.xml` with the following content, which adds a button (this is the body content you can see in *Figure 4.1*, with the closed navigation drawer):

   ```xml
   <?xml version="1.0" encoding="utf-8"?>
   <androidx.constraintlayout.widget.ConstraintLayout
   xmlns:android=
   ```

```xml
    "http://schemas.android.com/apk/res/android"
    xmlns:app="http://schemas.android.com/apk/res-auto"
    android:layout_width="match_parent"
    android:layout_height="match_parent">
    <TextView
        android:id="@+id/text_home"
        android:layout_width="match_parent"
        android:layout_height="wrap_content"
        android:layout_margin="8dp"
        android:text="@string/home_fragment"
        android:textAlignment="center"
        android:layout_gravity="center_horizontal"
        android:textSize="20sp"
        app:layout_constraintBottom_toBottomOf="parent"
        app:layout_constraintEnd_toEndOf="parent"
        app:layout_constraintStart_toStartOf="parent"
        app:layout_constraintTop_toTopOf="parent" />
    <Button
        android:id="@+id/button_home"
        android:layout_width="140dp"
        android:layout_height="140dp"
        android:layout_marginTop="16dp"
        android:text="@string/link_to_content_button"
        app:layout_constraintEnd_toEndOf="parent"
        app:layout_constraintStart_toStartOf="parent"
        app:layout_constraintTop_toBottomOf="@+id/text_home" />
</androidx.constraintlayout.widget.ConstraintLayout>
```

TextView is the same as what's specified in the other fragment layouts, except it has an ID (id) with which it constrains the button below it.

8. Create the navigation graph that will be used in the app. Select **File** | **New** | **Android Resource File** (making sure the res folder is selected in the Project see this option. Select **Navigation** as the **Resource Type** value and name it mobile_navigation.xml.

This creates the navigation graph:

Figure 4.3 – The Android Studio New Resource File dialog

9. Open the `mobile_navigation.xml` file in the `res/navigation` folder and update it with the code from the file in the following link. A truncated version of the code is shown here. Use this link to access the entire code: `https://packt.link/ZRDiT`.

mobile_navigation.xml

```xml
<?xml version="1.0" encoding="utf-8"?>
<navigation xmlns:android=
    "http://schemas.android.com/apk/res/android"
    xmlns:app="http://schemas.android.com/apk/res-auto"
    xmlns:tools="http://schemas.android.com/tools"
    android:id="@+id/mobile_navigation"
    app:startDestination="@+id/nav_home">

    <fragment
        android:id="@+id/nav_home"
        android:name="com.example.navigationdrawer.HomeFragment"
```

```
            android:label="@string/home"
            tools:layout="@layout/fragment_home">
            <action
                android:id="@+id/nav_home_to_content"
                app:destination="@id/nav_content"
                app:popUpTo="@id/nav_home" />
    </fragment>
...
```

This creates all the destinations in your app. However, it doesn't specify whether these are primary or secondary destinations. This should be familiar from the fragment Jetpack navigation exercise from the previous chapter.

The most important point to note here is `app:startDestination="@+id/nav_home`, which specifies what will be displayed to start with when the navigation loads and that there is an action available from within `HomeFragment` to move to the `nav_content` destination in the graph:

```
            <action
                android:id="@+id/nav_home_to_content"
                app:destination="@id/nav_content"
                app:popUpTo="@id/nav_home" />
```

You are now going to see how this is set up in `HomeFragment` and its layout.

10. Open `HomeFragment` and add two `import` statements for the `Button` and `Navigation` imports and update `onCreateView` to set up the button:

HomeFragment

```
    import android.widget.Button
    import androidx.navigation.Navigation

        override fun onCreateView(
            inflater: LayoutInflater,
            container: ViewGroup?,
            savedInstanceState: Bundle?
        ): View? {
            val view = inflater.inflate(R.layout.fragment_home, container, false)
            view.findViewById<Button> (R.id.button_home)?.setOnClickListener(
```

```
            Navigation.createNavigateOnClickListener
(R.id.nav_home_to_content, null)
        )
        return view
    }
```

This uses the `ClickListener` navigation to complete the `R.id.nav_home_to_content` action when `button_home` is clicked.

However, these changes will not do anything yet as you still need to set up the navigation host for your app and add all the other layout files, along with the navigation drawer.

11. Create a Nav host fragment by creating a new file in the layout folder called `content_main.xml`. This can be done by right-clicking on the `layout` folder in the `res` directory and then going to **File | New | Layout Resource File**. Once created, update it with `FragmentContainerView`:

```
<?xml version="1.0" encoding="utf-8"?>
<androidx.fragment.app.FragmentContainerView
    xmlns:android=
      "http://schemas.android.com/apk/res/android"
    xmlns:app="http://schemas.android.com/apk/res-auto"
    android:id="@+id/nav_host_fragment"
    android:name="androidx.navigation.fragment.
NavHostFragment"
    android:layout_width="match_parent"
    android:layout_height="match_parent"
    app:defaultNavHost="true"
    app:navGraph="@navigation/mobile_navigation" />
```

12. You'll notice that the navigation graph is set to the graph you just created:

```
app:navGraph="@navigation/mobile_navigation"
```

13. With that, the body of the app and its destination have been set up. Now, you need to set up the UI navigation. Create another layout resource file called `nav_header_main.xml` and add the following content:

```
<?xml version="1.0" encoding="utf-8"?>
<LinearLayout xmlns:android=
    "http://schemas.android.com/apk/res/android"
    xmlns:app="http://schemas.android.com/apk/res-auto"
    android:layout_width="match_parent"
```

```xml
        android:layout_height="176dp"
        android:background="@color/teal_700"
        android:gravity="bottom"
        android:orientation="vertical"
        android:paddingStart="16dp"
        android:paddingTop="16dp"
        android:paddingEnd="16dp"
        android:paddingBottom="16dp"
        android:theme="@style/ThemeOverlay.AppCompat.Dark">
        <ImageView
            android:id="@+id/imageView"
            android:layout_width="wrap_content"
            android:layout_height="wrap_content"
            android:contentDescription="@string/nav_header_desc"
            android:paddingTop= "8dp"
            app:srcCompat="@mipmap/ic_launcher_round" />
        <TextView
            android:layout_width="match_parent"
            android:layout_height="wrap_content"
            android:paddingTop= "8dp"
            android:text="@string/app_name"
            android:textAppearance= "@style/TextAppearance.AppCompat.Body1" />
</LinearLayout>
```

This is the layout that's displayed in the header of the navigation drawer.

14. Create the app bar with a toolbar layout file called `app_bar_main.xml`, and include the following content:

```xml
<?xml version="1.0" encoding="utf-8"?>
<androidx.coordinatorlayout.widget.CoordinatorLayout xmlns:android=
   "http://schemas.android.com/apk/res/android"
    xmlns:app="http://schemas.android.com/apk/res-auto"
    xmlns:tools="http://schemas.android.com/tools"
    android:layout_width="match_parent"
```

```
                android:layout_height="match_parent"
                tools:context=".MainActivity">
                <com.google.android.material.appbar.AppBarLayout
                    android:layout_width="match_parent"
                    android:layout_height="wrap_content"
                    android:theme= "@style/Theme.NavigationDrawer.
        AppBarOverlay">
                    <androidx.appcompat.widget.Toolbar
                        android:id="@+id/toolbar"
                        android:layout_width="match_parent"
                        android:layout_height="?attr/actionBarSize"
                        android:background="?attr/colorPrimary"
                        app:popupTheme= "@style/Theme.
        NavigationDrawer.PopupOverlay" />
                </com.google.android.material.appbar.AppBarLayout>
                <include layout="@layout/content_main" />
            </androidx.coordinatorlayout.widget.CoordinatorLayout>
```

This integrates the main body layout of the app with the app bar that appears above it. The remaining part is to create the items that will appear in the navigation drawer and create and populate the navigation drawer with these items.

15. To use icons with these menu items, you need to copy the vector assets in the drawable folder of the completed exercise to the drawable folder of your project. Vector assets use coordinates for points, lines, and curves to layout images with associated color information.

    They are significantly smaller when compared to PNG and JPG images, and vectors can be resized to different sizes without loss of quality. You can find them here: https://packt.link/CurtF.

    Copy the following drawables:

    - favorites.xml
    - archive.xml
    - recent.xml
    - home.xml
    - bin.xml

16. Create a menu with these items. To do this, go to **File** | **New** | **Android Resource File**, select **Menu** as the resource type, call it `activity_main_drawer`, and then populate it with the following content:

```xml
<?xml version="1.0" encoding="utf-8"?>
<menu xmlns:android=
    "http://schemas.android.com/apk/res/android"
    xmlns:tools="http://schemas.android.com/tools"
    tools:showIn="navigation_view">
    <group
        android:id="@+id/menu_top"
        android:checkableBehavior="single">
        <item
            android:id="@+id/nav_home"
            android:icon="@drawable/home"
            android:title="@string/home" />
        <item
            android:id="@+id/nav_recent"
            android:icon="@drawable/recent"
            android:title="@string/recent" />
        <item
            android:id="@+id/nav_favorites"
            android:icon="@drawable/favorites"
            android:title="@string/favorites" />
    </group>
    <group
        android:id="@+id/menu_bottom"
        android:checkableBehavior="single">
        <item
            android:id="@+id/nav_archive"
            android:icon="@drawable/archive"
            android:title="@string/archive" />
        <item
            android:id="@+id/nav_bin"
            android:icon="@drawable/bin"
            android:title="@string/bin" />
```

```
        </group>
</menu>
```

This sets up the menu items that will appear in the navigation drawer itself. The name of the IDs is the magic that ties up the menu items to the destinations within the navigation graph.

If the IDs of the menu items (in `activity_main_drawer.xml`) exactly match the IDs of the destinations in the navigation graph (which, in this case, are fragments within `mobile_navigation.xml`), then the destination is automatically loaded into the navigation host.

17. The layout for `MainActivity` ties the navigation drawer to all the layouts specified previously. Open `activity_main.xml` and update it with the following content:

```xml
<?xml version="1.0" encoding="utf-8"?>
<androidx.drawerlayout.widget.DrawerLayout
xmlns:android=
   "http://schemas.android.com/apk/res/android"
    xmlns:app="http://schemas.android.com/apk/res-auto"
    xmlns:tools="http://schemas.android.com/tools"
    android:id="@+id/drawer_layout"
    android:layout_width="match_parent"
    android:layout_height="match_parent"
    android:fitsSystemWindows="true"
    tools:openDrawer="start">
    <include
        layout="@layout/app_bar_main"
        android:layout_width="match_parent"
        android:layout_height="match_parent" />
    <com.google.android.material.navigation.NavigationView
        android:id="@+id/nav_view"
        android:layout_width="wrap_content"
        android:layout_height="match_parent"
        android:layout_gravity="start"
        android:fitsSystemWindows="true"
        app:headerLayout="@layout/nav_header_main"
        app:menu="@menu/activity_main_drawer" />
</androidx.drawerlayout.widget.DrawerLayout>
```

As you can see, an `include` is used to add `app_bar_main.xml`. The `<include>` element allows you to add layouts that will be replaced at compile time with the actual layout itself.

18. They allow us to encapsulate different layouts as they can be reused in multiple layout files within the app. `NavigationView` (the class that creates the navigation drawer) specifies the layout files you have just created to configure its header and menu items:

    ```
    app:headerLayout="@layout/nav_header_main"
    app:menu="@menu/activity_main_drawer"
    ```

19. Now that you have specified all the layout files, update `MainActivity` by adding the following interaction logic:

    ```kotlin
    package com.example.navigationdrawer
    import android.os.Bundle
    import androidx.appcompat.app.AppCompatActivity
    import androidx.navigation.findNavController
    import androidx.navigation.fragment.NavHostFragment
    import androidx.navigation.ui.*
    import com.google.android.material.navigation.NavigationView
    class MainActivity : AppCompatActivity() {
        private lateinit var appBarConfiguration: AppBarConfiguration
        override fun onCreate(savedInstanceState: Bundle?) {
            super.onCreate(savedInstanceState)
            setContentView(R.layout.activity_main)
            setSupportActionBar(findViewById(R.id.toolbar))
            val navHostFragment = supportFragmentManager.findFragmentById(R.id.nav_host_fragment) as NavHostFragment
            val navController = navHostFragment.navController
            //Creating top level destinations
            //and adding them to the draw
            appBarConfiguration = AppBarConfiguration(
                setOf(
                    R.id.nav_home, R.id.nav_recent, R.id.nav_favorites, R.id.nav_archive, R.id.nav_bin
                ), findViewById(R.id.drawer_layout)
            )
            setupActionBarWithNavController(navController, appBarConfiguration)
    ```

```
            findViewById<NavigationView>(R.id.nav_view)
?.setupWithNavController(navController)
        }
        override fun onSupportNavigateUp(): Boolean {
            val navController = findNavController(R.id.nav_
host_fragment)
            return navController.
navigateUp(appBarConfiguration) || super.
onSupportNavigateUp()
        }
}
```

Now, let's go through the preceding code. The `setSupportActionBar(toolbar)` line configures the toolbar used in the app by referencing it from the layout and setting it. Retrieving NavHostFragment is done with the following code:

```
        val navHostFragment = supportFragmentManager
.findFragmentById(R.id.nav_host_fragment) as
NavHostFragment
            val navController = navHostFragment.navController
```

Next, you add the menu items you want to display in the navigation drawer:

```
appBarConfiguration = AppBarConfiguration(
    setOf(
        R.id.nav_home, R.id.nav_recent, R.id.nav_
favorites, R.id.nav_archive, R.id.nav_bin
    ), findViewById(R.id.drawer_layout)
)
```

The `drawer_layout` is the container for the `nav_view`, the main app bar, and its included content.

This may seem like you are doing this twice as these items are displayed in the `activity_main_drawer.xml` menu for the navigation drawer. However, the function of setting these in `AppBarConfiguration` is that these primary destinations will not display an up arrow when they are selected as they are at the top level.

It also adds `drawer_layout` as the last parameter to specify which layout should be used when the hamburger menu is selected to display in the navigation drawer. The next line is as follows:

```
setupActionBarWithNavController(navController,
appBarConfiguration)
```

This sets up the app bar with the navigation graph so that any changes that are made to the destinations are reflected in the app bar:

```
findViewById<NavigationView> (R.id.nav_view)?.
setupWithNavController(navController)
```

This is the last statement in `onCreate`, and it specifies the item within the navigation drawer that should be highlighted when the user clicks on it. The next function in the class handles pressing the up button for the secondary destination, ensuring that it goes back to its parent primary destination:

```
override fun onSupportNavigateUp(): Boolean {
    val navController = findNavController(R.id.nav_host_fragment)
    return navController.navigateUp(appBarConfiguration)
|| super.onSupportNavigateUp()
}
```

The app bar can also display other menu items through the overflow menu, which, when configured, is displayed as three vertical dots at the top on the right-hand side. Let's create an overflow menu to display the **Settings** screen.

20. To add the overflow menu to the app bar, go to **File | New | Android Resource File**. Select **Menu** for the resource type and call the filename `main.xml`.

    Update it with the following content:

    ```
    <?xml version="1.0" encoding="utf-8"?>
    <menu xmlns:android=
        "http://schemas.android.com/apk/res/android"
        xmlns:app="http://schemas.android.com/apk/res-auto">
        <item
            android:id="@+id/nav_settings"
            android:title="@string/settings"
            app:showAsAction="never" />
    </menu>
    ```

    This configuration shows one item: `Settings`. Since it specifies the same ID as the `SettingsFragment` destination in the navigation graph, `android:id="@+id/nav_settings"` it will open the `SettingsFragment` fragment.

    The attribute being set to `app:showAsAction="never"` ensures it will stay as a menu option within the three dots overflow menu and will not appear on the app bar itself. There are other values for `app:showAsAction`, which set menu options to always appear on the app bar and if there is room.

See the full list here: https://developer.android.com/guide/topics/resources/menu-resource.

21. To add the overflow menu to the app bar, add the following to the MainActivity class:

    ```
    override fun onCreateOptionsMenu(menu: Menu): Boolean {
        menuInflater.inflate(R.menu.main, menu)
        return true
    }
    override fun onOptionsItemSelected(item: MenuItem):
    Boolean {
        return item.onNavDestinationSelected(findNavController
    (R.id.nav_host_fragment))
    }
    ```

    You will also need to add the following imports:

    ```
    import android.view.Menu
    import android.view.MenuItem
    ```

    The onCreateOptionsMenu function selects the menu to add to the app bar, while onOptionsItemSelected handles what to do when the item is selected using the item.onNavDestinationSelected(findNavController(R.id.nav_host_fragment)) navigation function. This is used to navigate to the destination within the navigation graph.

22. Run the app and navigate to a top-level destination using the navigation drawer. The following screenshot shows an example of navigating to the Recent destination:

Figure 4.4 – Recent menu item opened from the navigation drawer

23. When you open the navigation drawer again you will see that the Recent menu item is selected:

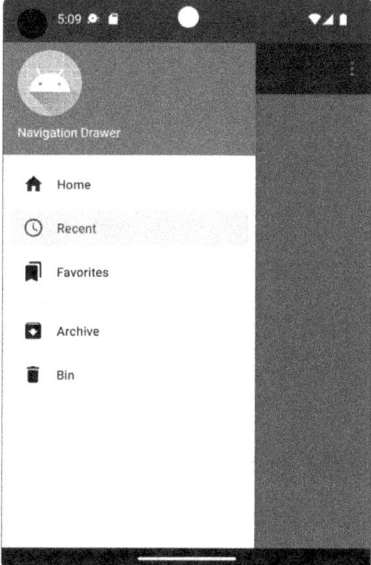

Figure 4.5 – The highlighted Recent menu item in the navigation drawer

24. Select the **Home** menu item again to display the button with the label **LINK TO CONTENT BUTTON**:

Figure 4.6 – The Home screen with a button for the secondary destination

25. Click this button to go to the secondary destination. You will see an up arrow displayed:

Figure 4.7 – Secondary destination with an up arrow displayed

In all the preceding screenshots, the overflow menu is displayed. After selecting it, you will see a **Settings** option appear. Upon pressing it, you will be taken to the **Settings** fragment with the up arrow displayed:

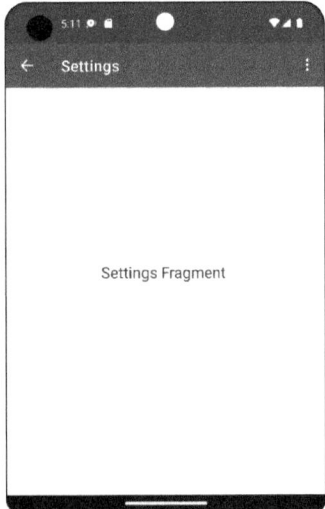

Figure 4.8 – Settings Fragment

Although there are quite a few steps to go through to set up an app with a navigation drawer, once created, it is very configurable. By adding a menu item entry to the drawer menu and a destination to the navigation graph, a new fragment can be created and set up for use immediately.

This removes a lot of the boilerplate code you needed to use fragments in the previous chapter. The next navigational pattern you'll explore is bottom navigation. This has become the most popular navigational pattern in Android, largely because it makes the main sections of the app easily accessible.

Bottom navigation

Bottom navigation is used when there are a limited number of top-level destinations, and these can range from three to five primary destinations that are not related to each other. Each item on the bottom navigation bar displays an icon and an optional text label.

This navigation allows quick access as the items are always available, no matter which secondary destination of the app the user navigates to.

Exercise 4.02 – adding bottom navigation to your app

Create a new app in Android Studio named Bottom Navigation using the **Empty Activity** project template, leaving all the other defaults as they are. Do not use the **Bottom Navigation Activity** project template, as we are going to use incremental steps to build the app.

You will build a loyalty app that provides offers, rewards, and so on for customers who have signed up to use it. Bottom navigation is quite common for this kind of app because there will typically be limited top-level destinations.

Let's get started:

1. Many of the steps are very similar to the previous exercise, as you will be using Jetpack navigation and defining destinations in a navigation graph and a corresponding menu.
2. Create a new project with an Empty Activity called Bottom Navigation.
3. Add the Gradle dependencies you will require to app/build.gradle:

   ```
   implementation
   'androidx.navigation:navigation-fragment-ktx:2.5.3'
   implementation
   'androidx.navigation:navigation-ui-ktx:2.5.3'
   ```

4. Append strings.xml in the res/values folder with the following values:

strings.xml

```
<!-- Bottom Navigation -->
<string name="home">Home</string>
<string name="tickets">Tickets</string>
```

```xml
<string name="offers">Offers</string>
<string name="rewards">Rewards</string>
<!-- Action Bar -->
<string name="settings">Settings</string>
<string name="cart">Shopping Cart</string>
<string name="content">Content</string>
<string name="home_fragment">Home Fragment</string>
<string name="tickets_fragment">Tickets Fragment</string>
<string name="offers_fragment">Offers Fragment</string>
<string name="rewards_fragment">Rewards Fragment</string>
<string name="settings_fragment"> Settings Fragment</string>
<string name="cart_fragment"> Shopping Cart Fragment</string>
<string name="content_fragment">Content Fragment</string>
<string name="link_to_content_button"> Link to Content Button</string>
```

5. Create eight fragments with the following names:

   - HomeFragment
   - ContentFragment
   - OffersFragment
   - RewardsFragment
   - SettingsFragment
   - TicketsFragment
   - CartFragment

6. Apply the same layout that you applied in the previous exercise for all the fragments adding the corresponding string resource except for `fragment_home.xml`. For this layout, use the same layout file that you used in *Exercise 4.01 – creating an App with a navigation drawer*.

7. Create the navigation graph as you did in the previous exercise and call it `mobile_navigation.xml`. Update it with the code from the following file provided in the link. A truncated version of the code is shown here. See the link for the entire code block you need to use: https://packt.link/Fwuyl.

mobile_navigation.xml

```
<navigation xmlns:android=
    "http://schemas.android.com/apk/res/android"
    xmlns:app="http://schemas.android.com/apk/res-auto"
    xmlns:tools="http://schemas.android.com/tools"
    android:id="@+id/mobile_navigation"
    app:startDestination="@+id/nav_home">
    <fragment
        android:id="@+id/nav_home"
        android:name="com.example.bottomnavigation.HomeFragment"
        android:label="@string/home"
        tools:layout="@layout/fragment_home">
        <action
            android:id="@+id/nav_home_to_content"
            app:destination="@id/nav_content"
            app:popUpTo="@id/nav_home" />
    </fragment>
    ...
```

8. Update the `onCreateView` function in `HomeFragment` to use the destination in the navigation graph to navigate to `ContentFragment`. You will also need to add the following imports:

```
import android.widget.Button
import androidx.navigation.Navigation

override fun onCreateView(
    inflater: LayoutInflater,
    container: ViewGroup?,
    savedInstanceState: Bundle?
): View? {
    val view = inflater.inflate(R.layout.fragment_home, container, false)
```

```
        view.findViewById<Button>(R.id.button_home)
?.setOnClickListener(
            Navigation.createNavigateOnClickListener (R.id.
nav_home_to_content, null)
        )
        return view
}
```

9. Now that the destinations have been defined in the navigation graph, create the menu in the bottom navigation to reference these destinations. First, however, you need to gather the icons that will be used in this exercise. Go to the completed exercise on GitHub and find the vector assets in the `drawable` folder: https://packt.link/pUXvC.

   Copy the following drawables to the drawable folder of your project:

   - `cart.xml`
   - `home.xml`
   - `offers.xml`
   - `rewards.xml`
   - `tickets.xml`

10. Create a `bottom_nav_menu.xml` file (right click on the `res` folder and select **Android Resource File**) and select **Menu** and then replace the contents with the following using all of these icons except the `cart.xml` vector asset, which will be used for the top toolbar. Notice that the IDs of the items match the IDs in the navigation graph:

bottom_nav_menu.xml

```xml
        <?xml version="1.0" encoding="utf-8"?>
        <menu xmlns:android=
            "http://schemas.android.com/apk/res/android">
            <item
                android:id="@+id/nav_home"
                android:icon="@drawable/home"
                android:title="@string/home" />
            <item
                android:id="@+id/nav_tickets"
                android:icon="@drawable/tickets"
                android:title="@string/tickets"/>
            <item
```

```xml
            android:id="@+id/nav_offers"
            android:icon="@drawable/offers"
            android:title="@string/offers" />
    <item
        android:id="@+id/nav_rewards"
        android:icon="@drawable/rewards"
        android:title="@string/rewards"/>
</menu>
```

11. Update the `activity_main.xml` file with the following content:

activity_main.xml

```xml
<?xml version="1.0" encoding="utf-8"?>
<androidx.constraintlayout.widget.ConstraintLayout
xmlns:android=
    "http://schemas.android.com/apk/res/android"
    xmlns:app="http://schemas.android.com/apk/res-auto"
    android:id="@+id/container"
    android:layout_width="match_parent"
    android:layout_height="match_parent"
    android:paddingTop="?attr/actionBarSize">
    <com.google.android.material.bottomnavigation.
BottomNavigationView
        android:id="@+id/nav_view"
        android:layout_width="0dp"
        android:layout_height="wrap_content"
        android:layout_marginStart="0dp"
        android:layout_marginEnd="0dp"
        android:background="?android:attr/windowBackground"
        app:layout_constraintBottom_toBottomOf="parent"
        app:layout_constraintStart_toStartOf="parent"
        app:layout_constraintEnd_toEndOf="parent"
        app:menu="@menu/bottom_nav_menu"
        app:labelVisibilityMode="labeled"/>
    <androidx.fragment.app.FragmentContainerView
        android:id="@+id/nav_host_fragment"
        app:layout_constraintStart_toStartOf="parent"
```

```
            app:layout_constraintEnd_toEndOf="parent"
            app:layout_constraintTop_toTopOf="parent"
            android:name ="androidx.navigation.fragment.NavHostFragment"
            android:layout_width="match_parent"
            android:layout_height="match_parent"
            app:defaultNavHost="true"
            app:navGraph="@navigation/mobile_navigation" />
    </androidx.constraintlayout.widget.ConstraintLayout>
```

The `BottomNavigation` view is configured with the menu you created previously, that is, `app:menu="@menu/bottom_nav_menu"`, while `FragmentContainerView` is configured with `app:navGraph="@navigation/mobile_navigation"`. As the bottom navigation in the app is not connected directly to the app bar, there are fewer layout files to set up.

12. Update `MainActivity` with the following content:

```
    package com.example.bottomnavigation
    import android.os.Bundle
    import androidx.appcompat.app.AppCompatActivity
    import androidx.navigation.findNavController
    import androidx.navigation.fragment.NavHostFragment
    import androidx.navigation.ui.*
    import com.google.android.material.bottomnavigation.BottomNavigationView
    class MainActivity : AppCompatActivity() {
        private lateinit var appBarConfiguration: AppBarConfiguration
        override fun onCreate(savedInstanceState: Bundle?) {
            super.onCreate(savedInstanceState)
            setContentView(R.layout.activity_main)
        val navHostFragment = supportFragmentManager.findFragmentById (R.id.nav_host_fragment) as NavHostFragment
        val navController = navHostFragment.navController
            //Creating top level destinations
            //and adding them to bottom navigation
            appBarConfiguration = AppBarConfiguration(setOf(
                R.id.nav_home, R.id.nav_tickets, R.id.nav_
```

```
offers, R.id.nav_rewards))
        setupActionBarWithNavController(navController,
appBarConfiguration)
        findViewById<BottomNavigationView>(R.id.nav_view)
?.setupWithNavController(navController)
    }
    override fun onSupportNavigateUp(): Boolean {
        val navController = findNavController(R.id.nav_
host_fragment)
        return navController.
navigateUp(appBarConfiguration) || super.
onSupportNavigateUp()
    }
}
```

The preceding code should be very familiar because it was explained in the previous exercise. The main change here is that instead of a NavigationView that holds the main UI navigation for the navigation drawer, it is now replaced with a BottomNavigationView. The configuration after this is the same.

13. Run the app. You should see the following output:

Figure 4.9 – Bottom navigation with Home selected

14. The display shows the four menu items you set up, with the **Home** item selected as the start destination. Click the square button to be taken to the secondary destination within **Home**:

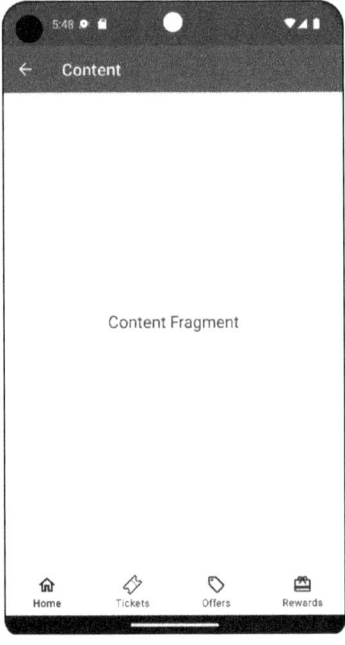

Figure 4.10 – Secondary destination within Home

15. The action that makes this possible is the nav_home_to_content action specified in the navigation graph:

mobile_navigation.xml (snippet)

```
<fragment
    android:id="@+id/nav_home"
    android:name="com.example.bottomnavigation.HomeFragment"
    android:label="@string/home"
    tools:layout="@layout/fragment_home">
    <action
        android:id="@+id/nav_home_to_content"
        app:destination="@id/nav_content"
        app:popUpTo="@id/nav_home" />
</fragment>
```

16. Since only a limited amount of items are added to the bottom navigation (typically three to five), sometimes action items (those that have a dedicated icon) are added to the app bar. Create another menu called `main.xml` and add the following content:

main.xml

```xml
<?xml version="1.0" encoding="utf-8"?>
<menu xmlns:android=
    "http://schemas.android.com/apk/res/android"
    xmlns:app="http://schemas.android.com/apk/res-auto">
    <item
        android:id="@+id/nav_cart"
        android:title="@string/cart"
        android:icon="@drawable/cart"
        app:showAsAction="always" />
    <item
        android:id="@+id/nav_settings"
        android:title="@string/settings"
        app:showAsAction="never" />
</menu>
```

17. This menu will be used in the overflow menu in the app bar. The overflow menu will be available when you click on the three dots. A `cart` vector asset will also be displayed on the top app bar because the `app:showAsAction` attribute is set to `always`. Configure the overflow menu within `MainActivity` by adding the following:

Add these two imports at the top of the file:

```
import android.view.Menu
import android.view.MenuItem
```

And then these two functions:

```
override fun onCreateOptionsMenu(menu: Menu): Boolean {
    menuInflater.inflate(R.menu.main, menu)
    return true
}
override fun onOptionsItemSelected(item: MenuItem): Boolean {
    super.onOptionsItemSelected(item)
    return item.onNavDestinationSelected(findNavController
```

```
            (R.id.nav_host_fragment))
        }
```

18. This will now display the main menu in the app bar. Run the app again, and you'll see the following:

Figure 4.11 – Bottom navigation with the overflow menu

Selecting the shopping cart takes you to the secondary destination we configured in the navigation graph:

Figure 4.12 – Bottom navigation with the overflow menu in the secondary destination

As you've seen in this exercise, setting up bottom navigation is quite straightforward. The navigation graph and the menu setup simplify linking the menu items to the fragments. Additionally, integrating the action bar and the overflow menu are also small steps to implement.

If you are developing an app that has very well-defined top-level destinations and switching between them is important, then the visibility of these destinations makes bottom navigation an ideal choice. The final primary navigation pattern to explore is tabbed navigation.

This is a versatile pattern as it can be used as an app's primary navigation and as secondary navigation with the other navigation patterns we've studied.

Tabbed navigation

Tabbed navigation is mostly used when you want to display related items. It is common to have fixed tabs if there's only a few of them (typically between two and five tabs) and scrolling horizontal tabs if you have more than five tabs. They are used mostly for grouping destinations that are at the same hierarchical level.

This can be the primary navigation if the destinations are related. This might be the case if the app you developed is in a narrow or specific subject field where the primary destinations are related, such as a news app. More commonly, it is used with bottom navigation to present secondary navigation that's available within a primary destination. The following exercise demonstrates using tabbed navigation for displaying related items.

Exercise 4.03 – using tabs for app navigation

Create a new app in Android Studio with an `Empty Activity` named `Tab Navigation`. You are going to build a skeleton movies app that displays the genres of movies. Let's get started:

1. Update the `strings.xml` with the following content:

strings.xml

```
<string name="action">Action</string>
<string name="comedy">Comedy</string>
<string name="drama">Drama</string>
<string name="sci_fi">Sci-Fi</string>
<string name="family">Family</string>
<string name="crime">Crime</string>
<string name="history">History</string>
<string name="dummy_text">
    Lorem ipsum dolor sit amet, consectetuer
```

```
        adipiscing elit. Aenean commodo ligula eget dolor.
        Aenean massa. Cum sociis natoque penatibus et magnis dis
        parturient montes, nascetur ridiculus mus. Donec quam
        felis, ultricies nec, pellentesque eu, pretium quis,
        sem. Nulla consequat massa quis enim. Donec pede justo,
        fringilla vel, aliquet nec, vulputate eget, arcu. In enim
        justo, rhoncus ut, imperdiet a, venenatis vitae, justo.
        Nullam dictum felis eu pede mollis pretium. Integer
        tincidunt. Cras dapibus. Vivamus elementum semper nisi.
        Aenean vulputate eleifend tellus. Aenean leo ligula,
        porttitor eu, consequat vitae, eleifend ac, enim. Aliquam
        lorem ante, dapibus in, viverra quis, feugiat a, tellus.
        Phasellus viverra nulla ut metus varius laoreet. Quisque
        rutrum. Aenean imperdiet. Etiam ultricies nisi vel augue.
        Curabitur ullamcorper ultricies nisi.
    </string>
```

The `<string name="dummy_text">` file specified provides some body text for each movie genre:

2. In order to be able to swipe through the tabs left and right, we need to use a `ViewPager` component. Add the following dependency to app `build.gradle`:

    ```
    implementation "androidx.viewpager2:viewpager2:1.0.0"
    ```

3. Create a new blank `MoviesFragment` fragment which will display some body text and replace the layout file content with the code following code snippet.

fragment_movies.xml

```xml
<?xml version="1.0" encoding="utf-8"?>
<androidx.constraintlayout.widget.ConstraintLayout
xmlns:android=
    "http://schemas.android.com/apk/res/android"
    xmlns:app="http://schemas.android.com/apk/res-auto"
    xmlns:tools="http://schemas.android.com/tools"
    android:layout_width="match_parent"
    android:layout_height="match_parent"
    tools:context=".MoviesFragment">

    <TextView
        android:layout_width="wrap_content"
        android:layout_height="wrap_content"
```

```
            android:textSize="16sp"
            android:text="@string/dummy_text"
            android:padding="16dp"
            app:layout_constraintStart_toStartOf="parent"
            app:layout_constraintEnd_toEndOf="parent"
            app:layout_constraintTop_toTopOf="parent" />
    </androidx.constraintlayout.widget.ConstraintLayout>
```

4. Update the `activity_main.xml` file with the following content:

```
<androidx.constraintlayout.widget.ConstraintLayout
xmlns:android="http://schemas.android.com/apk/res/
android" xmlns:app="http://schemas.android.com/apk/
res-auto"
    android:layout_width="match_parent"
    android:layout_height="match_parent"
    android:orientation="vertical">

    <com.google.android.material.tabs.TabLayout
        app:layout_constraintTop_toTopOf="parent"
        android:id="@+id/tab_layout"
        android:layout_width="match_parent"
        android:layout_height="wrap_content"
        app:tabMode="fixed"/>
    <androidx.viewpager2.widget.ViewPager2
      app:layout_constraintTop_toBottomOf="@id/tab_layout"
        android:id="@+id/view_pager"
        android:layout_width="match_parent"
        android:layout_height="wrap_content"/>
</androidx.constraintlayout.widget.ConstraintLayout>
```

The layout displays `TabLayout` at the top and notices that it sets the tabs to be fixed with the `app:tabMode="fixed"` attribute. To display the required content, you will use `ViewPager`, a swipeable layout that allows you to add multiple views or fragments so that when a user swipes to change one of the tabs, the body content displays the corresponding view or fragment. For this exercise, you are going to swipe between movie fragments.

The format of the tabs can be fixed so all are visible on the screen at the same time or scrollable, so some tabs will initially be off-screen if they don't fit within the horizontal screen space available.

Next, we need to provide the content for `ViewPager`. The component that provides the data that's used in `ViewPager` is called an adapter.

5. Create a simple adapter that will be used to display our movies. Go to **File** | **New** | **Kotlin Class/File** to create the class and call it `MovieGenresAdapter`:

```kotlin
package com.example.tabnavigation

import androidx.fragment.app.Fragment
import androidx.fragment.app.FragmentManager
import androidx.lifecycle.Lifecycle
import androidx.viewpager2.adapter.FragmentStateAdapter

val TAB_GENRES_SCROLLABLE = listOf(
    R.string.action,
    R.string.comedy,
    R.string.drama,
    R.string.sci_fi,
    R.string.family,
    R.string.crime,
    R.string.history
)
val TAB_GENRES_FIXED = listOf(R.string.action, R.string.comedy, R.string.drama)

class MovieGenresAdapter(fragmentManager: FragmentManager, lifecycle: Lifecycle) :
    FragmentStateAdapter(fragmentManager, lifecycle) {

    override fun getItemCount(): Int {
        return TAB_GENRES_FIXED.size
    }

    override fun createFragment(position: Int): Fragment {
        return MoviesFragment()
    }
}
```

First, look at the `MovieGenresAdapter` class header. It extends from `FragmentStateAdapter`, which is an adapter used to populate fragments within `ViewPager`.

The callback method's functions are as follows:

- `getItemCount()`: This returns the total number of fragments we will be inserting, which, as we are matching the number of pages to the number of tabs, is the size of the `TABS_GENRE_FIXED` constant.

- `createFragment(position Int)`: This creates the fragment to be displayed in `ViewPager` at the passed in argument position. Here we are setting this to be the same fragment, but in a real app, you would populate it with different fragments.

6. Update `MainActivity` so that it uses tabs with `ViewPager`:

    ```
    package com.example.tabnavigation

    import androidx.appcompat.app.AppCompatActivity
    import android.os.Bundle
    import androidx.viewpager2.widget.ViewPager2
    import com.google.android.material.tabs.TabLayout
    import com.google.android.material.tabs.TabLayoutMediator

    class MainActivity : AppCompatActivity() {
        override fun onCreate(savedInstanceState: Bundle?) {
            super.onCreate(savedInstanceState)
            setContentView(R.layout.activity_main)

            val viewPager = findViewById<ViewPager2>(R.id.view_pager)
            val tabLayout = findViewById<TabLayout>(R.id.tab_layout)

            val adapter = MovieGenresAdapter(supportFragmentManager, lifecycle)
            viewPager.adapter = adapter

            TabLayoutMediator(tabLayout, viewPager) { tab, position ->
                tab.text = resources.getString(TAB_GENRES_
    ```

```
            FIXED[position])
                }.attach()
        }
}
```

7. You then retrieve the Views from the layout and link the tabs to `ViewPager` with the `TabLayoutMediator`. The tab itself is exposed for you to customize. In this instance, we are just setting the text. The position is also available to link the tab position to the fragment position in `ViewPager`. Creating this tab navigation is simple and effective.

8. Run the app up, and you should see the following:

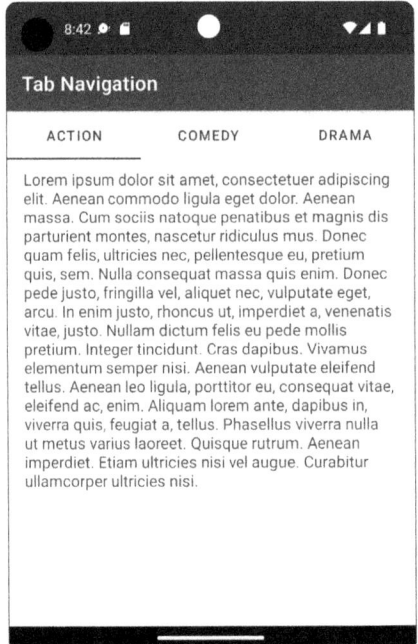

Figure 4.13 – Tab layout with fixed tabs

You can swipe left and right in the body of the page to go to each of the three tabs, and you can also select one of the respective tabs to perform the same action. Now, let's change the tab data that's being displayed and set the tabs so that they can be scrolled through.

9. First, change `MovieGenresAdapter` to use a few extra genres by updating the `getItemCount` function:

```
override fun getItemCount(): Int {
    return TAB_GENRES_SCROLLABLE.size
}
```

10. In `MainActivity`, set `TabLayoutMediator` to use the updated item count in the Adapter to set the tab text for these extra pages:

    ```
    TabLayoutMediator(tabLayout, viewPager) { tab, position ->
        tab.text = resources.getString(TAB_GENRES_SCROLLABLE[position])
    }.attach()
    ```

11. You will also need to change the `app:tabMode="fixed"` line to `app:tabMode="scrollable"` in the `activity_layout.xml` file.

12. Run the app now, and you should see the following:

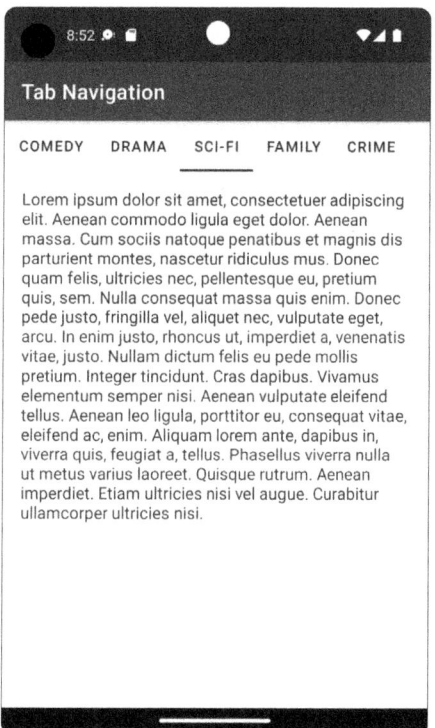

Figure 4.14 – The Tab Navigation layout with scrollable tabs

The list of tabs continues to display off the screen. The tabs can be swiped and selected, and the body content can also be swiped so that you can go left and right through the tab pages.

With this exercise, you learned how versatile tabs are when it comes to providing navigation in an app. Fixed-width tabs can be used for both primary and secondary navigation. At the same time, scrollable tabs can be used to group related items together for secondary navigation, so you also need to add primary navigation to the app.

In this example, the primary navigation has been omitted for simplicity, but for more real-world and complex apps, you can either add a navigation drawer or bottom navigation.

Activity 4.01 – building primary and secondary app navigation

You have been tasked with creating a sports app. It can have three or more top-level destinations. One of the primary destinations, however, must be called My Sports and should link to one or more secondary destinations, which are sports. You can use any one of the navigation patterns we have explored in this chapter or a combination of them, and you can also introduce any customizations that you feel are appropriate.

There are different ways of attempting this activity. One approach would be to use bottom navigation and add the individual secondary sports destinations to the navigation graph so that it can link to these destinations. It is fairly simple and delegates to the navigation graph using actions. Here is what the home screen should look like after using this approach:

Figure 4.15 – Bottom navigation for the My Sports app

> **Note**
> The solution to this activity can be found at: `https://packt.link/By7eE`.

Summary

This chapter covered the most important navigation techniques you need to know about in order to create clear and consistent navigation in your apps. You started off by learning how to create an Android Studio project with a navigation drawer to connect navigation menu items to individual fragments using Jetpack navigation. You then progressed to actions within Jetpack navigation to navigate to other secondary destinations in your app within the navigation graph.

The next exercise then used bottom navigation to display primary navigation destinations that are always visible on the screen. We followed this by looking at tabbed navigation, where you learned how to display both fixed and scrollable tabs. For each navigational pattern, you were shown when it might be more suitable, depending on the type of app you were building. We finished this chapter by building our own app using one or more of these navigational patterns and adding both primary and secondary destinations.

This chapter built upon the comprehensive introduction we provided to Android with Android Studio in *Chapter 1, Creating Your First App*, as well as what you learned about activities and fragments in *Chapter 2, Building User Screen Flows*, and *Chapter 3, Developing the UI with Fragments*. These chapters covered the knowledge, practice, and fundamental Android components you need to create apps. This chapter tied these previous chapters together by guiding you through the primary navigational patterns available to make your apps stand out and be easy to use.

The next chapter will build on these concepts and introduce you to more advanced ways of displaying app content. You will start off by learning about binding data with lists using `RecyclerView`. After that, you will explore the different mechanisms you can use to retrieve and populate content within apps.

Part 2: Displaying Network Calls

In this part, we will look at how we can integrate popular libraries and frameworks used for building Android apps. We will start with libraries used to fetch and process data from the internet, then continue with the `RecyclerView` library used for rendering lists.

Next, we will look at how we can handle permissions and use Google Maps, followed by performing tasks in the background with Services and `WorkManager`, and then displaying notifications to the user. Finally, we will look at Jetpack Compose and how we can use it to simplify the creation of user interfaces.

We will cover the following chapters in this section:

- *Chapter 5, Essential Libraries: Retrofit, Moshi, and Glide*
- *Chapter 6, Adding and Interacting with RecyclerView*
- *Chapter 7, Android Permissions and Google Maps*
- *Chapter 8, Services, WorkManager, and Notifications*
- *Chapter 9, Building User Interfaces Using Jetpack Compose*

5
Essential Libraries: Retrofit, Moshi, and Glide

In this chapter, we will cover the steps needed to present app users with dynamic content fetched from remote servers. You will be introduced to the different libraries required to retrieve and handle this dynamic data.

By the end of this chapter, you will be able to fetch data from a network endpoint using Retrofit, parse JSON payloads into Kotlin data objects using Moshi, and load images into `ImageView` using Glide.

In the previous chapter, we learned how to implement navigation in our app. In this chapter, we will learn how to present dynamic content to the user as they navigate around our app.

We will cover the following topics in this chapter:

- Introducing REST, API, JSON, and XML
- Fetching data from a network endpoint
- Parsing a JSON response
- Loading images from a remote URL

Technical requirements

The complete code for all the exercises and the activity in this chapter is available on GitHub at `https://packt.link/Uqtjm`

Introducing REST, API, JSON, and XML

Data presented to users can come from different sources. It can be hardcoded into an app, but that comes with limitations. To change hardcoded data, we have to publish an update to our app. Some data, such as currency exchange rates, the real-time availability of assets, and the current weather, cannot be hardcoded by its nature. Other data may become outdated, such as the terms of use of an app.

In such cases, you usually fetch the relevant data from a server. One of the most common architectures for serving such data is **representational state transfer** (**REST**) architecture. REST architecture is defined by a set of six constraints: client-server architecture, statelessness, cacheability, a layered system, code on demand (optional), and a uniform interface.

> **Note**
>
> To read more about REST, visit `https://packt.link/YsSRV`.

When applied to a web service **application programming interface** (**API**), we get a **Hypertext Transfer Protocol** (**HTTP**)-based RESTful API. The HTTP protocol is the foundation of data communication for the World Wide Web, hosted on and accessible via the **internet**. It is the protocol used by servers all around the world to serve websites to users in the form of HTML documents, images, style sheets, and so forth.

> **Note**
>
> An interesting article on this topic can be found at `https://developer.mozilla.org/en-US/docs/Web/HTTP/Overview`.

RESTful APIs rely on the standard HTTP methods – `GET`, `POST`, `PUT`, `DELETE`, and `PATCH` – to fetch and transform data. These methods allow us to fetch, store, delete, and update data entities on remote servers.

We can rely on the built-in Java `HttpURLConnection` class to execute these HTTP methods. Alternatively, we can use a library such as `OkHttp`, which offers additional features such as gzipping (compressing and decompressing), redirects, retries, and both synchronous and asynchronous calls. Interestingly, from Android 4.4, `HttpURLConnection` is just a wrapper around `OkHttp`. If we choose `OkHttp`, we might as well go for **Retrofit** (as we will in this chapter), the current industry standard. We can then benefit from its type-safety, which is better suited for handling REST calls.

Most commonly, data is represented by **JavaScript Object Notation** (**JSON**). JSON is a text-based data transfer format. As the name implies, it was derived from JavaScript. However, it has since become one of the most popular standards for data transfer, and its most modern programming languages have libraries that encode or decode data to or from JSON.

A simple JSON payload may look something like this:

```
{"employees":[
  {"name": "James", "email": "james.notmyemail@gmail.com"},
  {"name": "Lea", "email": "lea.dontemailme@gmail.com"}
]}
```

Another common data structure used by RESTful services is **Extensible Markup Language** (**XML**), which encodes documents in a format that is human- and machine-readable. XML is considerably more verbose than JSON. The same data structure as the previous in XML would look something like this:

```
<employees>
    <employee>
        <name>James</name>
        <email>james.notmyemail@gmail.com</email>
    </employee>
    <employee>
        <name>Lea</name>
        <email>lea.dontemailme@gmail.com</email>
    </employee>
</employees>
```

In this chapter, we will focus on JSON.

When obtaining a JSON payload, we essentially receive a string. To convert that string into a data object, we have a few options, the most popular ones being libraries such as **GSON**, **Jackson**, and **Moshi**, as well as the built-in `org.json` package. For its lightweight nature, we will focus on Moshi.

Finally, we will look into loading images from the web. Doing so will allow us to provide up-to-date images and load the right images for the user's device. It will also let us only load the images when we need them, thus keeping our APK size smaller.

Fetching data from a network endpoint

For the purpose of this section, we will use The Cat API (`https://thecatapi.com/`). This RESTful API offers us vast data about, well…cats.

To get started, we will create a new project. We then have to grant our app internet access permission. This is done by adding the following code to your `AndroidManifest.xml` file right before the `Application` tag:

```
<uses-permission
    android:name="android.permission.INTERNET" />
```

Next, we need to set up our app to include Retrofit. Retrofit is a type-safe library provided by Square, which is built on top of the `OkHttp` HTTP client. Retrofit helps us generate **Uniform Resource Locators** (**URLs**), which are the addresses of the server endpoints we want to access. It also makes the decoding of JSON payloads easier by providing integration with several parsing libraries. Sending data to the server is also easier with Retrofit, as it helps with encoding the requests.

> **Note**
>
> You can read more about Retrofit here `https://square.github.io/retrofit/`.

To add Retrofit to our project, we need to add the following code to the `dependencies` block of the `build.gradle` file of our app:

```
implementation 'com.squareup.retrofit2:retrofit:(insert latest version)'
```

> **Note**
>
> You can find the latest version here `https://github.com/square/retrofit`.

With Retrofit included in our project, we can proceed to set it up.

First, to access an HTTP(S) endpoint, we start by defining the contract with that endpoint. A contract to access the `https://api.thecatapi.com/v1/images/search` endpoint looks like this:

```
interface TheCatApiService {
    @GET("images/search")
    fun searchImages(
        @Query("limit") limit: Int,
        @Query("size") format: String
    ): Call<String>
}
```

There are a few things to note here. First, you will notice that the contract is implemented as an interface. This is how you define contracts for Retrofit. Next, you will notice that the name of the interface implies that this interface can, eventually, cover all calls made to `TheCatAPIService`.

It is a bit unfortunate that Square chose `Service` as the conventional suffix for these contracts, as the term service has a different meaning in the Android world, as you will see in *Chapter 8, Services, WorkManager, and Notifications*. Nevertheless, this is the convention.

To define our endpoint, we start by stating the method with which the call will be made using the appropriate annotation—in our case, `@GET`. The parameter passed to the annotation is the path of the endpoint to access. You'll notice that `https://api.thecatapi.com/v1/` is stripped from that path.

That is because this is the common address for all of the endpoints of `TheCatAPI`, and so it will be passed to our Retrofit instance at construction time instead. Next, we choose a meaningful name for our function—in this case, we'll call the image search endpoint, so `searchImages` seems appropriate. The parameters of the `searchImages` function define the values we can pass to the API when we make the calls.

There are different ways in which we can transfer data to the API. `@Query` allows us to define values added to the query of our request URL (that's the optional part of the URL that comes after the question mark). It takes a key-value pair (in our case, we have `limit` and `size`) and a data type. If the data type is not a string, the value of that type will be transformed into a string. Any value passed will be URL-encoded for us.

Another such way is using `@Path`. This annotation can be used to replace a token in our path wrapped in curly brackets with a provided value. The `@Header`, `@Headers`, and `@HeaderMap` annotations will allow us to add or remove HTTP headers from the request. `@Body` can be used to pass content in the body of the `POST` and `PUT` requests.

Lastly, we have a return type. To keep things simple at this stage, we will accept the response as a string. We wrapped our string in a `Call` interface. `Call` is Retrofit's mechanism for executing network requests synchronously (via `execute()`) or asynchronously (via `enqueue(Callback)`). When using coroutines, we can make the function a `suspend` function and omit the `Call` wrapper (see *Chapter 14, Coroutines and Flow*, for more information on coroutines).

With our contract defined, we can get Retrofit to implement our service interface:

```
val retrofit = Retrofit.Builder()
    .baseUrl("https://api.thecatapi.com/v1/").build()
val theCatApiService =
    retrofit.create(TheCatApiService::class.java)
```

If we try to run our app with this code, our app will crash with an `IllegalArgumentException`. This is because Retrofit needs us to tell the app how to process the server response to a string. This processing is done with what Retrofit calls **converters**. To set a `ConverterFactory` instance to our `retrofit` instance, we need to add the following:

```
val retrofit = Retrofit.Builder()
    .baseUrl("https://api.thecatapi.com/v1/")
    .addConverterFactory(ScalarsConverterFactory.create())
    .build()
```

For our project to recognize `ScalarsConverterFactory`, we need to update our app's `build.gradle` file by adding another dependency:

```
implementation 'com.squareup.retrofit2:converter-
scalars:(insert latest version)'
```

Now, we can obtain a `Call` instance by calling `val call = theCatApiService.searchImages(1, "full")`. With the instance obtained in this fashion, we can execute an async request by calling `call.enqueue(Callback)`.

Our `Callback` implementation will have two methods: `onFailure(Call, Throwable)` and `onResponse(Call, Response)`. Note that we are not guaranteed to have a successful response if `onResponse` is called. `onResponse` is called whenever we successfully receive *any* response from the server, and no unexpected exception occurs.

So, to confirm that the response is successful, we should check the `response.isSuccessful` property. The `onFailure` function will be called in the case of a network error or an unexpected exception somewhere along the way.

So, where should we implement the Retrofit code? In clean architecture, data is provided by repositories. Repositories, in turn, have data sources. One such data source can be a network data source. This is where we will implement our network calls. Our ViewModels will then request data from repositories via use cases.

In the case of **Model-View-ViewModel** (**MVVM**), the ViewModel is an abstraction of the view that exposes properties and commands.

For our implementation, we will simplify the process by instantiating Retrofit and the service in the Activity. This is not good practice. *Do not do this in a production app*. It does not scale well and is very difficult to test. Instead, adopt an architecture that decouples your views from your business logic and your data. See *Chapter 15, Architecture Patterns*, for some ideas.

Exercise 5.01 – reading data from an API

In the following chapters, we will develop an app for an imaginary secret agency with a worldwide network of agents saving the world from countless dangers. The secret agency in question is quite unique: it operates secret cat agents.

In this exercise, we will create an app that presents us with one random secret cat agent from The Cat API. Before you can present data from an API to your user, you first must fetch that data. Let's start with that here:

1. Create a new **Empty Activity** project (**File** | **New** | **New Project** | **Empty Activity**). Click **Next**.
2. Name your application `Cat Agent Profile`.
3. Make sure your package name is `com.example.catagentprofile`.
4. Set **Save location** to where you want to save your project.
5. Leave everything else at its default values and click **Finish**.
6. Make sure you are on the **Android** view in your **Project** pane:

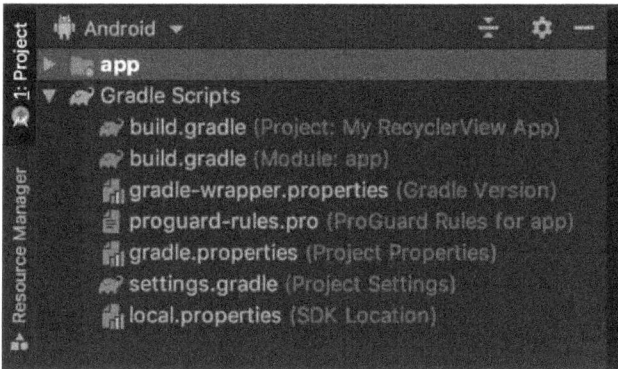

Figure 5.1 – The Android view in the Project pane

7. Open your `AndroidManifest.xml` file. Add internet permissions to your app like so:

   ```
   <manifest xmlns:android=
       "http://schemas.android.com/apk/res/android"
       package="com.example.catagentprofile">
       <uses-permission
       android:name="android.permission.INTERNET" />
       <application ...> ... </application>
   </manifest>
   ```

8. To add Retrofit and the scalars converter to your app, open the `build.gradle` app module, (Gradle Scripts | build.gradle (Module: app)), and add the following lines anywhere inside the `dependencies` block:

   ```
   dependencies {
       ...
       implementation
           'com.squareup.retrofit2:retrofit:2.9.0'
       implementation 'com.squareup.retrofit2:
           converter-scalars:2.9.0'
       ...
   }
   ```

   Your `dependencies` block should now look something like this:

   ```
   dependencies {
       implementation "org.jetbrains.kotlin:kotlin-stdlib
           :$kotlin_version"
   ```

```
    ...
    implementation
        'com.squareup.retrofit2:retrofit:2.9.0'
    implementation 'com.squareup.retrofit2:
        converter-scalars:2.9.0'
    ...
}
```

Between the time of writing and when you carry out this exercise, some dependencies may have changed. You should still only add the lines in bold from the preceding code block. These will add Retrofit and support for reading server responses as single strings.

> **Note**
>
> It is worth noting that Retrofit now requires, as a minimum, Android API 21 or Java 8.

9. Click the **Sync Project with Gradle Files** button in Android Studio.
10. Open your `activity_main.xml` file in the **Text** mode.
11. To be able to use your label to present the latest server response, you need to assign an ID to it:

    ```
    <TextView
        android:id="@+id/main_server_response"
        android:layout_width="wrap_content"
        android:layout_height="wrap_content"
        android:text="Hello World!"
        ... />
    ```

12. In the **Project** pane on the left, right-click on your app package (`com.example.catagentprofile`), then select **New | Package**.
13. Name your package `api`.
14. Now, right-click on the newly created package (`com.example.catagentprofile.api`), then select **New | Kotlin File/Class**.
15. Name your new file `TheCatApiService`. For **Kind**, choose **Interface**.
16. Add the following to the `interface` block:

    ```
    interface TheCatApiService {
        @GET("images/search")
        fun searchImages(
            @Query("limit") limit: Int,
    ```

```
            @Query("size") format: String
    ) : Call<String>
}
```

This defines the image search endpoint. Make sure to import all the required Retrofit dependencies.

17. Open your `MainActivity` file.
18. At the top of the `MainActivity` class block, add the following:

    ```
    class MainActivity : AppCompatActivity() {
        private val retrofit by lazy {
            Retrofit.Builder()
                .baseUrl("https://api.thecatapi.com/v1/")
                .addConverterFactory(
                    ScalarsConverterFactory.create()
                ).build()
        }
        private val theCatApiService by lazy {
            retrofit.create(TheCatApiService::class.java)
        }
        ...
    }
    ```

    This will instantiate Retrofit and the API service. We use `lazy` to make sure the instances are only created when needed.

19. Add `serverResponseView` as a field:

    ```
    class MainActivity : AppCompatActivity() {
        private val serverResponseView: TextView by lazy {
            findViewById(R.id.main_server_response)
        }
    ```

    This will look up the view with the `main_server_response` ID the first time `serverResponseView` is accessed and then keep a reference to it.

20. Now, add the `getCatImageResponse()` function after the `onCreate(Bundle?)` function:

    ```
    override fun onCreate(savedInstanceState: Bundle?) {
        ...
    }
    private fun getCatImageResponse() {
    ```

```kotlin
    val call = theCatApiService.searchImages(1,
      "full")
    call.enqueue(object : Callback<String> {
      override fun onFailure(call: Call<String>, t:
      Throwable) {
        Log.e("MainActivity", "Failed to get
        search results", t)
      }
      override fun onResponse(
        call: Call<String>, response: Response<String>
      ) {
        if (response.isSuccessful) {
          serverResponseView.text = response.body()
            } else {
          Log.e(
            "MainActivity",
            "Failed to get search results\n${
              response.errorBody()?.string().orEmpty()
            }"
          )
        }
      }
    })
  }
```

This function will fire off the search request and handle the possible outcomes—a successful response, an error response, and any other thrown exception.

21. Invoke a call to getCatImageResponse() in onCreate(). This will trigger the call as soon as the activity is created:

```kotlin
    override fun onCreate(savedInstanceState: Bundle?) {
        ...
        getCatImageResponse()
    }
```

22. Add the missing imports.

23. Run your app by clicking the **Run 'app'** button or pressing *Ctrl + R*. On the emulator, it should look like this:

Parsing a JSON response   213

Figure 5.2 – The app presenting the server response JSON

Because every time you run your app, a new call is made, and a random response is returned, your result will likely differ. However, whatever your result, if successful, it should be a JSON payload. Next, we will learn how to parse that JSON payload and extract the data we want from it.

Parsing a JSON response

Now that we have successfully retrieved a JSON response from an API, it is time to learn how to use the data we have obtained. To do so, we need to parse the JSON payload. This is because the payload is a plain string representing the data object, and we are interested in the specific properties of that object. If you look closely at *Figure 5.2*, you may notice that the JSON contains breed information, an image URL, and some other bits of information. However, for our code to use that information, first, we must extract it.

As mentioned in the introduction, multiple libraries exist that will parse a JSON payload for us. The most popular ones are Google's GSON (see https://github.com/google/gson) and, more recently, Square's Moshi (see https://github.com/square/moshi). Moshi is very lightweight, which is why we have chosen to use it in this chapter.

What do JSON libraries do? Basically, they help us convert data classes into JSON strings (serialization) and vice versa (deserialization). This helps us communicate with servers that understand JSON strings while allowing us to use meaningful data structures in our code.

To use Moshi with Retrofit, we need to add the Moshi Retrofit converter to our project. This is done by adding the following line to the `dependencies` block of our app's `build.gradle` file:

```
implementation 'com.squareup.retrofit2:converter-moshi:2.9.0'
```

Since we will no longer the responses as strings, we can go ahead and remove the scalars Retrofit converter. Next, we need to create a data class to map the server JSON response to. One convention is to suffix the names of API response data classes with `Data`—so we'll call our data class `ImageResultData`. Another common suffix is `Entity`.

When we design our server response data classes, we need to take two factors into account: the structure of the JSON response and our data requirements. The first will affect our data types and field names, while the second will allow us to omit fields we do not currently need. JSON libraries ignore data in fields we have not defined in our data classes.

One more thing JSON libraries do for us is automatically map JSON data to fields if they happen to have the exact same name. While this is a nice feature, it carries risk. If we rely solely on it, our data classes (and the code accessing them) will be tightly coupled to the API naming.

Because not all APIs are designed well, you might end up with meaningless field names, such as `fn` or `last`, or inconsistent naming. Luckily, there is a solution to this problem. Moshi provides us with an `@Json` annotation. It can be used to map a JSON field name to a meaningful field name:

```
data class UserData(
    @field:Json(name = "fn") val firstName: String,
    @field:Json(name = "last") val lastName: String
)
```

The `field:` prefix is used to ensure the generated Java field is annotated correctly.

Some consider it better practice to include the annotation even when the API name is the same as the field name for the sake of consistency. We prefer the conciseness of direct conversion when the field name is clear enough. This approach can be challenged when obfuscating our code. If we do, we must either exclude our data classes or make sure to annotate all fields.

While we are not always lucky enough to have properly documented APIs, when we do, it is best to consult the documentation when designing our model. Our model would be a data class into which the JSON data from all calls we make will be decoded. The documentation for the image search endpoint of The Cat API can be found at https://protect-eu.mimecast.com/s/d7uqCWwlKf56Q17s6kKb5?domain=developers.thecatapi.com/.

You will often find documentation is or inaccurate. If this happens to be the case, the best thing you can do is contact the owners of the API and request that they update the documentation. You may have to resort to experimenting with an endpoint, unfortunately. This is risky because undocumented fields or structures are not guaranteed to remain the same, so when possible, try and get the documentation updated.

Based on the response schema obtained from the preceding link, we can define our model as follows:

```
data class ImageResultData(
    @field:Json(name = "url") val imageUrl: String,
    val breeds: List<CatBreedData>
)
data class CatBreedData(
    val name: String,
    val temperament: String
)
```

Note that the response structure is that of a list of results. This means we need our responses mapped to `List<ImageResultData>`, not simply `ImageResultData`. Now, we need to update `TheCatApiService`. Instead of `Call<String>`, we can now have `Call<List<ImageResultData>>`.

Next, we need to update the construction of our Retrofit instance. Instead of `ScalarsConverterFactory`, we will now have `MoshiConverterFactory`. Lastly, we need to update our callback since it should no longer handle string calls but `List<ImageResultData>` instead:

```
@GET("images/search")
fun searchImages(
    @Query("limit") limit: Int,
    @Query("size") format: String
) : Call<List<ImageResultData>>
```

Exercise 5.02 – extracting the image URL from the API response

So, we have a server response as a string. Now, we want to extract the image URL from that string and present only that URL on the screen:

1. Open the app's `build.gradle` file and replace the scalars converter implementation with a Moshi converter one:

    ```
    implementation 'com.squareup.retrofit2: retrofit:2.9.0'
    implementation 'com.squareup.retrofit2:
    converter-moshi:2.9.0'
    testImplementation 'junit:junit:4.13.2'
    ```

2. Click the **Sync Project with Gradle Files** button.
3. Under your app package (com.example.catagentprofile), create a model package.
4. Within the com.example.catagentprofile.model package, create a new Kotlin file named CatBreedData.
5. Populate the newly created file with the following:

    ```
    package com.example.catagentprofile.model
    data class CatBreedData(
        val name: String,
        val temperament: String
    )
    ```

6. Next, create ImageResultData under the same package.
7. Set its contents to the following:

    ```
    package com.example.catagentprofile.model
    import com.squareup.moshi.Json
    data class ImageResultData(
        @field:Json(name = "url") val imageUrl: String,
        val breeds: List<CatBreedData>
    )
    ```

8. Open the TheCatApiService file and update the searchImages return type:

    ```
    @GET("images/search")
    fun searchImages(
        @Query("limit") limit: Int,
        @Query("size") format: String
    ) : Call<List<ImageResultData>>
    ```

9. Lastly, open MainActivity.
10. Update the Retrofit initialization block to use the Moshi converter to deserialize JSON:

    ```
    private val retrofit by lazy {
        Retrofit.Builder()
            .baseUrl("https://api.thecatapi.com/v1/")
            .addConverterFactory(MoshiConverterFactor
            .create())
    ```

```
            .build()
    }
```

11. Update the `getCatImageResponse()` function to handle the `List<ImageResultData>` requests and responses:

```
    private fun getCatImageResponse() {
      val call = theCatApiService.searchImages(1, "full")
      call.enqueue(object : Callback<List<ImageResultData>> {
        override fun onFailure(
          call: Call<List<ImageResultData>>,
          t: Throwable) {
          Log.e("MainActivity", "Failed to get search
          results", t)
        }
        override fun onResponse(
          call: Call<List<ImageResultData>>,
          response: Response<List<ImageResultData>>
        ) {
          if (response.isSuccessful) {
            val imageResults = response.body()
            val firstImageUrl =
              imageResults?.firstOrNull()?
              .imageUrl ?: "No URL"
            serverResponseView.text =
              "Image URL: $firstImageUrl"
          } else {
            Log.e("MainActivity", "Failed to get search
            results\n${response.errorBody()?.string().
    orEmpty()}"
          )
        }
      }
    })
   }
```

12. Now, you need to check for a successful response and that there is at least one `ImageResultData` instance. You can then read the `imageUrl` property of that instance and present it to the user.

13. Run your app. It should now look something like the following:

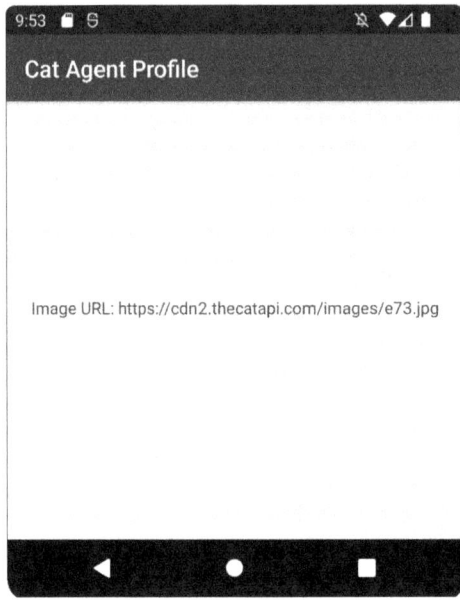

Figure 5.3 – The app presenting the parsed image URL

14. Again, due to the random nature of the API responses, your URL will likely be different.

You have now successfully extracted a specific property from an API response. Next, we will learn how to load the image from the URL provided to us by the API.

Loading images from a remote URL

We just learned how to extract data from an API response. That data often includes URLs to images we want to present to the user. There is quite a bit of work involved in achieving that. First, you must fetch the image as a binary stream from the URL. Then, you need to transform that binary stream into an image (it could be a GIF, JPEG, or one of a few other image formats).

Then, you need to convert it into a bitmap instance, potentially resizing it to use less memory. You may also want to apply other transformations to it at that point. Then, you need to set it to `ImageView`.

Sounds like a lot of work, doesn't it? Well, luckily for us, there are a few libraries that do all of that (and more) for us. The most commonly used libraries are Square's **Picasso** (see `https://square.github.io/picasso/`) and **Glide** by Bump Technologies (see `https://github.com/bumptech/glide`). Facebook's **Fresco** (see `https://frescolib.org/`) is somewhat less popular. A library that's gained traction recently is **Coil** (see `https://coil-kt.github.io/coil/`).

We will proceed with Glide because it is consistently the fastest of the two popular options for loading images, whether from the internet or the cache. However, it's worth noting that Picasso is more lightweight, so it is a trade-off, and both libraries are quite useful.

To include Glide in your project, add it to the `dependencies` block of your app's `build.gradle` file:

```
dependencies {
    implementation 'com.github.bumptech.glide:glide:4.14.2'
    ...
}
```

In fact, because we might change our minds at a later point, this is a great opportunity to abstract away the concrete library to have a simpler interface of our own. So, let's start by defining our `ImageLoader` interface:

```
interface ImageLoader {
    fun loadImage(imageUrl: String, imageView: ImageView)
}
```

This is a naïve implementation. In a production implementation, you might want to add arguments (or multiple functions) to support options such as different cropping strategies or loading states.

Our implementation of the interface will rely on Glide, so it will look something like this:

```
class GlideImageLoader(private val context: Context) :
ImageLoader {
    override fun loadImage(imageUrl: String, imageView:
    ImageView) {
        Glide.with(context)
            .load(imageUrl).centerCrop().into(imageView)
    }
}
```

We prefix our class name with `Glide` to differentiate it from other potential implementations. Constructing `GlideImageLoader` with `context` allows us to implement the clean `loadImage(String, ImageView)` interface without having to worry about the context, which Glide requires for image loading.

In fact, Glide is smart about the Android context. That means we could have separate implementations for the `Activity` and `Fragment` scopes, and Glide would know when an image-loading request went beyond the relevant scope.

Since we haven't yet added an `ImageView` to our layout, let's do that now:

```
<TextView
    ...
    app:layout_constraintBottom_toTopOf="@+id/
    main_profile_image"
    ... />
<ImageView
    android:id="@+id/main_profile_image"
    android:layout_width="150dp"
    android:layout_height="150dp"
    app:layout_constraintBottom_toBottomOf="parent"
    app:layout_constraintEnd_toEndOf="parent"
    app:layout_constraintStart_toStartOf="parent"
    app:layout_constraintTop_toBottomOf="@+id/
    main_server_response" />
```

This will add an `ImageView` with an ID of `main_profile_image` below our `TextView`.

We can now create an instance of `GlideImageLoader` in `MainActivity`:

```
private val imageLoader: ImageLoader by lazy {
    GlideImageLoader(this) }
```

In a production app, you would inject the dependency, rather than creating it inline.

Next, we tell our Glide loader to load the image and, once loaded, center-crop it inside the provided `ImageView`. This means the image will be scaled up or down to fully fill the `ImageView`, with any excess content cut off (cropped). Since we already obtained an image URL before, all we need to do is make the call:

```
val firstImageUrl = imageResults?.firstOrNull()?.imageUrl
    .orEmpty()
if (!firstImageUrl.isBlank()) {
    imageLoader.loadImage(firstImageUrl, profileImageView)
} else {
    Log.d("MainActivity", "Missing image URL")
}
```

We have to ensure the result contains a string that is not empty or made of spaces (isBlank() in the preceding code block). Then, we can safely load the URL into ImageView. And we're done. If we run our app now, we should see something like the following screenshot:

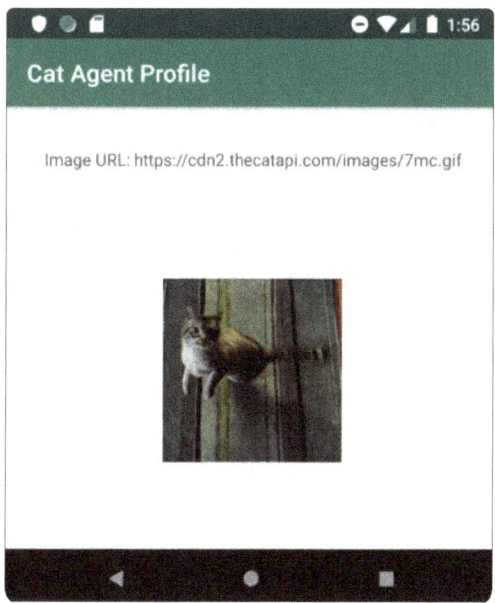

Figure 5.4 – Server response image URL with the actual image

Remember that the API returns random results, so the actual image is likely to be different. If we're lucky, we might even get an animated GIF, which we would then see animated.

Exercise 5.03 – loading the image from the obtained URL

In the previous exercise, we extracted the image URL from the API response. Now, we will use that URL to fetch an image from the web and display it in our app:

1. Open the app's `build.gradle` file and add the Glide dependency:

    ```
    dependencies {
        ...
        implementation 'com.github.bumptech.glide:glide:4.14.2'
        ...
    }
    ```

Synchronize your project with the Gradle files.

2. On the left **Project** panel, right-click on your project package name (`com.example.catagentprofile`) and select **New | Kotlin File/Class**.

3. Fill in `ImageLoader` in the **Name** field. For **Kind**, choose **Interface**.

4. Open the newly created `ImageLoader.kt` file and update it like so:

```
interface ImageLoader {
    fun loadImage(imageUrl: String, imageView: ImageView)
}
```

This will be your interface for any image loader in the app.

5. Right-click on the project package name again and select **New | Kotlin File/Class**.

6. Name the new file `GlideImageLoader` and select **Class** for **Kind**.

7. Update the newly created file:

```
class GlideImageLoader(private val context: Context) : ImageLoader {
    override fun loadImage(imageUrl: String,
    imageView: ImageView) {
        Glide.with(context).load(imageUrl)
            .centerCrop().into(imageView)
    }
}
```

8. Open `activity_main.xml` and update it like so:

```
<?xml version="1.0" encoding="utf-8"?>
<androidx.constraintlayout.widget.ConstraintLayout
...>
    <TextView
        android:id="@+id/main_server_response"
        android:layout_width="wrap_content"
        android:layout_height="wrap_content"
        android:text="Hello World!"
        app:layout_constraintBottom_toTopOf="@+id/
            main_profile_image"
        app:layout_constraintLeft_toLeftOf="parent"
```

```
                app:layout_constraintRight_toRightOf="parent"
                app:layout_constraintTop_toTopOf="parent" />
        <ImageView
            android:id="@+id/main_profile_image"
            android:layout_width="150dp"
            android:layout_height="150dp"
            app:layout_constraintBottom_toBottomOf="parent"
            app:layout_constraintEnd_toEndOf="parent"
            app:layout_constraintStart_toStartOf="parent"
            app:layout_constraintTop_toBottomOf=
                "@+id/main_server_response" />
    </androidx.constraintlayout.widget.ConstraintLayout>
```

   This will add an ImageView named main_profile_image below TextView.

9. Open the MainActivity.kt file.
10. Add a field for your newly added ImageView at the top of your class:

    ```
    private val serverResponseView: TextView by lazy {
        findViewById(R.id.main_server_response) }
    private val profileImageView: ImageView by lazy {
        findViewById(R.id.main_profile_image) }
    ```

11. Define ImageLoader just above the onCreate(Bundle?) function:

    ```
    private val imageLoader: ImageLoader by lazy {
        GlideImageLoader(this) }
    override fun onCreate(savedInstanceState: Bundle?) {
    ```

12. Update your getCatImageResponse() function like so:

    ```
    private fun getCatImageResponse() {
      val call = theCatApiService.searchImages(1, "full")
      call.enqueue(object : Callback<List<ImageResultData>> {
        override fun onFailure(...) { ... }
        override fun onResponse(...) {
          if (response.isSuccessful) {
            val imageResults = response.body()
            val firstImageUrl = imageResults
    ```

```
                    ?.firstOrNull()?.imageUrl.orEmpty()
                if (firstImageUrl.isNotBlank()) {
                  imageLoader.loadImage(
                    firstImageUrl, profileImageView)
                } else {
                  Log.d("MainActivity", "Missing image URL")
                }
                serverResponseView.text =
                  "Image URL: $firstImageUrl"
              } else {
                Log.e("MainActivity", "Failed to get search
                results\n${response.errorBody()?.string().
                orEmpty()}"
                )
              }
            }
          })
        }
```

13. Now, once you have a non-blank URL, it will be loaded into `profileImageView`.
14. Run the app:

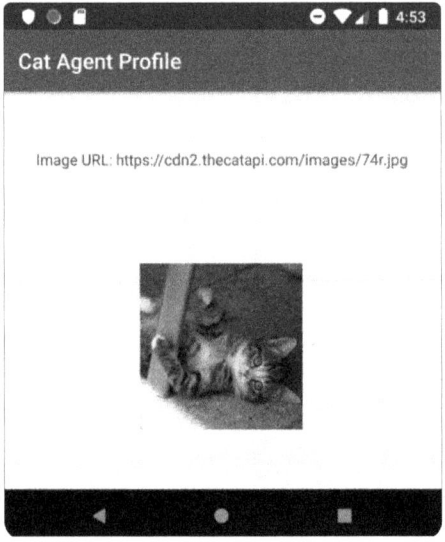

Figure 5.5 – Exercise outcome showing a random image and its source URL

The following are bonus steps.

15. Update your layout like so:

```xml
<?xml version="1.0" encoding="utf-8"?>
<androidx.constraintlayout.widget.ConstraintLayout
xmlns:android=
   "http://schemas.android.com/apk/res/android"
    xmlns:app="http://schemas.android.com/apk/res-auto"
    xmlns:tools="http://schemas.android.com/tools"
    android:layout_width="match_parent"
    android:layout_height="match_parent"
    tools:context=".MainActivity">
    <TextView
        android:id="@+id/main_agent_breed_label"
        android:layout_width="wrap_content"
        android:layout_height="wrap_content"
        android:padding="16dp"
        android:text="Agent breed:"
        app:layout_constraintStart_toStartOf="parent"
        app:layout_constraintTop_toTopOf="parent" />
    <TextView
        android:id="@+id/main_agent_breed_value"
        android:layout_width="wrap_content"
        android:layout_height="wrap_content"
        android:paddingTop="16dp"
        app:layout_constraintStart_toEndOf=
            "@+id/main_agent_breed_label"
        app:layout_constraintTop_toTopOf=
            "@+id/main_agent_breed_label" />
    <ImageView
        android:id="@+id/main_profile_image"
        android:layout_width="150dp"
        android:layout_height="150dp"
        android:layout_margin="16dp"
        app:layout_constraintStart_toStartOf="parent"
```

```
            app:layout_constraintTop_toBottomOf=
                "@+id/main_agent_breed_label" />
    </androidx.constraintlayout.widget.ConstraintLayout>
```

This will add an Agent breed label and tidy up the view layout. Now, your layout looks a bit more like a proper cat agent profile app.

16. In MainActivity.kt, locate the following lines:

    ```
    private val serverResponseView: TextView by lazy {
        findViewById(R.id.main_server_response) }
    ```

    Replace the preceding lines with the following to look up the new name field:

    ```
    private val agentBreedView: TextView by lazy {
        findViewById(R.id.main_agent_breed_value) }
    ```

17. Update getCatImageResponse() like so:

    ```
    private fun getCatImageResponse() {
      val call = theCatApiService.searchImages(1, "full")
      call.enqueue(object :
      Callback<List<ImageResultData>> {
        override fun onFailure(call:
          Call<List<ImageResultData>>, t: Throwable) {
          Log.e("MainActivity", "Failed to get search
          results", t)
        }
        override fun onResponse(
          call: Call<List<ImageResultData>>,
          response: Response<List<ImageResultData>>
        ) {
          if (response.isSuccessful) {
            val imageResults = response.body()
            val firstImageUrl = imageResults
              ?.firstOrNull()?.imageUrl.orEmpty()
            if (!firstImageUrl.isBlank()) {
              imageLoader.loadImage(
              firstImageUrl, profileImageView)
            } else {
              Log.d("MainActivity", "Missing image URL")
    ```

```
            }
            agentBreedView.text = imageResults
              ?.firstOrNull()?.breeds?.firstOrNull()
              ?.name ?: "Unknown"
          } else {
            Log.e("MainActivity", "Failed to get search
            results\n${response.errorBody()?.string().
            orEmpty()}")
          }
        }
      })
    }
```

This is done to load the first breed returned from the API into `agentNameView`, with a fallback to `Unknown`.

18. At the time of writing, not many pictures in The Cat API have breed data. However, if you run your app enough times, you will end up seeing something like this:

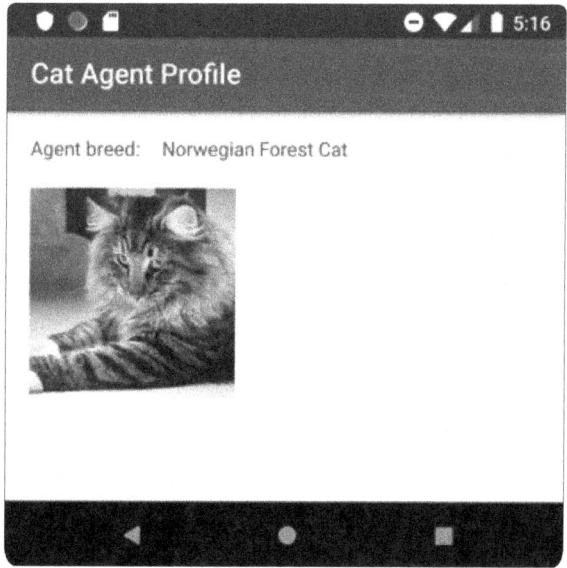

Figure 5.6 – Showing the cat agent image and breed

In this chapter, we learned how to fetch data from a remote API. We then learned how to process that data and extract the information we needed from it. Lastly, we learned how to present an image on the screen when given an image URL.

In the following activity, we will apply our knowledge to develop an app that tells the user the current weather in New York, presenting the user with a relevant weather icon.

Activity 5.01 – displaying the current weather

Let's say we want to build an app that shows the current weather in New York. Furthermore, we also want to display an icon representing the current weather.

This activity aims to create an app that polls an API endpoint for the current weather in JSON format, transforms that data into a local model, and uses that model to present the current weather. It also extracts the URL to an icon representing the current weather and fetches that icon to be displayed on the screen.

We will use the free OpenWeatherMap API for the purpose of this activity. The documentation can be found at `https://openweathermap.org/api`. To sign up for an API token, please go to `https://home.openweathermap.org/users/sign_up`. You can find your keys and generate new ones as needed at `https://home.openweathermap.org/api_keys`.

The steps for this activity are as follows:

1. Create a new app.
2. Grant internet permissions to the app in order to be able to make API and image requests.
3. Add Retrofit, the Moshi converter, and Glide to the app.
4. Update the app layout to support the presentation of the weather in a textual form (a short and long description) as well as a weather icon image.
5. Define the model. Create classes that will contain the server response.
6. Add the Retrofit service for the OpenWeatherMap API, `https://api.openweathermap.org/data/2.5/weather`.
7. Create a Retrofit instance with a Moshi converter.
8. Call the API service.
9. Handle the successful server response.
10. Handle the different failure scenarios.

The expected output is shown here:

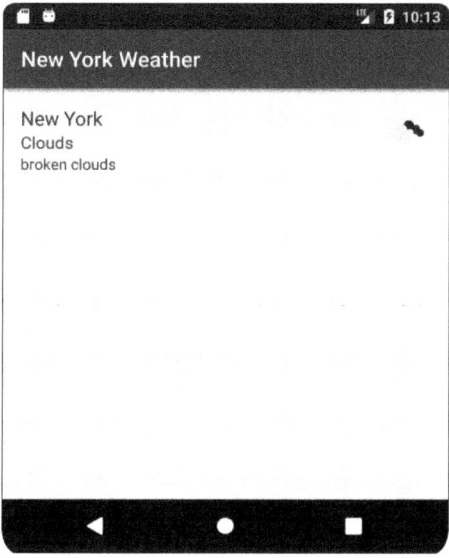

Figure 5.7 – The final weather app

> **Note**
> The solution for this activity can be found at https://packt.link/By7eE.

Summary

In this chapter, we learned how to fetch data from an API using Retrofit. We then learned how to handle JSON responses, as well as plain text responses, using Moshi. We also saw how different error scenarios could be handled.

We later learned how to load images from URLs using Glide and how to present them to the user via `ImageView`.

There are quite a few popular libraries for fetching data from APIs and for loading images. We only covered some of the most popular ones. You might want to try out some of the other libraries to find out which ones fit your purposes best.

In the next chapter, we will be introduced to `RecyclerView`, which is a powerful UI component that we can use to present our users with lists of items.

6
Adding and Interacting with RecyclerView

In this chapter, you will learn how to add lists and grids of items to your apps and effectively leverage the recycling power of `RecyclerView`. You'll also learn how to handle user interaction with the item views on the screen and support different item view types – for example, for titles. Later in the chapter, you'll add and remove items dynamically.

By the end of the chapter, you will have the skills required to present your users with interactive lists of rich items.

In the previous chapter, we learned how to fetch data, including lists of items and image URLs, from APIs, and how to load images from URLs. Combining that knowledge with the ability to display lists of items is the goal of this chapter.

Quite often, you will want to present your users with a list of items. For example, you might want to show them a list of pictures on their device or let them select their country from a list of all countries. To do that, you would need to populate multiple views, all sharing the same layout but presenting different content.

Historically, this was achieved by using `ListView` or `GridView`. While both are still viable options, they do not offer the robustness and flexibility of `RecyclerView`. For example, they do not support large datasets well, they do not support horizontal scrolling, and they do not offer rich divider customization.

> **Note**
> Customizing the divider between items in `RecyclerView` can be easily achieved using `RecyclerView.ItemDecorator`.

So, what does `RecyclerView` do? `RecyclerView` orchestrates the creation, population, and reuse (hence the name) of views representing lists of items. To use `RecyclerView`, you need to familiarize yourself with two of its dependencies – the adapter (and through it, the view holder) and the layout manager. These dependencies provide our `RecyclerView` with the content to show, as well as tell it how to present that content and lay it out on the screen.

The adapter provides `RecyclerView` with child views (nested Android views within `RecyclerView` used to represent individual data items) to draw on the screen, binds those views to data (via `ViewHolder` instances), and reports user interaction with those views.

The layout manager tells `RecyclerView` how to lay its children out. We are provided with three layout types by default – linear, grid, and staggered grid – managed by `LinearLayoutManager`, `GridLayoutManager`, and `StaggeredGridLayoutManager` respectively.

> **Note**
> This chapter relies on the use of the Jetpack RecyclerView library: `https://packt.link/FBX4d`.

In this chapter, we will develop an app that lists secret agents and whether they are currently active or sleeping (and, thus, unavailable). The app will then allow us to add new agents or delete existing ones by swiping them away. There is a twist, though – as you saw in *Chapter 5, Essential Libraries: Retrofit, Moshi, and Glide*, all our agents will be cats.

We will cover the following topics in this chapter:

- Adding `RecyclerView` to our layout
- Populating `RecyclerView`
- Responding to clicks in `RecylerView`
- Supporting different item types
- Swiping to remove items
- Adding items interactively

Technical requirements

The complete code for all the exercises and the activity in this chapter is available on GitHub at `https://packt.link/IJbeG`.

Adding RecyclerView to our layout

In *Chapter 3, Developing the UI with Fragments*, we saw how we can add views to our layouts to be inflated by activities, fragments, or custom views. `RecyclerView` is just another such view. To add it to our layout, we need to add the following tag to our layout:

```
<androidx.recyclerview.widget.RecyclerView
    android:id="@+id/recycler_view"
```

```
android:layout_width="match_parent"
android:layout_height="wrap_content"
tools:listitem="@layout/item_sample" />
```

You should already be able to recognize the `android:id` attribute, as well as the `android:layout_width` and `android:layout_height` ones.

We can use the optional `tools:listitem` attribute to tell Android Studio which layout to inflate as a list item in our preview toolbar. This will give us an idea of how `RecyclerView` might look in our app.

Adding a `RecyclerView` tag to our layout means we now have an empty container to hold the child views representing our list items. Once populated, it will handle the presenting, scrolling, and recycling of child views for us.

Exercise 6.01 – adding an empty RecyclerView to your main activity

To use `RecyclerView` in your app, you first need to add it to one of your layouts. Let's add it to the layout inflated by our main activity:

1. Start by creating a new empty activity project (**File** | **New** | **New Project** | **Empty Activity**). Name your application `My RecyclerView App`. Make sure your package name is `com.example.myrecyclerviewapp`.

2. Set the save location to where you want to save your project. Leave everything else at their default values and click **Finish**. Make sure you are on the **Android** view in your **Project** pane:

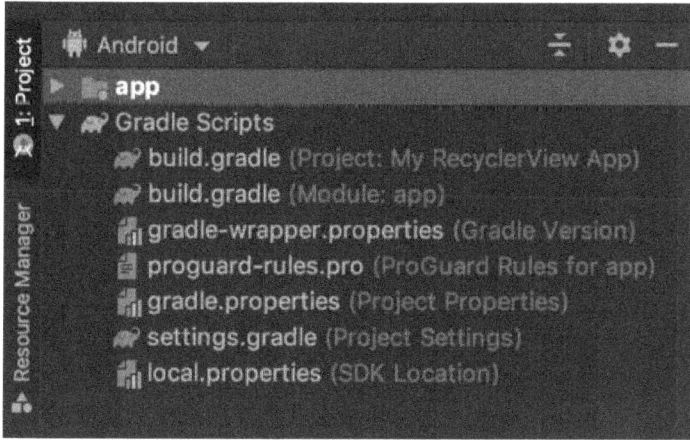

Figure 6.1 – The Android view in the Project pane

3. Open your `activity_main.xml` file in **Text** mode.

4. To turn your label into a title at the top of the screen under which you can add your `RecyclerView`, add an ID to `TextView` and align it to the top, like so:

   ```
   <TextView
       android:id="@+id/hello_label"
       android:layout_width="wrap_content"
       android:layout_height="wrap_content"
       android:text="Hello World!"
       app:layout_constraintLeft_toLeftOf="parent"
       app:layout_constraintRight_toRightOf="parent"
       app:layout_constraintTop_toTopOf="parent" />
   ```

5. Add the following after the `TextView` tag to add an empty `RecyclerView` element to your layout, constrained below your `hello_label` `TextView` title:

   ```
   <androidx.recyclerview.widget.RecyclerView
     android:id="@+id/recycler_view"
     android:layout_width="match_parent"
     android:layout_height="wrap_content"
     app:layout_constraintTop_toBottomOf="@+id/hello_label" />
   ```

   Your layout file should now look something like this:

   ```
   <?xml version="1.0" encoding="utf-8"?>
   <androidx.constraintlayout.widget.ConstraintLayout
       ...>
     <TextView
       android:id="@+id/hello_label"
       android:layout_width="wrap_content"
       android:layout_height="wrap_content"
       android:text="Hello World!"
       app:layout_constraintLeft_toLeftOf="parent"
       app:layout_constraintRight_toRightOf="parent"
       app:layout_constraintTop_toTopOf="parent" />
     <androidx.recyclerview.widget.RecyclerView
       android:id="@+id/recycler_view"
       android:layout_width="match_parent"
       android:layout_height="wrap_content"
   ```

```
        app:layout_constraintTop_toBottomOf="@+id/hello_label" />
    </androidx.constraintlayout.widget.ConstraintLayout>
```

6. Run your app by clicking the **Run app** button or pressing *Ctrl + R* (*Shift + F10* in Windows). On the emulator, it should look like this:

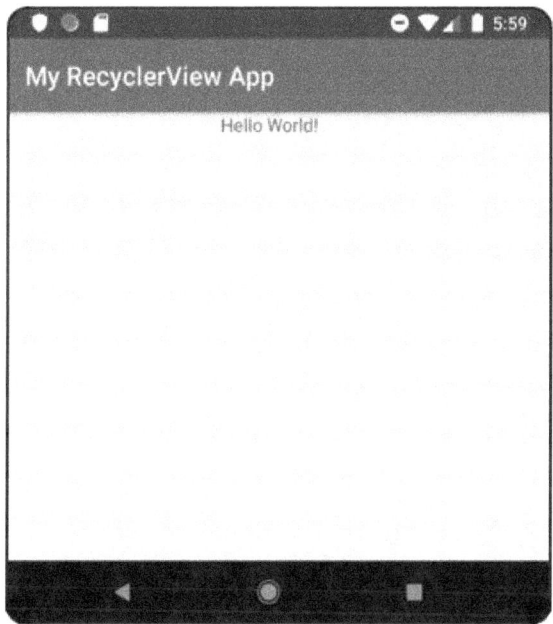

Figure 6.2 – The app with an empty RecyclerView (image cropped for space)

As you can see, our app runs, and our layout is presented on the screen. However, we do not see our `RecyclerView`. Why is that? At this stage, our `RecyclerView` has no content. `RecyclerView` with no content does not render by default – so, while our `RecyclerView` is indeed on the screen, it is not visible. This brings us to the next step – populating `RecyclerView` with content that we can actually see.

Populating RecyclerView

So, we added `RecyclerView` to our layout. For us to benefit from `RecyclerView`, we need to add content to it. Let's see how we go about doing that.

As we mentioned before, to add content to our `RecyclerView`, we would need to implement an adapter. An adapter binds our data to child views. In simpler terms, this means it tells `RecyclerView` how to plug data into views designed to present that data.

For example, let's say we want to present a list of employees.

First, we need to design our UI model. This will be a data object holding all the information needed by our view to present a single employee. Because this is a UI model, one convention is to suffix its name with `UiModel`:

```
data class EmployeeUiModel(
    val name: String,
    val biography: String,
    val role: EmployeeRole,
    val gender: Gender,
    val imageUrl: String
)
```

We will define `EmployeeRole` and `Gender` as follows:

```
enum class EmployeeRole {
    HumanResources,
    Management,
    Technology
}
enum class Gender {
    Female,
    Male,
    Unknown
}
```

The values are provided as an example, of course. Feel free to add more of your own!

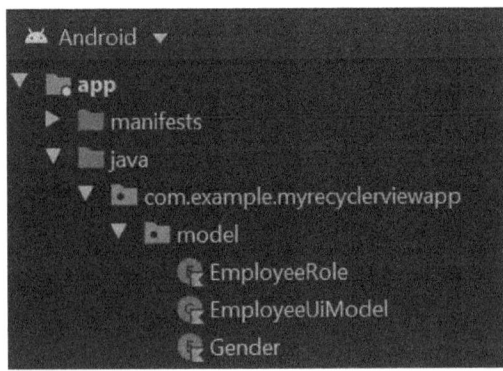

Figure 6.3 – The model's hierarchy

Now, we know what data to expect when binding to a view, so we can design our view to present this data (this is a simplified version of the actual layout, which we'll save as `item_employee.xml`). We'll start with `ImageView`:

```xml
<?xml version="1.0" encoding="utf-8"?>
<androidx.constraintlayout.widget.ConstraintLayout ...>
    <ImageView
        android:id="@+id/item_employee_photo"
        android:layout_width="60dp"
        android:layout_height="60dp"
        app:layout_constraintStart_toStartOf="parent"
        app:layout_constraintTop_toTopOf="parent"
        tools:background="@color/colorPrimary" />
```

Then, we will add a `TextView` for each field:

```xml
<TextView
   android:id="@+id/item_employee_name"
   android:layout_width="wrap_content"
   android:layout_height="wrap_content"
   android:layout_marginStart="16dp"
   android:layout_marginLeft="16dp"
   android:textStyle="bold"
   app:layout_constraintStart_toEndOf="@+id/item_employee_photo"
   app:layout_constraintTop_toTopOf="parent"
   tools:text="Oliver" />
<TextView
   android:id="@+id/item_employee_role"
   android:layout_width="wrap_content"
   android:layout_height="wrap_content"
   android:textColor="@color/colorAccent"
   app:layout_constraintStart_toStartOf="@+id/item_employee_name"
   app:layout_constraintTop_toBottomOf="@+id/
       item_employee_name"
   tools:text="Exotic Shorthair" />
<TextView
```

```
            android:id="@+id/item_employee_biography"
            android:layout_width="wrap_content"
            android:layout_height="wrap_content"
            app:layout_constraintStart_toStartOf="@+id/item_employee_role"
            app:layout_constraintTop_toBottomOf="@+id/item_employee_role"
            tools:text="Stealthy and witty. Better avoid in dark alleys." />
        <TextView
            android:id="@+id/item_employee_gender"
            android:layout_width="wrap_content"
            android:layout_height="wrap_content"
            android:textSize="30sp"
            app:layout_constraintEnd_toEndOf="parent"
            app:layout_constraintTop_toTopOf="parent"
            tools:text="&#9794;" />
</androidx.constraintlayout.widget.ConstraintLayout>
```

So far, there is nothing new. You should be able to recognize all of the different view types from *Chapter 2, Building User Screen Flows*:

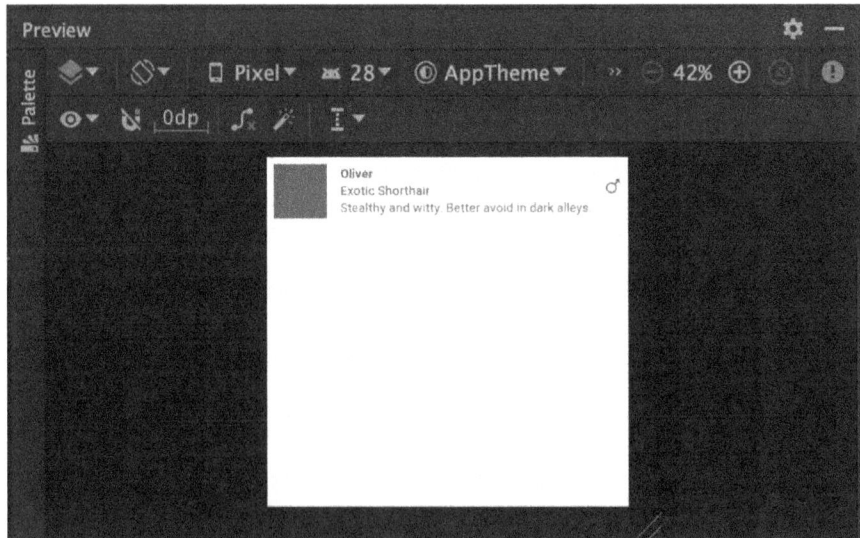

Figure 6.4 – A preview of the item_cat.xml layout file

With a data model and a layout, we now have everything we need to bind our data to the view. To do that, we will implement a view holder. Usually, a view holder has two responsibilities – it holds a reference to a view (as its name implies), but it also binds data to that view. We will implement our view holder as follows:

```kotlin
private const val FEMALE_SYMBOL = "\u2640"
private const val MALE_SYMBOL = "\u2642"
private const val UNKNOWN_SYMBOL = "?"
class EmployeeViewHolder(
  containerView: View,
  private val imageLoader: ImageLoader
) : ViewHolder(containerView) {
  private val employeeNameView: TextView by lazy {
    containerView.findViewById(R.id.item_employee_name) }
  private val employeeRoleView: TextView by lazy {
    containerView.findViewById(R.id.item_employee_role) }
  private val employeeBioView: TextView by lazy {
    containerView.findViewById(R.id.item_employee_bio) }
  private val employeeGenderView: TextView by lazy {
    containerView.findViewById(R.id.item_employee_gender) }
  fun bindData(employeeData: EmployeeUiModel) {
    imageLoader.loadImage(employeeData.imageUrl,
      employeePhotoView)
    employeeNameView.text = employeeData.name
    employeeRoleView.text = when (employeeData.role) {
      EmployeeRole.HumanResources -> "Human Resources"
      EmployeeRole.Management -> "Management"
      EmployeeRole.Technology -> "Technology"
    }
    employeeBioView.text = employeeData.biography
    employeeGenderView.text = when (employeeData.gender) {
      Gender.Female -> FEMALE_SYMBOL
      Gender.Male -> MALE_SYMBOL
      else -> UNKNOWN_SYMBOL
    }
  }
}
```

There are a few things worth noting in the preceding code. First, by convention, we suffixed the name of our view holder with `ViewHolder`. Second, note that `EmployeeViewHolder` needs to implement the abstract `RecyclerView.ViewHolder` class.

This is required so that the generic type of our adapter can be our view holder. Lastly, we lazily keep references to the views we are interested in. The first time `bindData(EmployeeUiModel)` is called, we will find these views in the layout and keep references to them.

Next, we introduced a `bindData(EmployeeUiModel)` function. This function will be called by our adapter to bind the data to the view held by the view holder. The last but most important thing to note is that we always make sure to set a state for all modified views for every possible input.

With our view holder set up, we can proceed to implement our adapter. We will start by implementing the minimum required functions, plus a function to set the data. Our adapter will look something like this:

```kotlin
class EmployeesAdapter(
  private val layoutInflater: LayoutInflater,
  private val imageLoader: ImageLoader
) : RecyclerView.Adapter<EmployeeViewHolder>() {
  private val employees = mutableListOf<EmployeeUiModel>()
  fun setData(newEmployees: List<EmployeeUiModel>) {
    employees.clear()
    employees.addAll(newEmployees)
    notifyDataSetChanged()
  }
  override fun onCreateViewHolder(parent: ViewGroup,
    viewType: Int): EmployeeViewHolder {
    val view = layoutInflater.inflate(
      R.layout.item_employee, parent, false)
    return EmployeeViewHolder(view, imageLoader)
  }
  override fun getItemCount() = employees.size
  override fun onBindViewHolder(
    holder: EmployeeViewHolder, position: Int) {
    holder.bindData(employees [position])
  }
}
```

Let's go over this implementation. First, we inject our dependencies into the adapter via its constructor. This will make testing our adapter much easier but will also allow us to change some of its behavior (for example, replace the image loading library) painlessly. In fact, we would not need to change the adapter at all in that case.

Then, we define a private mutable list of `EmployeeUiModel` to store the data currently provided by the adapter to `RecyclerView`. We also introduce a method (`setData`) to populate that list. Note that we keep a local list and set its contents, rather than allowing `employees` to be set directly.

This is mainly because Kotlin, just like Java, passes variables by reference. Passing variables by reference means changes to the content of the list passed into the adapter would change the list held by the adapter. So, for example, if an item was removed outside of the adapter, the adapter would have that item removed as well.

This becomes a problem because the adapter would not be aware of that change, and so would not be able to notify `RecyclerView`. There are other risks around a list being modified outside of the adapter, but covering them is beyond the scope of this book.

Another benefit of encapsulating the modification of data in a function is that we avoid the risk of forgetting to notify `RecyclerView` that a dataset has changed, which we do by calling `notifyDataSetChanged()`.

We proceed to implement the adapter's `onCreateViewHolder(ViewGroup, Int)` function. This function is called when `RecyclerView` needs a new `ViewHolder` to render data on a screen. It provides us with a `ViewGroup` container and a view type (we'll look into view types later in this chapter).

The function then expects us to return a view holder initialized with a view (in our case, an inflated one). So, we inflate the view we designed earlier, passing it to a new `EmployeeViewHolder` instance. Note that the last argument to the inflated function is `false`.

This makes sure we do not attach the newly inflated view to the parent. Attaching and detaching views will be managed by the layout manager. Setting the view to `true` or omitting it would result in `IllegalStateException` being thrown. Finally, we return the newly created `EmployeeViewHolder`.

To implement `getItemCount()`, we simply return the size of our `employees` list.

Lastly, we implement `onBindViewHolder(EmployeeViewHolder, Int)`. This is done by passing `EmployeeUiModel`, stored in `employees`, at the given position to the `bindData(EmployeeUiModel)` function of our view holder. Our adapter is now ready.

If we tried to plug our adapter into our `RecyclerView` at this point and run our app, we would still see no content. This is because we are still missing two small steps – setting data on our adapter and assigning a layout manager to our `RecyclerView`. The complete working code would look like this:

```kotlin
class MainActivity : AppCompatActivity() {
    private val employeesAdapter
        by lazy { EmployeesAdapter(layoutInflater,
        GlideImageLoader(this)) }
    private val recyclerView: RecyclerView
        by lazy { findViewById(R.id.main_recycler_view) }
    override fun onCreate(savedInstanceState: Bundle?) {
        super.onCreate(savedInstanceState)
        setContentView(R.layout.activity_main)
        recyclerView.adapter = employeesAdapter
        recyclerView.layoutManager = LinearLayoutManager(
            this, LinearLayoutManager.VERTICAL, false)
        employeesAdapter.setData(
            listOf(
                EmployeeUiModel(
                    "Robert",
                    "Rose quickly through the organization",
                    EmployeeRole.Management,
                    Gender.Male,
                    "https://images.pexels.com/photos/220453/
                    pexels-photo-220453.jpeg?
                    auto=compress&cs=tinysrgb&h=650&w=940"
                ),
                EmployeeUiModel(
                    "Wilma",
                    "A talented developer",
                    EmployeeRole.Technology,
                    Gender.Female,
                    "https://images.pexels.com/photos/3189024/
                    pexels-photo-3189024.jpeg?
                    auto=compress&cs=tinysrgb&h=650&w=940"
                ),
                EmployeeUiModel(
```

```
                          "Curious George",
                          "Excellent at retention",
                          EmployeeRole.HumanResources,
                          Gender.Unknown,
                          "https://images.pexels.com/photos/771742/
                          pexels-photo-771742.jpeg?
                          auto=compress&cs=tinysrgb&h=750&w=1260"
                )
            )
        )
    }
}
```

Running our app now, we would see a list of our employees.

Note that we hardcoded the list of employees. In a production app, following a **Model-View-View-Model** (**MVVM**) pattern (we will cover this pattern in *Chapter 15, Architecture Patterns*), you would be provided with data to present to your `ViewModel`. It is also important to note that we kept a reference to `employeesAdapter`.

This is so that we could, later on, set the data to different values. Some implementations rely on reading the adapter from `RecyclerView` itself – this can potentially result in unnecessary casting operations and unexpected states where the adapter is not yet assigned to `RecyclerView`, so this is generally not a recommended approach.

Lastly, note that we chose to use `LinearLayoutManager`, providing it with the activity for context, a `VERTICAL` orientation flag, and `false` to tell it that we do not want the order of the items in the list reversed.

Exercise 6.02 – populating your RecyclerView

`RecyclerView` is not very interesting without any content. It is time to populate `RecyclerView` by adding your secret cat agents to it.

A quick recap before you dive in – in the previous exercise, we introduced an empty list designed to hold a list of secret cat agents that users have at their disposal. In this exercise, you will be populating that list to present the users with the available secret cat agents in the agency:

1. To keep our file structure tidy, we will start by creating a model package. Right-click on the package name of our app, and then select **New** | **Package**:

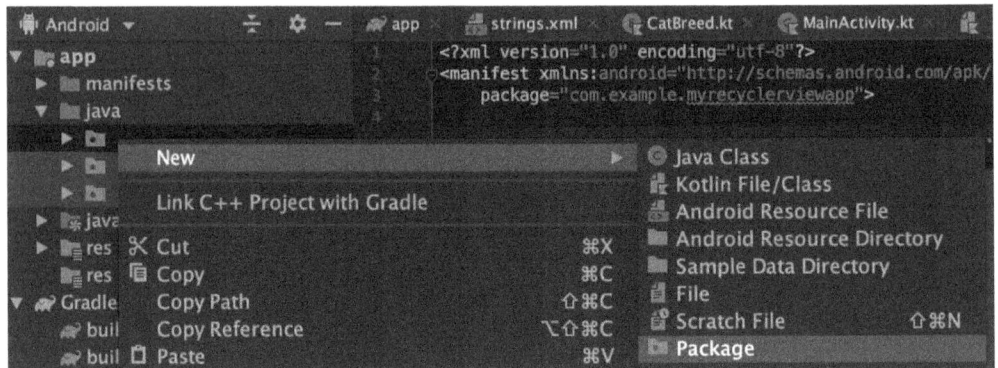

Figure 6.5 – Creating a new package

2. Name the new package `model`. Click **OK** to create the package.

3. To create our first model data class, right-click on the newly created model package, and then select **New | Kotlin File/Class**.

4. Under **Name**, fill in `CatUiModel`. Leave **kind** as **File** and click on **OK**. This will be the class holding the data we have about every individual cat agent.

5. Add the following to the newly created `CatUiModel.kt` file to define the data class with all the relevant properties of a cat agent:

   ```
   data class CatUiModel(
       val gender: Gender,
       val breed: CatBreed,
       val name: String,
       val biography: String,
       val imageUrl: String
   )
   ```

   For each cat agent, other than their name and photo, we want to know their gender, breed, and biography. This will help us choose the right agent for a mission.

6. Again, right-click on the model package, and then navigate to **New | Kotlin File/Class**.

7. This time, name the new file `CatBreed` and set `kind` to the Enum class. This class will hold our different cat breeds.

8. Update your newly created enum with some initial values, as follows:

   ```
   enum class CatBreed {
       AmericanCurl, BalineseJavanese, ExoticShorthair
   }
   ```

9. Repeat *step 6* and *step 7*, only this time call your file Gender. This will hold the accepted values for a cat agent's gender.

10. Update the Gender enum, like so:

    ```
    enum class Gender {
        Female, Male, Unknown
    }
    ```

11. Now, to define the layout of the view holding the data about each cat agent, create a new layout resource file by right-clicking on layout and then selecting **New** | **Layout resource file**:

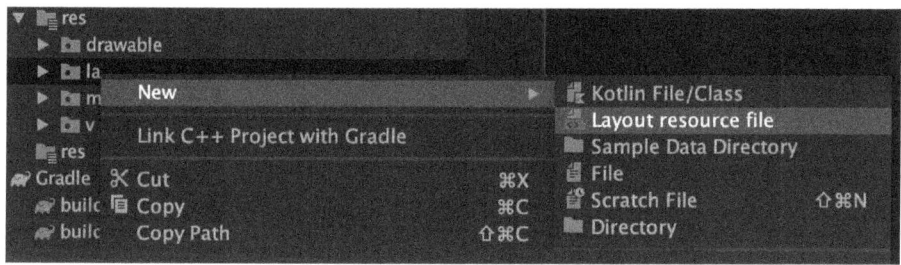

Figure 6.6 – Creating a new layout resource file

12. Name your resource item_cat. Leave all the other fields as they are and click **OK**.

13. Update the contents of the newly created item_cat.xml file (the following code block has been truncated for space, so use the following link to see the full code that you need to add):

item_cat.xml

```
<ImageView
    android:id="@+id/item_cat_photo"
    android:layout_width="60dp"
    android:layout_height="60dp"
    app:layout_constraintStart_toStartOf="parent"
    app:layout_constraintTop_toTopOf="parent"
    tools:background="@color/colorPrimary" />

<TextView
    android:id="@+id/item_cat_name"
    android:layout_width="wrap_content"
    android:layout_height="wrap_content"
    android:layout_marginStart="16dp"
```

```
                    android:layout_marginLeft="16dp"
                    android:textStyle="bold"
                    app:layout_constraintStart_toEndOf="@+id/
                        item_cat_photo"
                    app:layout_constraintTop_toTopOf="parent"
                    tools:text="Oliver" />
```

The complete code for this step can be found at http://packt.live/3sopUjo.

This will create a layout with an image and text fields for a name, breed, and biography to be used in our list.

14. You will need a copy of `ImageLoader.kt`, introduced in *Chapter 5, Essential Libraries: Retrofit, Moshi, and Glide*, so right-click on the package name of your app, navigate to **New | Kotlin File/Class**, then set the name to `ImageLoader` and **kind** to **Interface**, and click **OK**.

15. Similar to *Chapter 5, Essential Libraries: Retrofit, Moshi, and Glide*, you only need to add one function here:

```
interface ImageLoader {
    fun loadImage(imageUrl: String, imageView: ImageView)
}
```

Make sure to import `ImageView`.

16. Right-click on the package name of your app again, and then select **New | Kotlin File/Class**.
17. Call the new file `CatViewHolder`. Click **OK**.
18. To implement `CatViewHolder`, which will bind the cat agent data to your views, replace the contents of the `CatViewHolder.kt` file with the following:

```
private val FEMALE_SYMBOL = "\u2640"
private val MALE_SYMBOL = "\u2642"
private const val UNKNOWN_SYMBOL = "?"
class CatViewHolder(
    containerView: View,
    private val imageLoader: ImageLoader
) : ViewHolder(containerView) {
    private val catBiographyView: TextView by lazy {
        containerView
            .findViewById(R.id.item_cat_biography) }
    private val catBreedView: TextView by lazy {
        containerView.findViewById(R.id.item_cat_breed) }
```

```kotlin
    private val catGenderView: TextView by lazy {
      containerView.findViewById(R.id.item_cat_gender) }
    private val catNameView: TextView by lazy {
      containerView.findViewById(R.id.item_cat_name) }
    private val catPhotoView: ImageView by lazy {
      containerView.findViewById(R.id.item_cat_photo) }
    fun bindData(cat: CatUiModel) {
      imageLoader.loadImage(cat.imageUrl, catPhotoView)
      catNameView.text = cat.name
      catBreedView.text = when (cat.breed) {
        AmericanCurl -> "American Curl"
        BalineseJavanese -> "Balinese-Javanese"
        ExoticShorthair -> "Exotic Shorthair"
      }
      catBiographyView.text = cat.biography
      catGenderView.text = when (cat.gender) {
        Female -> FEMALE_SYMBOL
        Male -> MALE_SYMBOL
        else -> UNKNOWN_SYMBOL
      }
    }
}
```

19. Still under our app package name, create a new Kotlin file named `CatsAdapter`.

20. To implement `CatsAdapter`, which is responsible for storing the data for `RecyclerView`, as well as creating instances of your view holder and using them to bind data to views, replace the contents of the `CatsAdapter.kt` file with this:

```kotlin
package com.example.myrecyclerviewapp
import { ... }
class CatsAdapter(
    private val layoutInflater: LayoutInflater,
    private val imageLoader: ImageLoader
) : RecyclerView.Adapter<CatViewHolder>() {
    private val cats = mutableListOf<CatUiModel>()
    fun setData(newCats: List<CatUiModel>) {
        cats.clear()
```

```
            cats.addAll(newCats)
            notifyDataSetChanged()
        }
        override fun onCreateViewHolder(parent: ViewGroup,
           viewType: Int): CatViewHolder {
            val view = layoutInflater
                .inflate(R.layout.item_cat, parent, false)
            return CatViewHolder(view, imageLoader)
        }
        override fun getItemCount() = cats.size
        override fun onBindViewHolder(
            holder: CatViewHolder, position: Int) {
            holder.bindData(cats[position])
        }
    }
```

21. At this point, you need to include Glide in your project. Start by adding the following line of code to the dependencies block inside your app's `gradle.build` file:

    ```
    implementation 'com.github.bumptech.glide:glide:4.14.2'
    ```

22. Create a `GlideImageLoader` class in your app package path, containing the following:

    ```
    package com.example.myrecyclerviewapp
    [imports]
    class GlideImageLoader(context: Context) : ImageLoader {
        private val glide by lazy { Glide(context) }
        override fun loadImage(imageUrl: String,
        imageView: ImageView) {
            glide.load(imageUrl)
                .centerCrop().into(imageView)
        }
    }
    ```

    This is a simple implementation assuming the loaded image should always be center-cropped.

23. Update your `MainActivity` file:

    ```
    class MainActivity : AppCompatActivity() {
      private val recyclerView: RecyclerView by lazy {
    ```

```kotlin
      findViewById(R.id.recycler_view) }
  private val catsAdapter by lazy { CatsAdapter(
    layoutInflater, GlideImageLoader(this)) }
  override fun onCreate(savedInstanceState: Bundle?) {
    super.onCreate(savedInstanceState)
    setContentView(R.layout.activity_main)
    recyclerView.adapter = catsAdapter
    recyclerView.layoutManager = LinearLayoutManager(
      this, LinearLayoutManager.VERTICAL, false)
    catsAdapter.setData(
      listOf(
        CatUiModel(Gender.Male,
          CatBreed.BalineseJavanese, "Fred",
          "Silent and deadly",
          "https://cdn2.thecatapi.com/image/DBmIBhhyv.
          jpg"
        ),
        CatUiModel(Gender.Female,
          CatBreed.ExoticShorthair, "Wilma",
          "Cuddly assassin",
          "https://cdn2.thecatapi.com/images/KJF8fB_20.
          jpg"
        ),
        CatUiModel(Gender.Unknown,
          CatBreed.AmericanCurl, "Curious George",
          "Award winning investigator",
          "https://cdn2.thecatapi.com/images/vJB8rwfdX.
          jpg"
        )
      )
    )
  }
}
```

This will define your adapter, attach it to `RecyclerView`, and populate it with some hardcoded data.

24. In your `AndroidManifest.xml` file, add the following in the `manifest` tag before the application tag:

    ```xml
    <uses-permission android:name="android.permission.INTERNET" />
    ```

    Having this tag will allow your app to download images from the internet.

25. For some final touches, such as giving our title view a proper name and text, update your `activity_main.xml` file, like so:

    ```xml
    <?xml version="1.0" encoding="utf-8"?>
    <androidx.constraintlayout.widget.ConstraintLayout
        ...>
      <TextView
        android:id="@+id/main_label"
        android:layout_width="wrap_content"
        android:layout_height="wrap_content"
        android:text="@string/main_title"
        android:textSize="24sp"
        app:layout_constraintLeft_toLeftOf="parent"
        app:layout_constraintRight_toRightOf="parent"
        app:layout_constraintTop_toTopOf="parent" />
      <androidx.recyclerview.widget.RecyclerView
        android:id="@+id/recycler_view"
        android:layout_width="match_parent"
        android:layout_height="wrap_content"
        app:layout_constraintTop_toBottomOf="@+id/main_label"
    />
    </androidx.constraintlayout.widget.ConstraintLayout>
    ```

26. Also, update your `strings.xml` file to give your app a proper name and title:

    ```xml
    <resources>
      <string name="app_name">SCA - Secret Cat Agents</string>
      <string name="main_title">Our Agents</string>
    </resources>
    ```

27. Run your app. It should look like this:

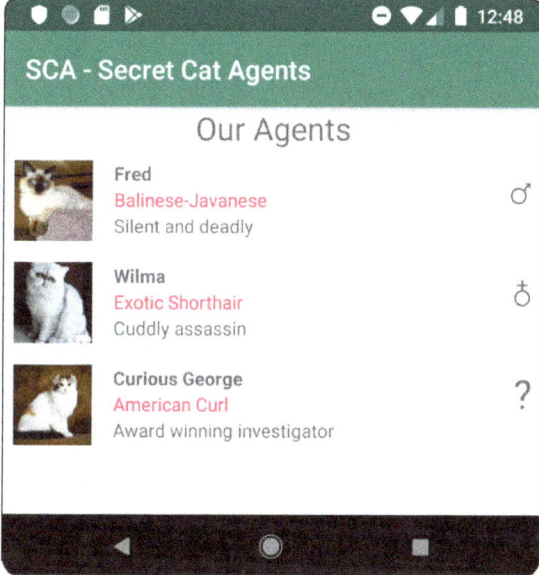

Figure 6.7 – RecyclerView with hardcoded secret cat agents

As you can see, `RecyclerView` now has content, and your app is starting to take shape. Note how the same layout is used to present different items based on the data bound to each instance. As you would expect, if you add enough items for them to go off screen, scrolling works. Next, we'll look into allowing a user to interact with the items inside our `RecyclerView`.

Responding to clicks in RecyclerView

What if we want to let our users select an item from a presented list? To achieve that, we need to communicate clicks back to our app.

The first step in implementing click interaction is to capture clicks on items at the `ViewHolder` level. To maintain separation between our view holder and the adapter, we define a nested `OnClickListener` interface in our view holder. We choose to define the interface within the view holder because that and the listener are tightly coupled.

The interface will, in our case, have only one function. The purpose of this function is to inform the owner of the view holder about the clicks. The owner of a view holder is usually a `Fragment` or an `Activity`. Since we know that a view holder can be reused, we know that it can be challenging to define it at construction time in a way that would tell us which item was clicked (since that item will change over time with reuse).

We work around that by passing the currently presented item back to the owner of the view holder on clicking. This means our interface would look like this:

```
interface OnClickListener {
    fun onClick(cat: CatUiModel)
}
```

We will also add this listener as a parameter to our `ViewHolder` constructor:

```
class CatViewHolder(
    containerView: View,
    private val imageLoader: ImageLoader,
    private val onClickListener: OnClickListener
) : ViewHolder(containerView) {
    ...
}
```

It will be used like this:

```
containerView.setOnClickListener {
    onClickListener.onClick(cat) }
```

Now, we want our adapter to pass in a listener. In turn, that listener will be responsible for informing the owner of the adapter about the click. This means our adapter, too, would need a nested listener interface, quite similar to the one we implemented in our view holder.

> **Note**
>
> While this seems like duplication that can be avoided by reusing the same listener, that is not a great idea, as it leads to tight coupling between the view holder and the adapter through the listener. What happens when you want your adapter to also report other events through the listener? You would have to handle those events coming from the view holder, even though they would not actually be implemented in the view holder.

Finally, to handle the click event and show a dialog, we define a listener in our activity and pass it to our adapter. We set that listener to show a dialog on clicking. In an MVVM implementation, you would be notifying the `ViewModel` of the click at this point instead. `ViewModel` would then update its state, telling the view (our activity) that it should display the dialog.

Exercise 6.03 – responding to clicks

Your app already shows the user a list of secret cat agents. It is time to allow your user to choose a secret cat agent by clicking on its view. Click events are delegated from the view holder to the adapter to the activity, as shown in *Figure 6.9*:

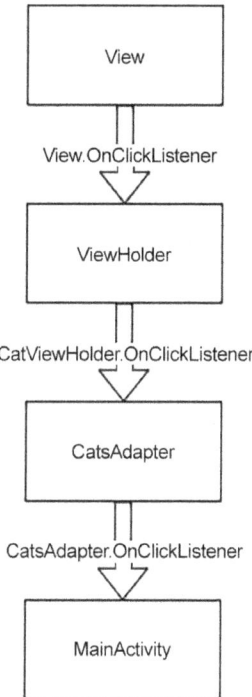

Figure 6.8 – The flow of click events

The following are the steps that you need to follow to complete this exercise:

1. Open your `CatViewHolder.kt` file. Add a nested interface to it right before the final closing curly bracket:

    ```
    interface OnClickListener {
        fun onClick(cat: CatUiModel)
    }
    ```

    This will be the interface that a listener will have to implement in order to register for click events on individual cat items.

2. Update the `CatViewHolder` constructor to accept `OnClickListener` and make `containerView` accessible:

   ```
   class CatViewHolder(
       private val containerView: View,
       private val imageLoader: ImageLoader,
       private val onClickListener: OnClickListener
   ) : ViewHolder(containerView) {
   ```

   Now, when constructing a `CatViewHolder` constructor, you also register for clicks on item views.

3. At the top of your `bindData(CatUiModel)` function, add the following to intercept clicks and report them to the provided listener:

   ```
   containerView.setOnClickListener {
       onClickListener.onClick(cat) }
   ```

4. Now, open your `CatsAdapter.kt` file. Add this nested interface right before the final closing curly bracket:

   ```
   interface OnClickListener {
       fun onItemClick(cat: CatUiModel)
   }
   ```

   This defines the interface that listeners will have to implement to receive item click events from the adapter.

5. Update the `CatsAdapter` constructor to accept a call implementing the `OnClickListener` adapter you just defined:

   ```
   class CatsAdapter(
       private val layoutInflater: LayoutInflater,
       private val imageLoader: ImageLoader,
       private val onClickListener: OnClickListener
   ) : RecyclerView.Adapter<CatViewHolder>() {
   ```

6. In `onCreateViewHolder(ViewGroup, Int)`, update the creation of the view holder, as follows:

   ```
   return CatViewHolder(view, imageLoader, object :
   CatViewHolder.OnClickListener {
   override fun onClick(cat: CatUiModel) =
   ```

```
    onClickListener.onItemClick(cat)
  })
```

This will add an anonymous class that delegates the `ViewHolder` click events to the adapter listener.

7. Finally, open your `MainActivity.kt` file. Update your `catsAdapter` construction as follows to provide the required dependencies to the adapter, in the form of an anonymous listener handling click events by showing a dialog:

```
private val catsAdapter by lazy {
  CatsAdapter(layoutInflater, GlideImageLoader(this),
    object : CatsAdapter.OnClickListener {
      override fun onClick(cat: CatUiModel) =
        onClickListener.onItemClick(cat)
    }
  )
}
```

8. Add the following function right before the final closing curly bracket:

```
private fun showSelectionDialog(cat: CatUiModel) {
  AlertDialog.Builder(this)
    .setTitle("Agent Selected")
    .setMessage("You have selected agent ${cat.name}")
    .setPositiveButton("OK") { _, _ -> }.show()
}
```

This function will show a dialog with the name of the cat whose data was passed in.

9. Make sure to import the right version of `AlertDialog`, which is `androidx.appcompat.app.AlertDialog`, not `android.app.AlertDialog`.

> **Note**
> The AppCompat version is usually a better choice because it offers backward compatibility.

10. Run your app. Clicking on one of the cats should now open a dialog:

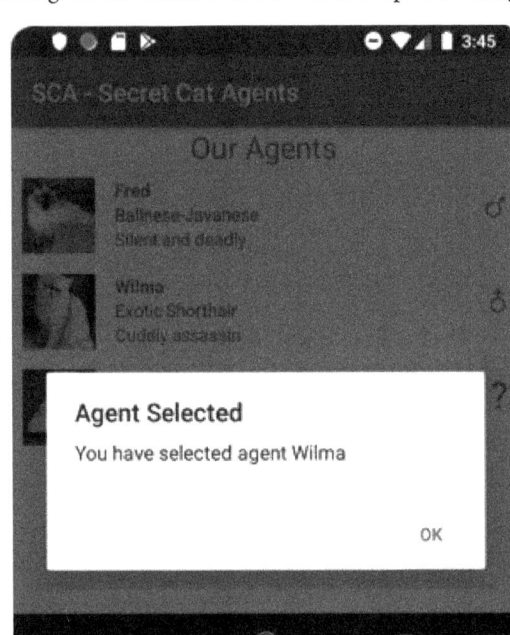

Figure 6.9 – A dialog showing an agent was selected

Try clicking the different items, and note the different messages presented. You now know how to respond to users clicking on items inside your `RecyclerView`. Next, we will look at how we can support different item types in our lists.

Supporting different Item types

In the previous sections, we learned how to handle a list of items of a single type (in our case, all our items were `CatUiModel`). What happens if you want to support more than one type of item? A good example of this would be having group titles on our list.

Let's say that instead of getting a list of cats, we get a list containing happy cats and sad cats. Each of the two groups of cats is preceded by a title of the corresponding group. Instead of a list of `CatUiModel` instances, our list would now contain `ListItem` instances. `ListItem` might look like this:

```
sealed class ListItem {
    data class Group(val name: String) : ListItem()
```

```
        data class Cat(val cat: CatUiModel) : ListItem()
}
```

Our list of items may look like this:

```
listOf(
    ListItem.Group("Happy Cats"),
    ListItem.Cat(
        CatUiModel(Gender.Female, CatBreed.AmericanCurl,
            "Kitty", "Kitty is warm and fuzzy.",
            "https://cdn2.thecatapi.com/images/..."
        )
    ),
    ListItem.Cat(
        CatUiModel(Gender.Male, CatBreed.ExoticShorthair,
            "Joey", "Loves to cuddle.",
            "https://cdn2.thecatapi.com/images/..."
        )
    ),
    ListItem.Group("Sad Cats"),
    ListItem.Cat(
        CatUiModel(Gender.Unknown, CatBreed.AmericanCurl,
            "Ginger", "Just not in the mood.",
            "https://cdn2.thecatapi.com/images/..."
        )
    ),
    ListItem.Cat(
        CatUiModel(Gender.Female, CatBreed.ExoticShorthair,
            "Butters", "Sleeps most of the time.",
            "https://cdn2.thecatapi.com/images/..."
        )
    )
)
```

In this case, having just one layout type will not do. Luckily, as you may have noticed in our earlier exercises, `RecyclerView.Adapter` provides us with a mechanism to handle this (remember the `viewType` parameter used in the `onCreateViewHolder(ViewGroup, Int)` function?).

To help the adapter determine which view type is needed for each item, we override its `getItemViewType(Int)` function. An example of an implementation that would do the trick for us is the following:

```
override fun getItemViewType(position: Int) =
  when (listData[position]) {
    is ListItem.Group -> VIEW_TYPE_GROUP
    is ListItem.Cat -> VIEW_TYPE_CAT
  }
```

Here, VIEW_TYPE_GROUP and VIEW_TYPE_CAT are defined as follows:

```
private const val VIEW_TYPE_GROUP = 0
private const val VIEW_TYPE_CAT = 1
```

This implementation maps the data type at a given position to a constant value, representing one of our known layout types. In our case, we know about titles and cats, thus the two types. The values we use can be any integer values as they're passed back to us, as is the case in the `onCreateViewHolder(ViewGroup, Int)` function. All we need to do is make sure not to repeat the same value more than once.

Now that we have told the adapter which view types are supported, we also need to tell it which view holder to use for each view type. This is done by implementing the `onCreateViewHolder(ViewGroup, Int)` function:

```
override fun onCreateViewHolder(parent: ViewGroup,
viewType: Int) = when (viewType) {
    VIEW_TYPE_GROUP -> {
        val view = layoutInflater.inflate(
            R.layout.item_title, parent, false)
        GroupViewHolder(view)
    }
    VIEW_TYPE_CAT -> {
        val view = layoutInflater.inflate(
            R.layout.item_cat, parent, false)
        CatViewHolder(view, imageLoader, object :
        CatViewHolder.OnClickListener {
            override fun onClick(cat: CatUiModel) =
                onClickListener.onItemClick(cat)
        })
```

```
    }
    else -> throw IllegalArgumentException(
        "Unknown view type requested: $viewType")
}
```

Unlike the earlier implementations of this function, we now take the value of `viewType` into account.

As we now know, `viewType` is expected to be one of the values we returned from `getItemViewType(Int)`.

For each of these values (`VIEW_TYPE_GROUP` and `VIEW_TYPE_CAT`), we inflate the corresponding layout and construct a suitable view holder. Note that we never expect to receive any other value, so we throw an exception if such a value is encountered.

> **Note**
> Depending on your needs, you could instead return a default view holder with a layout, showing an error or nothing at all. It may also be a good idea to log such values to allow you to investigate why you received them and decide how to handle them.

For our group title layout, a simple `TextView` may be sufficient. For a cat, the `item_cat.xml` layout can be used as is.

Now, let's move on to the view holder. We need to create a view holder for the group title. This means we will now have two different view holders. However, our adapter only supports one adapter type. The easiest solution is to define a common view holder that both `GroupViewHolder` and `CatViewHolder` will extend.

Let's call it `ListItemViewHolder`. The `ListItemViewHolder` class can be abstract, as we never intend to use it directly. To make it easy to bind data, we can also introduce a function in our abstract view holder – `abstract fun bindData(listItem: ListItemUiModel)`.

Our concrete implementations can expect to receive a specific type, and so we can add the following lines to both `GroupViewHolder` and `CatViewHolder` respectively:

```
require(listItem is ListItemUiModel.Group)
    { "Expected ListItemUiModel.Group" }
require(listItem is ListItemUiModel.Cat)
    { "Expected ListItemUiModel.Cat" }
```

Specifically, in `CatViewHolder`, thanks to some Kotlin magic, we can then use `define val cat = listItem.cat` and leave the rest of the class unchanged.

Having made those changes, we can now expect to see the `Happy Cats` and `Sad Cats` group titles, each followed by the relevant cats.

Exercise 6.04 – adding titles to RecyclerView

We now want to be able to present our secret cat agents in two groups – active agents that are available for us to deploy to the field and sleeper agents that cannot currently be deployed. We will do that by adding a title above the active agents and another above the sleeper agents:

1. Under `com.example.myrecyclerviewapp.model`, create a new Kotlin file called `ListItemUiModel`.

2. Add the following to the `ListItemUiModel.kt` file, defining our two data types – titles and cats:

   ```
   sealed class ListItemUiModel {
       data class Title(val title: String) :
           ListItemUiModel()
       data class Cat(val cat: CatUiModel) :
           ListItemUiModel()
   }
   ```

3. Create a new Kotlin file in `com.example.myrecyclerviewapp` named `ListItemViewHolder`. This will be our base view holder.

4. Populate the `ListItemViewHolder.kt` file with the following:

   ```
   abstract class ListItemViewHolder(containerView: View
   ) : RecyclerView.ViewHolder(containerView) {
       abstract fun bindData(listItem: ListItemUiModel)
   }
   ```

5. Open the `CatViewHolder.kt` file.

6. Make `CatViewHolder` extend `ListItemViewHolder`:

   ```
   class CatViewHolder(...) :
       ListItemViewHolder(containerView) {
   ```

7. Replace the `bindData(CatUiModel)` parameter with `ListItemUiModel` and make it override the `ListItemViewHolder` abstract function:

   ```
   override fun bindData(listItem: ListItemUiModel)
   ```

8. Add the following two lines to the top of the `bindData(ListItemUiModel)` function to enforce casting `ListItemUiModel` to `ListItemUiModel.Cat` and to fetch the cat data from it:

   ```
   require(listItem is ListItemUiModel.Cat)
       { "Expected ListItemUiModel.Cat" }
   val cat = listItem.cat
   ```

   Leave the rest of the file untouched.

9. Create a new layout file. Name your layout `item_title`.

10. Replace the default content of the newly created `item_title.xml` file with the following:

    ```xml
    <?xml version="1.0" encoding="utf-8"?>
    <TextView
      xmlns:android=
        "http://schemas.android.com/apk/res/android"
      xmlns:app="http://schemas.android.com/apk/res-auto"
      xmlns:tools="http://schemas.android.com/tools"
      android:id="@+id/item_title_title"
      android:layout_width="match_parent"
      android:layout_height="wrap_content"
      android:padding="8dp"
      android:textSize="16sp"
      android:textStyle="bold"
      app:layout_constraintStart_toStartOf="parent"
      app:layout_constraintTop_toTopOf="parent"
      tools:text="Sleeper Agents" />
    ```

This new layout, containing only a `TextView` with a 16 sp-sized bold font, will host our titles:

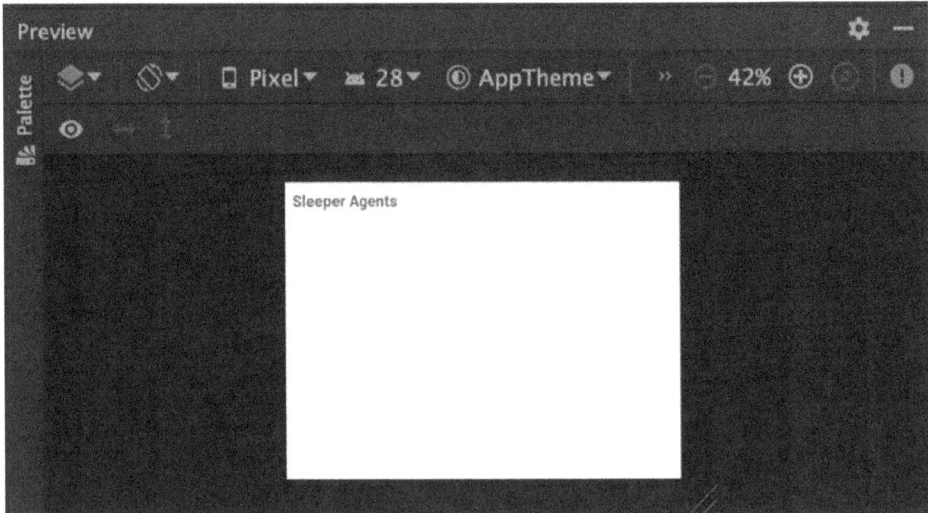

Figure 6.10 – A preview of the item_title.xml layout

11. Implement `TitleViewHolder` in a new file with the same name under `com.example.myrecyclerviewapp`:

    ```
    class TitleViewHolder(
      containerView: View
    ) : ListItemViewHolder(containerView) {
      private val titleView: TextView by lazy {
        containerView.findViewById(R.id.item_title_title)
      }
      override fun bindData(listItem: ListItemUiModel) {
        require(listItem is ListItemUiModel.Title)
          { "Expected ListItemUiModel.Title" }
        titleView.text = listItem.title
      }
    }
    ```

    This is very similar to `CatViewHolder`, but since we only set the text on `TextView`, it is also much simpler.

12. Now, to make things tidier, select `CatViewHolder`, `ListItemViewHolder`, and `TitleViewHolder`.

13. Move all the files to a new namespace; right-click on one of the files, and then select **Refactor | Move** (or press *F6*).
14. Append /viewholder to the prefilled **To directory** field. Leave **Search references** and **Update package directive (Kotlin files)** checked and **Open moved files in editor** unchecked. Click **OK**.
15. Open the CatsAdapter.kt file.
16. Now, rename CatsAdapter ListItemsAdapter. Right-click on the CatsAdapter class name in the code window, and then select **Refactor | Rename** (or *Shift + F6*).

> **Note**
> It is important to maintain the naming of variables, functions, and classes to reflect their actual usage to avoid future confusion.

17. When CatsAdapter is highlighted, type ListItemsAdapter and press *Enter*.
18. Change the adapter generic type to ListItemViewHolder:

```
class ListItemsAdapter(...) :
    RecyclerView.Adapter<ListItemViewHolder>() {
```

19. Update listData and setData(List<CatUiModel>) to handle ListItemUiModel instead:

```
private val listData = mutableListOf<ListItemUiModel>()
fun setData(newListData: List<ListItemUiModel>) {
    listData.clear()
    listData.addAll(newListData)
    notifyDataSetChanged()
}
```

20. Update onBindViewHolder(CatViewHolder) to comply with the adapter contract change:

```
override fun onBindViewHolder(holder:
    ListItemViewHolder, position: Int) {
    holder.bindData(listData[position])
}
```

21. At the top of the file, after the imports and before the class definition, add the view type constants:

```
private const val VIEW_TYPE_TITLE = 0
private const val VIEW_TYPE_CAT = 1
```

22. Implement `getItemViewType(Int)`, like so:

    ```
    override fun getItemViewType(position: Int) =
        when (listData[position]) {
            is ListItemUiModel.Title -> VIEW_TYPE_TITLE
            is ListItemUiModel.Cat -> VIEW_TYPE_CAT
        }
    ```

23. Lastly, change your `onCreateViewHolder(ViewGroup, Int)` implementation, as follows:

    ```
    override fun onCreateViewHolder(parent: ViewGroup,
      viewType: Int) = when (viewType) {
      VIEW_TYPE_TITLE -> {
        val view = layoutInflater.inflate(
          R.layout.item_title, parent, false)
        TitleViewHolder(view)
      }
      VIEW_TYPE_CAT -> {
        val view = layoutInflater.inflate(
          R.layout.item_cat, parent, false)
        CatViewHolder(
          view,
          imageLoader,
          object : CatViewHolder.OnClickListener {
            override fun onClick(cat: CatUiModel) =
              onClickListener.onItemClick(catData)
          })
      }
      else -> throw IllegalArgumentException("Unknown view type requested: $viewType")
    }
    ```

24. Update MainActivity to populate the adapter with appropriate data, replacing the previous catsAdapter.setData(List<CatUiModel>) call (note that the following code has been truncated for space; refer to the link after the code block to access the full code that you need to add):

MainActivity.kt

```
listItemsAdapter.setData(
  listOf(
      ListItemUiModel.Title("Sleeper Agents"),
      ListItemUiModel.Cat(
         CatUiModel(Gender.Male,
            CatBreed.ExoticShorthair, "Garvey",
            "Garvey is as a lazy, fat, and cynical orange cat.",
            "https://cdn2.thecatapi.com/images/FZpeiLi4n.jpg"
         )
      ),
      ListItemUiModel.Cat(
         CatUiModel(Gender.Unknown,
            CatBreed.AmericanCurl, "Curious George",
            "Award winning investigator",
            "https://cdn2.thecatapi.com/images/vJB8rwfdX.jpg"
         )
      ),
      ListItemUiModel.Title("Active Agents"),
```

The complete code for this step can be found at http://packt.live/3icCrSt.

25. Since catsAdapter is no longer holding CatsAdapter but ListItemsAdapter, rename it accordingly. Name it listItemsAdapter.

26. Run the app. You should see something similar to the following:

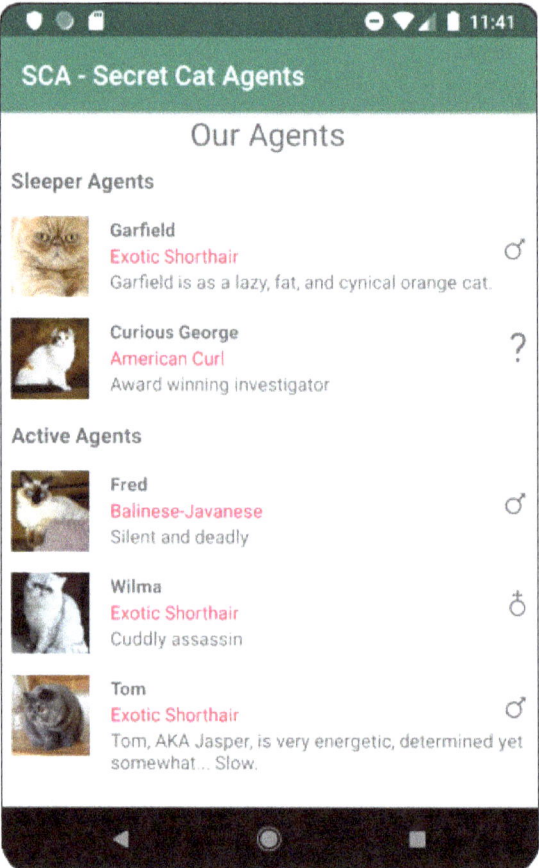

Figure 6.11 – RecyclerView with the Sleeper Agents and Active Agents header views

As you can see, we now have titles above our two agent groups. Unlike the **Our Agents** title, these titles will scroll with our content. Next, we will learn how to swipe an item to remove it from `RecyclerView`.

Swiping to remove Items

In the previous sections, we learned how to present different view types. However, up until now, we have worked with a fixed list of items. What if you want to be able to remove items from the list? There are a few common mechanisms to achieve that – fixed **delete** buttons on each item, swiping to delete, and long-clicking to select and then a tapping a **delete** button, to name a few. In this section, we will focus on the *swiping to delete* approach.

Let's start by adding the deletion functionality to our adapter. To tell the adapter to remove an item, we need to indicate which item we want to remove. The simplest way to achieve this is by providing the position of the item. In our implementation, this will directly correlate to the position of the item in our `listData` list. So, our `removeItem(Int)` function should look like this:

```
fun removeItem(position: Int) {
    listData.removeAt(position)
    notifyItemRemoved(position)
}
```

> **Note**
> Just like when setting data, we need to notify `RecyclerView` that the dataset has changed – in this case, an item was removed.

Next, we need to define swipe gesture detection. This is done by utilizing `ItemTouchHelper`, which handles certain touch events, namely dragging and swiping, by reporting them to us via a callback. We handle these callbacks by implementing `ItemTouchHelper.Callback`. Also, `RecyclerView` provides `ItemTouchHelper.SimpleCallback`, which takes away the writing of a lot of boilerplate code.

We want to respond to swipe gestures but ignore move gestures. More specifically, we want to respond to swipes to the right. Moving is used to reorder items, which is beyond the scope of this chapter. So, our implementation of `SwipToDeleteCallback` will look as follows:

```
inner class SwipeToDeleteCallback :
  ItemTouchHelper.SimpleCallback(0, ItemTouchHelper.RIGHT) {
  override fun onMove(
    recyclerView: RecyclerView,
    viewHolder: RecyclerView.ViewHolder,
    target: RecyclerView.ViewHolder
  ): Boolean = false
  override fun getMovementFlags(
    recyclerView: RecyclerView,
    viewHolder: RecyclerView.ViewHolder
  ) = if (viewHolder is CatViewHolder) {
    makeMovementFlags(
      ItemTouchHelper.ACTION_STATE_IDLE,
      ItemTouchHelper.RIGHT
    ) or makeMovementFlags(
```

```
            ItemTouchHelper.ACTION_STATE_SWIPE,
            ItemTouchHelper.RIGHT
        )
    } else { 0 }
    override fun onSwiped(viewHolder: RecyclerView.ViewHolder,
        direction: Int) {
        val position = viewHolder.adapterPosition
        removeItem(position)
    }
}
```

Because our implementation is tightly coupled to our adapter and its view types, we can comfortably define it as an inner class. The benefit we gain is the ability to directly call methods on the adapter.

As you can see, we return `false` from the `onMove(RecyclerView, ViewHolder, ViewHolder)` function. This means we ignore move events.

Next, we need to tell `ItemTouchHelper` which items can be swiped. We achieve this by overriding `getMovementFlags(RecyclerView, ViewHolder)`. This function is called when a user is about to start a drag or swipe gesture. `ItemTouchHelper` expects us to return the valid gestures for the provided view holder.

We check the `ViewHolder` class, and if it is `CatViewHolder`, we want to allow swiping; otherwise, we do not. We use `makeMovementFlags(Int, Int)`, which is a helper function used to construct flags in a way that `ItemTouchHelper` can decipher them.

Note that we define rules for `ACTION_STATE_IDLE`, which is the starting state of a gesture, thus allowing a gesture to start from the left or the right. We then combine it (using `or`) with the `ACTION_STATE_SWIPE` flags, allowing the ongoing gesture to swipe left or right. Returning 0 means neither swiping nor moving will occur for the provided view holder.

Once a swipe action is completed, `onSwiped(ViewHolder, Int)` is called. We then obtain the position from the passed-in view holder by calling `adapterPosition`. Now, `adapterPosition` is important because it is the only reliable way to obtain the real position of the item presented by the view holder.

With the correct position, we can remove the item by calling `removeItem(Int)` in the adapter.

To expose our newly created `SwipeToDeleteCallback` implementation, we define a read-only variable within our adapter named `swipeToDeleteCallback`, and set it to a new instance of `SwipeToDeleteCallback`.

Finally, to plug our `callback` mechanism to `RecyclerView`, we need to construct a new `ItemTouchHelper` and attach it to our `RecyclerView`. We should do this when setting up our

RecyclerView, which we do in the `onCreate(Bundle?)` function of our main activity. This is how the creation and attaching looks:

```
val itemTouchHelper =
  ItemTouchHelper(listItemsAdapter.swipeToDeleteCallback)
itemTouchHelper.attachToRecyclerView(recyclerView)
```

We can now swipe items to remove them from the list. Note how our titles cannot be swiped, just as we intended.

You may have noticed a small glitch – the last item is cut off as it animates upward. This is happening because `RecyclerView` shrinks to accommodate the new (smaller) number of items before the animation starts. A quick fix to this would be to fix the height of our `RecyclerView` by confining its bottom to the bottom of its parent.

Exercise 6.05 – adding swipe to delete functionality

We previously added `RecyclerView` to our app and then added items of different types to it. We will now allow users to delete some items (we want to let the users remove secret cat agents but not titles) by swiping them left or right:

1. To add item removal functionality to our adapter, add the following function to `ListItemsAdapter` right after the `setData(List<ListItemUiModel>)` function:

    ```
    fun removeItem(position: Int) {
        listData.removeAt(position)
        notifyItemRemoved(position)
    }
    ```

2. Next, right before the closing curly bracket of your `ListItemsAdapter` class, add the following `callback` implementation to handle the user swiping a cat agent left or right:

    ```
    inner class SwipeToDeleteCallback :
      ItemTouchHelper.SimpleCallback(0,
      ItemTouchHelper.LEFT or ItemTouchHelper.RIGHT) {
      override fun onMove(
        recyclerView: RecyclerView,
        viewHolder: RecyclerView.ViewHolder,
        target: RecyclerView.ViewHolder
      ): Boolean = false
        override fun getMovementFlags(
          recyclerView: RecyclerView,
    ```

```
              viewHolder: RecyclerView.ViewHolder
          ) = if (viewHolder is CatViewHolder) {
            makeMovementFlags(
              ItemTouchHelper.ACTION_STATE_IDLE,
              ItemTouchHelper.LEFT or ItemTouchHelper.RIGHT
            ) or makeMovementFlags(
              ItemTouchHelper.ACTION_STATE_SWIPE,
              ItemTouchHelper.LEFT or ItemTouchHelper.RIGHT
            )
          } else { 0 }
          override fun onSwiped(viewHolder:
            RecyclerView.ViewHolder, direction: Int) {
            val position = viewHolder.adapterPosition
            removeItem(position)
          }
      }
```

We have implemented an `ItemTouchHelper.SimpleCallback` instance, passing in the directions we were interested in – `LEFT` and `RIGHT`. Joining the values is achieved by using the `or` Boolean operator.

We have overridden the `getMovementFlags` function to make sure we have only handled swiping on a cat agent view and not on a title. Creating flags for both `ItemTouchHelper.ACTION_STATE_SWIPE` and `ItemTouchHelper.ACTION_STATE_IDLE` allows us to intercept both swipe and release events respectively.

Once a swipe is completed (the user has lifted their finger from the screen), `onSwiped` will be called, and in response, we remove the item at the position provided by the dragged view holder.

3. At the top of your adapter, expose an instance of the `SwipeToDeleteCallback` class you just created:

```
class ListItemsAdapter(...) :
  RecyclerView.Adapter<ListItemViewHolder>() {
  val swipeToDeleteCallback = SwipeToDeleteCallback()
```

4. Lastly, tie it all together by implementing `ItemViewHelper` and attaching it to our `RecyclerView`. Add the following code to the `onCreate(Bundle?)` function of your `MainActivity` file right after assigning the layout manager to your adapter:

```
recyclerView.layoutManager = ...
val itemTouchHelper = ItemTouchHelper(listItemsAdapter
```

```
        .swipeToDeleteCallback)
    itemTouchHelper.attachToRecyclerView(recyclerView)
```

5. To address the small visual glitch you would get when items are removed, scale `RecyclerView` to fit the screen by updating the code in `activity_main.xml`, as follows. The changes are in the `RecyclerView` tag, right before the `app:layout_constraintTop_toBottomOf` attribute:

   ```
   android:layout_height="0dp"
   app:layout_constraintBottom_toBottomOf="parent"
   app:layout_constraintTop_toBottomOf="@+id/main_label" />
   ```

   Note that there are two changes – we added a constraint at the bottom of the view to the bottom of the parent, and we set the layout height to 0dp. The latter change tells our app to calculate the height of `RecyclerView` based on its constraints:

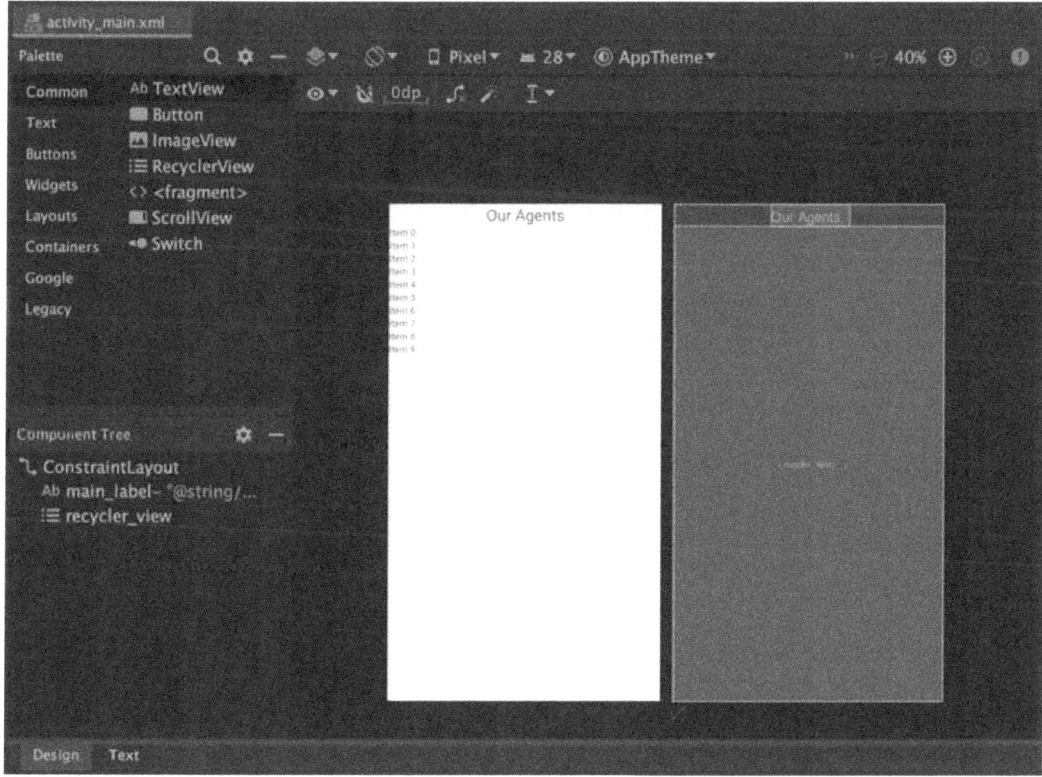

Figure 6.12 – RecyclerView taking the full height of the layout

6. Run your app. You should now be able to swipe secret cat agents left or right to remove them from the list. Note that `RecyclerView` handles the collapsing animation for us:

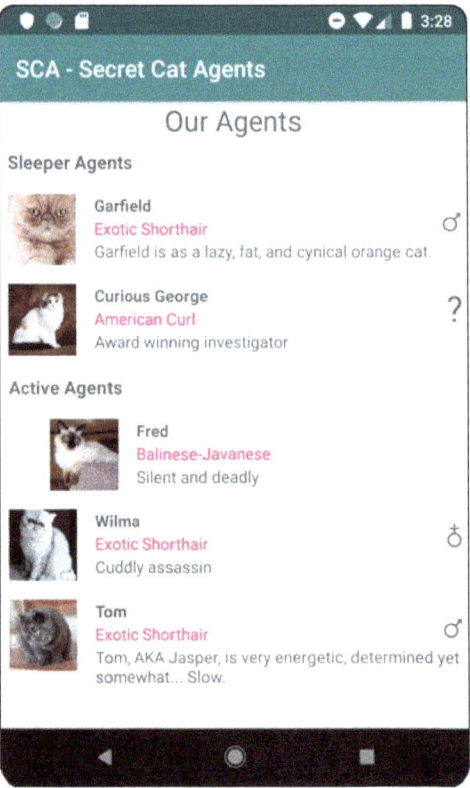

Figure 6.13 – A cat being swiped to the right

Note how even though titles are item views, they cannot be swiped. You have implemented a callback for swiping gestures that distinguishes between different item types and responds to a swipe by deleting the swiped item. Now, you know how to remove items interactively. Next, you will learn how to add new items as well.

Adding items interactively

We have just learned how to remove items interactively. What about adding new items? Let's look into it.

Similar to the way we implemented the removal of items, we start by adding a function to our adapter:

```
fun addItem(position: Int, item: ListItemUiModel) {
    listData.add(position, item)
```

```
        notifyItemInserted(position)
}
```

Note that the implementation is very similar to the `removeItem(Int)` function we implemented earlier. This time, we also receive an item to add and a position to add it to. We then add it to our `listData` list and notify `RecyclerView` that we added an item in the requested position.

To trigger a call to `addItem(Int, ListItemUiModel)`, we can add a button to our main activity layout. This button can be as follows:

```
<Button
    android:id="@+id/main_add_item_button"
    android:layout_width="match_parent"
    android:layout_height="wrap_content"
    android:text="Add A Cat"
    app:layout_constraintBottom_toBottomOf="parent" />
```

The app will now look like this:

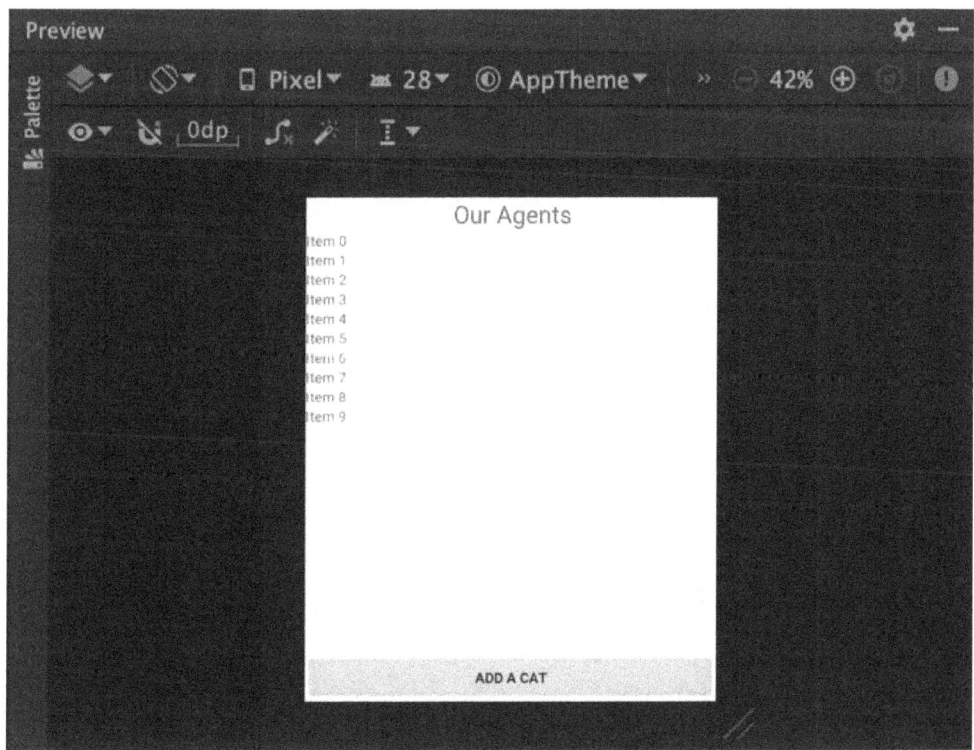

Figure 6.14 – The main layout with a button to add a cat

Don't forget to update your `RecyclerView` so that its bottom will be constrained to the top of this button. Otherwise, the button and `RecyclerView` will overlap.

In a production app, you could add a rationale about what a new item would be. For example, you could have a form for a user to fill in different details. For the sake of simplicity, in our example, we will always add the same dummy item – an anonymous female secret cat agent.

To add the item, we set `OnClickListener` on our button:

```
addItemButton.setOnClickListener {
  listItemsAdapter.addItem(1, ListItemUiModel.Cat(
    CatUiModel(Gender.Female, CatBreed.BalineseJavanese,
      "Anonymous", "Unknown",
      "https://cdn2.thecatapi.com/images/zJkeHza2K.jpg"
    ))
  )
}
```

And that is it. We add the item at position 1 so that it is added right below our first title, which is the item at position 0. In a production app, you could have logic to determine the correct place to insert an item. It could be below the relevant title or always be added at the top, bottom, or in the correct place to preserve some existing order.

We can now run the app. We will now have a new **Add a Cat** button. Every time we click the button, an anonymous secret cat agent will be added to `RecyclerView`. The newly added cats can be swiped away to be removed, just like the hardcoded cats before them.

Exercise 6.06 – implementing an Add A Cat button

Having implemented a mechanism to remove items, it is time we implemented a mechanism to add items:

1. Add a function to `ListItemsAdapter` to support adding items. Add it below the `removeItem(Int)` function:

    ```
    fun addItem(position: Int, item: ListItemUiModel) {
        listData.add(position, item)
        notifyItemInserted(position)
    }
    ```

2. Add a button to `activity_main.xml`, right after the `RecyclerView` tag:

    ```
    <Button
        android:id="@+id/main_add_item_button"
    ```

```
            android:layout_width="match_parent"
            android:layout_height="wrap_content"
            android:text="Add A Cat"
            app:layout_constraintBottom_toBottomOf="parent" />
```

3. Note that `android:text="Add A Cat"` is highlighted. If you hover your mouse over it, you will see that this is because of the hardcoded string. Click on the **Add** word to place the editor cursor over it.

4. Press *Option + Enter* (iOS) or *Alt + Enter* (Windows) to show the context menu, and then press *Enter* again to show the **Extract Resource** dialog.

5. Name the resource `add_button_label`. Press **OK**.

6. To change the bottom constraint on `RecyclerView` so that the button and `RecyclerView` do not overlap, within your `RecyclerView` tag, locate the following:

    ```
    app:layout_constraintBottom_toBottomOf="parent"
    ```

    Replace it with the following line of code:

    ```
    app:layout_constraintBottom_toTopOf="@+id/main_add_item_button"
    ```

7. Add a lazy field, holding a reference to the button at the top of the class, right after the definition of `recyclerView`:

    ```
    private val addItemButton: View by lazy {
        findViewById(R.id.main_add_item_button) }
    ```

    Note that `addItemButton` is defined as a view. This is because, in our code, we don't need to know the type of view to add a click listener to it. Choosing the more abstract type allows us to later change the type of view in the layout without having to modify this code.

8. Lastly, update `MainActivity` to handle the click. Find the line that says the following:

    ```
    itemTouchHelper.attachToRecyclerView(recyclerView)
    ```

    Right after it, add the following:

    ```
    addItemButton.setOnClickListener {
        listItemsAdapter.addItem(1,
            ListItemUiModel.Cat(CatUiModel(
                Gender.Female, CatBreed.BalineseJavanese,
                "Anonymous", "Unknown",
                "https://cdn2.thecatapi.com/images/zJkeHza2K.jpg"
    ```

```
            ) )
        )
    }
```

This will add a new item to `RecyclerView` every time the button is clicked.

9. Run the app. You should see a new button at the bottom of your app:

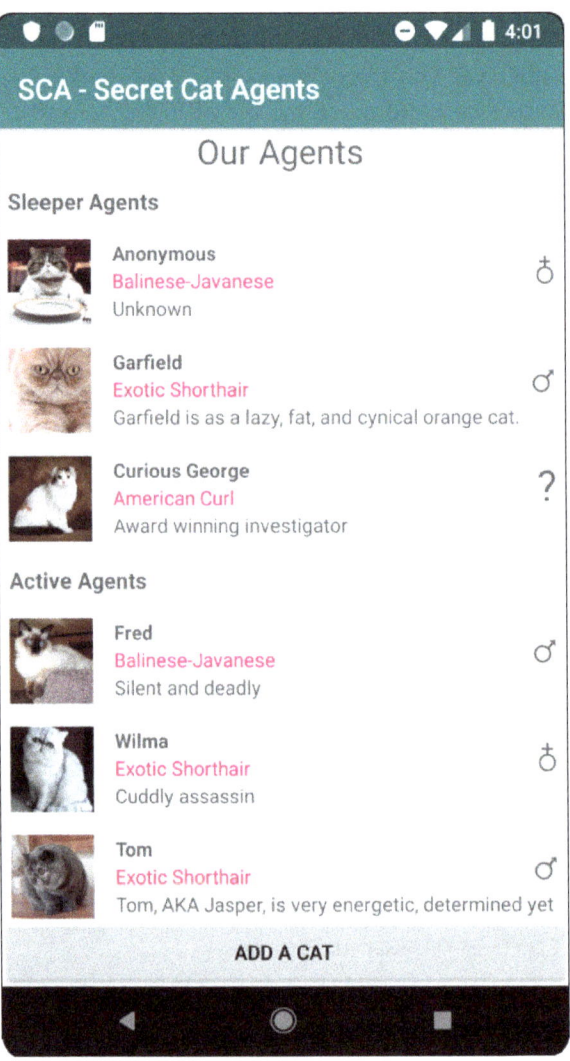

Figure 6.15 – An anonymous cat is added with the click of a button

10. Try clicking it a few times. Every time you click it, a new, anonymous secret cat agent is added to your `RecyclerView`. You can swipe away the newly added cats, just like you could with the hardcoded ones.

In this exercise, you added new items to `RecyclerView` in response to user interaction. You now know how to change the contents of `RecyclerView` at runtime. It is useful to know how to update lists at runtime because, quite often, the data you are presenting to your users changes while the app is running, and you want to present your users with a fresh, up-to-date state.

Activity 6.01 – managing a list of Items

Imagine you want to develop a recipe management app. Your app would support sweet and savory recipes. Users of your app could add new sweet or savory recipes, scroll through the list of added recipes – grouped by flavor (sweet or savory) – click a recipe to get information about it, and finally, delete recipes by swiping them aside.

The aim of this activity is to create an app with `RecyclerView` that lists the title of recipes, grouped by flavor. `RecyclerView` will support user interaction. Each recipe will have a title, a description, and a flavor. Interactions will include clicks and swipes.

A click will present a user with a dialog showing the description of the recipe. A swipe will remove the swiped recipe from the app. Finally, with two `EditText` fields (see *Chapter 3, Developing the UI with Fragments*) and two buttons, a user can add a new sweet or savory recipe respectively, with the title and description set to the values set in the `EditText` fields.

The steps to complete this are as follows:

1. Create a new empty activity app.
2. Add `RecyclerView` support to the app's `build.gradle` file.
3. Add `RecyclerView`, two `EditText` fields, and two buttons to the main layout. It should look something like this:

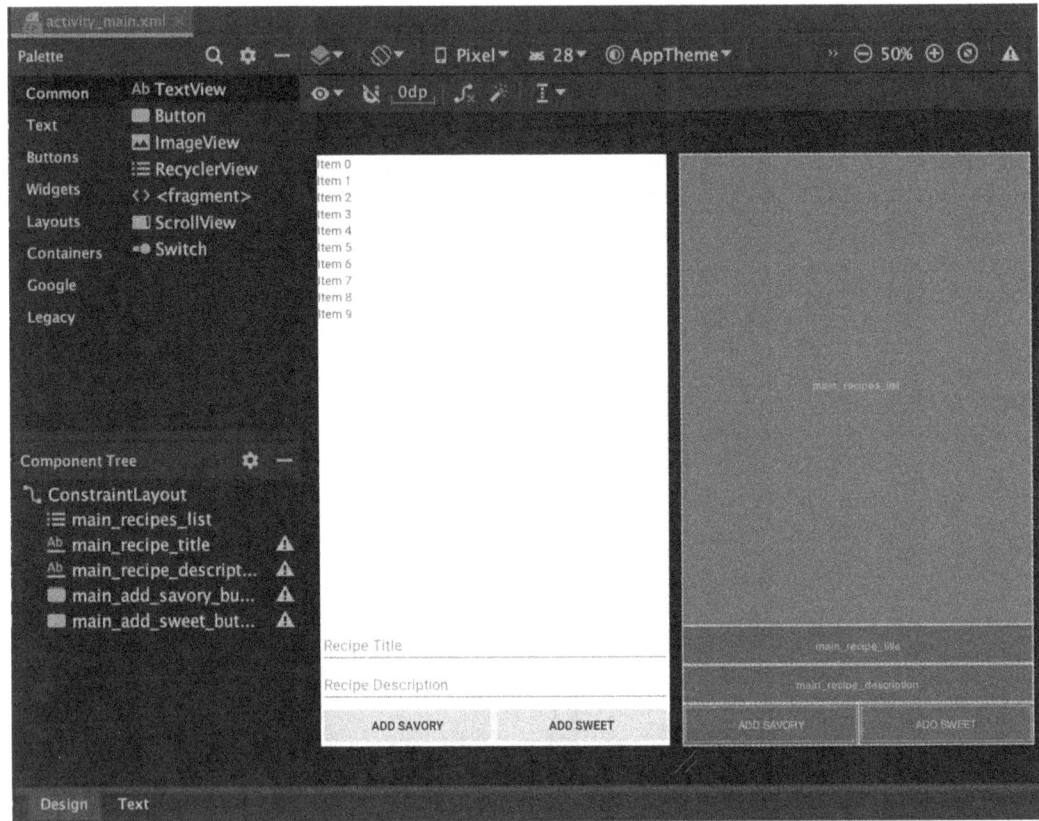

Figure 6.16 – The layout with RecyclerView, two EditText fields, and two buttons

4. Add models for the flavor titles and recipes, and an enum for flavor.
5. Add a layout for the flavor titles.
6. Add a layout for the recipe titles.
7. Add view holders for the flavor titles and recipe titles, as well as an adapter.
8. Add click listeners to show a dialog with recipe descriptions.
9. Update `MainActivity` to construct the new adapter, and hook up the buttons to add new savory and sweet recipes. Make sure the form is cleared after a recipe is added.
10. Add a swipe helper to remove items.

The final output will be as follows:

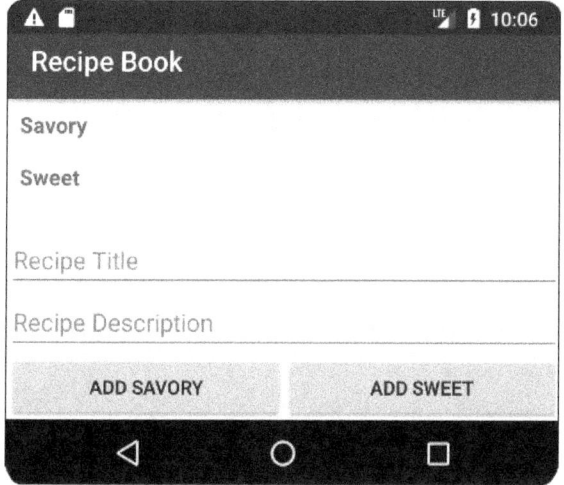

Figure 6.17 – The Recipe Book app

> **Note**
> The solution to this activity can be found at https://packt.link/By7eE.

Summary

In this chapter, we learned how to add `RecyclerView` support to our project. We also learned how to add `RecyclerView` to our layout and how to populate it with items. We went through adding different item types, which is particularly useful for titles. We covered interaction with `RecyclerView`, responding to clicks on individual items and responding to swipe gestures.

Lastly, we learned how to dynamically add and remove items to and from `RecyclerView`. The world of `RecyclerView` is very rich, and we have only scratched the surface. Going further would be beyond the scope of this book. However, it is strongly recommended that you investigate it on your own so that you can have carousels, designed dividers, and fancier swipe effects in your apps.

You can start your exploration here: https://packt.link/ClmMn.

In the next chapter, we will look into requesting special permissions on behalf of our app to enable it to perform certain tasks, such as accessing a user's contacts list or microphone. We will also look into using Google's Maps API and accessing a user's physical location.

7
Android Permissions and Google Maps

This chapter will teach you how to request and obtain app permissions in Android. You will gain a solid understanding of how to include local and global interactive maps in your app by using the Google Maps API and how to request permissions to use device features that provide richer functionality.

By the end of the chapter, you will be able to create permission requests for your app and handle missing permissions.

In the previous chapter, we learned how to present data in lists using `RecyclerView`. Then, we used that knowledge to present the user with a list of secret cat agents. In this chapter, we will learn how to find the user's location on the map and how to deploy cat agents to the field by selecting locations on the map.

First, we will explore the Android permissions system. Many Android features are not immediately available to us. These features are gated behind a permission system to protect the user. For us to access those features, we must ask the user to allow us to do so. Some such features include but are not limited to obtaining the user's location, accessing the user's contacts, accessing their camera, and establishing a Bluetooth connection. Different Android versions enforce different permission rules.

> **Note**
> When Android 6 (Marshmallow) was introduced in 2015, for example, several permissions you could silently obtain on installation were deemed insecure and became runtime permissions, requiring explicit user consent.

We will then look at the Google Maps API. This API allows us to present the user with a map of any desired location in the world. We will add data to that map and let the user interact with the map. The API also lets you show points of interest and render a street view of supported locations, though we will not explore these features in this book.

We will cover the following topics in this chapter:

- Requesting permission from the user
- Showing a map of the user's location
- Map clicks and custom markers

Technical requirements

The complete code for all the exercises and the activity in this chapter is available on GitHub at `https://packt.link/6ShZd`

Requesting permission from the user

Our app might want to implement certain features that Google deems dangerous. This usually means access to those features could risk the user's privacy. For example, some permissions may allow you to read users' messages or determine their current location.

Depending on the required permission and the target Android API level we are developing, we may need to request that permission from the user. If the device is running on Android 6 (Marshmallow, API level 23), and the target API of our app is 23 or higher (it almost certainly will be, as most devices by now will run newer versions of Android), there will be no alert for the user about any permissions requested by the app at install time. Instead, our app must ask the user to grant those permissions at runtime.

When we request permission, the user sees a dialog like the one shown in *Figure 7.1*.

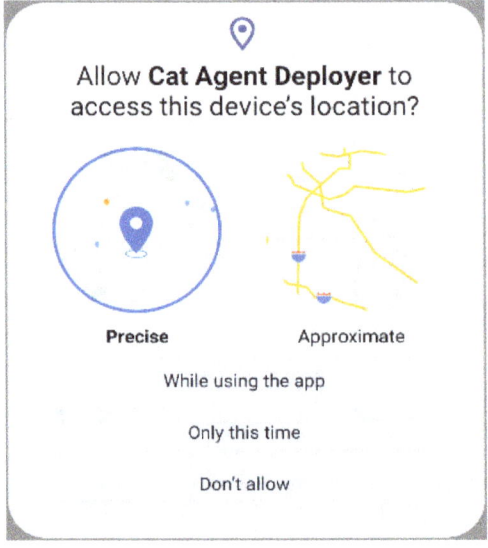

Figure 7.1 – Permission dialog for device location access

> **Note**
> For a full list of permissions and their protection level, see here: https://developer.android.com/reference/android/Manifest.permission.

We must include permissions in our manifest file when we intend to use them. A manifest with the SEND_SMS permission would look something like the following snippet:

```
<manifest xmlns:android=
    "http://schemas.android.com/apk/res/android"
  package="com.example.snazzyapp">
  <uses-permission android:name="android.permission.SEND_SMS" />
  <application …> ... </application>
</manifest>
```

Safe permissions (or normal permissions, as Google calls them) would be automatically granted to the user. However, dangerous ones would only be granted if explicitly approved by the user. If we fail to request permission from the user and try to execute an action that requires that permission, the result would be the action not running at best and our app crashing at worst.

We should first check whether the user has already granted us that permission before asking the user for permission. If the user has not yet granted us permission, we may need to check whether a rationale dialog should be shown prior to the permission request. This depends on how obvious the justification for the request would be to the user.

For example, if a camera app requests permission to access the camera, we can safely assume the reason would be clear to the user. However, some cases may not be as clear to the user, especially if the user is not tech-savvy. In those cases, we may have to justify the request to the user.

Google provides us with a function called shouldShowRequestPermissionRationale(Activity, String) for this purpose. Under the hood, this function checks whether the user has previously denied the permission but also whether the user has denied us permission before.

The idea is to allow us to justify our request to the user for permission before requesting it, thus increasing the likelihood of them approving the request. Once we determine whether the app should present the user with our rationale or whether no rationale was required, we can proceed to request the permission.

Let's see how we can request permission. First, we must include the Jetpack Activity and Fragment dependencies in our app `gradle` file:

```
implementation "androidx.activity:activity-ktx:1.6.1"
implementation "androidx.fragment:fragment-ktx:1.5.5"
```

This will provide us with `ActivityResultLauncher`, which we will use to launch the permission request dialog and handle the user's response.

The following is an example of an `Activity` class requesting the `Location` permission:

```
class MainActivity : AppCompatActivity() {
  private lateinit var requestPermissionLauncher:
    ActivityResultLauncher<String>
  override fun onCreate() {
    ...
    requestPermissionLauncher = registerForActivityResult(
      RequestPermission()) { isGranted ->
        if (isGranted) {... } else { ... }     }
```

When `Activity` is created, we register the launcher to handle permission request responses. We keep a reference to the result for later use. When `Activity` is resumed, we check the status of the permission and continue accordingly:

```
override fun onResume() {
  ...
  when {
    hasLocationPermission() -> getLastLocation()
    shouldShowRequestPermissionRationale(this,
    ACCESS_FINE_LOCATION) -> {
      showPermissionRationale {
        requestPermissionLauncher
          .launch(ACCESS_FINE_LOCATION)
      }
    }
    else -> requestPermissionLauncher.
            launch(ACCESS_FINE_LOCATION)
  }
}
```

We start by checking the location permission (`ACCESS_FINE_LOCATION`) by calling `getHasLocationPermissions()`:

```
private fun hasLocationPermission() =
  checkSelfPermission(this, Manifest.permission.ACCESS_FINE_
  LOCATION) == PERMISSION_GRANTED
```

This function checks whether the user has granted us the requested permissions by calling `checkSelfPermission(Context, String)` with the requested permission.

If the user hasn't granted us permission, we call `shouldShowRequestPermission Rationale(Activity, String)`, which we mentioned earlier, to check whether a rationale dialog should be presented to the user.

If showing our rationale is needed, we call `showPermissionRationale(() -> Unit)`, passing in a lambda that will use `requestPermissionLauncher` to launch the request dialog after the user dismisses our rationale dialog using the positive button. If no rationale is needed, we launch the dialog with `requestPermissionLauncher` directly. The following code is for presenting the rationale dialog:

```
private fun showPermissionRationale(
    positiveAction: () -> Unit) {
    AlertDialog.Builder(this)
        .setTitle("Location permission")
        .setMessage("We need your permission to find your current
            position")
        .setPositiveButton(android.R.string.ok) { _, _ ->
            positiveAction()
        }
        .setNegativeButton(android.R.string.cancel) { dialog, _ ->
            dialog.dismiss() }
        .create().show()
}
```

Our `showPermissionRationale` function presents the user with a dialog with a brief explanation of why we need their permission. The **OK** button will execute the positive action provided, and the **Cancel** one will dismiss the dialog:

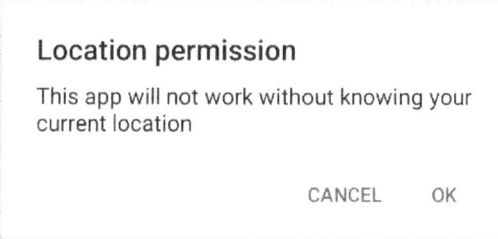

Figure 7.2 – Rationale dialog

Lastly, we request permission by calling `requestPermissionLauncher.launch(ACCESS_FINE_LOCATION)` using the request permission launcher we declared earlier.

If we've requested the location permission from the user, the response will be processed by the request permission launcher depending on the value returned via `isGranted`, as shown in the following code:

```
registerForActivityResult(RequestPermission()) { isGranted ->
  if (isGranted) { getLastLocation() }
  else {
    showPermissionRationale { requestPermissionLauncher
      .launch(ACCESS_FINE_LOCATION) }
  }
}
```

This code is the expansion of the code we observed earlier, added to the `onCreate` function of `Activity`. This chapter will take us through the development of an app that shows us our current location on a map and allows us to place a marker where we want to deploy our secret cat agent.

Let's start with our first exercise.

Exercise 7.01 – requesting the location permission

In this exercise, we will request that the user provide the location permission. We will first create a **Google Maps Activity** project. Then, we will define the permission required in the manifest file. To get started, let's implement the code required to request permission from the user to access their location:

1. Start by creating a new Google Maps Activity project (**File** | **New** | **New Project** | **Google Maps Activity**). We're not using Google Maps in this exercise. However, the Google Maps Activity is still a good choice in this case. It will save you a lot of boilerplate coding in the next exercise (*Exercise 7.02*). Don't worry; it will have no impact on your current exercise. Click **Next**, as shown in the following screenshot:

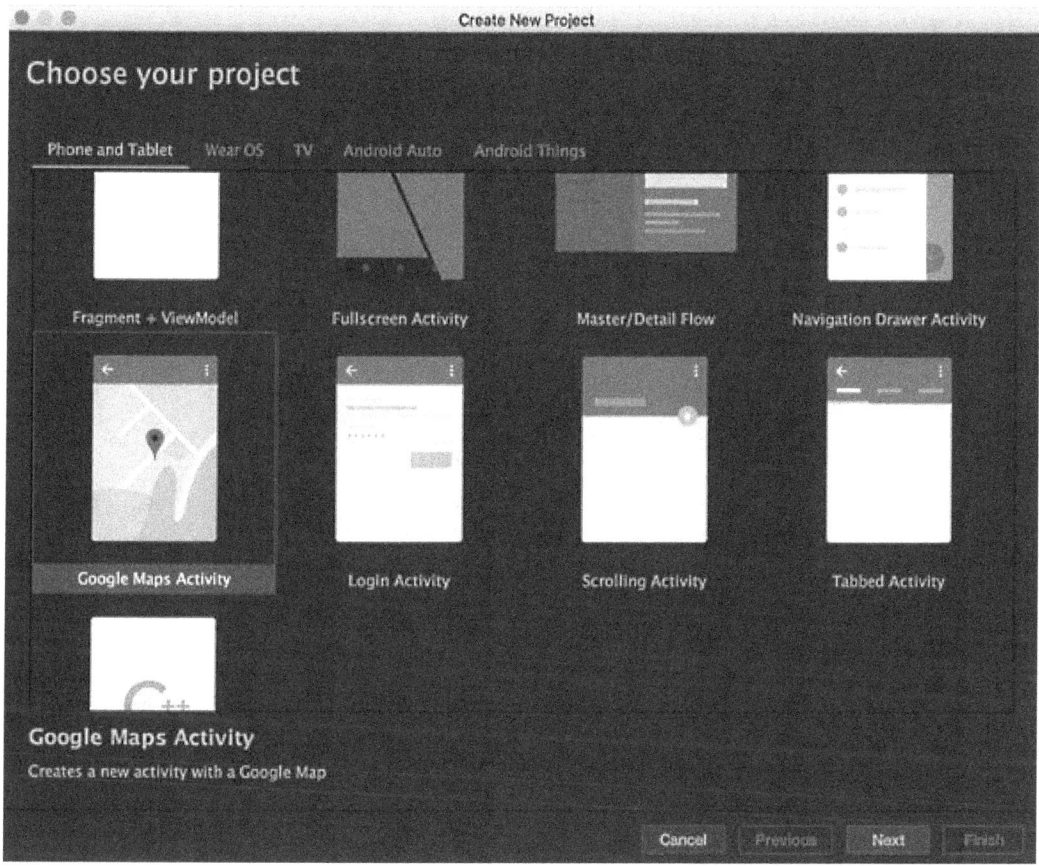

Figure 7.3 – Choose your project

2. Name your application `Cat Agent Deployer`.
3. Make sure your package name is `com.example.catagentdeployer`.
4. Set the save location to where you want to save your project.
5. Leave everything else at its default values and click **Finish**.

6. Make sure you are on the **Android** view in your **Project** pane:

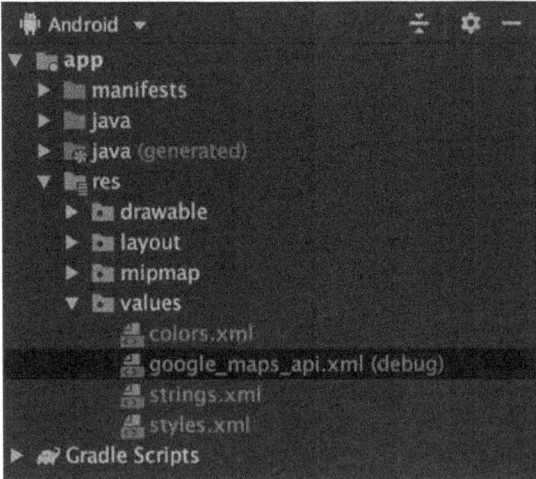

Figure 7.4 – The Android view

7. Open your `AndroidManifest.xml` file. Make sure the location permission was already added to your app:

```
<manifest ...>
    <uses-permission android:name=
        "android.permission.ACCESS_FINE_LOCATION" />
    <application ...> ... </application>
</manifest>
```

> **Note**
> `ACCESS_FINE_LOCATION` is the permission you will need to obtain the user's location based on GPS in addition to the less accurate Wi-Fi and mobile data-based location information you could obtain by using the `ACCESS_COARSE_LOCATION` permission.

8. Open your `MapsActivity.kt` file. At the bottom of the `MapsActivity` class block, add an empty `getLastLocation()` function:

```
class MapsActivity : ... {
    ...
    private fun getLastLocation() {
        Log.d("MapsActivity", "getLastLocation() called.")
```

```
        }
    }
```

This will be the function you will call when you have ensured the user has granted you the location permission.

9. Next, add the request permission launcher to the top of the `MapsActivity` class:

```
class MapsActivity : ... {
    private lateinit var requestPermissionLauncher:
        ActivityResultLauncher<String>
```

This is the variable through which we will launch the permission request and track user responses.

Now navigate to the bottom of the `onCreate()` function and register for activity results, storing the result in `requestPermissionLauncher`, which you declared in the previous step:

```
override fun onCreate(savedInstanceState: Bundle?) {
    ...
    requestPermissionLauncher =
      registerForActivityResult(RequestPermission()) {
      isGranted ->
        if (isGranted) { getLastLocation() }
        else {
          showPermissionRationale {
            requestPermissionLauncher
                .launch(ACCESS_FINE_LOCATION)
          }
        }
    }
}
```

10. To present users with the rationale for the permission request, implement the `showPermissionRationale(() -> Unit)` function right before the `getLastLocation()` function:

```
private fun showPermissionRationale(positiveAction: ()
-> Unit) {
  AlertDialog.Builder(this)
    .setTitle("Location permission")
    .setMessage("This app will not work without knowing
        your current location")
```

```
            .setPositiveButton(android.R.string.ok) { _, _ ->
                positiveAction() }
            .setNegativeButton(android.R.string.cancel) { dialog,
                _ -> dialog.dismiss() }
            .create().show()
    }
```

This function will present the user with a simple alert dialog explaining that the app will not work without knowing their current location, as shown in *Figure 7.1*. Tapping **OK** will execute the provided positiveAction lambda. Tapping **CANCEL** will dismiss the dialog.

11. To determine whether or not your app already has the location permission, introduce the following hasLocationPermission() function right before the requestPermissionWithRationaleIfNeeded() function:

```
    private fun hasLocationPermission() =
        ContextCompat.checkSelfPermission(
            this, Manifest.permission.ACCESS_FINE_LOCATION
        ) == PackageManager.PERMISSION_GRANTED
```

12. Finally, update the onMapReady() function of your MapsActivity class to determine the permission status and proceed accordingly:

```
    override fun onMapReady(googleMap: GoogleMap) {
      mMap = googleMap
      when {
        hasLocationPermission() -> getLastLocation()
        shouldShowRequestPermissionRationale(this,
          ACCESS_FINE_LOCATION) -> {
          showPermissionRationale {
            requestPermissionLauncher
              .launch(ACCESS_FINE_LOCATION)
          }
        }
        else -> requestPermissionLauncher
          .launch(ACCESS_FINE_LOCATION)
      }
    }
```

The when statement will check whether the permission was already granted. If not, it will check whether a rationale dialog should be presented. Then, if the rationale is accepted by the user or no rationale dialog is required, it will present a standard permission request dialog to the user (as shown in *Figure 7.1*), asking them to allow the app to access their location. You pass the requested permission you want the user to grant your app (`Manifest.permission.ACCESS_FINE_LOCATION`).

13. Run your app. You should now see a system permission dialog requesting you to allow the app to access the location of the device, as shown in *Figure 7.5*.

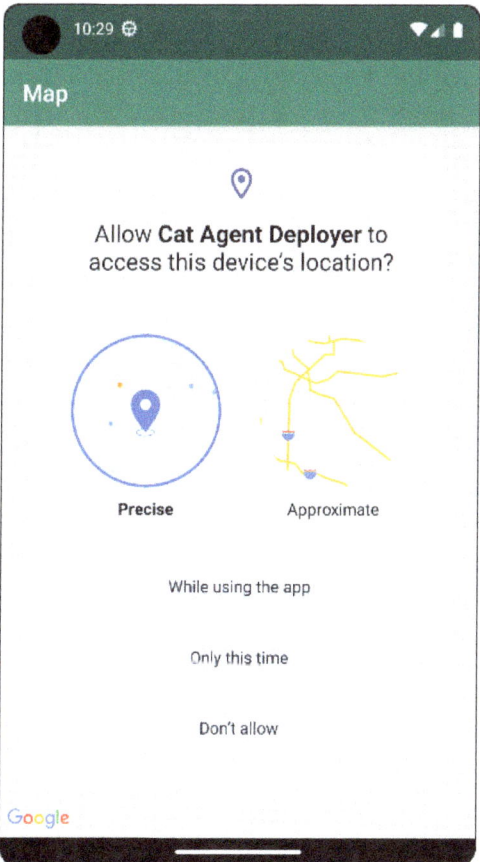

Figure 7.5 – App requesting the location permission

If the user denies the permission, the rationale dialog will appear. If the rationale is accepted, the system permission dialog will show again. Up until SDK 31, the user had the option to choose not to let the app ask for permission again (*Figure 7.6*). From SDK 31 onwards, not asking a third time is the default. Allowing it afterwards requires using the device settings.

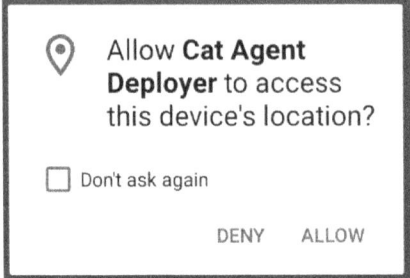

Figure 7.6 – The Don't ask again message

Once the user has allowed or permanently denied the permission, the dialog will never show again. To reset the state of your app permissions, you would have to manually grant it permission via the **App Info** interface.

Now that we can get the location permission, we will now look into obtaining the user's current location.

Showing a map of the user's location

Having successfully obtained permission from the user to access their location, we can now ask the user's device to provide us with its last known location. This is also usually the user's current location. We will then use this location to present the user with a map of their current location.

To obtain the user's last known location, Google has provided us with the Google Play Location service and, more specifically, with the `FusedLocationProviderClient` class. The `FusedLocationProviderClient` class helps us interact with Google's Fused Location Provider API, which is a location API that intelligently combines different signals from multiple device sensors to provide us with device location information.

To access the `FusedLocationProviderClient` class, we must first include the Google Play Location service library in our project. This simply means adding the following code snippet to the `dependencies` block of our `build.gradle` app:

```
implementation "com.google.android.gms:play-services-
location:21.0.1"
```

With the location service imported, we can now obtain an instance of the `FusedLocationProviderClient` class by calling `LocationServices.getFusedLocationProviderClient(this@MainActivity)`.

Once we have a fused location client, we can obtain the user's last location by calling `fusedLocationClient.lastLocation`.

> **Note**
> This is given that we have already received the location permission from the user.

Since this is an asynchronous call, we should also provide a success listener as a minimum. If we wanted to, we could also add listeners for cancellation, failure, and the completion of requests. Calling `lastLocation` returns `Task<Location>`. `Task` is a Google API abstract class whose implementations perform async operations.

In this case, that operation is returning a location. So, adding listeners is simply a matter of chaining calls. We will add the following code snippet to our call:

```
.addOnSuccessListener { location: Location? -> }
```

Note that the `location` parameter could be `null` if the client fails to obtain the user's current location. This is not very common but could happen if, for example, the user disabled their location services during the call.

Once the code inside our success listener block is executed, and `location` is not `null`, we have the user's current location in the form of a `Location` instance.

A `Location` instance holds a single coordinate on Earth, expressed using longitude and latitude.

> **Note**
> For our purpose, it is sufficient to know that each point on the surface of the Earth is mapped to a single pair of **longitude** (**Lng**) and **latitude** (**Lat**) values.

This is where it gets exciting. Google lets us present any location on an interactive map by using a `SupportMapFragment` class. All it takes is to sign up for a free API key. When you create your application with a Google Maps Activity, Android Studio immediately opens the `AndroidManifest.xml` file. A comment in the file sends us to https://packt.link/FK58V to obtain the required API key. You can copy that link to your browser or press *Ctrl/Command* + click on it.

On the page, follow the directions and click **Create**. Once you have a key, replace the `YOUR_API_KEY` string in the meta tag value with your newly obtained key.

At this point, if you run your app, you will already see an interactive map on your screen, as seen in *Figure 7.7*.

Figure 7.7 – Interactive map

To position the map based on our current location, we create a `LatLng` instance with the coordinates from our `Location` instance and call `moveCamera(CameraUpdate)` on the `GoogleMap` instance. To satisfy the `CameraUpdate` requirement, we call `CameraUpdateFactory.newLatLng(LatLng)`, passing in the `LatLng` parameter created earlier. The call would look something like this:

```
mMap.moveCamera(CameraUpdateFactory.newLatLng(latLng))
```

> **Note**
>
> To discover the rest of the available `CameraUpdateFactory` options, visit https://packt.link/EBRnt.

We could also call `newLatLngZoom(LatLng, Float)` to modify the zoom-in and zoom-out features of the map.

> **Note**
>
> Valid zoom values range between 2.0 (farthest) and 21.0 (closest). Values outside of that range are capped.

Some areas may not have tiles to render the closest zoom values.

We call `addMarker(MarkerOptions)` on the `GoogleMap` instance to add a marker at the user's coordinate. The `MarkerOptions` parameters are configured by chaining calls to a `MarkerOptions()` instance. We could call `position(LatLng)` and `title(String)` for a simple marker at our desired position. The call would look like the following:

```
mMap.addMarker(MarkerOptions().position(latLng)
    .title("Pin Label"))
```

The order in which we chain the calls does not matter.

Let's practice this in the following exercise.

Exercise 7.02 – obtaining the user's current location

Now that your app can be granted location permission, you can use the location permission to get the user's current location. You will then display the map and update it to zoom into the user's current location and show a pin at that location. To do this, perform the following steps:

1. First, add the Google Play location service to your `build.gradle` file. You should add it within the `dependencies` block:

    ```
    dependencies {
        implementation "com.google.android.gms: play-services-
        location:21.0.1"
        ...
    }
    ```

2. Click the **Sync Project with Gradle Files** button in Android Studio for Gradle to fetch the newly added dependency.

3. Obtain an API key: open the `AndroidManifest.xml` file (`app/src/main/AndroidManifest.xml`) and *Ctrl* / *Cmd* + click the link https://developers.google.com/maps/documentation/android-sdk/get-api-key.

4. Follow the instructions on the website until you have generated a new API key.

5. Update your `google_maps_api.xml` file by replacing YOUR_API_KEY with your new API key in the following code:

   ```
   <meta-data
   android:name="com.google.android.geo.API_KEY"
             android:value="YOUR_API_KEY" />
   ```

6. Open your `MapsActivity.kt` file. At the top of your `MapsActivity` class, define a lazily initialized fused location provider client:

   ```
   class MapsActivity : ... {
     private val fusedLocationProviderClient by lazy {
       LocationServices
         .getFusedLocationProviderClient(this)
     }
     override fun onCreate(savedInstanceState: Bundle?)
     { ... }
     ...
   }
   ```

   By making `fusedLocationProviderClient` initialize lazily, you are ensuring it is only initialized when needed, which essentially guarantees the `Activity` class will have been created before initialization.

7. Introduce an `updateMapLocation(LatLng)` function and an `addMarkerAtLocation(LatLng, String)` function immediately after the `getLastLocation()` function to zoom the map at a given location and add a marker at that location, respectively:

   ```
   private fun updateMapLocation(location: LatLng) {
       mMap.moveCamera(CameraUpdateFactory.newLatLngZoom(
           location, 7f))
   }
   ```

```kotlin
private fun addMarkerAtLocation(location: LatLng,
title: String) {
    mMap.addMarker(MarkerOptions().title(title)
        .position(location))
}
```

8. Now update your `getLastLocation()` function to retrieve the user's location:

```kotlin
private fun getLastLocation() {
  fusedLocationProviderClient.lastLocation
    .addOnSuccessListener { location: Location? ->
      location?.let {
        val userLocation = LatLng(
          location.latitude, location.longitude)
        updateMapLocation(userLocation)
        addMarkerAtLocation(userLocation, "You")
      }
    }
}
```

Your code requests the last location by calling `lastLocation` and then attaches a `lambda` function as an `OnSuccessListener` interface. Once a location is obtained, the `lambda` function is executed, updating the map location. The code then adds a marker at that location with the `You` title if a non-null location was returned.

9. Run your app. It should look like *Figure 7.8*.

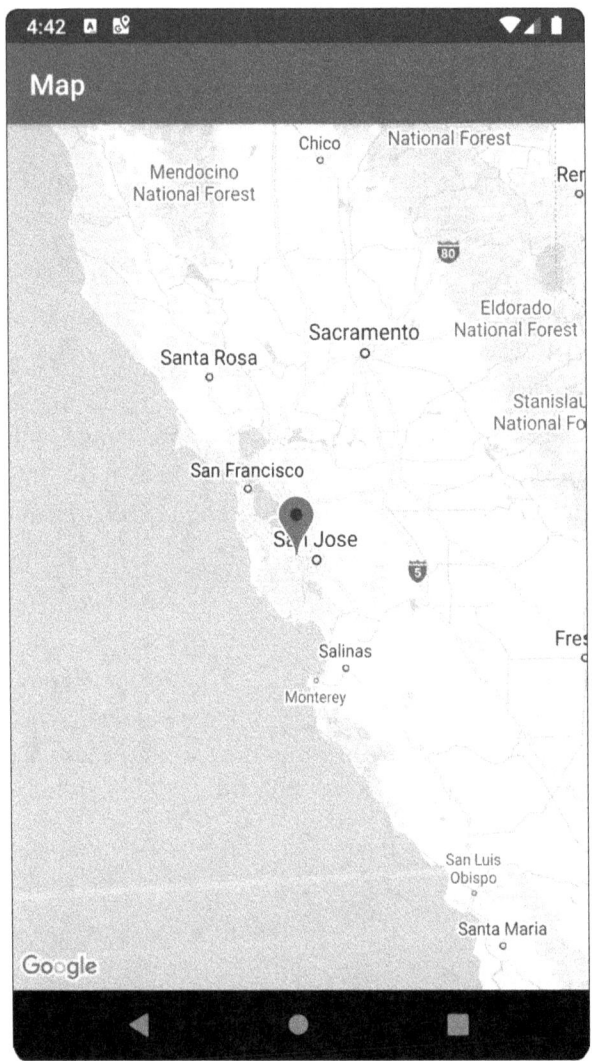

Figure 7.8 – Interactive map with a marker at the current location

Once the app has been granted permission, it can request the user's last location from the Google Play location service via the fused location provider client. This gives you an easy and concise way to fetch the user's current location. Remember to turn on the location on your device for the app to work.

With the user's location, your app can tell the map where to zoom and where to place a marker. If the user clicks on the marker, they will see the title you assigned to it (You in the exercise).

In the next section, we will learn how to respond to clicks on the map and how to move markers.

Map clicks and custom markers

With a map showing the user's current location by zooming in at the right location and placing a marker there, we have a rudimentary knowledge of how to render the desired map and how to obtain the required permissions and the user's current location.

In this section, we will learn how to respond to a user interacting with the map and how to use markers more extensively. We will learn how to move markers on the map and replace the default pin marker with custom icons. When we know how to let the user place a marker anywhere on the map, we can let them choose where to deploy the secret cat agent.

We need to add a listener to the `GoogleMap` instance to listen for clicks on the map. Looking at our `MapsActivity.kt` file, the best place to do so would be in `onMapReady(GoogleMap)`. A naïve implementation might look like this:

```
override fun onMapReady(googleMap: GoogleMap) {
    mMap = googleMap.apply {
        setOnMapClickListener { latLng ->
            addMarkerAtLocation(latLng, "Deploy here")
        }
    }
    ...
}
```

However, if we ran this code, we'd find that a new marker is added for every click on the map. This is not our desired behavior.

To control a marker on the map, we need to keep a reference to that marker. That is achieved easily enough by keeping a reference to the output of `GoogleMap.addMarker(MarkerOptions)`. The `addMarker` function returns a `Marker` instance. To move a marker on the map, we assign a new position value to it by calling its `position` setter.

To replace the default pin icon with a custom icon, we must provide a `BitmapDescriptor` to the marker or the `MarkerOptions()` instance. The `BitmapDescriptor` wrappers work around bitmaps used by `GoogleMap` to render markers and ground overlays, but we won't cover that in this book.

We obtain `BitmapDescriptor` by using `BitmapDescriptorFactory`. The factory will require an asset, which can be provided in a few ways. You can provide it with the name of a bitmap in the `assets` directory, a `Bitmap`, a filename of a file in the internal storage, or a resource ID.

The factory can also create default markers of different colors. We are interested in the `Bitmap` option because we intend to use a vector drawable, and the factory does not directly support those. In addition, when converting the drawable to a `Bitmap`, we can manipulate it to suit our needs (for example, we could change its color).

Android Studio offers us quite a wide range of free vector `Drawables` out of the box. For this example, we want the `paw` drawable. To do this, right-click anywhere in the left Android pane, and select **New** | **Vector Asset**.

Now, click the Android icon next to the **Clip Art** label for the list of icons (see *Figure 7.9*):

Figure 7.9 – Asset Studio

We'll now access a window to choose from the offered pool of clip art (*Figure 7.10*):

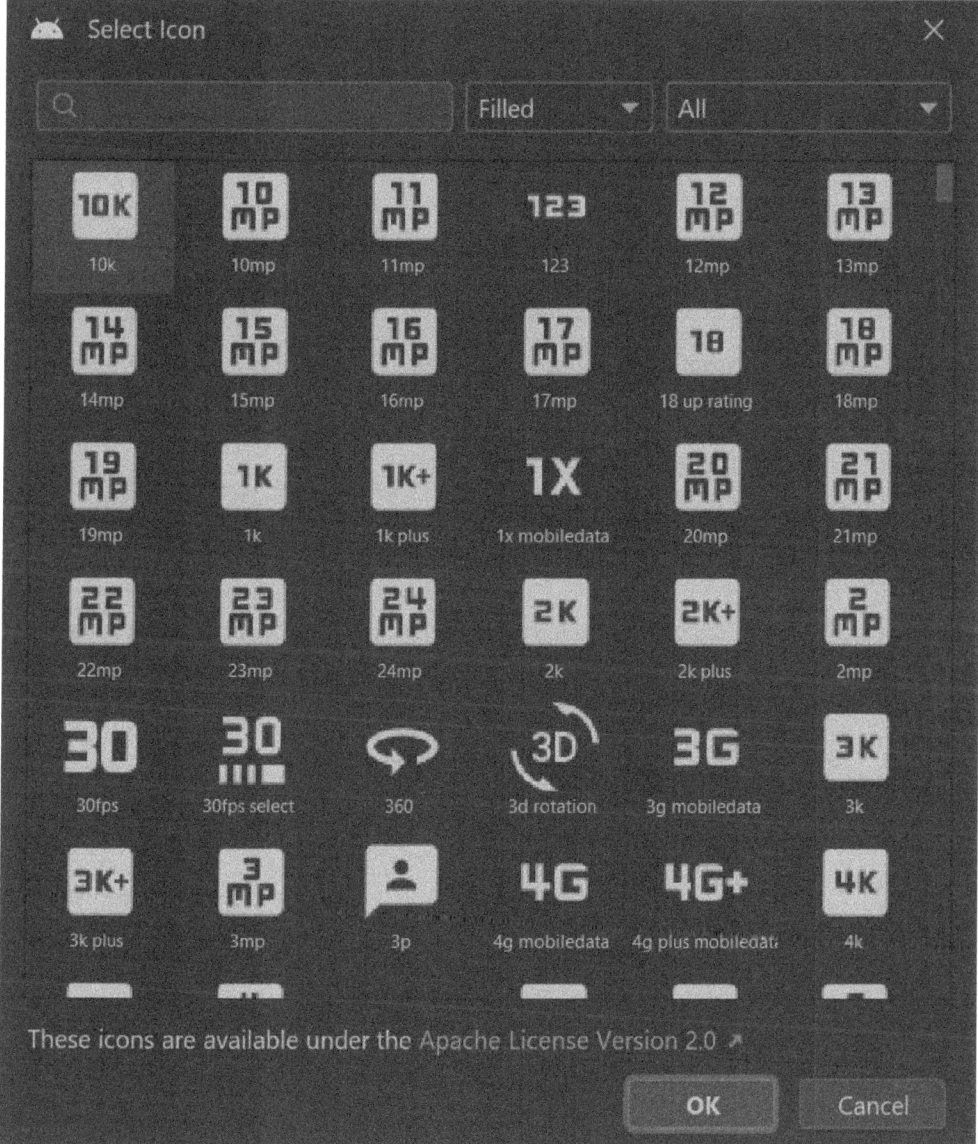

Figure 7.10 – Selecting an icon

Once we choose an icon, we can name it, and it will be created for us as a vector drawable XML file. We will name it `target_icon`.

To use the created asset, we must first get it as a `Drawable` instance. This is done by calling `ContextCompat.getDrawable(Context, Int)`, passing in the activity and `R.drawable.target_icon` as a reference to our asset. Next, we need to define bounds for the `Drawable` instance to draw in.

Calling `Drawable.setBound(Int, Int, Int, Int)` with `(0, 0, drawable.intrinsicWidth, drawable.intrinsicHeight)` will tell the drawable to draw within its intrinsic size.

To change the color of our icon, we can tint it. To tint a `Drawable` instance in a way that is supported by devices running APIs older than 21, we must first wrap our `Drawable` instance with `DrawableCompat` by calling `DrawableCompat.wrap(Drawable)`. The returned `Drawable` can then be tinted using `DrawableCompat.setTint(Drawable, Int)`.

Next, we need to create a `Bitmap` to hold our icon. Its dimensions can match those of the `Drawable` bounds, and we want its `Config` to be `Bitmap.Config.ARGB_8888`.

> **Note**
> **ARGB** stands for **Alpha, Red, Green, and Blue**. The 8 indicates the number of bits per channel. `ARGB_8888` means we want 8-bit red, green, blue, and alpha channels.

We then create a `Canvas` for the `Bitmap`, allowing us to draw our `Drawable` instance by calling… you guessed it, `Drawable.draw(Canvas)`:

```
private fun getBitmapDescriptorFromVector(@DrawableRes
vectorDrawableResourceId: Int): BitmapDescriptor? {
  val bitmap = ContextCompat.getDrawable(this,
    vectorDrawableResourceId)?.let { vectorDrawable ->
      vectorDrawable.setBounds(0, 0,
        vectorDrawable.intrinsicWidth,
        vectorDrawable.intrinsicHeight)
      val drawableWithTint = DrawableCompat.
        wrap(vectorDrawable)
      DrawableCompat.setTint(drawableWithTint, Color.RED)
      val bitmap = Bitmap.createBitmap(
        vectorDrawable.intrinsicWidth,
        vectorDrawable.intrinsicHeight,
        Bitmap.Config.ARGB_8888
      )
      val canvas = Canvas(bitmap)
```

```
        drawableWithTint.draw(canvas)
        bitmap
    }
    return BitmapDescriptorFactory.fromBitmap(bitmap)
        .also { bitmap?.recycle() }
}
```

With the `Bitmap` containing our icon, we are now ready to obtain a `BitmapDescriptor` instance from `BitmapDescriptorFactory`. Don't forget to recycle your `Bitmap` afterward. This will avoid a memory leak.

You have learned how to present the user with a meaningful map by centering it on their current location and showing their current location using a custom marker.

Exercise 7.03 – adding a custom marker where the map was clicked

In this exercise, you will respond to a user's map click by placing a red paw-shaped marker at the location on the map the user clicked:

1. In MapsActivity.kt (found under app/src/main/java/com/example/catagentdeployer), right below the definition of the mMap variable, define a nullable Marker variable to hold a reference to the paw marker on the map:

   ```
   private lateinit var mMap: GoogleMap
   private var marker: Marker? = null
   ```

2. Update addMarkerAtLocation(LatLng, String) to also accept a nullable BitmapDescriptor with a default value of null:

   ```
   private fun addMarkerAtLocation(
       location: LatLng, title: String,
       markerIcon: BitmapDescriptor? = null
   ) = mMap.addMarker(
       MarkerOptions().title(title).position(location)
           .apply { markerIcon?.let { icon(markerIcon) } }
   )
   ```

   If the markerIcon provided is not null, the app sets it to MarkerOptions. The function now returns the marker it added to the map.

3. Create a getBitmapDescriptorFromVector(Int): BitmapDescriptor? function below your addMarkerAtLocation(LatLng, String, BitmapDescriptor?): Marker function to provide BitmapDescriptor given a Drawable resource ID:

```
private fun getBitmapDescriptorFromVector(@DrawableRes
vectorDrawableResourceId: Int): BitmapDescriptor? {
  val bitmap = ContextCompat.getDrawable(this,
    vectorDrawableResourceId)?.let { vectorDrawable ->
      vectorDrawable.setBounds(0, 0,
        vectorDrawable.intrinsicWidth,
        vectorDrawable.intrinsicHeight)
      val drawableWithTint = DrawableCompat
        .wrap(vectorDrawable)
      DrawableCompat.setTint(drawableWithTint, Color.RED)
      val bitmap = Bitmap.createBitmap(
        vectorDrawable.intrinsicWidth,
        vectorDrawable.intrinsicHeight,
        Bitmap.Config.ARGB_8888
      )
      val canvas = Canvas(bitmap)
      drawableWithTint.draw(canvas)
      bitmap
    }
  return BitmapDescriptorFactory.fromBitmap(bitmap)
    .also { bitmap?.recycle() }
}
```

This function first obtains a drawable using `ContextCompat` by passing in the provided resource ID. It then sets the drawing bounds for the drawable, wraps it in `DrawableCompat`, and sets its tint to red.

Then, it creates a `Bitmap` and a `Canvas` for that `Bitmap`, upon which it draws the tinted drawable. The bitmap is then returned to be used by `BitmapDescriptorFactory` to build `BitmapDescriptor`. Lastly, `Bitmap` is recycled to avoid a memory leak.

4. Before you can use the `Drawable` instance, you must first create it. Right-click on the Android pane, and then select **New | Vector Asset**.

5. In the window that opens, click on the Android icon next to the **Clip Art** label to select a different icon (*Figure 7.11*):

Figure 7.11 – Asset Studio

6. From the list of icons, select the **pets** icon. You can type `pets` into the search field if you can't find the icon. Once you select the **pets** icon, click **OK**:

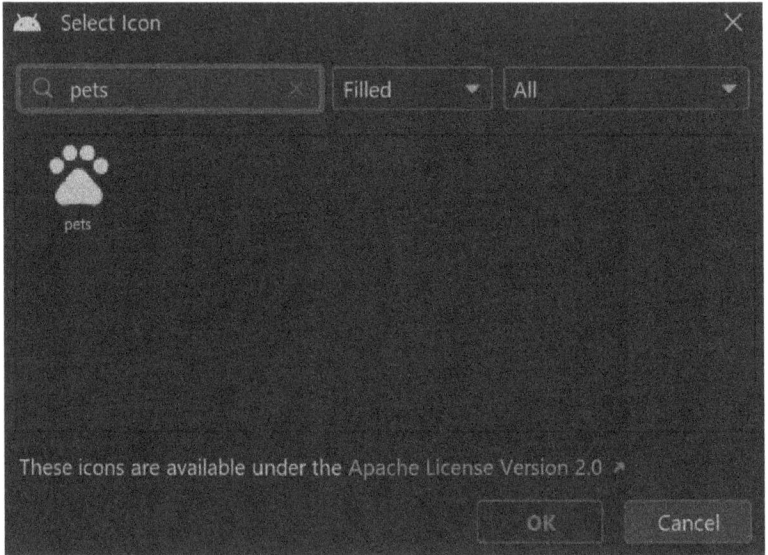

Figure 7.12 – Selecting an icon

7. Name your icon `target_icon`. Click **Next** and **Finish**.

8. Define an `addOrMoveSelectedPositionMarker(LatLng)` function to create a new marker or move it to the provided location if one has already been created. Add it after the `getBitmapDescriptorFromVector(Int)` function:

    ```
    private fun addOrMoveSelectedPositionMarker(latLng:
    LatLng) {
      if (marker == null) {
        marker = addMarkerAtLocation(latLng,
          "Deploy here", getBitmapDescriptorFromVector(
            R.drawable.target_icon)
        )
      } else { marker?.apply { position = latLng } }
    }
    ```

9. Update your `onMapReady(GoogleMap)` function to set an `OnMapClickListener` event on `mMap`, which will add a marker to the clicked location or move the existing marker to the clicked location:

    ```
    override fun onMapReady(googleMap: GoogleMap) {
        mMap = googleMap.apply {
            setOnMapClickListener { latLng ->
    ```

```
            addOrMoveSelectedPositionMarker(latLng)
        }
    }
    if (hasLocationPermission()) { ... }
}
```

Run your app. It should look like *Figure 7.13*.

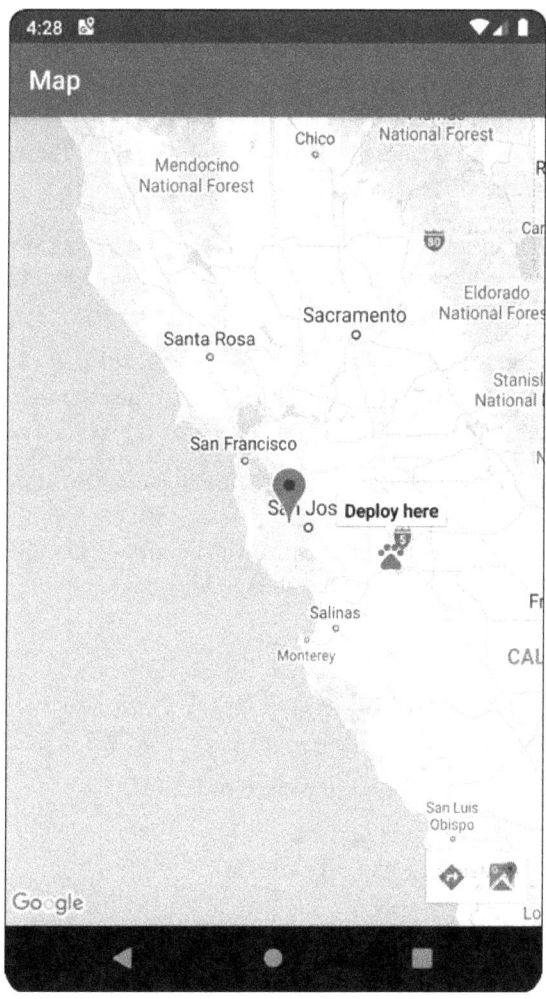

Figure 7.13 – The complete app

Clicking anywhere on the map will now move the paw icon to that location. Clicking the paw icon will show the **Deploy here** label.

> **Note**
> The location of the paw is a geographical one, not a screen one. That means if you drag your map or zoom in, the paw will move with the map and remain in the same geographical location.

You now know how to respond to user clicks on the map and add and move markers around. You also know how to customize the appearance of markers.

Activity 7.01 – creating an app to find the location of a parked car

Some people often forget where it was that they parked their car. Let's say you want to help these people by developing an app that lets the user store the last place they parked. The app will show a pin at the car's location when the user launches the app. The user can click an **I'm parked here** button to update the pin location to the current location the next time they park.

Your goal in this activity is to develop an app that shows the user a map of the current location. The app must first ask the user for permission to access their location. Make sure to also provide a rationale dialog, if needed, according to the SDK.

The app will show a car icon where the user last told it the car was. The user can click a button labeled **I'm parked here** to move the car icon to the current location. When the user relaunches the app, it will show the user's current location and the car icon where the car was last parked.

As a bonus feature of your app, you can choose to add functionality that stores the car's location so that it can be restored after the user has killed and then re-opened the app. This bonus functionality relies on using `SharedPreferences`; a concept that will be covered in *Chapter 10*, *Persisting Data*. As such, *steps 9* and *10* here will give you the required implementation.

The following steps will help you complete the activity:

1. Create a Google Maps Activity app.
2. Obtain an API key for the app and update your `google_maps_api.xml` file with that key.
3. Show a button at the bottom with an **I'm parked here** label.
4. Include the location service in your app.
5. Request the user's permission to access their location.
6. Obtain the user's location and place a pin on the map at that location.
7. Add a car icon to your project.
8. Add functionality to move the car icon to the user's current location.

9. Bonus step: store the selected location in `SharedPreferences`. This function, placed in your activity, will help you do this:

    ```
    private fun saveLocation(latLng: LatLng) =
      getPreferences(MODE_PRIVATE)?.edit()?.apply {
        putString("latitude", latLng.latitude.toString())
        putString("longitude", latLng.longitude.toString())
        apply()
      }
    ```

10. Bonus step: restore any saved location from `SharedPreferences`. You can use the following function:

    ```
    val latitude = sharedPreferences
        .getString("latitude", null)
        ?.toDoubleOrNull() ?: return null
    val longitude = sharedPreferences
        .getString("longitude", null)
        ?.toDoubleOrNull() ?: return null
    ```

With this activity completed, you have demonstrated your understanding of requesting permissions in an Android app. You have also shown that you can present the user with a map and control pins on that map. Finally, you have also demonstrated your knowledge of obtaining the user's current location. Well done.

> **Note**
>
> The solution to this activity can be found at `https://packt.link/By7eE`.

Summary

In this chapter, we learned about Android permissions. We touched on the reasons for having them and saw how we could request the user's permission to perform certain tasks. We also learned how to use Google's Maps API and how to present the user with an interactive map.

Lastly, we leveraged our knowledge of presenting a map and requesting permissions to find out the user's current location and present it on the map. Of course, there is a lot more that can be done with the Google Maps API, and you could explore a lot more possibilities with certain permissions.

You should now have enough understanding of the foundations to explore further. To read more about permissions, visit `https://packt.link/57BdN`. To read more about the Maps API, visit `https://packt.link/8akrP`.

In the next chapter, we will learn how to perform background tasks using `Services` and `WorkManager`. We will also learn how to present the user with notifications, even when the app is not running. These are powerful tools to have in your arsenal as a mobile developer.

8
Services, WorkManager, and Notifications

This chapter will introduce you to the concepts of managing long-running tasks in the background of an app. By the end of this chapter, you will be able to trigger a background task, create a notification for the user when a background task is complete, and launch an application from a notification. This chapter will give you a solid understanding of how to manage background tasks and keep the user informed about the progress of these tasks.

In the previous chapter, we learned how to request permissions from the user and use Google's Maps API. With that knowledge, we obtained the user's location and allowed them to deploy an agent on a local map. In this chapter, we will learn how to track a long-running process and report its progress to the user.

We will build an example app where we will assume that **Secret Cat Agents** (**SCAs**) get deployed in a record time of 15 seconds. When a cat successfully deploys, we will notify the user and let them launch the app, presenting them with a successful deployment message.

> **Note**
> We will go with 15 seconds because that way, we will avoid having to wait for very long before our background task completes.

Ongoing background tasks are quite common in the mobile world. Background tasks run even when an application is not active. Examples of long-running background tasks include the downloading of files, resource clean-up jobs, playing music, and tracking the user's location.

Historically, Google offered Android developers multiple ways of executing such tasks: `Services`, `JobScheduler`, and Firebase's `JobDispatcher` and `AlarmManager`. With the fragmentation that existed in the Android world, it was quite a mess to cope with. Luckily for us, since March 2019, we have had a better (more stable) option.

With the introduction of `WorkManager`, Google has abstracted the logic of choosing a background executing mechanism based on the API version away for us. We still use a foreground service, which is a special kind of service, for certain tasks that should be known to the user while running, such as playing music or tracking the user's location in a running app.

> **Note**
> **Services** are application components designed to run in the background, even when an app is not running. Except for foreground services, which are tied to a notification, services have no user interface.

Before we proceed, take a quick step back. We have mentioned services, and we will be focusing on foreground services, but we haven't quite explained what services are. Services are application components designed to run in the background, even when an app is not running.

Services have no user interface, with foreground services being the exception. It is important to note that services run on the main thread of their hosting process. This means that their operations can block the app. It is up to us to start a separate thread from within a service to avoid that.

Let's get started and look at the implementation of the multiple approaches available in Android for managing a background task.

We will cover the following topics in this chapter:

- Starting a background task using `WorkManager`
- Background operations noticeable to the user – using a Foreground Service

Technical requirements

The complete code for all the exercises and the activity in this chapter is available on GitHub at `https://packt.link/i8IRQ`.

Starting a background task using WorkManager

The first question we will address here is should we opt for `WorkManager` or a foreground service? To answer that, a good rule of thumb is to ask whether you need the action to be tracked by the user in real time.

If the answer is yes (for example, if you have a task such as responding to the user's location or playing music in the background), then you should use a foreground service with its attached notification to give the user a real-time indication of state. When the background task can be delayed or does not require user interaction (for example, downloading a large file), use `WorkManager`.

> **Note**
> Starting with version `2.3.0-alpha02` of `WorkManager`, you can launch a foreground service via the `WorkManager` singleton by calling `setForegroundAsync(ForegroundInfo)`. Our control over that foreground service is quite limited. It does allow you to attach a (predefined) notification to the work, which is why it is worth mentioning.

In our example app, we will track the SCAs' preparation for deployment. Before an agent can head out, they need to stretch, groom their fur, visit the litter box, and suit up. Each one of these tasks takes some time. Because you can't rush a cat, the agent will finish each step in its own time. All we can do is wait (and let the user know when the task is done). `WorkManager` is perfect for such a scenario.

To use `WorkManager`, we need to familiarize ourselves with its four main classes:

- `WorkManager`: This receives work and enqueues it based on provided arguments and constraints (such as internet connectivity and the device charging).
- `Worker`: This is a wrapper around the work that needs doing. It has one function, `doWork()`, which we override to implement the background work code. The `doWork()` function will be executed in a background thread.
- `WorkRequest`: This class binds a `Worker` class to arguments and constraints. There are two types of `WorkRequest`: `OneTimeWorkRequest`, which runs the work once, and `PeriodicWorkRequest`, which can be used to schedule work to run at a fixed interval.
- `ListenableWorker.Result`: You probably guessed it, but this is the class holding the result of the executed work. The result can be one of `Success`, `Failure`, or `Retry`.

Other than these four classes, we also have the `Data` class, which holds data passed to and from the worker.

Let's get back to our example. We want to define four tasks that need to occur in sequential order: the cat needs to stretch, then it needs to groom its fur, then visit the litter box, and, finally, it needs to suit up.

Before we can start using `WorkManager`, we have to first include its dependency in our app `build.gradle` file:

```
implementation "androidx.work:work-runtime:2.8.0"
```

With `WorkManager` included in our project, we'll go ahead and create our workers. The first worker will look something like this:

```
class CatStretchingWorker(
  context: Context, workerParameters: WorkerParameters
) : Worker(context, workerParameters) {
  override fun doWork(): Result {
    val catAgentId =
```

```
      inputData.getString(INPUT_DATA_CAT_AGENT_ID)
    Thread.sleep(3000L)
    val outputData = Data.Builder()
      .putString(OUTPUT_DATA_CAT_AGENT_ID, catAgentId)
      .build()
    return Result.success(outputData)
  }
  companion object {
    const val INPUT_DATA_CAT_AGENT_ID = "id"
    const val OUTPUT_DATA_CAT_AGENT_ID = "id"
  }
}
```

We start by extending `Worker` and overriding its `doWork()` function. We then read the SCA ID from the input data. Then, because we have no real sensors to track the progress of the cat stretching, we fake our wait by introducing a 3-second (3,000 milliseconds) `Thread.sleep(Long)` call. Finally, we construct an output `data` class with the ID we received in our input and return it with the successful result.

Once we've created workers for all our tasks (`CatStretchingWorker`, `CatFurGroomingWorker`, `CatLitterBoxSittingWorker`, and `CatSuitUpWorker`), similarly to how we created the first one, we can call `WorkManager` to chain them. Let's also assume we can't tell the progress of the agent unless we're connected to the internet. Our call would look something like this:

```
val catStretchingInputData = Data.Builder()
  .putString(CatStretchingWorker.INPUT_DATA_CAT_AGENT_ID,
  "catAgentId").build()
val catStretchingRequest = OneTimeWorkRequest
  .Builder(CatStretchingWorker::class.java)
val catStretchingRequest = OneTimeWorkRequest.Builder(
  CatStretchingWorker::class.java)
    .setConstraints(networkConstraints)
    .setInputData(catStretchingInputData)
    .build()
...
WorkManager.getInstance(this)
  .beginWith(catStretchingRequest)
  .then(catFurGroomingRequest)
```

```
.then(catLitterBoxSittingRequest)
.then(catSuitUpRequest)
.enqueue()
```

In the preceding code, we first construct a `Constraints` instance declaring we need to be connected to the internet for the work to execute. We then define our input data, setting it to the SCA ID. Next, we bind the constraints and input data to our `Worker` class by constructing `OneTimeWorkRequest`.

The construction of the other `WorkRequest` instances has been left out, but they are almost identical to the one shown here. We can now chain all the requests and enqueue them in the `WorkManager` class. You can enqueue a single `WorkRequest` instance by passing it directly to the `WorkManager` `enqueue()` function, or you can have multiple `WorkRequest` instances run in parallel by passing them all to the `WorkManager` `enqueue()` function as a list.

Our tasks will be executed by `WorkManager` when the constraints are met.

Each `Request` instance has a unique identifier. `WorkManager` exposes a `LiveData` property for each request, allowing us to track the progress of its work by passing its unique identifier, as shown in the following code:

```
workManager.getWorkInfoByIdLiveData(
  catStretchingRequest.id
).observe(this) { info ->
  if (info.state.isFinished) { doSomething() }
}
```

The state of work can be `BLOCKED` (there is a chain of requests, and this one is not next in the chain), `ENQUEUED` (there is a chain of requests, and this work is next), `RUNNING` (the work in `doWork()` is executing), and `SUCCEEDED`. Work can also be canceled, leading to a `CANCELLED` state, or it can fail, leading to a `FAILED` state.

Finally, there's `Result.retry`. Returning this result tells the `WorkManager` class to enqueue the work again. The policy governing when to run the work again is defined by a `backoff` criteria set on `WorkRequest Builder`. The default `backoff` policy is exponential, but we can set it to be linear instead. We can also define the initial `backoff` time.

Let's put the knowledge gained so far into practice in the following exercise. In this section, we will track our SCA from the moment we fire off the command to deploy it to the field to the moment it arrives at its destination.

Exercise 8.01 – executing background work with the WorkManager class

In this first exercise, we will track the SCA as it prepares to head out by enqueuing the chained `WorkRequest` classes:

1. Start by creating a new `Empty Activity` project (**File** | **New** | **New Project** | **Empty Activity**). Click **Next**.
2. Name your application `Cat Agent Tracker`.
3. Make sure your package name is `com.example.catagenttracker`.
4. Set the save location to where you want to save your project.
5. Leave everything else at its default values and click **Finish**.
6. Make sure you are on the Android view in your **Project** pane.
7. Open your app's `build.gradle` file. In the `dependencies` block, add the `WorkManager` dependency:

   ```
   dependencies {
       ...
       implementation "androidx.work:work-runtime:2.8.0"
       ...
   }
   ```

   This will allow you to use `WorkManager` and its dependencies in your code.

8. Create a new package under your app package (right-click on `com.example.catagenttracker`, then **New** | **Package**). Name the new package `com.example.catagenttracker.worker`.
9. Create a new class under `com.example.catagenttracker.worker` named `CatStretchingWorker` (right-click on `worker`, then **New** | **New Kotlin File/Class**). Under **Kind**, choose **Class**.
10. To define a `Worker` instance that will sleep for 3 seconds, update the new class like so:

    ```
    package com.example.catagenttracker.worker
    [imports]
    class CatStretchingWorker(
      context: Context, workerParameters: WorkerParameters
    ) : Worker(context, workerParameters) {
      override fun doWork(): Result {
        val catAgentId =
          inputData.getString(INPUT_DATA_CAT_AGENT_ID)
    ```

```kotlin
      Thread.sleep(3000L)
      val outputData = Data.Builder()
        .putString(OUTPUT_DATA_CAT_AGENT_ID, catAgentId)
        .build()
      return Result.success(outputData)
    }
    companion object {
      const val INPUT_DATA_CAT_AGENT_ID = "inId"
      const val OUTPUT_DATA_CAT_AGENT_ID = "outId"
    }
  }
```

This will add the required dependencies for a `Worker` implementation and then extend the `Worker` class. To implement the actual work, you will override `doWork(): Result`, making it read the Cat Agent ID from the input, sleep for 3 seconds (3,000 milliseconds), construct an output data instance with the Cat Agent ID, and pass it via a `Result.success` value.

11. Repeat *steps 9* and *10* to create three more identical workers named `CatFurGroomingWorker`, `CatLitterBoxSittingWorker`, and `CatSuitUpWorker`.

12. Open `MainActivity`. Right before the end of the class, add the following:

```kotlin
    private fun getCatAgentIdInputData(
      catAgentIdKey: String, catAgentIdValue: String) =
      Data.Builder()
        .putString(catAgentIdKey, catAgentIdValue).build()
```

This helper function constructs an input `Data` instance for you with the Cat Agent ID.

13. Add the following to the `onCreate(Bundle?)` function:

```kotlin
    override fun onCreate(savedInstanceState: Bundle?) {
      super.onCreate(savedInstanceState)
      setContentView(R.layout.activity_main)
      val networkConstraints = Constraints.Builder()
        .setRequiredNetworkType(NetworkType.CONNECTED)
        .build()
      val catAgentId = "CatAgent1"
      val catStretchingRequest = OneTimeWorkRequest.
        Builder(CatLitterBoxSittingWorker::class.java)
        .setConstraints(networkConstraints)
        .setInputData(
```

```
                getCatAgentIdInputData(CatStretchingWorker
                    .INPUT_DATA_CAT_AGENT_ID, catAgentId)
                ).build()
            val catFurGroomingRequest = OneTimeWorkRequest.
                Builder(CatFurGroomingWorker::class.java)
                .setConstraints(networkConstraints)
                .setInputData(getCatAgentIdInputData(
                    CatFurGroomingWorker.INPUT_DATA_CAT_AGENT_ID,
                    catAgentId)
                ).build()
            val catLitterBoxSittingRequest = OneTimeWorkRequest.
                Builder(CatLitterBoxSittingWorker::class.java)
                .setConstraints(networkConstraints)
                .setInputData(getCatAgentIdInputData(
        CatLitterBoxSittingWorker.INPUT_DATA_CAT_AGENT_ID,
                    catAgentId)
                ).build()
            val catSuitUpRequest = OneTimeWorkRequest.Builder(
                CatSuitUpWorker::class.java
            ).setConstraints(networkConstraints)
                .setInputData(getCatAgentIdInputData(CatSuitUpWorker.
                    INPUT_DATA_CAT_AGENT_ID, catAgentId)
                ).build()
        }
```

The first line added defines a network constraint. It tells the `WorkManager` class to wait for an internet connection before executing work. Then, you define your Cat Agent ID. Finally, you define four requests, passing in your `Worker` classes, the network constraints, and the Cat Agent ID in the form of input data.

14. At the top of the class, define your `WorkManager`:

    ```
    private val workManager = WorkManager.getInstance(this)
    ```

15. Add a chained `enqueue` request right below the code you just added, still within the `onCreate` function:

    ```
    val catSuitUpRequest = ...
    workManager.beginWith(catStretchingRequest)
        .then(catFurGroomingRequest)
    ```

```
            .then(catLitterBoxSittingRequest)
            .then(catSuitUpRequest)
            .enqueue()
```

Your `WorkRequests` are now enqueued to be executed in sequence when their constraints are met and the `WorkManager` class is ready to execute them.

16. Define a function to show a toast with a provided message. It should look like this:

    ```
    private fun showResult(message: String) {
        Toast.makeText(this, message, LENGTH_SHORT).show()
    }
    ```

17. To track the progress of the enqueued `WorkRequest` instances, add the following after the enqueue call:

    ```
    workManager.beginWith(catStretchingRequest)
        ...
        .enqueue()
    workManager.getWorkInfoByIdLiveData(
        catStretchingRequest.id).observe(this) { info ->
            if (info.state.isFinished) {
                showResult("Agent done stretching")
            }
        }
    workManager.getWorkInfoByIdLiveData(
        catFurGroomingRequest.id).observe(this) { info ->
            if (info.state.isFinished) {
                showResult("Agent done grooming its fur")
            }
        }
    workManager.getWorkInfoByIdLiveData(
        catLitterBoxSittingRequest.id).observe(this) { info ->
            if (info.state.isFinished) {
                showResult("Agent done sitting in litter box")
            }
        }
    workManager.getWorkInfoByIdLiveData(
        catSuitUpRequest.id).observe(this) { info ->
    ```

```
            if (info.state.isFinished) {
                showResult("Agent done suiting up. Ready to go!")
            }
    }
```

The preceding code observes a `WorkInfo` observable provided by the `WorkManager` class for each `WorkRequest`. When each request is finished, a toast is shown with a relevant message.

18. Run your app:

Figure 8.1 – Toasts showing in order

You should now see a simple `Hello World!` screen. However, if you wait a few seconds, you will start seeing toasts informing you of the progress of your SCA preparing to deploy to the field. You will notice that the toasts follow the order in which you enqueued the requests and execute their delays sequentially.

Background operations noticeable to the user – using a Foreground Service

With our SCA all suited up, they are now ready to get to the assigned destination. To track the SCA, we will periodically poll the location of the SCA using a Foreground Service and update the sticky notification (a notification that cannot be dismissed by the user) attached to that service with the new location.

> **Note**
>
> For the sake of simplicity, we will fake the location. Following what you learned in *Chapter 7, Android Permissions and Google Maps*, you could later replace this implementation with a real one that uses a map.

Foreground services are another way of performing background operations. The name may be a bit counterintuitive. It is meant to differentiate these services from the base Android (background) services. The former are tied to a notification, while the latter run in the background with no user-facing representation built in.

Another important difference between foreground services and background services is that the latter are candidates for termination when the system is low on memory, while the former are not.

As of Android 9 (Pie, or API level 28), we have to request the `FOREGROUND_SERVICE` permission to use foreground services. Since it is a normal permission, it will be granted to our app automatically.

Before we can launch a foreground service, we must first create one. A foreground service is a subclass of the Android abstract `Service` class. If we do not intend to bind to the service, and in our example, we do not, we can simply override `onBind(Intent)` so that it returns `null`.

As a side note, binding is one of the ways for interested clients to communicate with a service. We will not focus on this approach in this book, as there are other, easier approaches, as you will discover.

A Foreground Service must be tied to a notification. On Android 8 (Oreo, or API level 26) and above, if a Foreground Service is not tied to one within the **Application Not Responding** (**ANR**) time window (around five seconds), the service is stopped, and the app is declared as not responding.

Because of this requirement, it is best if we tie the service to a notification as soon as we can. The best place to do that would be in the `onCreate()` function of the service. A quick implementation would look something like this:

```kotlin
private fun onCreate() {
  val channelId = if (Build.VERSION.SDK_INT >= Build.VERSION_CODES.O) {
    val newChannelId = "ChannelId"
    val channelName = "My Background Service"
    val channel = NotificationChannel(newChannelId,
      channelName, NotificationManager.IMPORTANCE_DEFAULT)
    val service = getSystemService(
      Context.NOTIFICATION_SERVICE) as NotificationManager
    service.createNotificationChannel(channel)
    newChannelId
  } else { "" }
    val flag = if (Build.VERSION.SDK_INT >=
      Build.VERSION_CODES.S) FLAG_IMMUTABLE else 0
  val pendingIntent = Intent(this,
    MainActivity::class.java).let { notificationIntent ->
      PendingIntent.getActivity(this, 0,
        notificationIntent, flag)
    }
  val notification =
    NotificationCompat.Builder(this, channelId)
      .setContentTitle("Content title")
      .setContentText("Content text")
      .setSmallIcon(R.drawable.notification_icon)
      .setContentIntent(pendingIntent)
      .setTicker("Ticker message").build()
  startForeground(NOTIFICATION_ID, notificationBuilder.build())
}
```

Let's break this down.

We start by defining the channel ID. This is only required for Android Oreo or above and is ignored in earlier versions of Android. In Android Oreo, Google introduced the concept of channels. Channels are used to group notifications and allow users to filter out unwanted notifications.

Next, we define `pendingIntent`. This will be the `Intent` launched if the user taps on the notification. In this example, the main activity would be launched. It is constructed by wrapping an `Intent` activity launching in a `PendingIntent`. The request code is set to 0 because this example doesn't expect a result, so the code will not be used.

The flag is set to 0 for APIs older than S (31). Otherwise, it is set to the recommended `PendingIntent.FLAG_IMMUTABLE`, which means that additional arguments passed to the `Intent` on send will be ignored.

With the channel ID and `pendingIntent`, we can construct our notification. We use `NotificationCompat`, which takes away some of the boilerplate around supporting older API levels. We pass in the context and the channel ID. We define the title, text, small icon, `Intent`, and ticker message and build the notification to complete the builder:

```
val notification =
  NotificationCompat.Builder(this, channelId)
    .setContentTitle("Content title")
    .setContentText("Content text")
    .setSmallIcon(R.drawable.notification_icon)
    .setContentIntent(pendingIntent)
    .setTicker("Ticker message")
    .build()
```

To start a service in the foreground, attaching the notification to it, we call it the `startForeground(Int, Notification)` function, passing in a notification ID (any unique int value to identify this service, which must not be 0) and a notification, which must have its priority set to `PRIORITY_LOW` or higher. In our case, we have not specified the priority, which sets it to `PRIORITY_DEFAULT`:

If launched, our service will now show a sticky notification. Clicking on the notification would launch our main activity.

Currently, our service doesn't perform any operations aside from showing a notification. To add some functionality to it, we need to override `onStartCommand(Intent?, Int, Int)`. This function gets called when the service is launched via an Intent.

This also gives us the opportunity to read any extra data passed via that Intent. It also provides us with flags (which may be set to `START_FLAG_REDELIVERY` or `START_FLAG_RETRY`) and a unique request ID. We will get to reading the extra data later in this chapter. You don't need to worry about the flags or the request ID in a simple implementation.

It is important to note that `onStartCommand(Intent?, Int, Int)` gets called on the UI thread, so don't perform any long-running operations here, or your app will freeze, giving the user a poor experience. Instead, we could create a new handler using a new `HandlerThread` (a thread with a looper, a class used to run a message loop for a thread) and post our work to it.

This means we'll have an infinite loop running, waiting for us to post to it via a `Handler`. When we receive a `start` command, we can post the work we want to be done to it. That work will then be executed on that thread.

When our long-running work is done, there are a few things we may want to happen. First, we may want to inform whoever is interested (our main activity, if it is running, for example) that we are done. Then, we probably want to stop running in the foreground. Lastly, if we do not expect to require the service again, we could stop it.

An app has several ways to communicate with a service: binding, using broadcast receivers, using a bus architecture, or using a result receiver, to name just a few. For our example, we will use Google's `LiveData`.

Before we proceed, it is worth touching on broadcast receivers. Broadcast receivers allow our app to send and receive messages using a pattern much like the *publish-subscribe design pattern*.

The system broadcasts events such as the device booting up or charging having started. Our services can broadcast status updates as well. For example, they can broadcast a long calculation result on completion. If our app registers to receive a certain message, the system will inform it when that message is broadcast.

This used to be a common way to communicate with services, but the `LocalBroadcastManager` class is now deprecated as it was an application-wide event bus that encouraged anti-patterns.

Having said that, broadcast receivers are still useful for system-wide events. We first define a class overriding the `BroadcastReceiver` abstract class:

```
class ToastBroadcastReceiver : BroadcastReceiver() {
    override fun onReceive(context: Context, intent: Intent) {
        StringBuilder().apply {
            append("Action: ${intent.action}\n")
            append(
                "URI:${intent.toUri(Intent.URI_INTENT_SCHEME)}\n")
            toString().let { eventText ->
                Toast.makeText(context, eventText,
                    Toast.LENGTH_LONG).show()
            }
        }
    }
}
```

When an event is received by `ToastBroadcastReceiver`, it will show a toast showing the action and URI of the event.

We can register our receiver via the `Manifest.xml` file:

```xml
<receiver android:name=".ToastBroadcastReceiver"
android:exported="true">
  <intent-filter>
    <action android:name=
       "android.intent.action.ACTION_POWER_CONNECTED" />
  </intent-filter>
</receiver>
```

Specifying `android:exported="true"` tells the system that this receiver can receive messages from outside of the application. The action defines the message we are interested in. We can specify multiple actions. In this example, we listen for when the device starts charging. Keep in mind that setting this value to `true` allows other apps, including malicious ones, to activate this receiver.

We can also register for messages in code:

```
val filter = IntentFilter(ConnectivityManager.CONNECTIVITY_
ACTION).apply { addAction(Intent.ACTION_POWER_CONNECTED) }
registerReceiver(ToastBroadcastReceiver(), filter)
```

Adding this code to an activity or in our custom application class would register a new instance of our receiver as well. This receiver will live so long as the context (activity or application) is valid. So, correspondingly, if the activity or application is destroyed, our receiver will be freed to be garbage collected.

Now back to our implementation. `LiveData` already comes bundled with `androidx.appcompat`, saving us the trouble of having to manually include it in our project. We can define a `LiveData` instance in the companion object of the service, like so:

```
companion object {
   private val mutableWorkCompletion = MutableLiveData<String>()
   val workCompletion: LiveData<String> = mutableWorkCompletion
}
```

Note that we hide the `MutableLiveData` instance behind a `LiveData` interface. This is so that consumers can only read the data. We can now use the `mutableWorkCompletion` instance to report completion by assigning it a value. However, we must remember that values can only be assigned to the `LiveData` instances on the main thread.

This means once our work is done, we must switch back to the main thread. We can easily achieve that—all we need is a new handler with the main `Looper` (obtained by calling `Looper.getMainLooper()`), to which we can post our update.

Now that our service is ready to do some work, we can finally launch it. Before we do, we must make sure we added the service to our `AndroidManifest.xml` file within the `<application></application>` block, as shown in the following code:

```
<application ...>
    <service android:name="ForegroundService" />
</application>
```

To launch the service we just added to our manifest, we create `Intent`, passing in any extra data required, as shown in the following code:

```
val serviceIntent = Intent(this, ForegroundService::class.
java).apply {
    putExtra("ExtraData", "Extra value")
}
```

Then, we call `ContextCompat.startForegroundService(Context, Intent)` to fire off `Intent` and launch the service.

Exercise 8.02 – tracking your SCA's work with a Foreground Service

In the first exercise, you tracked the SCA as it was preparing to head out using the `WorkManager` class and multiple Worker instances showing toasts. In this exercise, you will track the SCA as it deploys to the field and moves toward the assigned target by showing a sticky notification counting down the time to arrival at the destination.

This notification will be driven by a Foreground Service, which will present and continuously update it. Clicking the notification at any time will launch your main activity if it's not already running and will always bring it to the foreground:

1. Start by adding the `WorkManager` dependency to your app's `build.gradle` file:

    ```
    implementation "androidx.work:work-runtime:2.8.0"
    ```

2. Create a new class called `RouteTrackingService`, extending the abstract `Service` class:

    ```
    class RouteTrackingService : Service() {
      override fun onBind(intent: Intent): IBinder? = null
    }
    ```

You will not rely on binding in this exercise, so it is safe to simply return `null` in the `onBind(Intent)` implementation.

3. In the newly created service, define some constants that you will later need, as well as the `LiveData` instance used to observe progress:

```
companion object {
    const val NOTIFICATION_ID = 0xCA7
    const val EXTRA_SECRET_CAT_AGENT_ID = "scaId"
    private val mutableTrackingCompletion =
        MutableLiveData<String>()
    val trackingCompletion: LiveData<String> =
        mutableTrackingCompletion
}
```

`NOTIFICATION_ID` has to be a unique identifier for the notification owned by this service and must not be 0. Now, `EXTRA_SECRET_CAT_AGENT_ID` is the constant you will use to pass data to the service. `mutableTrackingCompletion` is private and is used to allow you to post completion updates internally via `LiveData` without exposing the mutability outside of the service. `trackingCompletion` is then used to expose the `LiveData` instance for observation in an immutable fashion.

4. Add a function to your `RouteTrackingService` class to provide `PendingIntent` to your sticky notification:

```
private fun getPendingIntent(): PendingIntent {
  val flag = if (Build.VERSION.SDK_INT >= Build.VERSION_
    CODES.S) FLAG_IMMUTABLE else 0
  return PendingIntent.getActivity(this, 0, Intent(this,
    MainActivity::class.java), flag)
```

This will launch `MainActivity` whenever the user clicks on `Notification`. You call `PendingIntent.getActivity()`, passing a context, no request code (0), `Intent` that will launch `MainActivity`, and the `FLAG_IMMUTABLE` flag if available, otherwise no flags (0) to it. You get back a `PendingIntent`, which will launch that activity.

5. Add another function to create `NotificationChannel` on devices running Android Oreo or newer, and return a channel ID:

```
private fun createNotificationChannel(): String =
  if (Build.VERSION.SDK_INT >= Build.VERSION_CODES.O) {
    val newChannelId = "CatDispatch"
    val channelName = "Cat Dispatch Tracking"
```

```
            val channel = NotificationChannel(
              newChannelId, channelName,
              NotificationManager.IMPORTANCE_DEFAULT
            )
            val service = requireNotNull(
              ContextCompat.getSystemService(this,
                NotificationManager::class.java)
            )
            service.createNotificationChannel(channel)
            newChannelId
        } else { "" }
```

You start by checking the Android version. You only need to create a channel if it's Android O or later. Otherwise, you can return an empty string. For Android O, you define the channel ID. This needs to be unique for a package. Next, you define a channel name that will be visible to the user.

This can (and should) be localized. We skipped that part for the sake of simplicity. A `NotificationChannel` instance is then created with the importance set to `IMPORTANCE_DEFAULT`. The importance dictates how disruptive the notifications posted to this channel are.

Lastly, a channel is created using `Notification Service` with the data provided in the `NotificationChannel` instance. The function returns the channel ID so that it can be used to construct the `Notification`.

6. Create a function to provide you with `Notification.Builder`:

```
    private fun getNotificationBuilder(pendingIntent:
    PendingIntent, channelId: String) =
        NotificationCompat.Builder(this, channelId)
          .setContentTitle("Agent approaching destination")
          .setContentText("Agent dispatched")
          .setSmallIcon(R.drawable.ic_launcher_foreground)
          .setContentIntent(pendingIntent)
          .setTicker("Agent dispatched, tracking movement")
          .setOngoing(true)
```

This function takes the `pendingIntent` and `channelId` instances generated from the functions you created earlier and constructs a `NotificationCompat.Builder` class.

The builder lets you define a title (the first row), text (the second row), a small icon (size differs based on the device) to use, the `Intent` to be triggered when the user clicks on **Notification**, and a ticker (used for accessibility; before Android Lollipop, this showed before the notification was presented).

Setting the notification to ongoing prevents users from dismissing them. This also prevents Android from muting the notification due to frequent updates.

You can set other properties, too. Explore the `NotificationCompat.Builder` class. In a real project, remember to use string resources from `strings.xml` rather than hardcoded strings.

7. Implement the following code to introduce a function to start the foreground service:

   ```
   private fun startForegroundService():
   NotificationCompat.Builder {
     val pendingIntent = getPendingIntent()
     val channelId = if (Build.VERSION.SDK_INT >=
       Build.VERSION_CODES.O) {
         createNotificationChannel()
     } else { "" }
     val notificationBuilder = getNotificationBuilder(
       pendingIntent, channelId)
     startForeground(NOTIFICATION_ID, notificationBuilder.build())
     return notificationBuilder
   }
   ```

   You first get `PendingIntent` using the `getPendingIntent` function you introduced earlier. Then, depending on the API level of the device, you create a notification channel and get its ID or set an empty ID.

   You pass `pendingIntent` and `channelId` to the function that constructs a `NotificationCompat.Builder` and start the service as a Foreground Service, providing it with `NOTIFICATION_ID` and a notification built using the builder. The function returns `NotificationCompat.Builder`, to be used later to update the notification.

8. Define two fields in your service—one to hold a reusable `NotificationCompat.Builder` class, and another to hold a reference to `Handler`, which you will later use to post work in the background:

   ```
   private lateinit var notificationBuilder:
   NotificationCompat.Builder
   private lateinit var serviceHandler: Handler
   ```

9. Next, override `onCreate()` to start the service as a foreground service and keep a reference of the `Notification.Builder`, and create `serviceHandler`:

   ```
   override fun onCreate() {
     super.onCreate()
     notificationBuilder = startForegroundService()
     val handlerThread =
       HandlerThread("RouteTracking").apply { start() }
     serviceHandler = Handler(handlerThread.looper)
   }
   ```

   Note that to create the `Handler` instance, you must first initialize and start `HandlerThread`.

10. Define a call that tracks your deployed SCA as it approaches its designated destination:

    ```
    private fun trackToDestination(notificationBuilder:
    NotificationCompat.Builder) {
      val notificationManager = getSystemService(NOTIFICATION_
        SERVICE) as NotificationManager
      for (i in 10 downTo 0) {
        Thread.sleep(1000L)
        notificationBuilder.setContentText(
          "$i seconds to destination").setSilent(true)
        notificationManager.notify(NOTIFICATION_ID,
          notificationBuilder.build())
      }
    }
    ```

    This will first obtain a reference to the `NotificationManager`. Then, it will count down from 10 to 0, sleeping for 1 second between updates and then updating the notification with the remaining time. Note we set the notification to be silent. This avoids the notification playing a sound every second.

11. Add a function to notify observers of completion on the main thread:

    ```
    private fun notifyCompletion(agentId: String) {
        Handler(Looper.getMainLooper()).post {
            mutableTrackingCompletion.value = agentId
        }
    }
    ```

By posting on a handler using the main `Looper`, you make sure that updates occur on the main (UI) app thread. When setting the value to the agent ID, you are notifying all observers that that agent ID has reached its destination.

12. Override `onStartCommand(Intent?, Int, Int)` like so:

    ```
    override fun onStartCommand(intent: Intent?, flags:
    Int, startId: Int): Int {
      val returnValue = super.onStartCommand(intent,
        flags, startId)
      val agentId = intent?.getStringExtra(EXTRA_SECRET_CAT_
        AGENT_ID)
        ?: throw IllegalStateException("Agent ID must be
        provided")
      serviceHandler.post {
        trackToDestination(notificationBuilder)
        notifyCompletion(agentId)
        stopForeground(true)
        stopSelf()
      }
      return returnValue
    }
    ```

    You first delegate the call to `super`, which internally calls `onStart()` and returns a backward-compatible state you could return. You store this returned value. Next, you obtain the SCA ID from the extras passed via the `Intent`. This service will not work without an agent ID, so you throw an exception if one is not provided.

    Next, you switch to the background thread defined in `onCreate` to track the agent to its destination in a blocking way. When tracking is done, you notify observers that the task is complete, stop the foreground service (removing the notification by passing `true`), and stop the service itself, as you don't expect to require it again soon. You then return the earlier stored return value from `super`.

13. Update your `AndroidManifest.xml` file to request the necessary permissions and introduce the service:

    ```
    <manifest ...>
      <uses-permission android:name="android.permission.
    FOREGROUND_SERVICE" />
      <uses-permission android:name="android.permission.POST_
    NOTIFICATIONS" />
    ```

```xml
<application ...>
  <service
    android:name=".RouteTrackingService"
    android:enabled="true"
    android:exported="true" />
  <activity ...>
```

First, we declare that our app requires the FOREGROUND_SERVICE permission. Unless we do so, the system will block our app from using foreground services. We also request the POST_NOTIFICATIONS permission, without which we cannot present notifications on SDK 33+. Next, we declare the service. Setting android:enabled="true" tells the system it can instantiate the service.

The default is "true", so this is optional. Defining the service with android:exported="true" tells the system that other applications can start the service. In our case, we don't need this extra functionality, but we have added it just so that you are aware of this capability.

14. Back to your MainActivity. Introduce a function to launch RouteTrackingService:

    ```kotlin
    private fun launchTrackingService() {
      RouteTrackingService.trackingCompletion.observe(
        this) { agentId ->
          showResult("Agent $agentId arrived!")
      }
      val serviceIntent = Intent(this,
        RouteTrackingService::class.java).apply {
          putExtra(EXTRA_SECRET_CAT_AGENT_ID, "007")
      }
      ContextCompat.startForegroundService(this,
        serviceIntent)
    }
    ```

    This function first observes the trackingCompletion LiveData for completion updates, showing a result on completion. Then, it defines an Intent for launching the service, setting the SCA ID as an extra parameter for that Intent. It then launches the service as a foreground service using ContextCompat, which hides away compatibility-related logic for you.

15. Now, extract the logic from onCreate() (everything that comes after the setContentView call) to a private function named dispatchCat.

16. Update `dispatchCat` to start tracking the SCA as soon as it is suited up and ready to go:

    ```
    workManager.getWorkInfoByIdLiveData(
      catSuitUpRequest.id).observe(this) { info ->
      if (info.state.isFinished) {
        showResult("Agent done suiting up. Ready to go!")
        launchTrackingService()
      }
    }
    ```

17. Create a new private function called `ensurePermissionGrantedAndDispatchCat`. In this function, make sure you have the `POST_NOTIFICATIONS` permission. Request it if you don't. Refer to *Chapter 7, Android Permissions and Google Maps*, for the implementation details. If or when you have the permission, call `dispatchCat`.

18. Launch the app:

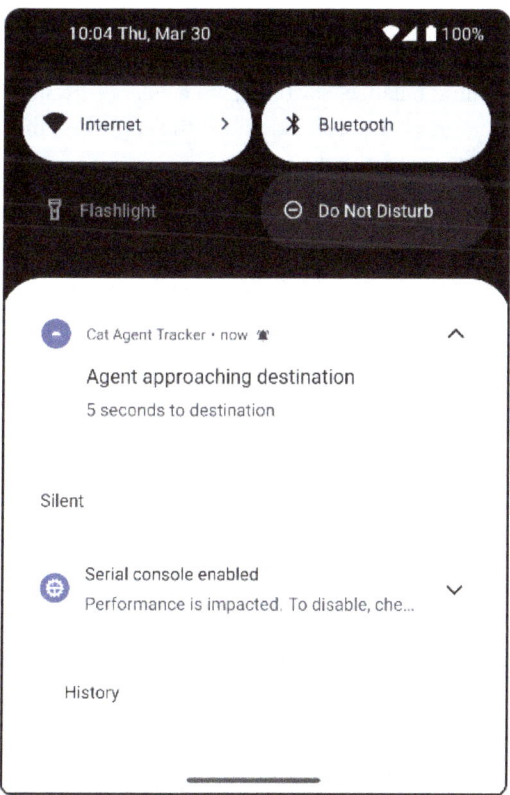

Figure 8.2 – The notification counting down

After the notifications informing you of the SCA's preparation steps, you should see a notification in your status bar. That notification should then count down from 10 to 0, disappear, and be replaced by a toast informing you that the agent arrived at its destination. Seeing that last toast tells you that you managed to communicate the SCA ID to the service as well as get it back on completion of the background task.

With all the knowledge gained from this chapter, let's complete the following activity.

Activity 8.01 – reminder to drink water

The average human loses about 2,500 **milliliters** (**ml**) of water per day. To stay healthy, we need to consume as much water as we lose. However, due to the busy nature of modern life, a lot of us forget to stay hydrated regularly.

> **Note**
> See https://packt.link/90nbQ for more information on this.

Suppose you wanted to develop an app that keeps track of your water loss (statistically) and gives you a constant update on your fluid balance. Starting from a balanced state, the app would gradually decrease the user's tracked water level. The user could tell the app when they drank a glass of water, and it would update the water level accordingly.

The continuous updating of the water level will leverage your knowledge of running a background task, and you will also utilize your knowledge of communicating with a service to update a balance in response to user interaction.

The following steps will help you complete the activity:

1. Create an empty activity project and name your app My Water Tracker.
2. Add the foreground service and post notifications permissions to your AndriodManifest.xml file.
3. Create a new service.
4. Define a variable in your service to track the water level.
5. Define constants for a notification ID and for an extra Intent data key.
6. Set up the creation of the notification from the service.
7. Add functions to request the notification permissions (if required), start the Foreground Service and update the water level.
8. Set the water level to decrease by 0.144 ml every 5 seconds.
9. Handle the addition of fluids from outside the service.

10. Make sure the service cleans up callbacks and messages when destroyed.
11. Register the service in the `Manifest.xml` file.
12. Start the service from `MainActivity` when the activity is created, after the notification permission is granted, if needed.
13. Add a button to the main activity layout with a **Drank a Glass of Water** label.
14. When the user clicks the button, notify the service that it needs to increment the water level by 250 ml.

> **Note**
> The solution to this activity can be found at `https://packt.link/By7eE`.

Summary

In this chapter, we learned how to execute long-running background tasks using `WorkManager` and foreground services. We discussed how to communicate progress to the user, and how to get the user back into an app once a task is finished executing. All the topics covered in this chapter are quite broad, and you could explore communicating with services, building notifications, and using the `WorkManager` class further.

For most common scenarios, you now have the tools you need. Common use cases include background downloads, the background cleaning up of cached assets, playing media while the app is not running in the foreground, and, combined with the knowledge we gained from *Chapter 7, Android Permissions and Google Maps*, tracking the user's location over time.

In the next chapter, we will look into making our apps more robust and maintainable by writing unit and integration tests. This is particularly helpful when the code you write runs in the background, and it is not immediately evident when something goes wrong.

9
Building User Interfaces Using Jetpack Compose

In this section, you will learn how to use Jetpack Compose to create user interfaces using Kotlin code, how Compose revolutionized the way we built user interfaces, and how we can translate existing applications to Jetpack Compose. By the end of the chapter, you will be familiar with the most common UI elements in Compose and how to handle user actions.

We will cover the following topics in this chapter:

- What is Jetpack Compose?
- Handling user actions
- Theming in Compose
- Adding Compose to existing projects

Technical requirements

The complete code for all the exercises and the activity in this chapter is available on GitHub at https://packt.link/kb5FW

What is Jetpack Compose?

In past chapters, you learned how to set data into the Android View hierarchy and learned how to use different types of views for different purposes. That approach to user interface building is referred to as the **imperative approach**.

In the imperative approach, when we want to change the state of the user interface, we will need to manually change each user interface element until we reach the desired outcome.

Let's assume that because of a user action, we want our `TextView` to change the text and text color. This means that we would need to invoke `setText` and `setTextColor` to achieve our desired effect.

As an alternative to the imperative approach, we have the **declarative approach**, in which we would need to describe the final state that we want our user interface to reach, and internally, the required invocations would be performed.

This means that our `TextView` would instead have the text and text color as attributes, and we could define different objects holding the states we want. In Jetpack Compose, this would look like the following example:

```
@Composable
fun MyTextDisplay(myState: MyState) {
    Text(text = myState.text, color = myState.color)
}
data class MyState(
    val text: String,
    val color: Color
)
```

In the preceding example, we define a `@Composable` function that will display a `Text` element. The values for `text` and `color` would be held in a separate data class that will represent the state of the text.

This mechanism allows Jetpack Compose to redraw the user interface for any change that occurs in `MyState`. This process is called **recomposition**, and the way it works is that Compose will invoke the `@Composable` functions that had a change.

If we want to create a new screen in Compose, we have the choice of `Row` or `Column` functions to arrange elements. For vertical arrangement, we can use `Column`, and for horizontal alignment, we use Row:

```
@Composable
fun MyScreen(){
    Column {
        Text(text = "My Static Text")
        TextField(value = "My Text Field", onValueChange = {

        })
        Button(onClick = { }) { }
        Icon(painter = painterResource(R.drawable.icon),
            contentDescription = stringResource(id =
            R.string.icon_content_description))
```

         }
}

In the preceding example, using `Column`, we show different user interface elements above each other:

- `Text` will show a simple label with the `My Static Text` label as `text`.
- `TextField` will show an input field that's pre-filled with the `My Text Field` text through the `value` parameter. The `onValueChange` lambda will pick up any text that the user inserts.
- `Button` will show a button, and through the `onClick` lambda, we can pick up the click events on it.
- `Icon` will display a particular icon from the `drawable` or `mipmap` folders and will have a content description set from the string resources.

If we want to display lists of items, we can use the following approach when the number of items is known and small enough to fit on the screen of the device:

```
@Composable
fun MyList(items: List<String>) {
    Column {
        items.forEach { item -> Text(text = item) }
    }
}
```

In the preceding example, we iterate over each item in the list and show a `Text` for each item. If the number of items is unknown and large enough to require scrolling, then we would run into performance issues because the recomposition would go over both visible and not visible items. For this scenario, we could use the following approach:

```
@Composable
fun MyList(items: List<String>) {
    LazyColumn {
        item { Text(text = "Header") }
        items(items) { item-> Text(text = item) }
        item { Text(text = "Footer") }
    }
}
```

Here, we are using a `LazyColumn` function and wrapping our collection with the `items` function. We are also able to add static items as a header and footer of the list through the `item` function. Like `Column`, we can use `LazyRow` to display a list of items with a horizontal scroll.

If we want to add spacings between the UI elements, then `Modifiers` become useful:

```
@Composable
fun MyScreen(){
    Column {
        Text(text = "My Static Text")
        TextField(value = "My Text Field", onValueChange =
        { }
    }
}
```

In the preceding example, we've added `16.dp` padding to the entire column of items in all directions. If we wanted different paddings in different directions, we would've had something like the following:

```
@Composable
fun MyScreen() {
    Column(
        modifier = Modifier.padding(
            top = 5.dp, bottom = 5.dp,
            start = 10.dp, end = 10.dp)
    ) {
    }
}
```

In the preceding example, we've set `5.dp` vertical padding and `10.dp` horizontal padding. Because the values are repeated, we can instead use the following approach:

```
@Composable
fun MyScreen() {
    Column(
        modifier = Modifier.padding(
            vertical = 5.dp,
            horizontal = 10.dp)
    ) {
    }
}
```

Here, we've used the `vertical` and `horizontal` parameters to set the vertical and horizontal paddings. If we want to make the row clickable, then we can use the following modifier:

```
@Composable
fun MyScreen() {
    Column(
        modifier = Modifier.padding(
            vertical = 5.dp,
            horizontal = 10.dp
        ).clickable { }
    ) { }
}
```

In the preceding example, we've used the `clickable` method from `Modifier`. This will register a click listener on the entire `Column`.

When it comes to the relationship between Activities and Fragments and screens built in Compose, we have the following extensions for Compose:

```
class MainActivity : ComponentActivity() {
    override fun onCreate(savedInstanceState: Bundle?) {
        super.onCreate(savedInstanceState)
        setContent { MyScreen() }
    }
}
```

In the preceding example, we have the `setContent` extension function defined for `ComponentActivity`, which will set a `@Composable` function as the content for an `Activity`. The same is true for Fragments; however, when starting a new project, the current recommendation is to have a single Activity and make all your additional screens separate `@Composable` functions.

> **Note**
>
> In Android Studio, we have the possibility of having previews of our `@Composable` functions with the `@Preview` annotation.

Exercise 9.01 – first Compose screen

Create an Android application that has one screen defined using `@Composable` functions. The screen should have the following elements:

- `Text`, which will say `"Enter a number"`
- `TextField`, which will only accept integers
- `Button`, which has the text `"Click Me"`
- `LazyColumn`, which will display a list of 100 items with the following format for each row: `"Item c"`

Perform the following steps to complete the exercise:

1. Create a new Android Studio project and select **Empty Compose Activity**.
2. Add the following to `strings.xml` in the `res/values` folder:

   ```
   <string name="enter_number">Enter a
       number</string>
   <string name="click_me">Click Me</string>
   <string name="item_format">Item %s</string>
   ```

3. Define the user interface of the exercise:

   ```
   @Composable
   fun MyScreen(
       items: List<String>
   ) {
       LazyColumn {
           item {
               Column(modifier = Modifier.padding(16.dp))
               {
                   Text(text = stringResource(id =
                   R.string.enter_number))
                   TextField(
                       value = "",
                       keyboardOptions =
                       KeyboardOptions(keyboardType =
                       KeyboardType.Number),
                       onValueChange = {
   ```

```
                })
                Button(onClick = { }) {
                    Text(text = stringResource(id =
                    R.string.click_me))
                }
            }
        }
        items(items) { item ->
            Column(modifier =
            Modifier.padding(vertical = 4.dp)) {
                Text(text = item)
            }
        }
    }
}
```

We have chosen to place everything inside the `LazyColumn` block. This will make the entire content scrollable, including `Text`, `TextField`, and `Button`, and not just the list of items.

To make the keyboard only accept numeric input, we have used the `keyboardOptions` parameter of the `TextField` function. For `Button`, to add text to it, we needed to use the content parameter of the function and place a new `Text` inside the lambda.

4. Finally, modify the `MainActivity` code to use the function we've just defined:

```
class MainActivity : ComponentActivity() {
    override fun onCreate(savedInstanceState: Bundle?) {
        super.onCreate(savedInstanceState)
        setContent {
            val items = (1..100).toList().map {
                stringResource(id =
                R.string.item_format, formatArgs =
                arrayOf("$it"))
            }
            MyScreen(items)
        }
    }
}
```

Here, we generate a new list of items that will say "`Item [count]`" and invoke the `@Composable` function, `MyScreen`.

If we run the preceding example, we will see the following output:

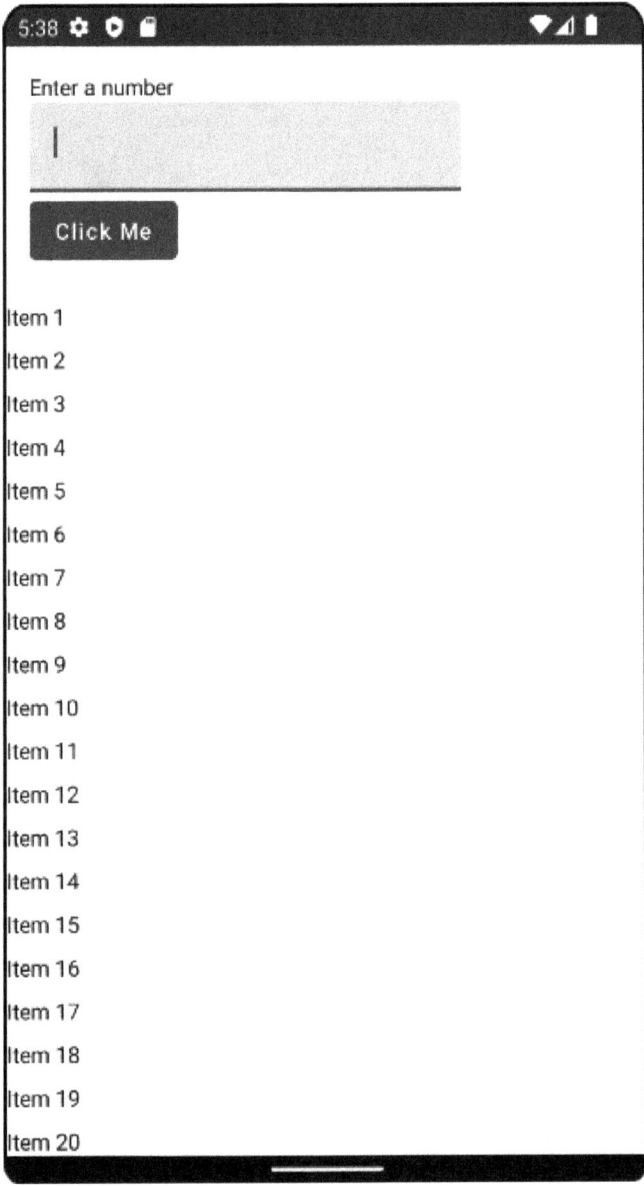

Figure 9.1 – Output of Exercise 9.01

We can see in the preceding figure that the screen was built according to the exercise specifications. One thing to note is the fact that we aren't allowed to input any text in the input field. This is because we have set the value of `TextField` to an empty string. We will look at how we can properly handle this aspect in further sections of the book.

In this section, we have looked at how we can use Jetpack Compose to build user interfaces in a simple way without the involvement of other languages and syntaxes such as Kotlin. Next, we will continue the exploration of Compose and how we can handle user actions and manage states.

Handling user actions

In the previous section, we learned how to build user interfaces using Jetpack Compose. In the exercise, we were unable to collect the data that the user sets in `TextField`. In this section, we will learn how to handle the user input as well as the state of the user interface.

Let's assume we have the following example:

```
@Composable
fun MyScreen() {
    Column { TextField(value = "", onValueChange = {}) }
}
```

In this example, we define a `TextField` that is empty and has no handling for the change of the value. As we've seen, this won't let us introduce any new input from the keyboard because it will always set the text to an empty string.

For us to be able to introduce a new text, we will need to create a mutable variable to store the text inside and for it to survive recomposition. In Jetpack Compose, we can use the `@Composable` function called `remember` to define a `MutableState` that will hold our text:

```
@Composable
fun MyScreen() {
    var text by remember { mutableStateOf("") }
    Column { TextField(value = text, onValueChange = {}) }
}
```

In the preceding example, we define a mutable variable called `text`, which we then set in `TextField`. The `text` variable is initialized through the `remember` function, which will hold a `MutableState` that has the initial value set to an empty string. This still isn't complete; we will now need to connect the state with the change in value for `TextField`:

```
@Composable
fun MyScreen() {
```

```
        var text by remember { mutableStateOf("") }
        Column {
            TextField(value = text, onValueChange = { text = it })
        }
    }
}
```

Here, we modify the `onValueChange` lambda to change the `text` state with the latest text that the user inserted.

> **Note**
> We can use the `rememberSaveable` function to instead retain the value of objects across configuration changes such as Activity recreation.

When dealing with states, we tend to turn `@Composable` functions from being stateless (doesn't manage a state) to stateful (manages one or more states). As a guideline, we should try to keep our functions as stateless as possible through a pattern called **state hoisting**. This involves moving the state management to the caller of our `@Composable` function. This would turn the preceding example into the following:

```
@Composable
fun MyScreen() {
    var text by rememberSaveable { mutableStateOf("") }
    MyScreenContent(text = text, onTextChange = { text = it })
}


@Composable
fun MyScreenContent(text: String, onTextChange: (String) -> Unit) {
    Column {
        TextField(value = text, onValueChange =
            onTextChange)
    }
}
```

In the preceding example, we've split our functions in two. The `MyScreen` function will manage the state of the text and invoke the `MyScreenContent` function, which is now stateless. This approach introduces multiple benefits such as the *reusability* of our stateless functions, *decoupled* state management, and *a single source of truth* for our state.

> **Note**
> When you use the Jetpack Compose State and `MutableState` objects, you might need to manually import the following two methods for getting and setting a state: `androidx.compose.runtime.getValue` and `androidx.compose.runtime.setValue`.

When dealing with states in Jetpack Compose, when a change occurs in a state, then the recomposition process is triggered. This might cause problems when we want to show one-off events such as `Snackbar` and `Toast`. To achieve this, we can use `LaunchedEffect`:

```
@Composable
fun MyScreenContent() {
    val context = LocalContext.current
    LaunchedEffect(anObjectToChange) {
        Toast.makeText(context, "Toast text",
        Toast.LENGTH_SHORT).show()
    }
}
```

The preceding example will show a `Toast` message every time `anObjectToChange` takes on a different value. If we replace `anObjectToChange` with `Unit`, then the `LaunchedEffect` block will be executed only once.

Exercise 9.02 – handling user inputs

Modify *Exercise 9.01 – first Compose screen* such that when the user introduces a number in `TextField` and clicks the button, then a list of items as big as the number introduced will be generated and the list below the button will be populated with them. The text will be the same for each item as before.

To represent the state of the user interface, a data class will be created that will hold the number of items, with 0 as a default, and the list of items, which will be empty by default.

Perform the following steps to complete the exercise:

1. Create the `MyScreenState` data class, which will hold the state of the user interface:

    ```
    data class MyScreenState(
        val itemCount: String = "",
        val items: List<String> = emptyList()
    )
    ```

2. Create the `@Composable` method called `MyScreenContent`, which will have `MyScreenState` as a parameter and render the state:

```
@Composable
fun MyScreenContent(
    myScreenState: MyScreenState,
    onItemCountChange: (String) -> Unit,
    onButtonClick: @Composable () -> Unit
) {
    LazyColumn {
        item {
            Column(modifier = Modifier.padding(16.dp))
            {
                Text(text = stringResource(id =
                R.string.enter_number))
                TextField(
                    value = myScreenState.itemCount,
                    keyboardOptions =
                    KeyboardOptions(keyboardType =
                    KeyboardType.Number),
                    onValueChange =
                    onItemCountChange
                )
                Button(onClick = onButtonClick) {
                    Text(text = stringResource(id =
                    R.string.click_me))
                }
            }
        }
        items(myScreenState.items) { item ->
            Column(modifier =
            Modifier.padding(vertical = 4.dp)) {
                Text(text = item)
            }
        }
    }
}
```

In the preceding example, we set `itemCount` from `myScreenState` in our `TextField` and `items` from `myScreenState` as items in the list. We've also added our text change listener and button listener as parameters to the function, making it stateless.

3. Modify the `MyScreen` function such that it will call `MyScreenContent` and handle the listener for the text change and for the button click:

```
@Composable
fun MyScreen() {
    var state by remember {
        mutableStateOf(MyScreenState())
    }
    val context = LocalContext.current
    MyScreenContent(state, {
        state = state.copy(itemCount = it)
    }, {
        state = state.copy(items =
        (1..state.itemCount.toInt()).toList().map {
            context.getString(R.string.item_format,
            "$it")
        })
    })
}
```

Here, we are creating a new `MutableState` that will hold `MyScreenState` with its defaults. We will then invoke `MyScreenContent` in which we pass the state. When the text changes, we set the state to be the copy of the existing state with the new text, and when the button is clicked, we generate a new list of items up until the current `itemCount` and update the state.

4. Update the `MainActivity` class to invoke the `MyScreen` function without any parameters:

```
class MainActivity : ComponentActivity() {
    override fun onCreate(savedInstanceState: Bundle?) {
        super.onCreate(savedInstanceState)
        setContent {
            MyScreen()
        }
    }
}
```

If we run the exercise and insert a number, then we should see the following screen:

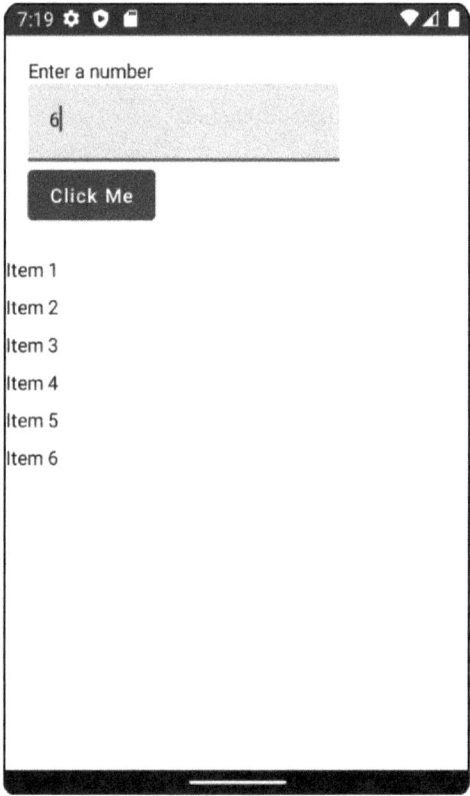

Figure 9.2 – Output of Exercise 9.02

When the app is first launched, we should see the `TextField` empty and no elements in the list below the button. When a number is set, then the state is changed to reflect the new text, and when the button is clicked, then the list of items is displayed with the size being that of the inserted number.

In this section, we have looked at how we can handle user input, keep it in a state, and manage that state across recomposition. In the section that follows, we will look at how we can further decorate our user interface elements.

Theming in Compose

In the previous section, we learned how to handle user actions and how to manage the state of a particular screen. But how do we keep our application's user interface elements consistent across the entire application? In this section, we will look at how we can create reusable elements that are linked to the application's theme.

You may have noticed, when you carried out the previous exercises, that Android Studio created some files in a `ui.theme` package. This is because Jetpack Compose is built upon the Material Design library and will assign a theme to your application that is built on Material Design. The approach taken is the following:

1. In the `Color.kt` file, all the colors of the application are declared:

    ```
    val Purple200 = Color(0xFFBB86FC)
    val Purple500 = Color(0xFF6200EE)
    val Purple700 = Color(0xFF3700B3)
    val Teal200 = Color(0xFF03DAC5)
    ```

    In the preceding example, we have the color hexadecimal names.

2. In `Shape.kt`, the following code is generated:

    ```
    val Shapes = Shapes(
        small = RoundedCornerShape(4.dp),
        medium = RoundedCornerShape(4.dp),
        large = RoundedCornerShape(0.dp)
    )
    ```

    This will indicate what the size of the icons you use in the application should be.

3. In `Type.kt`, the following code is generated:

    ```
    val Typography = Typography(
        body1 = TextStyle(
            fontFamily = FontFamily.Default,
            fontWeight = FontWeight.Normal,
            fontSize = 16.sp
        )
    )
    ```

    This will represent how the text in your application is rendered. The `Typography` class holds configurations for how your headings, subtitles, paragraphs, buttons, and captions text should be.

4. In the `Theme.kt` file, we have two color palettes defined:

    ```
    private val DarkColorPalette = darkColors(
        primary = Purple200,
        primaryVariant = Purple700,
        secondary = Teal200
    )
    ```

```
private val LightColorPalette = lightColors(
    primary = Purple500,
    primaryVariant = Purple700,
    secondary = Teal200
)
```

Here, there's a light and dark color palette defined and the `primary`, `primaryVariant`, and `secondary` colors are set. The rest of the colors in the `lightColors` and `darkColors` functions will remain with their default values.

5. In the same file, the application's theme is generated:

```
@Composable
fun MyApplicationTheme(
    darkTheme: Boolean = isSystemInDarkTheme(),
    content: @Composable () -> Unit
) {
    val colors = if (darkTheme) {
        DarkColorPalette
    } else {
        LightColorPalette
    }
    MaterialTheme(
        colors = colors,
        typography = Typography,
        shapes = Shapes,
        content = content
    )
}
```

Here, a check will be made to see whether the device has light or dark mode enabled and use the appropriate set of colors for each mode. It will also set the typography you configured and how the shapes in your application should be. Just because it is set in the theme, that doesn't mean that our user interface elements will automatically inherit it.

6. When the `MainActivity` class is generated, it will have the following structure:

```
class MainActivity : ComponentActivity() {
    override fun onCreate(savedInstanceState: Bundle?) {
        super.onCreate(savedInstanceState)
        setContent {
```

```
                MyApplicationTheme {
                    Surface(color =
                    MaterialTheme.colors.background) { }
                }
            }
        }
    }
```

When `setContent` is called, then your application's theme will be called, and the `Surface` function will set your application's background.

We can now use the preceding setup as a starting point to define a theme for our application and start to create reusable user interface components. Let's assume we want all the paragraphs in the application to use the same typography and have the same color; in this case, we will use `MaterialTheme.typography.body1` and `MaterialTheme.colors.onBackground`:

```
@Composable
fun ParagraphText(text: String) {
    Text(
        text = text,
        style = MaterialTheme.typography.body1,
        color = MaterialTheme.colors.onBackground
    )
}
```

In the preceding example, we've defined the `ParagraphText` function, which will set the style and color of the text from `MaterialTheme`. We might now have a problem if we want the same style and a different text color, where we need to duplicate the style attribute for each. Another solution is to create two functions – one for the style and the other on top of it for the color:

```
@Composable
fun OnBackgroundParagraphText(text: String) {
    ParagraphText(text = text, color =
    MaterialTheme.colors.onBackground)
}
@Composable
fun ParagraphText(text: String, color: Color) {
    Text(
        text = text,
        style = MaterialTheme.typography.body1,
```

```
        color = color
    )
}
```

In the preceding example, we've moved the color to be a parameter in the `ParagraphText` function and then created a new function called `OnBackgroundParagraphText`, which allows us to set `MaterialTheme.colors.onBackground` to the Text defined in `ParagraphText`. If we want to use our new function, then we can do the following:

```
@Composable
fun MyScreen() {
    OnBackgroundParagraphText(text = "My text")
}
```

This is a simple function call, just like using the `Text` function.

Now, let's assume we are using this text across our entire application and the application goes through a redesign where instead of using `MaterialTheme.typography.body1`, we need to use `MaterialTheme.typography.body2`, and the text color needs to be red. In this case, we would then modify the `ParagraphText` function as follows:

```
@Composable
fun ParagraphText(text: String, color: Color) {
    Text(
        text = text,
        style = MaterialTheme.typography.body2,
        color = color
    )
}
```

Here, we have changed the style of the `Text` function to use `MaterialTheme.typography.body2`. To change the color, we could modify `OnBackgroundParagraphText`, but the color currently used is recommended to be used on top of the current background, so we could also change the value of `MaterialTheme.colors.onBackground`. For this, we could go into `Theme.kt` and do the following:

```
private val DarkColorPalette = darkColors(
    primary = Purple200,
    primaryVariant = Purple700,
    secondary = Teal200,
    onBackground = Color.Red
```

```
)
private val LightColorPalette = lightColors(
    primary = Purple500,
    primaryVariant = Purple700,
    secondary = Teal200,
    onBackground = Color.Red
)
```

Here, we changed the value of `onBackground` to red, which will impact all the user interface elements that reference `onBackground`. We can now see how we can easily apply this across all the user interface elements in the application without touching the code where those elements are used.

If we want to have multiple screens in our application, we can connect Compose with the `navigation` library, which is available here:

```
implementation "androidx.navigation:navigation-compose:2.5.3
```

Let's now assume we have two screens defined in Jetpack Compose:

```
@Composable
fun Screen1(onButtonClick: () -> Unit) {
    Button(onClick = onButtonClick) {
        Text(text = "Click Me")
    }
}
@Composable
fun Screen2(input1: String, input2: String) {
    Text(text = "My inputs are $input1 and $input2")
}
```

`Screen1` will display one button and `Screen2` has two inputs that will be displayed. We now want to connect the screens so that when the button is clicked on `Screen1`, `Screen2` opens with two hardcoded inputs passed. This would look like the following:

```
@Composable
fun MyApp(navController: NavHostController) {
    NavHost(navController = navController,
        startDestination = "screen1") {
        composable("screen1") {
            Screen1 { navController.navigate
```

```
                ("screen2/Input1?input2=Input2") }
        }
        composable(
            "screen2/{input1}?input2={input2}",
            arguments = listOf(navArgument("input1") {
            type = NavType.StringType },
            navArgument("input2") { type =
            NavType.StringType }
            )
        ) {
            Screen2(
                input1 = it.arguments?
                .getString("input1").orEmpty(),
                input2 = it.arguments?
                .getString("input2").orEmpty()
            )
        }
    }
}
```

We have defined a new @Composable function called MyApp, which uses NavHost to keep all the screens in the application. NavHost will open Screen1 as a default through the screen1 URL. In the onButtonClick lambda from Screen1, we navigate to Screen2 and we pass the input1 and input2 strings.

This is done through the screen2/{input1}?input2={input2} URL. This is also how we will pass arguments between the two screens, through either the path parameter (input1) or the argument (input2). For each input, we will need to specify that we will be expecting a string as a type.

Screen2 will then be opened, and the input extracted through the it variable, which is a NavBackStackEntry type. We can call this function from the setContent method of the Activity:

```
class MainActivity : ComponentActivity() {
    override fun onCreate(savedInstanceState: Bundle?) {
        super.onCreate(savedInstanceState)
        setContent {
            MyApplicationTheme {
                Surface(color =
                MaterialTheme.colors.background) {
```

```
                    val navController =
                        rememberNavController()
                    MyApp(navController)
                }
            }
        }
    }
}
```

Here, we hoist the state of `NavHostController` and then call the `MyApp` function.

Exercise 9.03 – applying themes

Modify *Exercise 9.02 – handling user inputs* such that `MyScreen`, `MyScreenContent`, and `MyScreenState` are split into two screens, with `ItemCountScreen`, `ItemCountScreenContent`, and `ItemCountScreenState` on one side, which will hold `Text`, `TextField`, and `Button`, and `ItemScreen`, `ItemScreenContent`, and `ItemScreenState` on the other, which will hold the item list.

The two screens will be saved in `ItemCountScreen.kt` and `ItemScreen.kt` files. `ItemCountScreen` will be shown first, and when the button is clicked, then `ItemScreen` is shown with the number of items set in the previous screen.

New functions will also be created to represent `Text` used across the application: one for the `"Enter a number"` text, which will be `MaterialTheme.typography.h5`; the `"Click Me"` text will be `MaterialTheme.typography.button`; and `"Item [count]"` will be `MaterialTheme.typography.body1`.

The colors will be set to `MaterialTheme.colors.onBackground` for the text, and `Color.red` for the button text.

Perform the following steps to complete the exercise:

1. In the `app/build.gradle` file, add the `navigation` library dependency:

    ```
    implementation "androidx.navigation:navigation-
    compose:2.5.3"
    ```

2. In the `ui.theme` package, create a Kotlin file called `Elements`.
3. In the `Elements.kt` file, add the functions for the title text on the first screen:

    ```
    @Composable
    fun OnBackgroundTitleText(text: String) {
    ```

```
        TitleText(text = text, color =
            MaterialTheme.colors.onBackground)
}
@Composable
fun TitleText(text: String, color: Color) {
    Text(text = text, style =
        MaterialTheme.typography.h5, color = color)
}
```

4. In the same file, add the functions for the "`Item [count]`" text:

```
@Composable
fun OnBackgroundItemText(text: String) {
    ItemText(text = text, color =
        MaterialTheme.colors.onBackground)
}
@Composable
fun ItemText(text: String, color: Color) {
    Text(text = text, style =
        MaterialTheme.typography.body1, color = color)
}
```

5. In the same file, add the functions for the button text:

```
@Composable
fun PrimaryTextButton(text: String, onClick: () ->
Unit) {
    TextButton(text = text, textColor = Color.Red,
        onClick = onClick)
}
@Composable
fun TextButton(text: String, textColor: Color,
onClick: () -> Unit) {
    Button(
        onClick = onClick, colors = ButtonDefaults
        .buttonColors(contentColor = textColor)
    ) {
        Text(text = text, style =
```

```
            MaterialTheme.typography.button)
        }
    }
```

In this example, as the button sets the color for the content in a different way, we had to use `contentColor` from the `ButtonColors` class.

6. Create a new Kotlin file called `ItemCountScreen`.

7. In this file, create a new class called `ItemCountScreenState`:

   ```
   data class ItemCountScreenState(
       val itemCount: String = ""
   )
   ```

8. In the same file, create a new function called `ItemCountScreenContent`, which will hold the newly created `OnBackgroundTitleText` and `PrimaryTextButton` functions:

   ```
   @Composable
   fun ItemCountScreenContent(
       itemCountScreenState: ItemCountScreenState,
       onItemCountChange: (String) -> Unit,
       onButtonClick: () -> Unit
   ) {
       Column {
           OnBackgroundTitleText(text = stringResource(id
               = R.string.enter_number))
           TextField(
               value = itemCountScreenState.itemCount,
               keyboardOptions = KeyboardOptions(
               keyboardType = KeyboardType.Number),
               onValueChange = onItemCountChange
           )
           PrimaryTextButton(text = stringResource(id =
           R.string.click_me), onClick = onButtonClick)
       }
   }
   ```

9. In the same file, create a new function called `ItemCountScreen`:

```
@Composable
fun ItemCountScreen(onButtonClick: (String) -> Unit) {
    var state by remember {
        mutableStateOf(ItemCountScreenState())
    }
    ItemCountScreenContent(state, {
        state = state.copy(itemCount = it)
    }, {
        onButtonClick(state.itemCount)
    })
}
```

10. Create a new Kotlin file called `ItemScreen`.

11. In that file, create a new class called `ItemScreenState`:

```
data class ItemScreenState(
    val items: List<String> = emptyList()
)
```

12. In the same file, create a new function called `ItemScreenContent`, which will use `OnBackgroundItemText`:

```
@Composable
fun ItemScreenContent(
    itemScreenState: ItemScreenState
) {
    LazyColumn {
        items(itemScreenState.items) { item ->
            Column(modifier =
            Modifier.padding(vertical = 4.dp)) {
                OnBackgroundItemText(text = item)
            }
        }
    }
}
```

13. In the same file, create a new function called `ItemScreen`:

    ```
    @Composable
    fun ItemScreen(itemCount: String) {
        ItemScreenContent(itemScreenState =
        ItemScreenState((1..itemCount.toInt()).toList()
        .map {
            stringResource(id = R.string.item_format,
            formatArgs = arrayOf("$it"))
        }))
    }
    ```

14. In the `MainActivity` file, create the `MyApp` function, which will manage our two screens defined previously:

    ```
    @Composable
    fun MyApp(navController: NavHostController) {
        NavHost(navController = navController,
        startDestination = "itemCountScreen") {
            composable("itemCountScreen") {
                ItemCountScreen { navController.navigate(
                "itemScreen/?itemCount=$it") }
            }
            composable(
                "itemScreen/?itemCount={itemCount}",
                arguments =
                listOf(navArgument("itemCount") {type =
                NavType.StringType })
            ) {
                ItemScreen(
                    it.arguments?.getString("itemCount")
                    .orEmpty()
                )
            }
        }
    }
    ```

15. Finally, modify the `setContent` function so that `MyApp` will be called:

```
class MainActivity : ComponentActivity() {
    override fun onCreate(savedInstanceState: Bundle?)
    {
        super.onCreate(savedInstanceState)
        setContent {
            MyApplicationTheme {
                Surface(color =
                MaterialTheme.colors.background) {
                    val navController =
                        rememberNavController()
                    Column(modifier =
                        Modifier.padding(16.dp)) {
                            MyApp(navController)
                    }
                }
            }
        }
    }
}
```

If we run the application, we should see the following output:

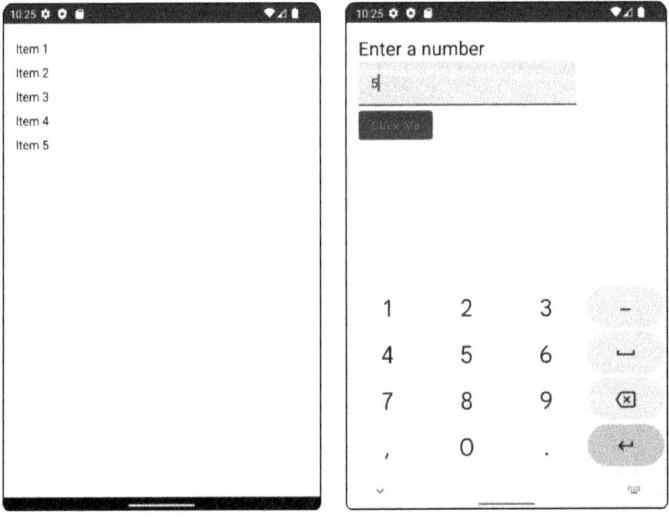

Figure 9.3 – Output of Exercise 9.03

We should be able to see the screens split in two, and on entering a number in one screen, we should transition to the other screen with a generated list of items. We should also see the newly defined styles for the Text functions. We should be able to control these styles only from the Elements class and cause no modifications to the screens themselves.

In this section, we have learned how to apply theming to an application and how we can create multiple screens and navigate between them using Jetpack Compose. In the section that follows, we will look at how we can integrate Compose into an existing project and how well it can be integrated with other popular libraries.

Adding Compose to existing projects

In this section, we will look at what options we have in terms of introducing Jetpack Compose into an existing Android application and how to get Compose to work with different libraries.

When using Jetpack Compose, you should ideally have a small number of activities, or one if possible, and have all your screens built using Compose. For an existing project to be able to achieve this, it would need to start at the bottom of the View hierarchy, meaning that your existing views should start being migrated to be built in Compose.

To facilitate this transition, Jetpack Compose offers the possibility of using ComposeView in your XML layout, as in the following example:

```
<?xml version="1.0" encoding="utf-8"?>
<LinearLayout
    xmlns:android=
        "http://schemas.android.com/apk/res/android"
    android:orientation="vertical"
    android:layout_width="match_parent"
    android:layout_height="match_parent">
    <androidx.compose.ui.platform.ComposeView
        android:id="@+id/compose_view"
        android:layout_width="match_parent"
        android:layout_height="match_parent" />
</LinearLayout>
```

Here, we have an existing layout that will need to include a view that was defined using Jetpack Compose. In the layout XML file, we can put a ComposeView placeholder of where our view would live, and then in the Kotlin code, we can include the Compose user interface element:

```
class MyFragment : Fragment() {
    override fun onCreateView(
```

```kotlin
            inflater: LayoutInflater,
            container: ViewGroup?,
            savedInstanceState: Bundle?
    ): View? {
        return inflater.inflate(
        R.layout.my_fragment_layout, container).apply {
            findViewById<ComposeView>(R.id.compose_view)
            .apply {
                setViewCompositionStrategy(
                ViewCompositionStrategy
                .DisposeOnViewTreeLifecycleDestroyed)
                setContent {
                    MaterialTheme {
                        Text("My Text")
                    }
                }
            }
        }
    }
}
```

In this example, `Fragment` inflates the XML layout, looks up `ComposeView`, and marks the Compose content to be destroyed when the `View` of `Fragment` is also destroyed to prevent any leaks, and then sets the content of `ComposeView` to be `Text`.

When we want to go the opposite route and add Android views into the Compose code, then we have the option of using `AndroidView`:

```kotlin
@Composable
fun MyCustomisedElement(text: String) {
    AndroidView(factory = { context ->
        TextView(context).apply {
            this.text = text
        }
    })
}
```

In this example, we have defined a new @Composable function called MyCustomisedElement, which will invoke AndroidView, which in turn will create a TextView on which it will set the text we have defined as a parameter.

As we've seen in previous sections, we can use LocalContext.current to obtain a Context reference. This allows us to perform actions such as starting activities and services and showing Toasts.

Compose is also able to interact with other libraries that are useful when building Android applications. We will analyze these libraries in the chapters that follow, but now, we will look at how they work with Jetpack Compose:

- The ViewModel library is useful for keeping data across configuration changes in our Activities and Fragments and helps to make our code more testable. Compose can obtain references to ViewModel objects through the @Composable function called viewModel:

    ```
    @Composable
    fun MyScreen(viewModel: MyViewModel = viewModel()) {
        Text(text = viewModel.myText)
    }
    ```

    Here, we call viewModel to obtain a reference to MyViewModel and set Text with the value that viewModel was holding.

- Data stream libraries are useful in combination with the ViewModel library, as when we want to load data asynchronously from the internet or the local filesystem, we will need to notify the user interface that the data was loaded.

    Common data stream libraries are LiveData, RxJava, and Coroutines and Flows. We've seen that Compose uses a State object when we want to manage the state of the user interface. For each of the three libraries, Compose provides extension libraries in which a stream of data is converted into a State object:

    ```
    @Composable
    fun MyScreen(viewModel: MyViewModel = viewModel()) {
        viewModel.myLiveData.observeAsState()?.let{
        myLiveDataText->
            Text(text = myLiveDataText)
        }
        viewModel.myObservable.subscribeAsState()?.let{
         myObservableText->
            Text(text = myObservableText)
        }
        viewModel.myFlow.collectAsState()?.let{
    ```

```
            myFlowText->
                Text(text = myFlowText)
        }
    }
```

In this example, our `viewModel` object would have each of the data streams that would hold a string. For each of the streams, Compose calls the equivalent method to subscribe and monitor for changes in the value of the string. When a new value is emitted for each stream, then Compose sets it in the `Text`.

- Hilt is a dependency injection library designed for Android app development. If the `navigation` library is not present in the project, then using the `viewModel` function described previously should be enough to obtain a reference to your `ViewModel`; however, if the `navigation` library is present, then a library that makes `hilt` and `navigation` work together also needs to be included:

```
implementation 'androidx.hilt:hilt-navigation-compose:1.0.0'
```

To obtain references to `ViewModel` objects in the Compose code, we will need to replace the invocation to `viewModel` with an invocation to `hiltViewModel`.

You can find more information about integrating Jetpack Compose into your Android application at https://developer.android.com/jetpack/compose, and about compatibility with other libraries here: https://developer.android.com/jetpack/compose/libraries.

In this section, we have looked at how we can integrate the Jetpack Compose library and make it work with existing `View` objects and existing libraries in the project.

Activity 9.01 – first Compose app

Create a new app using Jetpack Compose, which will have three screens:

- The **insert rows** screen will have a title, a text field, and a button where a number can be inserted. When the button is clicked, then the user navigates to the next screen.

- The **insert columns** screen will have a title, a text field where a number can be inserted, and a button. When the button is clicked, then the user navigates to the next screen.

- A grid screen will display a grid with the number of rows and the number of columns inserted above. Each row will independently scroll using `LazyRow`, and for the columns, `LazyColumn` will be used. Each grid item will display the text `"Item [row][column]"`.

The first two screens will have their user interface elements using the same styling for the titles, text fields, and buttons, and the third screen will have a style for displaying the text in the grid.

To complete this activity, you need to take the following steps:

1. Create a new Android Studio project using an empty Compose Activity.
2. Add the `navigation` library dependency to the `app/build.gradle` file.
3. In the `ui.theme` package, create a new Kotlin file called `Elements`.
4. In that file, create `@Composable` functions for the titles used in the application.
5. In the same file, create `@Composable` functions for the text fields used in the application.
6. In the same file, create `@Composable` functions for the grid items.
7. In the same file, create `@Composable` functions for the buttons.
8. Create a new Kotlin file called `InsertRowsScreen`.
9. Create `InsertRowsScreenState`, `InsertRowsScreenContent`, and `InsertRowsScreen`, which will be responsible for holding the screen state and the screen content and for managing the screen state.
10. Create a new Kotlin file called `InsertColumnsScreen`.
11. Create `InsertColumnsScreenState`, `InsertColumnsScreenContent`, and `InsertColumnsScreen`, which will be responsible for holding the screen state and the screen content and for managing the screen state.
12. Create a new Kotlin file called `GridScreen`.
13. Create `GridScreenState`, `GridScreenContent`, and `GridScreen`, which will be responsible for holding the screen state and the screen content and for managing the screen state.
14. In `MainActivity`, create a new function that will set up the navigation between your screens.
15. In `MainActivity`, modify the `setContent` method block to invoke the function created previously.

> **Note**
> The solution to this activity can be found at `https://packt.link/Le1jE`.

Summary

In this chapter, we looked at how we can build user interfaces using Jetpack Compose. We started by creating simple user interface elements, and we looked at how we can make an entire screen using `@Composable` functions without any XML code.

Then, we analyzed state management and how we can handle user input, and looked at patterns such as state hoisting, in which we keep our functions as stateless as possible to increase reusability. We then looked at how we can define our own user interface elements and apply themes and styles to them, which allows us to change the entire look of an application without modifying the screens that use the changed elements.

Finally, we looked at how we can add Compose to an existing project and how Compose interacts with popular libraries used for app development. In the chapter's activity, we applied all these concepts and created an application with a consistent user interface definition with multiple screens defined in Compose.

In the next chapter, we will analyze how we can test our code on Android and look at some popular libraries we can use to achieve this.

Part 3: Testing and Code Structure

In this part, we will look at how we can structure our code to make it testable and the types of testing we can do in the code base. The Android Architecture Components will be used to assist in code structuring by separating code that performs tasks that can be tested from code that interacts with the user interface, which is harder to test.

We will then look at the available options we have with regard to saving data on the device. Finally, we will explore how we can manage the dependencies inside the application with the help of dependency injection.

We will cover the following chapters in this section:

- *Chapter 10, Unit Tests and Integration Tests with JUnit, Mockito, and Espresso*
- *Chapter 11, Android Architecture Components*
- *Chapter 12, Persisting Data*
- *Chapter 13, Dependency Injection with Dagger, Hilt, and Koin*

10
Unit Tests and Integration Tests with JUnit, Mockito, and Espresso

In this chapter, you will learn about testing on the Android platform and how to create unit tests, integration tests, and UI tests. You will see how to create each of these types of tests, analyze how they run, and work with frameworks such as JUnit, Mockito, Robolectric, and Espresso.

You will also learn about test-driven development (TDD), a software development practice that prioritizes tests over implementation. By the end of this chapter, you will be able to combine your new testing skills to work on a realistic project.

In previous chapters, you learned about how to load background data and display it in the UI and how to set up API calls to retrieve data. But how can you be sure that things work well? What if you're in a situation where you have to fix a bug in a project that you haven't interacted much with in the past? How can you know that the fix you are applying won't trigger another bug? The answer to these questions is through tests.

In this chapter, we will analyze the types of tests developers can write and we will look at available testing tools to ease the testing experience. The first issue that arises is the fact that desktops or laptops (which have different operating systems) are used to develop mobile applications. This implies that the tests must also be run on the device or an emulator, which will slow the tests down.

To solve this issue, we are presented with two types of tests: **local tests**, which are located in the `test` folder and will run on your machine, and **instrumented tests**, which are located in the `androidTest` folder and will run on the device or emulator.

We will cover the following topics in this chapter:

- JUnit
- Android Studio testing tips

- Mockito
- Integration tests
- UI tests
- TDD

Technical requirements

The complete code for all the exercises and the activity in this chapter is available on GitHub at `https://packt.link/pNbuk`

Types of testing

Both tests rely on the Java **JUnit** library, which helps developers set up their tests and group them into different categories. It also provides different configuration options, as well as extensions that other libraries can build upon. We will also investigate the testing pyramid, which helps guide developers as to how to structure their tests.

We will start at the bottom of the pyramid, which is represented by **unit tests**, move upward through **integration tests**, and finally, reach the top, which is represented by **end-to-end tests** (UI tests). You'll have the opportunity to learn about the tools that aid in writing each of these types of tests:

- **Mockito** and `mockito-kotlin`, which help mainly in unit tests and are useful for creating mocks or test doubles in which we can manipulate inputs so that we can assert different scenarios. (A mock or test double is an object that mimics the implementation of another object. Every time a test interacts with mocks, you can specify the behavior of these interactions.)

- **Robolectric**, which is an open source library that brings the Android framework onto your machine, allows you to test activities and fragments locally and not on the emulator. This can be used for both unit tests and integration tests.

- **Espresso**, which allows developers to create interactions (clicking buttons, inserting text in `EditText` components, and so on) and assertions (verifying that views display certain text, are currently being displayed to the user, are enabled, and so on) on an app's UI in an instrumented test.

In this chapter, we will also look at **TDD**. This is a software development process where tests take priority. A simple way of describing it is writing the test first. We will analyze how this approach is taken when developing features for Android applications. One of the things to keep in mind is that for an application to be properly tested, its classes must be properly written. One way to do this is by clearly defining the boundaries between your classes and splitting them based on the tasks you want them to accomplish.

Once you have achieved this, you can also rely on the **dependency inversion** and **dependency injection** principles when writing your classes. When these principles are applied properly, you should be able to inject fake objects into the subjects of your tests and manipulate the input to suit your testing scenario.

Dependency injection also helps when writing instrumented tests to help you swap modules that make network calls with local data in order to make your tests independent of external factors, such as networks. Instrumented tests are tests that run on a device or an emulator. The `instrument` keyword comes from the instrumentation framework, which assembles these tests and then executes them on the device.

Ideally, each application should have three types of tests:

- **Unit tests**: These are local tests that validate individual classes and methods. They should represent most of your tests and they should be fast, easy to debug, and easy to maintain. They are also known as small tests.
- **Integration tests**: These are either local tests with Robolectric or instrumented tests that validate interactions between your app's modules and components. These are slower and more complex than unit tests. The increase in complexity is due to the interaction between the components. These are also known as medium tests.
- **UI tests (end-to-end tests)**: These are instrumented tests that verify complete user journeys and scenarios. This makes them more complex and harder to maintain; they should represent the smallest number of your total test number. These are also known as large tests.

In the following figure, you can observe the **testing pyramid**. The recommendation from Google is to keep a ratio of 70:20:10 (unit tests: integration tests: UI tests) for your tests:

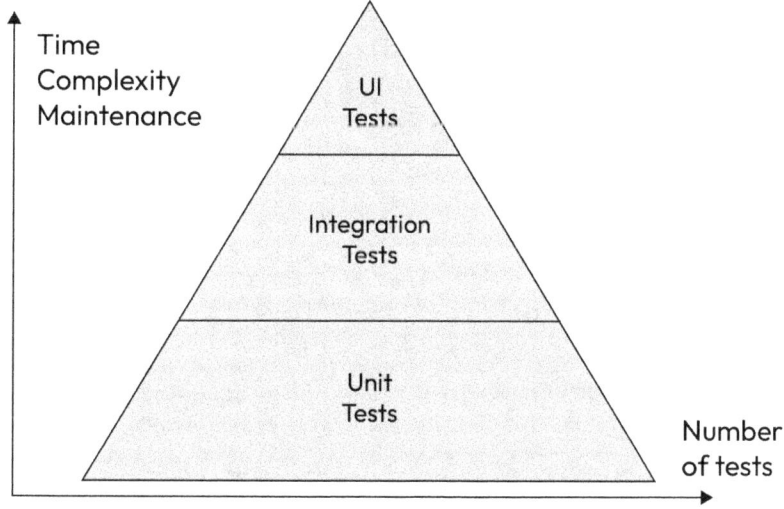

Figure 10.1 – Testing pyramid

As mentioned in the previous section, a unit test is a test that verifies a small portion of your code, and most of your tests should be unit tests that cover all sorts of scenarios (success, errors, limits, and more). Ideally, these tests should be local, but there are a few exceptions where you can make them instrumented. Those cases are rare and should be limited to when you want to interact with specific hardware of the device.

JUnit

JUnit is a framework for writing unit tests both in Java and Android. It is responsible for how tests are executed, allowing developers to configure their tests. It offers a multitude of features, such as the following:

- **Setup and teardown**: These are called before and after each test method is executed, allowing developers to set up relevant data for the test and clear it once the test is executed. They are represented by the `@Before` and `@After` annotations.
- **Assertions**: These are used to verify the result of an operation against an expected value.
- **Rules**: These allow developers to set up inputs that are common for multiple tests.
- **Runners**: Using these, you can specify how the tests can be executed.
- **Parameters**: These allow a test method to be executed with multiple inputs.
- **Orderings**: These specify in which order the tests should be executed.
- **Matchers**: These allow you to define patterns that can then be used to validate the results of the subject of your tests, or help you control the behavior of mocks.

In Android Studio, when a new project is created, the `app` module comes with the JUnit library in Gradle. This should be visible in `app/build.gradle`:

```
testImplementation 'junit:junit:4.13.2'
```

Let's look at the following class that we need to test:

```
class MyClass {
    fun factorial(n: Int): Int {
        return IntArray(n) {
            it+1
        }.reduce { acc, i ->
            acc * i
        }
    }
}
```

This method should return the factorial of the number n. We can start with a simple test that checks the value. To create a new unit test, you will need to create a new class in the test directory of your project.

The typical convention most developers follow is to add the Test suffix to your class name and place it under the same package in the test directory. For example, com.mypackage.ClassA will have the test in com.mypackage.ClassATest:

```
import org.junit.Assert.assertEquals
import org.junit.Test
class MyClassTest {
    private val myClass = MyClass()
    @Test
    fun computesFactorial() {
        val n = 3
        val result = myClass.factorial(n)
        assertEquals(6, result)
    }
}
```

In this test, you can see that we initialize the class under test, and the test method itself is annotated with the @Test annotation. The test method itself will assert that (3!)==6. The assertion is done using the assertEquals method from the JUnit library. A common practice in development is to split the test into three areas, also known as **Arrange-Act-Assert (AAA)**:

- **Arrange**: Where the input is initialized
- **Act**: Where the method under test is called
- **Assert**: Where the verification is done

We can write another test to make sure that the value is correct, but we will end up duplicating the code. We can now attempt to write a parameterized test. To do this, we will need to use the parameterized test runner. The preceding test has its own built-in runner provided by JUnit.

The parameterized runner will run the test repeatedly for different values that we provide, and it will look like the following – please note that import statements have been removed for brevity:

```
@RunWith(Parameterized::class)
class MyClassTest(
    private val input: Int,
    private val expected: Int
) {
    companion object {
```

```
            @Parameterized.Parameters
            @JvmStatic
            fun getData(): Collection<Array<Int>> = listOf(
                arrayOf(0, 1),
                arrayOf(1, 1),
                arrayOf(2, 2),
                arrayOf(3, 6),
                arrayOf(4, 24),
                arrayOf(5, 120)
            )
        }
        private val myClass = MyClass()
        @Test
        fun computesFactorial() {
            val result = myClass.factorial(input)
            assertEquals(expected, result)
        }
    }
```

This will run six tests. The usage of the `@Parameterized` annotation tells JUnit that this is a test with multiple parameters and allows us to add a constructor for the test that will represent the input value for our `factorial` function and the output. We then defined a collection of parameters with the use of the `@Parameterized.Parameters` annotation.

Each parameter for this test is a separate list containing the input and the expected output. When JUnit runs this test, it will run a new instance for each parameter and then execute the test method. This will produce five successes and one failure when we test 0!, meaning that we have found a bug.

We never accounted for a situation when n = 0. Now, we can go back to our code to fix the failure. We can do this by replacing the `reduce` function, which doesn't allow us to specify an initial value, with a `fold` function, which allows us to give the initial value of 1:

```
fun factorial(n: Int): Int {
    return IntArray(n) {
        it + 1
    }.fold(1, { acc, i -> acc * i })
}
```

Running the tests now, they will all pass. But that doesn't mean we are done here. There are many things that can go wrong. What happens if n is a negative number? Since we are dealing with factorials, we may get large numbers. We are working with integers in our examples, which means that the integer will overflow after 12!.

Normally, we would create new test methods in the `MyClassTest` class, but since the parameterized runner is used, all our new methods will be run multiple times, which will cost us time, so we will create a new test class to check our errors:

```
class MyClassTest2 {
    private val myClass = MyClass()
    @Test(expected =
        MyClass.FactorialNotFoundException::class)
    fun computeNegatives() {
        myClass.factorial(-10)
    }
}
```

This would lead to the following change in the class that was tested:

```
class MyClass {
    @Throws(FactorialNotFoundException::class)
    fun factorial(n: Int): Int {
        if (n < 0) {
            throw FactorialNotFoundException
        }
        return IntArray(n) {
            it + 1
        }.fold(1, { acc, i -> acc * i })
    }
    object FactorialNotFoundException : Throwable()
}
```

Let's solve the issue with very large factorials. We can use the `BigInteger` class, which can hold large numbers. We can update the test as follows (`import` statements not shown):

```
@RunWith(Parameterized::class)
class MyClassTest(
    private val input: Int,
    private val expected: BigInteger
) {
    companion object {
```

```
            @Parameterized.Parameters
            @JvmStatic
            fun getData(): Collection<Array<Any>> = listOf(
                arrayOf(0, BigInteger.ONE),
                arrayOf(1, BigInteger.ONE),
                arrayOf(2, BigInteger.valueOf(2)),
                arrayOf(3, BigInteger.valueOf(6)),
                arrayOf(4, BigInteger.valueOf(24)),
                arrayOf(5, BigInteger.valueOf(120)),
                arrayOf(13, BigInteger("6227020800")),
                arrayOf(25, BigInteger(
                    "15511210043330985984000000"))
            )
        }
        private val myClass = MyClass()
        @Test
        fun computesFactorial() {
            val result = myClass.factorial(input)
            assertEquals(expected, result)
        }
    }
```

The class under test now looks like this:

```
    @Throws(FactorialNotFoundException::class)
    fun factorial(n: Int): BigInteger {
        if (n < 0) {
            throw FactorialNotFoundException
        }
        return IntArray(n) {
            it + 1
        }.fold(BigInteger.ONE, { acc, i -> acc *
            i.toBigInteger() })
    }
```

In the preceding example, we implemented the factorial with the help of `IntArray`. This implementation is based more on Kotlin's ability to chain methods together, but it has one drawback: the fact that it uses memory for the array when it doesn't need to.

We only care about the factorial and not storing all the numbers from 1 to n. We can change the implementation to a simple `for` loop and use the tests to guide us during the refactoring process.

We can observe here two benefits of having tests in your application:

- They serve as updated documentation of how the features should be implemented
- They guide us when refactoring code by maintaining the same assertion and detecting whether new changes to the code broke it

Let's update the code to get rid of `IntArray`:

```
@Throws(FactorialNotFoundException::class)
fun factorial(n: Int): BigInteger {
    if (n < 0) {
        throw FactorialNotFoundException
    }
    var result = BigInteger.ONE
    for (i in 1..n) {
        result = result.times(i.toBigInteger())
    }
    return result
}
```

If we modify the `factorial` function, as in the preceding example, and run the tests, we should see them all passing.

In certain situations, your tests will use a resource that is common to the test or the application (databases, files, and so on). Ideally, this shouldn't happen for unit tests, but there can always be exceptions to this.

Let's analyze that scenario and see how JUnit can aid us with it. We will add a `companion` object, which will store the result, to simulate this behavior:

```
companion object {
    var result: BigInteger = BigInteger.ONE
}
@Throws(FactorialNotFoundException::class)
fun factorial(n: Int): BigInteger {
    if (n < 0) {
        throw FactorialNotFoundException
    }
    for (i in 1..n) {
```

```
            result = result.times(i.toBigInteger())
        }
        return result
    }
```

If we execute the tests for the preceding code, we will start seeing that some will fail. That's because after the first tests execute the `factorial` function, the result will have the value of the executed tests, and when a new test is executed, the result of the factorial will be multiplied by the previous value of the result.

Normally, this would be good because the tests tell us that we are doing something wrong and we should remedy this, but for this example, we will address the issue directly in the tests:

```
@Before
fun setUp() {
    MyClass.result = BigInteger.ONE
}
@After
fun tearDown() {
    MyClass.result = BigInteger.ONE
}
@Test
fun computesFactorial() {
    val result = myClass.factorial(input)
    assertEquals(expected, result)
}
```

In the tests, we've added two methods with the `@Before` and `@After` annotations. When these methods are introduced, JUnit will change the execution flow as follows: all methods with the `@Before` annotation will be executed, a method with the `@Test` annotation will be executed, and then all methods with the `@After` annotation will be executed. This process will repeat for every `@Test` method in your class.

If you find yourself repeating the same statements in your `@Before` method, you can consider using `@Rule` to remove the repetition. We can set up a test rule for the preceding example. Test rules should be in the `test` or `androidTest` packages, as their usage is only limited to testing. They tend to be used in multiple tests, so you can place your rules in a `rules` package (`import` statements not shown):

```
class ResultRule : TestRule {
    override fun apply(
        base: Statement,
```

```
            description: Description?
    ): Statement? {
        return object : Statement() {
            @Throws(Throwable::class)
            override fun evaluate() {
                MyClass.result = BigInteger.ONE
                try {
                    base.evaluate()
                } finally {
                    MyClass.result = BigInteger.ONE
                }
            }
        }
    }
}
```

In the preceding example, we can see that the rule will implement `TestRule`, which, in turn, comes with the `apply()` method. We then create a new `Statement` object that will execute the `base` statement (the test itself) and reset the value of the result before and after the statement. We can now modify the test as follows:

```
    @JvmField
    @Rule
    val resultRule = ResultRule()
    private val myClass = MyClass()
    @Test
    fun computesFactorial() {
        val result = myClass.factorial(input)
        assertEquals(expected, result)
    }
```

To add the rule to the test, we use the `@Rule` annotation. Since the test is written in Kotlin, we are using `@JvmField` to avoid generating getters and setters because `@Rule` requires a public field and not a method.

In this section, we have learned how we can use JUnit to write tests that can verify small units of our code, by verifying the results, errors, or behavior for different parameters. We've also learned how each test is run when they are part of a testing class and the order of operations being invoked. In the next section, we will look at how we can use Android Studio to understand how we can run tests and view the results.

Android Studio testing tips

Android Studio comes with a good set of shortcuts and visual tools to help with testing. If you want to create a new test for your class or go to existing tests for your class, you can use the *Ctrl + Shift + T* (Windows) or *Command + Shift + T* (Mac) shortcut. You will need to make sure that the contents of your class are currently in focus in the editor for the keyboard shortcut to take effect.

In order to run tests, there are multiple options: right-click your file or the package and select the **Run 'Tests in...'** option, or if you want to run a test independently, you can go to the particular test method and select the green icon at the top of the class, which will execute all the tests in the class.

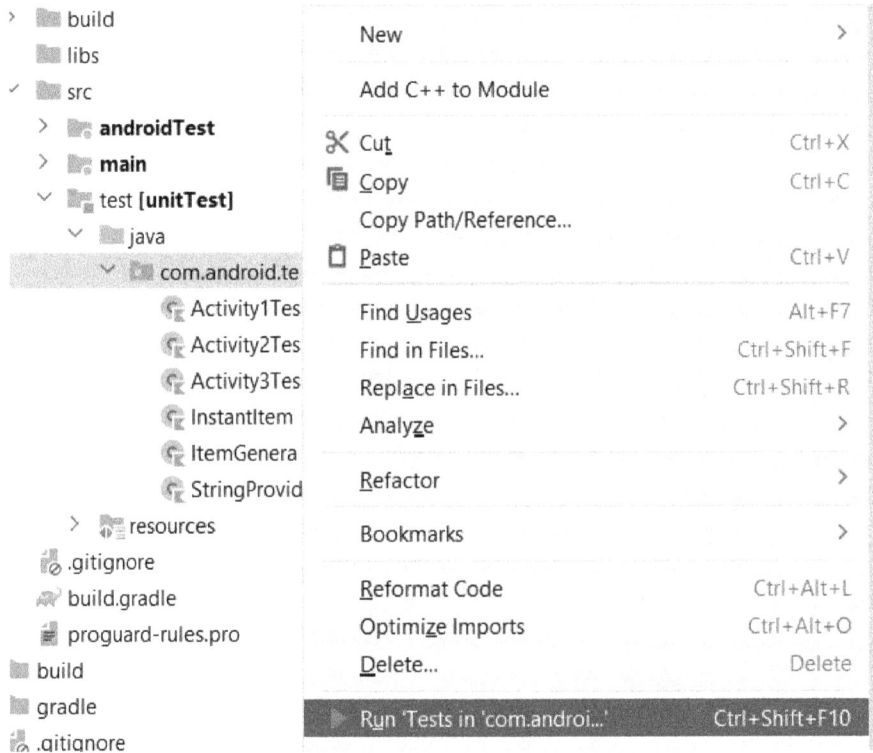

Figure 10.2 – Running a group of tests

For an individual test, you can click the green icon next to the `@Test` annotated methods.

```kotlin
@RunWith(MockitoJUnitRunner::class)
class StringProviderTest {

    @InjectMocks
    lateinit var stringProvider: StringProvider

    @Mock
    lateinit var context: Context

    @Test
    fun provideItemString() {
        val number = 5
        val expected = "expected"
        whenever("Item {5}").thenReturn(expected)

        val result = stringProvider.provideItemString(number)

        assertEquals(expected, result)
    }
}
```

Figure 10.3 – Icons for running individual tests

This will trigger the test execution, which will be displayed in the **Run** tab, as shown in the following screenshot. When the tests are completed, they will become either red or green, depending on their success state:

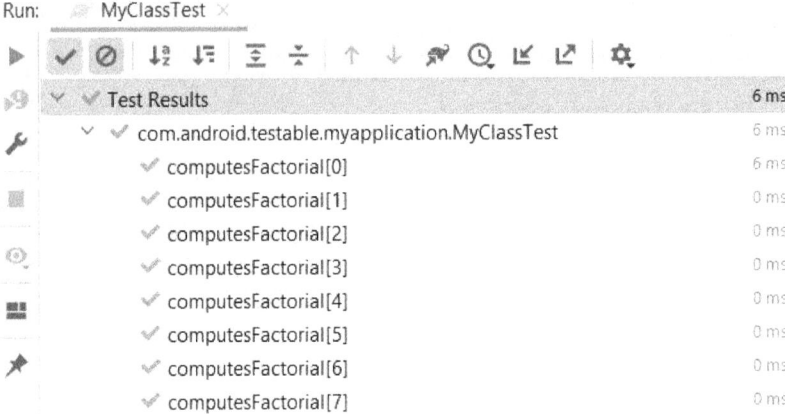

Figure 10.4 – Test output in Android Studio

Another important feature that can be found in tests is the debug one. This is important because you can debug both the test and the method under test, so if you find problems in fixing an issue, you can use this to view what the test used as input and how your code handles the input. The third feature you can find in the green icon next to a test is the **Run With Coverage** option.

This helps developers identify which lines of code are covered by the test and which ones are skipped. The higher the coverage, the higher the chances of finding crashes and bugs:

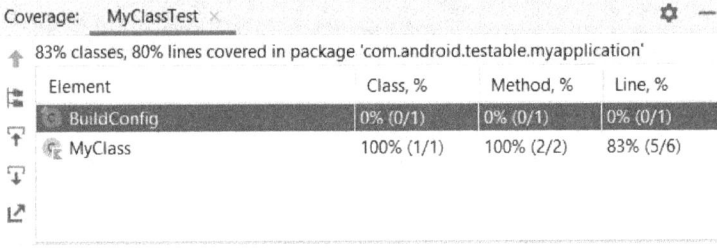

Figure 10.5 – Test coverage in Android Studio

In the preceding figure, you can see the coverage of our class broken down into the number of classes under test, the number of methods under test, and the number of lines under test.

Another way to run tests for your Android app is through the command line. This is usually handy in situations where your project has **continuous integration** set up, meaning that every time you upload your code to a repository in the cloud, a set of scripts will be triggered to test it and ensure functionality.

Since this is done in the cloud, there is no need for Android Studio to be installed. For simplicity, we will be using the **Terminal** tab in Android Studio to emulate that behavior. The **Terminal** tab is usually located in the bottom bar in Android Studio near the **Logcat** tab.

In every Android Studio project, a file called `gradlew` is present. This is an executable file that allows developers to execute Gradle commands. To run your local unit tests, you can use the following:

- `gradlew.bat test` (for Windows)
- `./gradlew test` (for macOS and Linux)

Once that command is executed, the app will be built and tested. You can find a variety of commands that you can input in **Terminal** in the **Gradle** tab located on the right-hand side of Android Studio.

If you see the message saying **Task list has not been built**, click it and uncheck **Do not build Gradle task list during Gradle Sync**, click **OK**, and then sync the project's Gradle files. The task list should then appear in the list.

The output of the tests, when executed from either the **Terminal** or **Gradle** tab, can be found in the `app/build/reports` folder.

Figure 10.6 – Gradle commands in Android Studio

In this section, we have learned about the various options for testing that Android Studio provides and how we can visualize testing results. In the section that follows, we will look at how we can mock objects in tests and how we can use Mockito to do so.

Mockito

In the preceding examples, we looked at how to set up a unit test and how to use assertions to verify the result of an operation. What if we want to verify whether a certain method was called? Or what if we want to manipulate the test input to test a specific scenario? In these types of situations, we can use **Mockito**.

This is a library that helps developers set up dummy objects that can be injected into the objects under test and allows them to verify method calls, set up inputs, and even monitor the test objects themselves.

The library should be added to your `test` Gradle setup, as follows:

```
testImplementation 'org.mockito:mockito-core:4.5.1'
```

Now, let's look at the following code example (please note that, for brevity, `import` statements have been removed from the following code snippets):

```
class StringConcatenator(private val context: Context) {
    fun concatenate(@StringRes stringRes1: Int,
      @StringRes stringRes2: Int): String {
        return context.getString(stringRes1).plus(context
          .getString(stringRes2))
    }
}
```

Here, we have the `Context` object, which normally cannot be unit-tested because it's part of the Android framework. We can use `mockito` to create a test double and inject it into the `StringConcatenator` object. Then, we can manipulate the call to `getString()` to return whatever input we chose. This process is referred to as **mocking**:

```
class StringConcatenatorTest {
    private val context = Mockito.mock(Context::class.java)
    private val stringConcatenator =
        StringConcatenator(context)
    @Test
    fun concatenate() {
        val stringRes1 = 1
```

```
        val stringRes2 = 2
        val string1 = "string1"
        val string2 = "string2"
        Mockito.`when`(context.getString(stringRes1))
            .thenReturn(string1)
        Mockito.`when`(context.getString(stringRes2))
            .thenReturn(string2)
        val result =
            stringConcatenator.concatenate(stringRes1,
            stringRes2)
        assertEquals(string1.plus(string2), result)
    }
}
```

In the test, we created a mock context. When the concatenate method was tested, we used Mockito to return a specific string when the getString() method was called with a particular input. This allowed us to then assert the result.

> **Note**
>
> ` is an escape character present in Kotlin and should not be confused with a quote mark. It allows the developer to give methods any name that they want, including special characters or reserved words.

Mockito is not limited to mocking Android framework classes only. We can create a SpecificStringConcatenator class, which will use StringConcatenator to concatenate two specific strings from strings.xml:

```
class SpecificStringConcatenator(private val
stringConcatenator: StringConcatenator) {
    fun concatenateSpecificStrings(): String {
        return stringConcatenator.concatenate(
            R.string.string_1, R.string.string_2)
    }
}
```

We can write the test for it as follows:

```
class SpecificStringConcatenatorTest {
    private val stringConcatenator = Mockito
```

```
            .mock(StringConcatenator::class.java)
        private val specificStringConcatenator =
            SpecificStringConcatenator(stringConcatenator)
        @Test
        fun concatenateSpecificStrings() {
            val expected = "expected"
            Mockito.'when'(stringConcatenator.concatenate(
                R.string.string_1, R.string.string_2))
                .thenReturn(expected)
            val result = specificStringConcatenator
                .concatenateSpecificStrings()
            assertEquals(expected, result)
        }
    }
```

Here, we are mocking the previous `StringConcatenator` and instructing the mock to return a specific result. If we run the test, it will fail because Mockito is not able to mock final classes. Here, it encounters a conflict with Kotlin that makes all classes *final* unless we specify them as *open*.

Luckily, there is a configuration we can apply that solves this dilemma without making the classes under test *open*:

1. Create a folder named `resources` in the `test` package.
2. In `resources`, create a folder named `mockito-extensions`.
3. In the `mockito-extensions` folder, create a file named `org.mockito.plugins.MockMaker`.
4. Inside the file, add the following line:

    ```
    mock-maker-inline
    ```

In situations where you have callbacks or asynchronous work and cannot use the JUnit assertions, you can use `mockito` to verify the invocation on the callback or lambdas:

```
class SpecificStringConcatenator(private val
stringConcatenator: StringConcatenator) {
    fun concatenateSpecificStrings(): String {
        return stringConcatenator.concatenate(
        R.string.string_1, R.string.string_2)
    }
    fun concatenateWithCallback(callback: Callback) {
```

```kotlin
        callback.onStringReady(concatenateSpecificStrings())
    }
    interface Callback {
        fun onStringReady(input: String)
    }
}
```

In the preceding example, we have added the `concatenateWithCallback` method, which will invoke the callback with the result of the `concatenateSpecificStrings` method. The test for this method would look something like this:

```kotlin
    @Test
    fun concatenateWithCallback() {
        val expected = "expected"
        Mockito.`when`(stringConcatenator.concatenate(
            R.string.string_1, R.string.string_2))
            .thenReturn(expected)
        val callback = Mockito.mock(
            SpecificStringConcatenator.Callback::class.java
        )
        specificStringConcatenator.concatenateWithCallback(
            callback)
        Mockito.verify(callback).onStringReady(expected)
    }
```

Here, we create a mock `Callback` object, which we can then verify at the end with the expected result. Notice that we had to duplicate the setup of the `concatenateSpecificStrings` method to test the `concatenateWithCallback` method. You should never mock the objects you are testing; however, you can use spy to change their behavior. We can spy the `stringConcatenator` object to change the outcome of the `concatenateSpecificStrings` method:

```kotlin
    @Test
    fun concatenateWithCallback() {
        val expected = "expected"
        val spy = Mockito.spy(specificStringConcatenator)
        Mockito.`when`(spy.concatenateSpecificStrings())
            .thenReturn(expected)
        val callback =
          Mockito.mock(SpecificStringConcatenator.Callback::
```

```
            class.java)
        specificStringConcatenator.concatenateWithCallback(
            callback)
        Mockito.verify(callback).onStringReady(expected)
}
```

Mockito also relies on dependency injection to initialize class variables and has a custom build JUnit test runner. This can simplify the initialization of our variables, as follows:

```
@RunWith(MockitoJUnitRunner::class)
class SpecificStringConcatenatorTest {
    @Mock
    lateinit var stringConcatenator: StringConcatenator
    @InjectMocks
    lateinit var specificStringConcatenator:
        SpecificStringConcatenator
}
```

In the preceding example, `MockitoRunner` will inject the variables with the `@Mock` annotation with mocks. Next, it will create a new non-mocked instance of the field with the `@InjectMocks` annotation. When this instance is created, Mockito will try to inject the mock objects that match the signature of the constructor of that object.

In this section, we have looked at how we can mock objects when we write tests and how we can use Mockito to do so. In the section that follows, we will look at a specialized library for Mockito that is better suited to be used with the Kotlin programming language, mockito-kotlin.

You may have noticed in the preceding example that the `when` method from Mockito has escaped. This is because of a conflict with the Kotlin programming language. Mockito is built mainly for Java, and when Kotlin was created, it introduced the `this` keyword. Conflicts like this are escaped using the ` character.

This, along with some other minor issues, causes some inconvenience when using Mockito in Kotlin. A few libraries were introduced to wrap Mockito and provide a nicer experience when using it. One of those is `mockito-kotlin`. You can add this library to your module using the following command:

```
testImplementation "org.mockito.kotlin:
mockito-kotlin:4.1.0"
```

A big visible change this library adds is replacing the `when` method with `whenever`. Another useful change is replacing the `mock` method to rely on generics, rather than class objects. The rest of the syntax is like the Mockito syntax.

We can now update the previous tests with the new library, starting with `StringConcatenatorTest` (`import` statements have been removed for brevity):

```
class StringConcatenatorTest {
    private val context = mock<Context>()
    private val stringConcatenator =
        StringConcatenator(context)
    @Test
    fun concatenate() {
        val stringRes1 = 1
        val stringRes2 = 2
        val string1 = "string1"
        val string2 = "string2"
        whenever(context.getString(stringRes1)).thenReturn(
            string1)
        whenever(context.getString(stringRes2)).thenReturn(
            string2)
        val result = stringConcatenator.concatenate(
            stringRes1, stringRes2)
        assertEquals(string1.plus(string2), result)
    }
}
```

As you can observe, the ` character has disappeared, and our mock initialization for the `Context` object has been simplified. We can apply the same thing for the `SpecificStringConcatenatorTest` class (`import` statements have been removed for brevity):

```
@RunWith(MockitoJUnitRunner::class)
class SpecificStringConcatenatorTest {
    @Mock
    lateinit var stringConcatenator: StringConcatenator
    @InjectMocks
    lateinit var specificStringConcatenator:
        SpecificStringConcatenator
    @Test
    fun concatenateSpecificStrings() {
        val expected = "expected"
        whenever(stringConcatenator.concatenate(
```

```
                R.string.string_1, R.string.string_2))
            .thenReturn(expected)
        val result = specificStringConcatenator
            .concatenateSpecificStrings()
        assertEquals(expected, result)
    }
    @Test
    fun concatenateWithCallback() {
        val expected = "expected"
        val spy = spy(specificStringConcatenator)
        whenever(spy.concatenateSpecificStrings())
            .thenReturn(expected)
        val callback =
            mock<SpecificStringConcatenator.Callback>()
        specificStringConcatenator.concatenateWithCallback(
            callback)
        verify(callback).onStringReady(expected)
    }
}
```

In this section, we have looked at how we can use the `mockito-kotlin` library and how it can simplify the Mockito functions in Kotlin. In what follows, we will do an exercise on how we can write unit tests with JUnit and Mockito.

Exercise 10.01 – testing the sum of numbers

Using JUnit, Mockito, and `mockito-kotlin`, write a set of tests for the following class that should cover the following scenarios:

- Assert the values for 0, 1, 5, 20, and `Int.MAX_VALUE`
- Assert the outcome for a negative number
- Fix the code and replace the sum of numbers with the formula $n*(n+1)/2$

> **Note**
>
> Throughout this exercise, `import` statements are not shown. To see the full code files, refer to https://packt.link/rv8C2.

The code to test is as follows:

```
class NumberAdder {
    @Throws(InvalidNumberException::class)
    fun sum(n: Int, callback: (BigInteger) -> Unit) {
        if (n < 0) {
            throw InvalidNumberException
        }
        var result = BigInteger.ZERO
        for (i in 1..n){
            result = result.plus(i.toBigInteger())
        }
        callback(result)
    }
    object InvalidNumberException : Throwable()
}
```

Perform the following steps to complete this exercise:

1. Let's make sure the necessary libraries are added to the app/build.gradle file:

    ```
    testImplementation 'junit:junit:4.13.2'
    testImplementation 'org.mockito:mockito-core:4.5.1'
    testImplementation 'org.mockito.kotlin:mockito-kotlin:4.1.0'
    ```

2. Create a class named NumberAdder and copy the preceding code inside it.

3. Move the cursor inside the newly created class and, with *Command* + *Shift* + *T* or *Ctrl* + *Shift* + *T*, create a test class called NumberAdderParameterTest.

4. Create a parameterized test inside this class that will assert the outcomes for the 0, 1, 5, 20, and Int.MAX_VALUE values:

    ```
    @RunWith(Parameterized::class)
    class NumberAdderParameterTest(
        private val input: Int,
        private val expected: BigInteger
    ) {
        companion object {
            @Parameterized.Parameters
    ```

```
            @JvmStatic
            fun getData(): List<Array<out Any>> = listOf(
                arrayOf(0, BigInteger.ZERO),
                arrayOf(1, BigInteger.ONE),
                arrayOf(5, 15.toBigInteger()),
                arrayOf(20, 210.toBigInteger()),
                arrayOf(Int.MAX_VALUE, BigInteger(
                    "2305843008139952128"))
            )
        }
        private val numberAdder = NumberAdder()
        @Test
        fun sum() {
            val callback = mock<(BigInteger) -> Unit>()
            numberAdder.sum(input, callback)
            verify(callback).invoke(expected)
        }
    }
```

5. Create a separate test class that handles the exception thrown when there are negative numbers, named NumberAdderErrorHandlingTest:

```
    @RunWith(MockitoJUnitRunner::class)
    class NumberAdderErrorHandlingTest {
        @InjectMocks
        lateinit var numberAdder: NumberAdder
        @Test(expected =
            NumberAdder.InvalidNumberException::class)
        fun sum() {
            val input = -1
            val callback = mock<(BigInteger) -> Unit>()
            numberAdder.sum(input, callback)
        }
    }
```

6. Since $1 + 2 + ...n = n * (n + 1) / 2$, we can use the formula in the code, and this would make the execution of the method run faster:

```
class NumberAdder {
    @Throws(InvalidNumberException::class)
    fun sum(n: Int, callback: (BigInteger) -> Unit) {
        if (n < 0) {
            throw InvalidNumberException
        }
        callback(n.toBigInteger()
        .times((n.toBigInteger() +
        1.toBigInteger())).divide(2.toBigInteger()))
    }
    object InvalidNumberException : Throwable()
}
```

7. Run the tests by right-clicking the package in which the tests are located and selecting **Run all in [package_name]**. An output similar to the following will appear, signifying that the tests have passed:

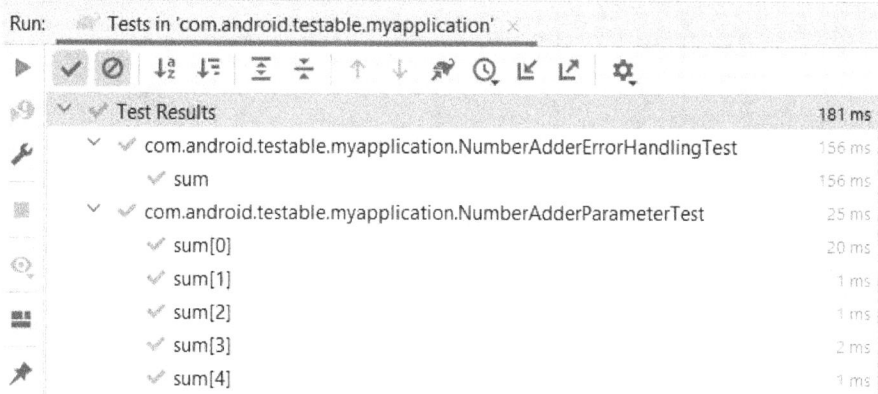

Figure 10.7 – Output of Exercise 10.01

By completing this exercise, we have taken the first steps into unit testing, managed to create multiple test cases for a single operation, taken the first steps into understanding Mockito, and used tests to guide us on how to refactor code without introducing any new issues.

Integration tests

Let's assume your project is covered by unit tests where a lot of your logic is held. You now have to add these tested classes to an activity or a fragment and require them to update your UI. How can you be certain that these classes will work well with each other? The answer to that question is through integration testing.

The idea behind this type of testing is to ensure that different components within your application integrate well with each other. Some examples include the following:

- Ensuring that your API-related components parse the data well and interact well with your storage components
- The storage components are capable of storing and retrieving the data correctly
- The UI components load and display the appropriate data
- The transition between different screens in your application

To aid with integration testing, the requirements are sometimes written in the format `Given - When - Then`. These usually represent acceptance criteria for a user story. Take the following example:

```
Given I am not logged in
And I open the application
When I enter my credentials
And click Login
Then I see the Main screen
```

We can use these steps to approach how we can write the integration tests for the feature we are developing.

On the Android platform, integration testing can be achieved with two libraries:

- **Robolectric**: This library gives developers the ability to test Android components as unit tests – that is, executing integration tests without an actual device or emulator
- **Espresso**: This library is helpful in instrumentation tests on an Android device or emulator

We'll have a look at these libraries in detail in the next sections.

Robolectric

Robolectric started as an open source library, which was meant to give users the ability to unit test classes from the Android framework as part of their local tests instead of the instrumented tests. Recently, it has been endorsed by Google and has been integrated with AndroidX Jetpack components.

One of the main benefits of this library is the simplicity of testing activities and fragments. This is a benefit when it comes to integration tests because we can use this feature to make sure that our components integrate well with each other.

Some of Robolectric's features are as follows:

- The possibility to instantiate and test the activity and fragment lifecycle
- The possibility to test view inflation
- The possibility to provide configurations for different Android APIs, orientations, screen sizes, layout directions, and so on
- The possibility to change the `Application` class, which then helps to change the modules to permit data mocks to be inserted

To add Robolectric, along with the AndroidX integration, we will need the following libraries:

```
testImplementation 'org.robolectric:robolectric:4.9'
testImplementation 'androidx.test.ext:junit:1.1.4'
```

The second library will bring a set of utility methods and classes required for testing Android components.

Let's assume we have to deliver a feature in which we display the text `Result x`, where x is the `factorial` function for a number that the user will insert in the `EditText` element. We will assume that we will use an Activity with an `EditText`, a `TextView`, and a `Button`. When the button is clicked, then we display in the `TextView` the factorial result of the number entered in the `EditText`.

To achieve this, we have two classes, one that computes the factorial and another that concatenates the word `Result` with the factorial if the number is positive, or it will return the text `Error` if the number is negative.

The `factorial` class will look something like this (throughout this example, `import` statements have been removed for brevity):

```
class FactorialGenerator {
    @Throws(FactorialNotFoundException::class)
    fun factorial(n: Int): BigInteger {
        if (n < 0) {
            throw FactorialNotFoundException
        }
        var result = BigInteger.ONE
        for (i in 1..n) {
            result = result.times(i.toBigInteger())
```

```
        }
        return result
    }
    object FactorialNotFoundException : Throwable()
}
```

The TextFormatter class will look like this:

```
class TextFormatter(
    private val factorialGenerator: FactorialGenerator,
    private val context: Context
) {
    fun getFactorialResult(n: Int): String {
        return try {
            context.getString(R.string.result,
                factorialGenerator.factorial(n).toString())
        } catch (e: FactorialGenerator
            .FactorialNotFoundException) {
            context.getString(R.string.error)
        }
    }
}
```

We can combine these two components in our activity and have something like this:

```
class MainActivity : AppCompatActivity() {
    private lateinit var textFormatter: TextFormatter
    override fun onCreate(savedInstanceState: Bundle?) {
        super.onCreate(savedInstanceState)
        setContentView(R.layout.activity_main)
        textFormatter = TextFormatter(FactorialGenerator(),
            applicationContext)
        findViewById<Button>(R.id.button)
            .setOnClickListener {
                findViewById<TextView>(R.id.text_view)
                    .text = textFormatter.getFactorialResult(
                    findViewById<EditText>(R.id.edit_text).text
                        .toString()
```

```
                .toInt())
        }
    }
}
```

We can observe three components interacting with each other in this case. We can use Robolectric to test our activity. By testing the activity that creates the components, we can also test the interaction between all three of the components. We can write a test that looks like this:

```
@RunWith(AndroidJUnit4::class)
class MainActivityTest {
    private val context =
        getApplicationContext<Application>()
    @Test
    fun `show factorial result in text view`() {
        val scenario = launch(MainActivity::class.java)
        scenario.moveToState(Lifecycle.State.RESUMED)
        scenario.onActivity { activity ->
            activity.findViewById<EditText>(R.id.edit_text)
                .setText(5.toString())
            activity.findViewById<Button>(R.id.button)
                .performClick()
            assertEquals(
                context.getString(R.string.result, "120"),
                activity.findViewById<TextView>(
                R.id.text_view).text
            )
        }
    }
}
```

In the preceding example, we can see the AndroidX support for the activity test. The AndroidJUnit4 test runner will set up Robolectric and create the necessary configurations, while the launch method will return a scenario object, which we can then play with to achieve the necessary conditions for the test. We can also observe how we can use the ` character to provide longer names to our functions, in which we can include whitespace characters.

If we want to add configurations for the test, we can use the `@Config` annotation both on the class and on each of the test methods:

```
@Config(
    sdk = [Build.VERSION_CODES.TIRAMISU],
    minSdk = Build.VERSION_CODES.KITKAT,
    maxSdk = Build.VERSION_CODES.TIRAMISU,
    application = Application::class,
    assetDir = "/assetDir/"
)
@RunWith(AndroidJUnit4::class)
class MainActivityTest
```

We can also specify global configurations in the `test/resources` folder in the `robolectric.properties` file, like so:

```
sdk=33
minSdk = 14
maxSdk = 33
```

Another important feature that has recently been added to Robolectric is support for the Espresso library. This allows developers to use the syntax from Espresso to interact with views and make assertions on the views.

Another library that can be used in combination with Robolectric is `FragmentScenario`, which allows the possibility to test fragments. These libraries can be added in Gradle using the following:

```
testImplementation 'androidx.fragment:
fragment-testing:1.5.5'
testImplementation 'androidx.test.espresso:
espresso-core:3.5.0'
```

Testing fragments is like activities using the `scenario` setup:

```
val scenario = launchFragmentInContainer<MainFragment>()
scenario.moveToState(Lifecycle.State.CREATED)
```

Espresso

Espresso is a library designed to perform interactions and assertions in a concise way. It was initially designed to be used in instrumented tests, and now it has migrated to be used with Robolectric as well. The typical usage for performing an action is as follows:

```
onView(Matcher<View>).perform(ViewAction)
```

For verification, we can use the following:

```
onView(Matcher<View>).check(ViewAssertion)
```

We can provide custom `ViewMatchers` if none can be found in the `ViewMatchers` class. Some of the most common ones are `withId` and `withText`. These two allow us to identify views based on their `R.id.myId` identifier or text identifier. Ideally, the first one should be used to identify a particular view.

Another interesting aspect of Espresso is the reliance on the `Hamcrest` library for matchers. This is a Java library that aims to improve testing. This allows multiple matchers to be combined if necessary. Let's say that the same ID is present in different views on your UI. You can narrow your search for a specific view using the following expression:

```
onView(allOf(withId(R.id.edit_text),
  withParent(withId(R.id.root))))
```

The `allOf` expression will evaluate all of the other operators and will pass only if all of the operators inside pass. The preceding expressions will translate to *Find the view with id=edit_text that has the parent with id=R.id.root*. Other `Hamcrest` operators may include `anyOf`, `both`, `either`, `is`, `isA`, `hasItem`, `equalTo`, `any`, `instanceOf`, `not`, `null`, and `notNull`.

`ViewActions` have a similar approach to `ViewMatchers`. We can find common ones in the `ViewActions` class. Common ones include `typeText`, `click`, `scrollTo`, `clearText`, `swipeLeft`, `swipeRight`, `swipeUp`, `swipeDown`, `closeSoftKeyboard`, `pressBack`, `pressKey`, `doubleClick`, and `longClick`. If you have custom views and certain actions are required, then you can implement your own `ViewAction` element by implementing the `ViewAction` interface.

Similar to the preceding examples, `ViewAssertions` have their own class. Typically, the `matches` method is used, where you can then use `ViewMatchers` and `Hamcrest` matchers to validate the result:

```
onView(withId(R.id.text_view)).check(matches(withText(
  "My text"))))
```

The preceding example will verify that the view with the `text_view` ID will contain the text My text:

```
onView(withId(R.id.button)).perform(click())
```

This will click the view with the ID button.

We can now rewrite the Robolectric test and add Espresso, which will give us this (the `import` statement is not shown):

```
@RunWith(AndroidJUnit4::class)
class MainActivityTest {
    @Test
    fun `show factorial result in text view`() {
        val scenario = launch(MainActivity::class.java)
        scenario.moveToState(Lifecycle.State.RESUMED)
        scenario.onActivity { activity ->
            onView(withId(R.id.edit_text)).perform(
                typeText("5"))
            onView(withId(R.id.button)).perform(click())
            onView(withId(R.id.text_view))
                .check(matches(withText(activity.getString(
                R.string.result, "120"))))
        }
    }
}
```

In the preceding code sample, we can observe how, using Espresso, we input the number 5 into `EditText`, then click on the button, and then assert the text displayed in `TextView` with the help of the `onView()` method to obtain a reference to the view, and then execute actions using `perform()` or make assertions using `check()`.

> **Note**
>
> For the following exercise, you will need an emulator or a physical device with USB debugging enabled. You can do so by selecting **Tools** | **AVD Manager** in Android Studio. Then, you can create one with the **Create Virtual Device** option by selecting the type of emulator, clicking **Next**, and then selecting an x86 image. Any image larger than Lollipop should be alright for this exercise. Next, you can give your image a name and click **Finish**.

Exercise 10.02 – double integration

Develop an application that observes the following requirements:

```
Given I open the application
And I insert the number n
When I press the Calculate button
Then I should see the text "The sum of numbers from 1 to n is
[result]"
Given I open the application
And I insert the number -n
When I press the Calculate button
Then I should see the text "Error: Invalid number"
```

You should implement both unit tests and integration tests using Robolectric and Espresso and migrate the integration tests to become instrumentation tests.

> **Note**
>
> Throughout this exercise, `import` statements are not shown. To see the full code files, refer to https://packt.link/EcmiV.

Implement the following steps to complete this exercise:

1. Let's start by adding the necessary test libraries to `app/build.gradle`:

    ```
    testImplementation 'junit:junit:4.13.2'
    testImplementation
        'org.mockito:mockito-core:4.5.1'
    testImplementation
        'org.mockito.kotlin:mockito-kotlin:4.1.0'
    testImplementation
        'org.robolectric:robolectric:4.9'
    testImplementation 'androidx.test.ext:junit:1.1.4'
    testImplementation
        'androidx.test.espresso:espresso-core:3.5.0'
    androidTestImplementation
        'androidx.test.ext:junit:1.1.4'
    androidTestImplementation
    ```

```
            'androidx.test.espresso:espresso-core:3.5.0'
        androidTestImplementation
            'androidx.test:rules:1.5.0'
```

2. For Robolectric, we will need to add extra configurations, the first of which is to add the following line to app/build.gradle in the android closure:

   ```
   testOptions.unitTests.includeAndroidResources = true
   ```

3. Create a resources directory in the test package. You will need to switch your Android Studio project view from **Android** to **Project**.

4. Add the robolectric.properties file and add the following configuration to that file:

   ```
   sdk=32
   ```

5. In resources, create a folder named mockito-extensions.

6. In the mockito-extensions folder, create a file named org.mockito.plugins.MockMaker, and inside the file, add the following line:

   ```
   mock-maker-inline
   ```

7. Create the NumberAdder class. This is similar to the one in *Exercise 10.01*:

   ```
   class NumberAdder {
       @Throws(InvalidNumberException::class)
       fun sum(n: Int, callback: (BigInteger) -> Unit) {
           if (n < 0) {
               throw InvalidNumberException
           }
           callback(n.toBigInteger().times((n.toLong() +
           1).toBigInteger()).divide(2.toBigInteger()))
       }
       object InvalidNumberException : Throwable()
   }
   ```

8. Create the tests for NumberAdder in the test folder. First, create NumberAdderParameterTest:

   ```
   @RunWith(Parameterized::class)
   class NumberAdderParameterTest(
       private val input: Int,
       private val expected: BigInteger
   ```

```kotlin
) {
    private val numberAdder = NumberAdder()

    @Test
    fun sum() {
        val callback = mock<(BigInteger) -> Unit>()
        numberAdder.sum(input, callback)
        verify(callback).invoke(expected)
    }
}
```

The complete code for this step can be found at https://packt.link/ghcTs.

9. Then, create the NumberAdderErrorHandlingTest test:

```kotlin
@RunWith(MockitoJUnitRunner::class)
class NumberAdderErrorHandlingTest {
    @InjectMocks
    lateinit var numberAdder: NumberAdder
    @Test(expected =
        NumberAdder.InvalidNumberException::class)
    fun sum() {
        val input = -1
        val callback = mock<(BigInteger) -> Unit>()
        numberAdder.sum(input, callback)
    }
}
```

10. In the main folder in the root package, create a class that will format the sum and concatenate it with the necessary strings:

```kotlin
class TextFormatter(
    private val numberAdder: NumberAdder,
    private val context: Context
) {
    fun getSumResult(n: Int, callback: (String) ->
    Unit) {
```

```kotlin
            try {
                numberAdder.sum(n) {
                    callback(context.getString(
                        R.string
                        .the_sum_of_numbers_from_1_to_is,
                        n, it.toString())
                    )
                }
            } catch (
              e: NumberAdder.InvalidNumberException) {
                callback(context.getString(
                    R.string.error_invalid_number))
            }
        }
    }
```

11. Unit-test this class for both the success and error scenarios. Start with the success scenario:

```kotlin
@RunWith(MockitoJUnitRunner::class)
class TextFormatterTest {
    @InjectMocks
    lateinit var textFormatter: TextFormatter
    @Mock
    lateinit var numberAdder: NumberAdder
    @Mock
    lateinit var context: Context
    @Test
    fun getSumResult_success() {
        val n = 10
        val sumResult = BigInteger.TEN
        val expected = "expected"
        whenever(numberAdder.sum(eq(n),
        any())).thenAnswer {
            (it.arguments[1] as (BigInteger) ->
            Unit).invoke(sumResult)
        }
        whenever(context.getString(
```

```
            R.string.the_sum_of_numbers_from_1_to_is,
            n, sumResult.toString())
        ).thenReturn(expected)
        val callback = mock<(String) -> Unit>()
        textFormatter.getSumResult(n, callback)
        verify(callback).invoke(expected)
    }
}
```

Then, create the test for the error scenario:

```
@Test
fun getSumResult_error() {
    val n = 10
    val expected = "expected"
    whenever(numberAdder.sum(eq(n),
        any())).thenThrow(NumberAdder
        .InvalidNumberException)
        whenever(context.getString(
        R.string.error_invalid_number))
        .thenReturn(expected)
    val callback = mock<(String) -> Unit>()
    textFormatter.getSumResult(n, callback)
    verify(callback).invoke(expected)
}
```

12. In main/res/values/strings.xml, add the following strings:

```
<string name="the_sum_of_numbers_from_1_to_is">
    The sum of numbers from 1 to %1$d is:
    %2$s</string>
<string name="error_invalid_number">Error: Invalid
    number</string>
<string name="calculate">Calculate</string>
```

13. Create the layout for activity_main.xml in the main/res/layout folder:

```
<EditText
    android:id="@+id/edit_text"
    android:layout_width="match_parent"
```

```
            android:layout_height="wrap_content"
            android:inputType="number" />
    <Button
        android:id="@+id/button"
        android:layout_width="wrap_content"
        android:layout_height="wrap_content"
        android:layout_gravity="center_horizontal"
        android:text="@string/calculate" />
```

The complete code for this step can be found at https://packt.link/hxZ0I.

14. In the main folder in the root package, create the MainActivity class, which will contain all the other components:

```
class MainActivity : AppCompatActivity() {
    private lateinit var textFormatter: TextFormatter
    override fun onCreate(savedInstanceState: Bundle?)
    {
        super.onCreate(savedInstanceState)
        setContentView(R.layout.activity_main)
        textFormatter = TextFormatter(NumberAdder(),
            applicationContext)
        findViewById<Button>(R.id.button)
        .setOnClickListener {
            textFormatter.getSumResult(findViewById
            <EditText>(R.id.edit_text).text.toString()
            .toIntOrNull() ?: 0) {
                findViewById<TextView>(R.id.text_view)
                .text = it
            }
        }
    }
}
```

15. Create a test for `MainActivity` and place it in the `test` directory. It will contain two test methods, one for success and one for errors:

    ```
    @RunWith(AndroidJUnit4::class)
    class MainActivityTest {
        @Test
        fun `show sum result in text view`() {
            val scenario =
                launch(MainActivity::class.java)
            scenario.moveToState(Lifecycle.State.RESUMED)
            scenario.onActivity { activity ->
                onView(withId(R.id.edit_text))
                    .perform(replaceText("5"))
                onView(withId(R.id.button)).perform(click(
                ))
                onView(withId(R.id.text_view))
                  .check(matches(withText(
                  activity.getString(
                  R.string.the_sum_of_numbers_from_1_to_is
                  , 5, "15"))))
            }
        }
    }
    ```

    The complete code for this step can be found at `https://packt.link/fZI3u`.

    If you run the tests by right-clicking the package in which the tests are located and selecting **Run all in [package_name]**, then an output like the following will appear:

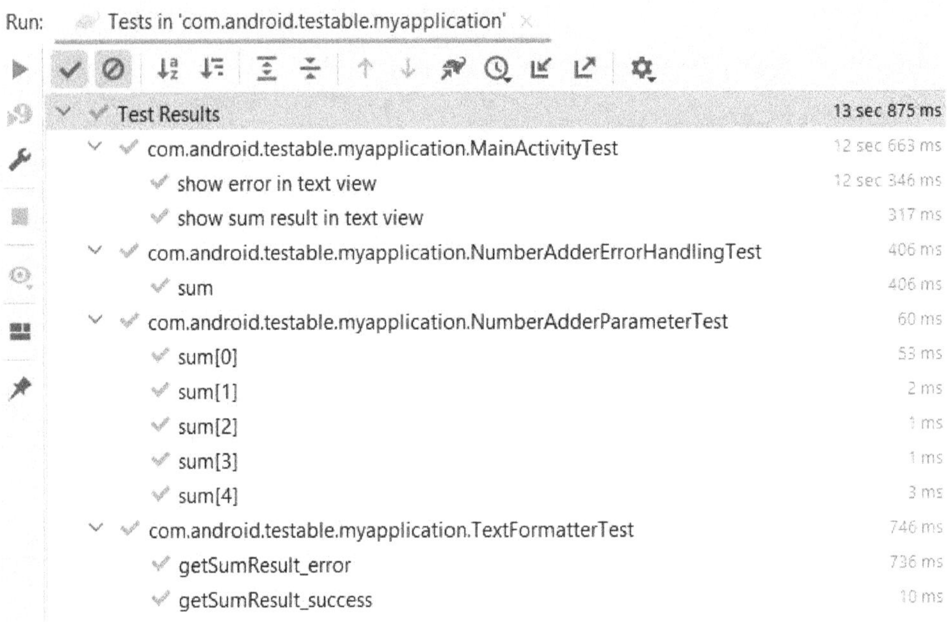

Run:	Tests in 'com.android.testable.myapplication'	
Test Results		13 sec 875 ms
com.android.testable.myapplication.MainActivityTest		12 sec 663 ms
show error in text view		12 sec 346 ms
show sum result in text view		317 ms
com.android.testable.myapplication.NumberAdderErrorHandlingTest		406 ms
sum		406 ms
com.android.testable.myapplication.NumberAdderParameterTest		60 ms
sum[0]		53 ms
sum[1]		2 ms
sum[2]		1 ms
sum[3]		1 ms
sum[4]		3 ms
com.android.testable.myapplication.TextFormatterTest		746 ms
getSumResult_error		736 ms
getSumResult_success		10 ms

Figure 10.8 – Result of executing the tests in the test folder for Exercise 10.02

If you execute the preceding tests, you should see an output like *Figure 10.8*. The Robolectric test is executed in the same way as a regular unit test; however, there is an increase in the execution time.

16. Let's now migrate the preceding test to an instrumented integration test. To do this, we will copy the preceding test from the `test` package into the `androidTest` package and remove the code related to scenarios from our tests. Make sure that in the `androidTest` folder, there is a Java folder that contains a package with the same name as the `main/java` folder. You will need to move your tests to this package.

17. After copying the file, we will use `ActivityTestRule`, which will launch our activity before every test is executed. We will also need to rename the class to avoid duplicates and rename the test methods because the syntax is not supported for instrumented tests:

```
@RunWith(AndroidJUnit4::class)
class MainActivityUiTest {
    @Test
    fun showSumResultInTextView() {
        val scenario =
            launch(MainActivity::class.java)
        scenario.moveToState(Lifecycle.State.RESUMED)
        onView(withId(R.id.edit_text)).perform(
```

```
            replaceText("5"))
        onView(withId(R.id.button)).perform(click())
        onView(withId(R.id.text_view)).check(matches(
            withText(getApplicationContext<Application>
            ().getString(R.string
            .the_sum_of_numbers_from_1_to_is, 5, "15"))
        ))
    }
}
```

The complete code for this step can be found at https://packt.link/hNB4A.

If you run the tests by right-clicking the package in which the tests are located and selecting **Run all in [package_name]**, then an output like the following will appear:

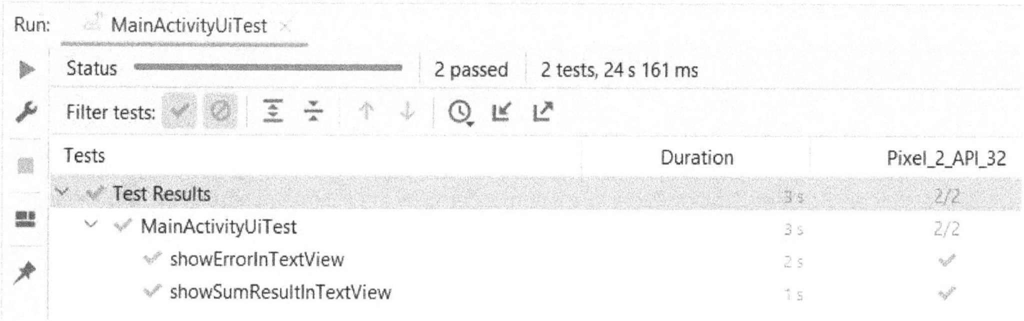

Figure 10.9 – Result of executing the tests in the androidTest folder for Exercise 10.02

In *Figure 10.9*, we can see what Android Studio displays as an output for the result. If you pay attention to the emulator while the tests are executing, you can see that for each test, your activity will be opened, the input will be set in the field, and the button will be clicked.

Both of our integration tests (on the workstation and the emulator) try to match the accepted criteria of the requirement. The integration tests verify the same behavior, the only difference is that one checks it locally and the other checks it on an Android device or emulator. The main benefit here is the fact that Espresso was able to bridge the gap between them, making integration tests easier to set up and execute.

In this section, we have implemented an exercise in which we have written tests with the Robolectric library combined with the Espresso library and looked at how we can migrate our Robolectric tests from the test folder to the androidTest folder. In the section that follows, we will look at how we can build upon the existing testing suite with instrumented tests that run on physical devices or emulators.

UI tests

UI tests are instrumented tests where developers can simulate user journeys and verify the interactions between different modules of the application. They are also referred to as end-to-end tests. For small applications, you can have one test suite, but for larger applications, you should split your test suites to cover user journeys (logging in, creating an account, setting up flows, and so on).

Because they are executed on the device, you will need to write them in the `androidTest` package, which means they will run with the **Instrumentation** framework. Instrumentation works as follows:

- The app is built and installed on the device
- A testing app will also be installed on the device that will monitor your app
- The testing app will execute the tests on your app and record the results

One of the drawbacks of this is the fact that the tests will share persisted data, so if a test stores data on the device, then the second test can have access to that data, which means that there is a risk of failure. Another drawback is that if a test comes across a crash, this will stop the entire testing because the application under test is stopped.

These issues were solved in the Jetpack updates with the introduction of the **orchestrator** framework. Orchestrators give you the ability to clear the data after each test is executed, sparing developers the need to make any adjustments. The orchestrator is represented by another application that will manage how the testing app will coordinate the tests and the data between the tests.

In order to add it to your project, you need a configuration similar to this in the `app/build.gradle` file:

```
android {
    ...
    defaultConfig {
        ...
        testInstrumentationRunner
            "androidx.test.runner.AndroidJUnitRunner"
        testInstrumentationRunnerArguments
            clearPackageData: 'true'
    }
    testOptions {
        execution 'ANDROIDX_TEST_ORCHESTRATOR'
```

```
        }
    }
    dependencies {
        ...
        androidTestUtil 'androidx.test:orchestrator:1.4.2'
    }
```

You can execute the orchestrator test on a connected device using Gradle's `connectedCheck` command, either from Terminal or from the list of Gradle commands.

In the configuration, you will notice the following line: `testInstrumentationRunner`. This allows us to create a custom configuration for the test, which gives us the opportunity to inject mock data into the modules:

```
testInstrumentationRunner "com.android.CustomTestRunner"
```

`CustomTestRunner` looks like this (`import` statements are not shown in the following code snippets):

```
class CustomTestRunner: AndroidJUnitRunner() {
    @Throws(Exception::class)
    override fun newApplication(
        cl: ClassLoader?,
        className: String?,
        context: Context?
    ): Application? {
        return super.newApplication(cl,
            MyApplication::class.java.name, context)
    }
}
```

The test classes themselves can be written by applying the JUnit4 syntax with the help of the `androidx.test.ext.junit.runners.AndroidJUnit4` test runner:

```
@RunWith(AndroidJUnit4::class)
class MainActivityUiTest {
}
```

The `@Test` methods themselves run in a dedicated test thread, which is why a library such as Espresso is helpful. Espresso will automatically move every interaction with a view on the UI thread. Espresso can be used for UI tests in a similar way as it is used with Robolectric tests:

```
@Test
fun myTest() {
    onView(withId(R.id.edit_text)).perform(replaceText
        ("5"))
    onView(withId(R.id.button)).perform(click())
    onView(withId(R.id.text_view)).check(matches(
        withText("my test")))
}
```

Typically, in UI tests, you will find interactions and assertions that may get repetitive. In order to avoid duplicating multiple scenarios in your code, you can apply a pattern called **Robot**. Each screen will have an associated `Robot` class in which the interactions and assertions can be grouped into specific methods. Your test code will use the robots and assert them. A typical robot will look something like this:

```
class MyScreenRobot {
    fun setText(): MyScreenRobot {
        onView(ViewMatchers.withId(R.id.edit_text))
            .perform(ViewActions.replaceText("5"))
        return this
    }
    fun pressButton(): MyScreenRobot {
        onView(ViewMatchers.withId(R.id.button))
            .perform(ViewActions.click())
        return this
    }
    fun assertText(): MyScreenRobot {
        onView(ViewMatchers.withId(R.id.text_view))
            .check(ViewAssertions.matches(ViewMatchers
            .withText("my test")))
        return this
    }
}
```

The test will look like this:

```
@Test
fun myTest() {
    MyScreenRobot()
        .setText()
        .pressButton()
        .assertText()
}
```

Because apps can be multithreaded and sometimes it takes a while to load data from various sources (internet, files, local storage, and so on), the UI tests will have to know when the UI is available for interactions. One way to implement this is through the use of idling resources.

These are objects that can be registered to Espresso before the test and injected into your application's components where multithreaded work is done. The apps will mark them as non-idle when the work is in progress and idle when the work is done. It is at this point that Espresso will then start executing the test. One of the most commonly used ones is CountingIdlingResource.

This specific implementation uses a counter that should be incremented when you want Espresso to wait for your code to complete its execution and decremented when you want to let Espresso verify your code. When the counter reaches 0, Espresso will resume testing. An example of a component with an idling resource looks something like this:

```
class MyHeavyliftingComponent(private val
countingIdlingResource:CountingIdlingResource) {
    fun doHeavyWork() {
        countingIdlingResource.increment()
        // do work
        countingIdlingResource.decrement()
    }
}
```

The Application class can be used to inject the idling resource, like this:

```
class MyApplication : Application(){
    val countingIdlingResource =
        CountingIdlingResource("My heavy work")
    val myHeavyliftingComponent =
        MyHeavyliftingComponent(countingIdlingResource)
}
```

Then, in the test, we can access the `Application` class and register the resource to Espresso:

```
@RunWith(AndroidJUnit4::class)
class MyTest {
    @Before
    fun setUp() {
        val myApplication =
            getApplicationContext<MyApplication>()
        IdlingRegistry.getInstance()
            .register(myApplication.countingIdlingResource)
    }
}
```

Espresso comes with a set of extensions that can be used to assert different Android components. One extension is intents testing. This is useful when you want to test an activity in isolation (more appropriate for integration tests). In order to use this, you need to add the library to Gradle:

```
androidTestImplementation 'androidx.test.espresso:
espresso-intents:3.5.0'
```

After you add the library, you need to set up the necessary intent monitoring using the `init` method from the `Intents` class, and to stop monitoring, you can use the `release` method from the same class. These operations can be done in the `@Before` and `@After` annotated methods of your test class.

To assert the values of the intent, you need to trigger the appropriate action and then use the `intended` method:

```
onView(withId(R.id.button)).perform(click())
intended(allOf(hasComponent(hasShortClassName
    (".MainActivity")), hasExtra(MainActivity
    .MY_EXTRA, "myExtraValue")))
```

The `intended` method works in a similar way to the `onView` method. It requires a matcher that can be combined with a `Hamcrest` matcher. The intent-related matchers can be found in the `IntentMatchers` class. This class contains methods to assert different methods of the `Intent` class: extras, data, components, bundles, and so on.

Another important extension library comes to the aid of `RecyclerView`. The `onData` method from Espresso is only capable of testing `AdapterViews` such as `ListView` and isn't capable of asserting `RecyclerView`. In order to use the extension, you need to add the following library to your project:

```
androidTestImplementation
'com.android.support.test.espresso:espresso-contrib:3.5.0'
```

This library provides a `RecyclerViewActions` class, which contains a set of methods that allow you to perform actions on items inside `RecyclerView`:

```
onView(withId(R.id.recycler_view))
.perform(RecyclerViewActions.actionOnItemAtPosition(0,
click()))
```

The preceding statement will click the item at position 0:

```
onView(withId(R.id.recycler_view)).perform(
RecyclerViewActions.scrollToPosition<RecyclerView
.ViewHolder>(10))
```

This will scroll to the 10th item in the list:

```
onView(withText("myText")).check(matches(isDisplayed()))
```

The preceding code will check whether a view with the `myText` text is displayed, which will also apply to `RecyclerView` items.

Testing in Jetpack Compose

Jetpack Compose offers the ability to test `@Composable` functions with a similar approach to Espresso. If we are using Robolectric, we can write our testing code in the `test` folder, and if not, we can use the `androidTest` folder and our tests will be viewed as instrumented tests. The testing library is the following:

```
androidTestImplementation "androidx.compose.ui:
ui-test-junit4:1.1.1"
```

If we want to also test the `Activity` that sets the `@Composable` function as content, then we will also need to add the following library:

```
debugImplementation "androidx.compose.ui:
ui-test-manifest:1.1.1"
```

To test, we would need to use a test rule that provides a set of methods used for interacting with the `@Composable` elements and performing assertions on them. We have multiple ways of obtaining that rule through the following approaches:

```
class MyTest {
    @get:Rule
    var composeTestRuleForActivity =
        createAndroidComposeRule(MyActivity::class.java)
    @get:Rule
    var composeTestRuleForNoActivity = createComposeRule()
    @Test
    fun testNoActivityFunction(){
        composeTestRuleForNoActivity.setContent {
            // Set method you want to test here
        }
    }
}
```

In the preceding snippet, we have two test rules. The first one, `composeTestRuleForActivity`, will start the `Activity` that holds the `@Composable` function that we want to test and will hold all the nodes we want to assert.

The second one, `composeTestRuleForNoActivity`, provides the ability to set as content the function we want to test. This will then allow the rule to have access to all the `@Composable` elements.

If we want to identify elements from our function, we have the following methods:

```
composeTestRule.onNodeWithText("My text")
composeTestRule.onNodeWithContentDescription(
    "My content description")
composeTestRule.onNodeWithTag("My test tag")
```

In the preceding snippet, we have the `onNodeWithText` method, which will identify a particular UI element using a text label that's visible to the user. The `onNodeWithContentDescription` method will identify an element using the content description set, and `onNodeWithTag` will identify an element using the test tag, which is set using the `Modifier.testTag` method.

Like Espresso, once we identify the element we want to interact with or perform assertions on, we have similar methods for both situations. For interacting with the element, we have methods such as the following:

```
composeTestRule.onNodeWithText("My text")
    .performClick()
    .performScrollTo()
    .performTextInput("My new text")
    .performGesture {
    }
```

In the preceding snippet, we perform a click, scroll, text insertion, and gesture into the element. For assertions, some examples are as follows:

```
composeTestRule.onNodeWithText("My text")
    .assertIsDisplayed()
    .assertIsNotDisplayed()
    .assertIsEnabled()
    .assertIsNotEnabled()
    .assertIsSelected()
    .assertIsNotSelected()
```

In the preceding example, we assert whether an element is displayed, not displayed, enabled, not enabled, selected, or not selected.

If our user interface has multiple elements with the same text, we have the option to extract all of them using the following:

```
composeTestRule.onAllNodesWithText("My text")
composeTestRule.onAllNodesWithContentDescription(
    "My content description")
composeTestRule.onAllNodesWithTag("My test tag")
```

Here, we extract all the nodes that have `My text` as a text, `My content description` as a content description, and `My test tag` as a test tag. The return is a collection, which allows us to assert each element of the collection individually, like so:

```
composeTestRule.onAllNodesWithText("My text")[0]
    .assertIsDisplayed()
```

Here, we assert that the first element that has My text is displayed. We also can perform assertions on the collection, like so:

```
composeTestRule.onAllNodesWithText("My text")
    .assertCountEquals(3)
    .assertAll(SemanticsMatcher.expectValue(
    SemanticsProperties.Selected, true))
    .assertAny(SemanticsMatcher.expectValue(
    SemanticsProperties.Selected, true))
```

Here, we assert that the number of elements that have My text as a text set is three, assert whether all elements match a SemanticsMatcher, or assert whether any of the elements match a SemanticMatcher. In this case, it would assert that all the elements are selected and at least one element is selected.

Another similarity to Espresso that we have when testing Jetpack Compose is the usage of IdlingResource. Compose provides its own IdlingResource abstraction, which is separate from Espresso and can be registered to our test rule as follows:

```
@Before
fun setUp() {
    composeTestRule.registerIdlingResource(
    idlingResource)
}
@After
fun tearDown() {
    composeTestRule.unregisterIdlingResource(
    idlingResource)
}
```

In the preceding snippet, we register IdlingResource in the @Before annotated method and unregister it in the @After method.

Exercise 10.03 – random waiting times

Write an application that will have two screens. The first screen will have a button. When the user presses the button, it will wait a random time between 1 and 5 seconds and then launch the second screen, which will display the text **Opened after x seconds**, where **x** is the number of seconds that passed. Write a UI test that will cover this scenario with the following features adjusted for the test:

- The random function will return a value of 1 when the test is run
- CountingIdlingResource will be used to indicate when the timer has stopped

> **Note**
>
> Throughout this exercise, import statements are not shown. To see the full code files, refer to https://packt.link/GG32r.

Take the following steps to complete this exercise:

1. Create a new Android Studio Project with no Activity.

2. Add the following libraries to app/build.gradle:

   ```
   implementation 'androidx.test.espresso:
       espresso-idling-resource:3.5.1'
   testImplementation 'junit:junit:4.13.2'
   androidTestImplementation
       'androidx.test.espresso:espresso-core:3.5.1'
   androidTestImplementation
       'androidx.test.ext:junit:1.1.5'
   androidTestImplementation
       'androidx.test:rules:1.5.0'
   ```

3. In the main folder in the root package, create a class; start with a Randomizer class:

   ```
   open class Randomizer(private val random: Random) {
       open fun getTimeToWait(): Int {
           return random.nextInt(5) + 1
       }
   }
   ```

4. In the main folder in the root package, create a class; create a Synchronizer class, which will use Randomizer and Timer to wait for the random time interval. It will also use CountingIdlingResource to mark the start of the task and the end of the task:

   ```
   class Synchronizer(
       private val randomizer: Randomizer,
       private val timer: Timer,
       private val countingIdlingResource:
           CountingIdlingResource
   ) {
       fun executeAfterDelay(callback: (Int) -> Unit) {
           val timeToWait = randomizer.getTimeToWait()
   ```

```kotlin
            countingIdlingResource.increment()
            timer.schedule(CallbackTask(callback,
                timeToWait), timeToWait * 1000L)
    }
    inner class CallbackTask(
        private val callback: (Int) -> Unit,
        private val time: Int
    ) : TimerTask() {
        override fun run() {
            callback(time)
            countingIdlingResource.decrement()
        }
    }
}
```

5. Now, create an `Application` class, which will be responsible for creating all the instances of the preceding classes:

```kotlin
class MyApplication : Application() {
    val countingIdlingResource =
        CountingIdlingResource("Timer resource")
    val randomizer = Randomizer(Random())
    val synchronizer =
        Synchronizer(randomizer, Timer(),
        countingIdlingResource)
}
```

6. Add the `MyApplication` class to `AndroidManifest` in the `application` tag with the `android:name` attribute.

7. Create an `activity_1` layout file, which will contain a parent layout and a button:

```xml
<?xml version="1.0" encoding="utf-8"?>
<LinearLayout
xmlns:android=
  "http://schemas.android.com/apk/res/android"
    android:layout_width="match_parent"
    android:layout_height="match_parent"
    android:orientation="vertical">
```

```xml
    <Button
        android:id="@+id/activity_1_button"
        android:layout_width="wrap_content"
        android:layout_height="wrap_content"
        android:layout_gravity="center"
        android:text="@string/press_me" />
</LinearLayout>
```

8. Create an `activity_2` layout file, which will contain a parent layout and `TextView`:

   ```xml
   <?xml version="1.0" encoding="utf-8"?>
   <LinearLayout
   xmlns:android=
     "http://schemas.android.com/apk/res/android"
       android:layout_width="match_parent"
       android:layout_height="match_parent"
       android:orientation="vertical">
       <TextView
           android:id="@+id/activity_2_text_view"
           android:layout_width="wrap_content"
           android:layout_height="wrap_content"
           android:layout_gravity="center" />
   </LinearLayout>
   ```

9. Create the `Activity1` class, which will implement the logic for the button click:

   ```kotlin
   class Activity1 : AppCompatActivity() {
       override fun onCreate(savedInstanceState: Bundle?)
       {
           super.onCreate(savedInstanceState)
           setContentView(R.layout.activity_1)
           findViewById<Button>(R.id.activity_1_button)
           .setOnClickListener {
               (application as MyApplication)
               .synchronizer.executeAfterDelay {
                   startActivity(Activity2.newIntent(this
                   , it))
               }
   ```

                }
            }
        }

10. Create the `Activity2` class, which will display the received data through the intent:

    ```
    class Activity2 : AppCompatActivity() {
        companion object {
            private const val EXTRA_SECONDS =
                "extra_seconds"
            fun newIntent(context: Context, seconds: Int)
                = Intent(context, Activity2::class.java)
                    .putExtra(EXTRA_SECONDS, seconds)
        }
        override fun onCreate(savedInstanceState: Bundle?)
        {
            super.onCreate(savedInstanceState)
            setContentView(R.layout.activity_2)
            findViewById<TextView>(
                R.id.activity_2_text_view).text =
                getString(R.string.opened_after_x_seconds,
                intent.getIntExtra(EXTRA_SECONDS, 0))
        }
    }
    ```

11. Make sure that the relevant strings are added to `strings.xml`:

    ```
    <string name="press_me">Press Me</string>
    <string name="opened_after_x_seconds">Opened after
    %d seconds</string>
    ```

12. Make sure that the two activities are added to `AndroidManifest.xml`:

    ```
    <application
        ... >
        <activity
            android:name=".Activity1"
            android:exported="true">
    </application>
    ```

The complete code for this step can be found at https://packt.link/TkEX9.

13. Create a `FlowTest` class in the `androidTest` directory, which will register `IdlingResource` from the `MyApplication` object and will assert the outcome of the click:

    ```
    @RunWith(AndroidJUnit4::class)
    @LargeTest
    class FlowTest {

        @Test
        fun verifyFlow() {
            onView(withId(R.id.activity_1_button))
                .perform(click())
            onView(withId(R.id.activity_2_text_view))
                .check(matches(withText(myApplication
                    .getString(R.string.opened_after_x_seconds,
                    1))))
        }
    }
    ```

    The complete code for this step can be found at https://packt.link/711Vw.

14. Run the test multiple times and check the test results. Notice that the test will have a 20% chance of success, but it will wait until the button from `Activity1` is clicked. This means that the idling resource is working. Another thing to observe is that there is an element of randomness here.

15. Tests don't like randomness, so we need to eliminate it by making the `Randomizer` class open and creating a subclass in the `androidTest` directory. We can do the same for the `MyApplication` class and provide a different randomizer called `TestRandomizer`:

    ```
    class TestRandomizer(random: Random) :
    Randomizer(random) {
        override fun getTimeToWait(): Int {
            return 1
        }
    }
    ```

16. Now, modify the `MyApplication` class in a way in which we can override the randomizer from a subclass:

    ```
    open class MyApplication : Application() {
        val countingIdlingResource =
            CountingIdlingResource("Timer resource")
        lateinit var synchronizer: Synchronizer
        override fun onCreate() {
            super.onCreate()
            synchronizer =
                Synchronizer(createRandomizer(), Timer(),
                countingIdlingResource)
        }
        open fun createRandomizer() = Randomizer(Random())
    }
    ```

17. In the `androidTest` directory, create `TestMyApplication`, which will extend `MyApplication` and override the `createRandomizer` method:

    ```
    class TestMyApplication : MyApplication() {
        override fun createRandomizer(): Randomizer {
            return TestRandomizer(Random())
        }
    }
    ```

18. Finally, in the `androidTest/java` folder in the root package, create an instrumentation test runner that will use this new `Application` class inside the test:

    ```
    class MyApplicationTestRunner : AndroidJUnitRunner() {
        @Throws(Exception::class)
        override fun newApplication(
            cl: ClassLoader?,
            className: String?,
            context: Context?
        ): Application? {
            return super.newApplication(cl,
                TestMyApplication::class.java.name, context)
        }
    }
    ```

19. Add the new test runner to the Gradle configuration:

    ```
    android {
        ...
        defaultConfig {
            ...
            testInstrumentationRunner
                "com.android.testable.myapplication
                .MyApplicationTestRunner"
        }
    }
    ```

    If we run the test now, the test should pass; however, we have a couple of problems with our dependencies. For the `Randomizer` class, we had to make our class open so that it could be extended in the `androidTest` folder.

    Another issue is the fact that our application code contains references to idling resources that are part of the testing libraries. To solve both problems, we will need to define abstractions for the `Randomizer` and `Synchronizer` classes.

20. In the `main/java` folder in the root package, create an interface called `Randomizer`:

    ```
    interface Randomizer {
        fun getTimeToWait(): Int
    }
    ```

21. Rename the previous `Randomizer` class `RandomizerImpl` and implement the `Randomizer` interface as follows:

    ```
    class RandomizerImpl(private val random: Random) :
    Randomizer {
        override fun getTimeToWait(): Int {
            return random.nextInt(5) + 1
        }
    }
    ```

22. In `MyApplication`, modify the `createRandomizer` method to have the `Randomizer` return type, which will return an instance of `RandomizerImpl`:

    ```
    open class MyApplication : Application() {
        ...
        open fun createRandomizer() : Randomizer =
    ```

```
            RandomizerImpl(Random())
    }
```

23. Modify `TestRandomizer` to implement the `Randomizer` interface:

    ```
    class TestRandomizer : Randomizer {
        override fun getTimeToWait(): Int {
            return 1
        }
    }
    ```

24. Modify `TestMyApplication` to correct the compile errors:

    ```
    class TestMyApplication : MyApplication() {
        override fun createRandomizer(): Randomizer {
            return TestRandomizer()
        }
    }
    ```

25. In `app/build.gradle`, make the idling resource dependency `androidTestImplementation`:

    ```
    androidTestImplementation 'androidx.test.espresso:
    espresso-idling-resource:3.5.1'
    ```

26. In the `main/java` folder in the root package, create an interface called `Synchronizer`:

    ```
    interface Synchronizer {
        fun executeAfterDelay(callback: (Int) -> Unit)
    }
    ```

27. Rename the previous `Synchronizer` class `SynchronizerImpl`, implement the `Synchronizer` interface, and remove the usages of `CountingIdlingResource`:

    ```
    class SynchronizerImpl(
        private val randomizer: Randomizer,
        private val timer: Timer
    ) : Synchronizer {
        override fun executeAfterDelay(callback: (Int) ->
        Unit) {
            val timeToWait = randomizer.getTimeToWait()
                timer.schedule(CallbackTask(callback,
    ```

```
                    timeToWait), timeToWait * 1000L)
        }
        inner class CallbackTask(
            private val callback: (Int) -> Unit,
            private val time: Int
        ) : TimerTask() {
            override fun run() {
                callback(time)
            }
        }
    }
}
```

28. Modify `MyApplication` so that it will open the ability to provide different `Synchronizer` instances from the `TestMyApplication` class:

    ```
    open class MyApplication : Application() {
        lateinit var synchronizer: Synchronizer
        override fun onCreate() {
            super.onCreate()
            synchronizer = createSynchronizer()

        }
        open fun createRandomizer(): Randomizer =
            RandomizerImpl(Random())
        open fun createSynchronizer(): Synchronizer =
            SynchronizerImpl(createRandomizer(), Timer())
    }
    ```

29. In the `androidTest` folder, create a class called `TestSynchronizer`, which will wrap a `Synchronizer`, and then use `CountingIdlingResource` to increment and decrement a counter when `executeAfterDelay` is started and finished:

    ```
    class TestSynchronizer(
        private val synchronizer: Synchronizer,
        private val countingIdlingResource:
            CountingIdlingResource
    ) : Synchronizer {
        override fun executeAfterDelay(callback: (Int) -> Unit) {
    ```

```
            countingIdlingResource.increment()
            synchronizer.executeAfterDelay {
                callback(it)
                countingIdlingResource.decrement()
            }
        }
    }
```

In the preceding example, we have a reference to a Synchronizer instance. When executeAfterDelay is called, then we inform Espresso to wait. We then invoke the actual Synchronizer instance, and when it finishes the execution, we then inform Espresso to resume.

30. Modify TestMyApplication to provide an instance of TestSynchronizer:

```
    class TestMyApplication : MyApplication() {

        val countingIdlingResource =
            CountingIdlingResource("Timer resource")
        override fun createRandomizer(): Randomizer {
            return TestRandomizer()
        }
        override fun createSynchronizer(): Synchronizer {
            return
                TestSynchronizer(super.createSynchronizer(),
                countingIdlingResource)
        }
    }
```

In the preceding snippet, we create a new TestSynchronizer that wraps the Synchronizer defined in MyApplication and adds the CountingIdlingResource.

31. In FlowTest, change the reference to MyApplication with TestMyApplication:

```
    private val myApplication =
        getApplicationContext<TestMyApplication>()
```

When running the test now, everything should pass, as shown in *Figure 10.10*:

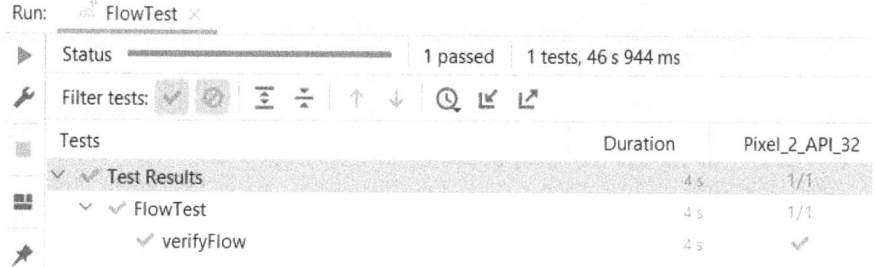

Figure 10.10 – Output of Exercise 10.03

This type of exercise shows how to avoid randomness in a test and provides concrete and repeatable input to make our tests reliable. Similar approaches are taken with dependency injection frameworks, where entire modules can be replaced in the test suite to ensure the test's reliability.

One of the most common things to be replaced is API communication. Another issue this approach solves is the decrease in waiting time. If this type of scenario were to have been repeated across your tests, then the execution time of them would have increased because of this.

In this exercise, we have looked at how we can write instrumented tests and execute them on an emulator or physical device. We have also analyzed how we can decorate our objects with `CountingIdlingResources` to be able to monitor asynchronous operations, and how we can switch dependencies that cause flakiness and provide stub data instead.

TDD

Let's assume that you are tasked with building an activity that displays a calculator with the add, subtract, multiply, and divide options. You must also write tests for your implementation. Typically, you would build your UI and your activity and a separate `Calculator` class. Then, you would write the unit tests for your `Calculator` class and then for your `activity` class.

If you were to translate the TDD process to implementing features on an Android app, you would have to write your UI test with your scenarios first. To achieve this, you can create a skeleton UI to avoid compile-time errors. After your UI test, you would need to write your `Calculator` test. Here, you would also need to create the necessary methods in the `Calculator` class to avoid compile-time errors.

If you ran your tests in this phase, they would fail. This would force you to implement your code until the tests pass. Once your `Calculator` tests pass, you can connect your calculator to your UI until your UI tests pass. While this seems like a counter-intuitive approach, it solves two issues once the process is mastered:

- Less time will be spent writing code because you will ensure that your code is testable, and you need to write only the amount of code necessary for the test to pass
- Fewer bugs will be introduced because developers will be able to analyze different outcomes

Have a look at the following diagram, which shows the TDD cycle:

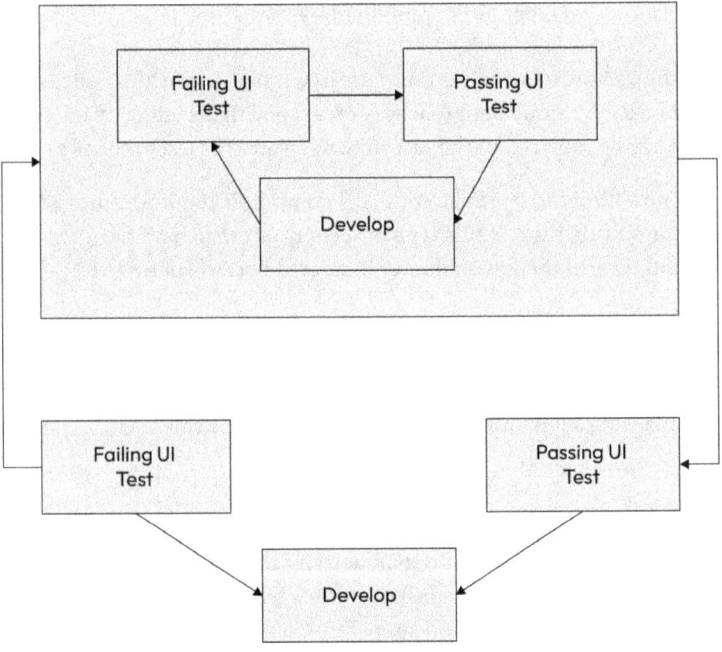

Figure 10.11 – TDD cycle

In the preceding figure, we can see the development cycle in a TDD process. You should start from a point where your tests are failing. Implement changes for the tests to pass. When you update or add new features, you can repeat the process.

Going back to our factorial examples, we started with a `factorial` function that didn't cover all our scenarios and had to keep updating the function every time a new test was added. TDD is built with that idea in mind. You start with an empty function. You start defining your testing scenarios: what are the conditions for success? What's the minimum? What's the maximum? Are there any exceptions to the main rule? What are they? These questions can help developers define their test cases. Then, these cases can be written. Let's now see how this can be done practically through the next exercise.

Exercise 10.04 – using TDD to calculate the sum of numbers

Write a function that has as input the integer n and will return the sum of numbers from 1 to n. The function should be written with a TDD approach, and the following criteria should be satisfied:

- For n<=0, the function will return the value -1
- The function should be able to return the correct value for Int.MAX_VALUE
- The function should be quick, even for Int.MAX_VALUE

Perform the following steps to complete this exercise:

1. Create a new Android Studio Project with No Activity.
2. Make sure that the following library is added to app/build.gradle:

    ```
    testImplementation 'junit:junit:4.13.2'
    ```

3. In the main/java folder in the root package, create an Adder class with the sum method, which will return 0, to satisfy the compiler:

    ```
    class Adder {
        fun sum(n: Int): Int = 0
    }
    ```

4. Create an AdderTest class in the test directory and define our test cases. We will have the following test cases: n=1, n=2, n=0, n=-1, n=10, n=20, and n=Int.MAX_VALUE. We can split the successful scenarios into one method and the unsuccessful ones into a separate method:

    ```
    class AdderTest {
        private val adder = Adder()
        @Test
        fun sumSuccess() {
            assertEquals(1, adder.sum(1))
            assertEquals(3, adder.sum(2))
            assertEquals(55, adder.sum(10))
            assertEquals(210, adder.sum(20))
            assertEquals(2305843008139952128L,
                adder.sum(Int.MAX_VALUE))
    ```

```
        }
        @Test
        fun sumError(){
            assertEquals(-1, adder.sum(0))
            assertEquals(-1, adder.sum(-1))
        }
    }
```

If we run the tests for the `AdderTest` class, we will see an output like the following figure, meaning that all our tests failed:

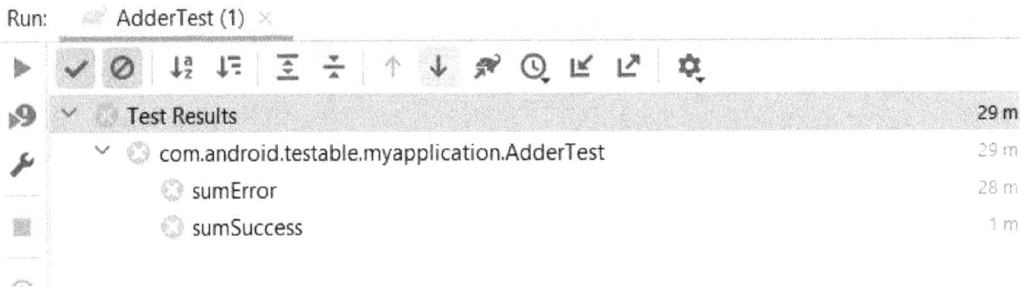

Figure 10.12 – Initial test status for Exercise 10.04

5. Let's first address the success scenarios by implementing the sum in a loop from 1 to n:

```
    class Adder {
        fun sum(n: Int): Long {
            var result = 0L
            for (i in 1..n) {
                result += i
            }
            return result
        }
    }
```

If we run the tests now, you will see that one will pass and the other will fail, like the following figure:

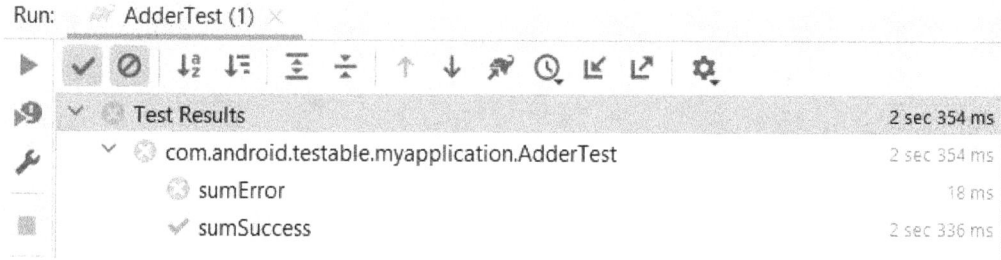

Figure 10.13 – Test status after resolving the success scenario for Exercise 10.04

6. If we look at the time it took to execute the successful test, it seems a bit long. This can add up when thousands of unit tests are present in one project. We can now optimize our code to deal with the issue by applying the *n(n+1)/2* formula:

```
class Adder {
    fun sum(n: Int): Long {
        return (n * (n.toLong() + 1)) / 2
    }
}
```

Running the tests now will drastically reduce the execution time to a few milliseconds.

7. Now, let's focus on solving our failure scenarios. We can do this by adding a condition for when n is smaller than or equal to 0:

```
class Adder {
    fun sum(n: Int): Long {
        return if (n > 0) (n * (n.toLong() + 1)) / 2
    else -1
    }
}
```

If we run the tests now, we should see them all passing, like the following figure:

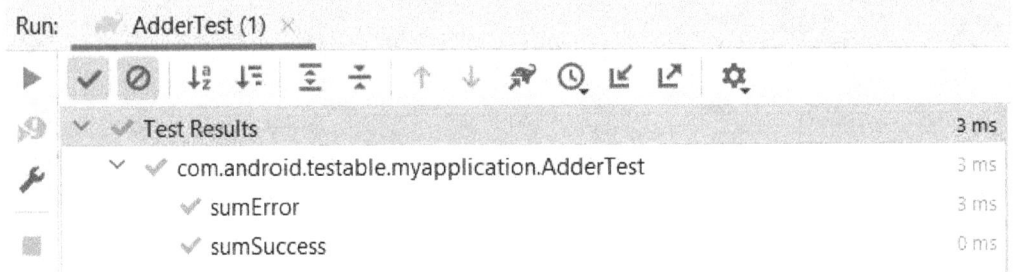

Figure 10.14 – Passing tests for Exercise 10.04

In this exercise, we have applied the concept of TDD to a very small example to demonstrate how the technique can be used. We have observed how, starting from the skeleton code, we can create a suite of tests to verify our conditions, and how by constantly running tests, we improved the code until a point where all the tests pass. As you have probably noticed, the concept isn't an intuitive one. Some developers find it hard to define how big skeleton code should be to start creating the test cases, while others, out of habit, focus on writing the code first and then developing the test. In either case, developers will need a lot of practice with the technique until it's properly mastered.

Activity 10.01 – developing with TDD

Using the TDD approach, develop an application that contains three activities and works as follows:

- In activity 1, you will display a numeric `EditText` element and a button. When the button is clicked, the number in `EditText` will be passed to activity 2.
- Activity 2 will generate a list of items asynchronously. The number of items will be represented by the number passed from activity 1. You can use the `Timer` class with a delay of 1 second. Each item in the list will display the text `Item x` where x is the position in the list. When an item is clicked, you should pass the clicked item to activity 3.
- Activity 3 will display the text `You clicked y`, where y is the text of the item the user has clicked.

The tests for the app will be the following:

- Unit tests with Mockito and `mockito-kotlin` annotated with `@SmallTest`
- Integration tests with Robolectric and Espresso annotated with `@MediumTest`
- UI tests with Espresso annotated with `@LargeTest` and using the `Robot` pattern

Run the test commands from the command line. To complete this activity, you need to take the following steps:

1. You will need Android Studio 4.1.1 or higher with Kotlin 1.4.21 or higher for the Parcelize Kotlin plugin.
2. Create the three activities and the UI for each of them.
3. In the `androidTest` folder, create three robots, one for each activity:

   - Robot 1 will contain the interaction with `EditText` and the button
   - Robot 2 will assert the number of items on the screen and interact with an item in the list
   - Robot 3 will assert the text displayed in `TextView`

4. Create an instrumented test class that will have one test method using the preceding robots.
5. Create an `Application` class that will hold instances of all the classes that will be unit-tested.
6. Create three classes representing integration tests, one for each of the activities. Each of these classes will contain one test method for interactions and data loading. Each integration test will assert the intents passed between the activities.
7. Create a class that will provide the text required for the UI. It will have a reference to a `Context` object and will contain two methods that will provide the text for the UI, which will return an empty string.
8. Create the test for the preceding class in which the two methods are tested.
9. Implement the class for the preceding tests to pass.
10. Create a class that will be responsible for loading the list in `Activity2`, and provide an empty method for loading. The class will have a reference to the timer and the idling resource. Here, you should also create a data class that will represent the model for `RecyclerView`.
11. Create a unit test for the preceding class.
12. Create the implementation for the preceding class and run the unit tests until they pass.
13. In the `Application` class, instantiate the classes that were unit-tested and start using them in your activities. Do this until your integration tests pass.
14. Provide `IntegrationTestApplication`, which will return a new implementation of the class responsible for loading. This is to avoid making your integration test for `Activity2` wait until loading is complete.
15. Provide `UiTestApplication`, which will again reduce the loading time of your models and connect the idling resource to Espresso. Implement the remaining work for the UI test to pass.

> **Note**
> The solution to this activity can be found at `https://packt.link/Ma4tD`.

Summary

In this chapter, we looked at the different types of testing and the frameworks available for implementing these tests. We also looked at the testing environment and how to structure it for each environment, as well as structuring your code into multiple components that can be individually unit-tested.

We analyzed different ways to test code, how we should approach testing, and how, by looking at different test results, we can improve our code. With TDD, we learned that by starting with testing, we can write our code faster and ensure it is less error-prone.

The activity is where all these concepts came together into building a simple Android application, and we can observe how, by adding tests, the development time increases, but this pays off in the long term by eliminating possible bugs that appear when the code is modified.

The frameworks we have studied are some of the most common ones, but there are others that build on top of these and are used by developers in their projects, such as Mockk (a mocking library designed for Kotlin that takes advantage of a lot of the features of the language) and Barista (written on top of Espresso and simplifies the syntax of UI tests), to note a few.

Think of all the concepts presented here as building blocks that fit into two processes present in the software engineering world: automation and continuous integration. Automation takes redundant and repetitive work out of the hands of developers and puts it into the hands of machines.

Instead of having a team of quality assurance people testing your application to make sure the requirements are met, you can instruct a machine through a variety of tests and test cases to test the application instead and just have one person reviewing the results of the tests.

Continuous integration builds on the concept of automation to verify your code the moment you submit it for review from other developers. A project with continuous integration would have a setup along the following lines: a developer submits work for review in a source control repository such as GitHub.

A machine in the cloud would then start executing the tests for the entire project, making sure that nothing was broken, and the developer can move on to a new task. If the tests pass, then the rest of the developers can review the code, and when it is correct, it can be merged, and a new build can be created in the cloud and distributed to the rest of the team and the testers.

All of this takes place while the initial developer can safely work on something else. If anything fails in the process, then they can pause the new task and go and address any issues in their work. The continuous integration process can then be expanded into continuous delivery, where similar automation can be set up when preparing a submission to Google Play that can be handled almost entirely by machines with minor involvement from developers.

In the chapters that follow, you will learn about how to organize your code when building more complex applications that use the storage capabilities of the device and connect to the cloud to request data. Each of those components can be individually unit-tested, and you can apply integration tests to assert a successful integration of multiple components.

11
Android Architecture Components

In this chapter, you will learn about the key components of the Android Jetpack libraries and what benefits they bring to the standard Android framework. You will also learn how to structure your code and give different responsibilities to your classes with the help of Jetpack components. Finally, you'll improve the test coverage of your code.

By the end of this chapter, you'll be able to create applications that handle the lifecycles of activities and fragments with ease. You'll also know more about how to persist data on an Android device using Room and how to use ViewModels to separate your logic from your Views.

In the previous chapters, you learned how to write unit tests. The question is: what can you unit-test? Can you unit-test activities and fragments? It is hard to unit-test activities and fragments on your machine because of the way they are built. Testing would be easier if you could move the code away from activities and fragments.

Also, consider the situation where you are building an application that supports different orientations, such as landscape and portrait, and supports multiple languages. What tends to happen in these scenarios by default is that when the user rotates the screen, the activities and fragments are recreated for the new display orientation.

Now, imagine that happens while your application is in the middle of processing data. You have to keep track of the data you are processing, keep track of what the user was doing to interact with your screens, and avoid causing a context leak.

> **Note**
> A **context leak** occurs when your destroyed activity cannot be garbage-collected because it is referenced in a component with a longer lifecycle – such as a thread that is currently processing your data.

We will cover the following topics in this chapter:

- ViewModel
- Data streams
- Room

Technical requirements

The complete code for all the exercises and the activity in this chapter is available on GitHub at `https://packt.link/89BCi`

Android components background

In many situations, you have to use `onSaveInstanceState` to save the current state of your activity/fragment, and then in `onCreate` or `onRestoreInstanceState`, you need to restore the state of your activity/fragment. This adds extra complexity to your code and makes it repetitive, especially if the processing code is part of your activity or fragment.

These scenarios are where `ViewModel` and `LiveData` come in. `ViewModels` are components built with the express goal of holding data in case of lifecycle changes. They also separate the logic from the Views, which makes them very easy to unit-test. `LiveData` is a component used to hold data and notify observers when changes occur while taking their lifecycle into account.

In simpler terms, the fragment only deals with the Views, `ViewModel` does the heavy lifting, and `LiveData` deals with delivering the results to the fragment, but only when the fragment is there and ready.

If you've ever used WhatsApp or a similar messaging app and you've turned off the internet, you'll have noticed that you are still able to use the application. The reason for this is that the messages are stored locally on your device. This is achieved with a database file called `SQLite` in most cases.

The Android Framework already allows you to use this feature for your application. However, this requires a lot of boilerplate code to read and write data. Every time you want to interact with the local storage, you must write a SQL query. When you read the SQLite data, you must convert it into a Java/Kotlin object.

All of this requires a lot of code, time, and unit testing. What if someone else were to handle the SQLite connection, and all you had to do was focus on the code part? This is where **Room** comes in. This is a library that is a wrapper over SQLite. All you need to do is define how your data should be saved and let the library take care of the rest.

Let's say you want your activity to know when there is an internet connection and when the internet drops. You can use something called `BroadcastReceiver` for this. A slight problem with this is that every time you register `BroadcastReceiver` in an activity, you must unregister it when the activity is destroyed.

You can use `Lifecycle` to observe the state of your activity, thereby allowing your receiver to be registered in the desired state and unregistered in the complementary one (for example, RESUMED-PAUSED, STARTED-STOPPED, or CREATED-DESTROYED).

`ViewModels`, `LiveData`, and `Room` are all part of the Android architecture components, which are part of the Android Jetpack libraries. The architecture components are designed to help developers structure their code, write testable components, and help reduce boilerplate code.

Other architecture components include `Databinding` (which binds views with models or `ViewModels`, allowing the data to be directly set in the Views), `WorkManager` (which allows developers to handle background work with ease), `Navigation` (which allows developers to create visual navigation graphs and specify relationships between activities and fragments), and `Paging` (which allows developers to load paginated data, which helps in situations where infinite scrolling is required).

ViewModel

The `ViewModel` component is responsible for holding and processing data required by the **user interface** (UI). It has the benefit of surviving configuration changes that destroy and recreate fragments and activities, which allows it to retain the data that can then be used to re-populate the UI.

It will eventually be destroyed when the activity or fragment is destroyed without being recreated or when the application process is terminated. This allows `ViewModel` to serve its responsibility and to have garbage collected when it is no longer necessary. The only method `ViewModel` has is the `onCleared()` method, which is called when `ViewModel` terminates. You can overwrite this method to terminate ongoing tasks and deallocate resources that will no longer be required.

Migrating data processing from the activities into `ViewModel` helps create better and faster unit tests. Testing an activity requires an Android test to be executed on a device. Activities also have states, which means that your test should get the activity into the proper state for the assertions to work. `ViewModel` can be unit-tested locally on your development machine and can be stateless, meaning that your data processing logic can be tested individually.

One of the most important features of `ViewModel` is that it allows communication between fragments. To communicate between fragments without `ViewModel`, you must make your fragment communicate with the activity, which will then call the fragment you wish to communicate with.

To achieve this with `ViewModel`, you can just attach it to the parent activity and use the same `ViewModel` in the fragment you wish to communicate with. This will reduce the boilerplate code that was required previously.

In the following diagram, you can see that `ViewModel` can be created at any point in an activity's lifecycle (in practice, they are normally initialized in `onCreate` for Activities and `onCreateView` or `onViewCreated` for Fragments because these represent the points where the views are created and ready to be updated), and that once created, it will live as long as the activity does:

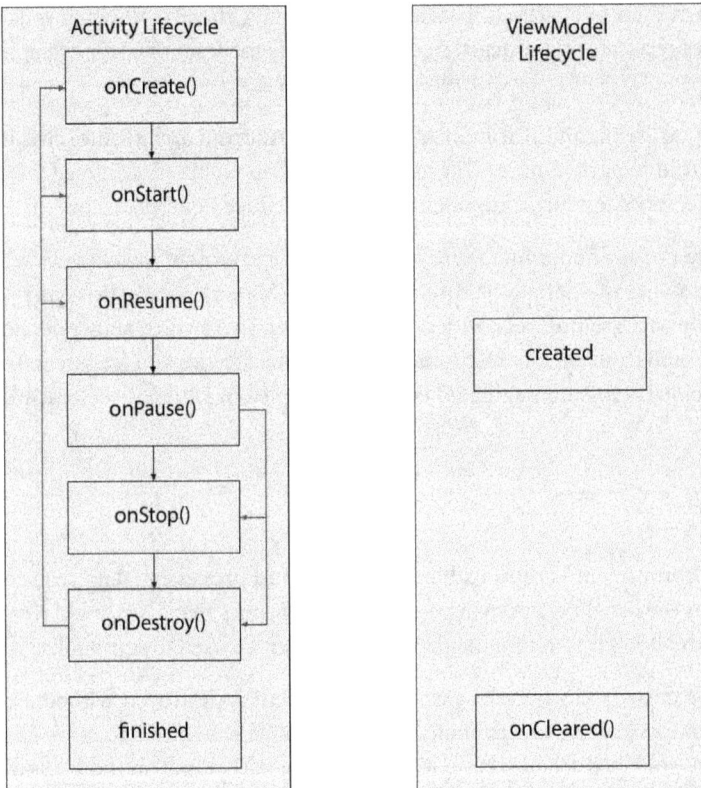

Figure 11.1 – The lifecycle of an activity compared to the ViewModel lifecycle

In the preceding diagram, we can see how the lifecycle of `Activity` compares to that of `ViewModel`. The red lines indicate what happens when the `Activity` is recreated, starting from the `onPause` method, ending in `onDestroy`, and then going from `onCreate` to `onResume` in a new instance of `Activity`.

The following diagram shows how `ViewModel` connects to a fragment:

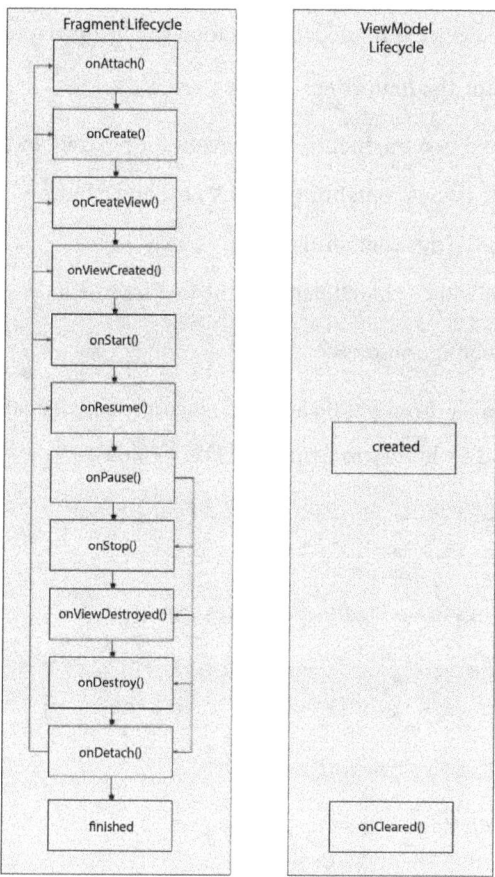

Figure 11.2 – The lifecycle of a fragment compared to the ViewModel lifecycle

In the preceding diagram, we can see how the lifecycle of `Fragment` compares to that of `ViewModel`. The red lines indicate what happens when `Fragment` is recreated, starting from the `onPause` method, ending in `onDetach`, and then going from `onAttach` to `onResume` in a new instance of `Fragment`.

In this section, we learned what a ViewModel is and the benefits it provides with regard to testing and performing logic, which survives the recreation of the activity and fragment.

Exercise 11.01 – shared ViewModel

You have been tasked with building an app with one screen split vertically into two when in portrait mode and horizontally when in landscape mode. The first half contains some text, and below it is a button.

The second half contains only text. When the screen is opened, the text in both halves displays **Total: 0**. When the button is clicked, the text will change to **Total: 1**. When clicked again, the text will change to **Total: 2**, and so on. When the device is rotated, the last total will be displayed in the new orientation.

To solve this task, we will define the following:

- An activity that will hold two fragments – one layout for portrait and another for landscape
- One fragment with one layout containing `TextView` and a button
- One fragment with one layout containing `TextView`
- One `ViewModel` that will be shared between the two fragments

Let's begin by setting up our configurations:

1. Create a new project in Android Studio and add an empty activity called `SplitActivity`.
2. Let's add the `ViewModel` library to `app/build.gradle`:

    ```
    implementation "androidx.lifecycle:
    lifecycle-viewmodel-ktx:2.5.1"
    ```

3. Add the following strings to `values/strings.xml`:

    ```
    <string name="press_me">Press Me</string>
    <string name="total">Total %d</string>
    ```

4. Create and define `SplitFragmentOne`:

    ```
    class SplitFragmentOne : Fragment() {
        override fun onCreateView(
            inflater: LayoutInflater,
            container: ViewGroup?,
            savedInstanceState: Bundle?
        ): View? {
            return inflater.inflate(
            R.layout.fragment_split_one, container, false)
        }
        override fun onViewCreated(view: View,
            savedInstanceState: Bundle?) {
            super.onViewCreated(view, savedInstanceState)
            view.findViewById<TextView>
                (R.id.fragment_split_one_text_view)
    ```

```kotlin
                    .text = getString(R.string.total, 0)
        }
    }
```

5. Add the `fragment_split_one.xml` file to the `res/layout` folder:

```xml
<?xml version="1.0" encoding="utf-8"?>
<LinearLayout xmlns:android=
    "http://schemas.android.com/apk/res/android"
    android:layout_width="match_parent"
    android:layout_height="match_parent"
    android:gravity="center"
    android:orientation="vertical">
    <TextView
        android:id="@+id/fragment_split_one_text_view"
        android:layout_width="wrap_content"
        android:layout_height="wrap_content" />
    <Button
        android:id="@+id/fragment_split_one_button"
        android:layout_width="wrap_content"
        android:layout_height="wrap_content"
        android:text="@string/press_me" />
</LinearLayout>
```

6. Now, let's create and define `SplitFragmentTwo`:

```kotlin
class SplitFragmentTwo : Fragment() {
    override fun onCreateView(
        inflater: LayoutInflater,
        container: ViewGroup?,
        savedInstanceState: Bundle?
    ): View? {
        Return inflater.inflate(
        R.layout.fragment_split_two, container, false)
    }
    override fun onViewCreated(view: View,
        savedInstanceState: Bundle?) {
        super.onViewCreated(view, savedInstanceState)
```

```
            view.findViewById<TextView> (
            R.id.fragment_split_two_text_view).text =
            getString(R.string.total, 0)
    }
}
```

7. Add the fragment_split_two.xml file to the res/layout folder:

   ```
   <?xml version="1.0" encoding="utf-8"?>
   <LinearLayout xmlns:android =
       "http://schemas.android.com/apk/res/android"
       android:layout_width="match_parent"
       android:layout_height="match_parent"
       android:gravity="center"
       android:orientation="vertical">
       <TextView
           android:id="@+id/fragment_split_two_text_view"
           android:layout_width="wrap_content"
           android:layout_height="wrap_content" />
   </LinearLayout>
   ```

8. Define SplitActivity:

   ```
   class SplitActivity : AppCompatActivity() {
       override fun onCreate(savedInstanceState: Bundle?)
       {
           super.onCreate(savedInstanceState)
           setContentView(R.layout.activity_split)
       }
   }
   ```

9. Create the activity_split.xml file in the res/layout folder:

   ```
   <?xml version="1.0" encoding="utf-8"?>
   <LinearLayout>
       <androidx.fragment.app.FragmentContainerView
           android:id="@+id/activity_fragment_split_1"
           android:name="{package.path}.SplitFragmentOne"
           android:layout_width="match_parent"
   ```

```
        />
        <androidx.fragment.app.FragmentContainerView
            android:id="@+id/activity_fragment_split_2"
            android:name="{package.path}.SplitFragmentTwo"
            android:layout_width="match_parent"
            />
</LinearLayout>
```

The complete code for this step can be found at https://packt.link/HPy9p.

Replace {package.path} with the name of the package in which your Fragments are located.

10. Next, let's create a layout-land folder in the res folder. Then, in the layout-land folder, we'll create an activity_split.xml file with the following layout:

```
<?xml version="1.0" encoding="utf-8"?>
<LinearLayout>
    <androidx.fragment.app.FragmentContainerView
        android:id="@+id/activity_fragment_split_1"
        android:name="{package.path}.SplitFragmentOne"
        />
    <androidx.fragment.app.FragmentContainerView
        android:id="@+id/activity_fragment_split_2"
        android:name="{package.path}.SplitFragmentTwo"
        />
</LinearLayout>
```

The complete code for this step can be found at https://packt.link/1zRQa.

Replace {package.path} with the name of the package in which your Fragments are located. Notice the same android:id attribute in both the activity_split.xml files. This allows the operating system to correctly save and restore the fragment's state during rotation.

11. In the main/java folder in the root package, create a TotalsViewModel that looks like this:

```
class TotalsViewModel : ViewModel() {
    var total = 0
```

```
        fun increaseTotal(): Int {
            total++
            return total
        }
    }
```

Notice that we extended from the `ViewModel` class, which is part of the lifecycle library. In the `ViewModel` class, we defined one method that increases the total value and returns the updated value.

12. Now, add the `updateText` and `prepareViewModel` methods to the `SplitFragment1` fragment:

    ```
    class SplitFragmentOne : Fragment() {
        ...
        override fun onViewCreated(view: View,
        savedInstanceState: Bundle?) {
            ...
            prepareViewModel()
        }

        private fun prepareViewModel() {
        }

        private fun updateText(total: Int) {
            view?.findViewById<TextView>
            (R.id.fragment_split_one_text_view)?.text =
            getString(R.string.total, total)
        }
    }
    ```

13. In the `prepareViewModel()` function, let's start adding our `ViewModel`:

    ```
    private fun prepareViewModel() {
        val totalsViewModel = ViewModelProvider(this)
            .get(TotalsViewModel::class.java)
    }
    ```

    This is how the `ViewModel` instance is accessed. `ViewModelProvider(this)` will bind `TotalsViewModel` to the lifecycle of the fragment. `.get(TotalsViewModel::class.java)` will retrieve the `TotalsViewModel` instance that we defined previously.

If the fragment is being created for the first time, it will produce a new instance, while if the fragment is recreated after a rotation, it will provide the previously created instance. We pass the class as an argument because a fragment or activity can have multiple ViewModels, and the class serves as an identifier for the type of `ViewModel` we want.

14. Now, set the last known value on the view:

    ```
    private fun prepareViewModel() {
        val totalsViewModel = ViewModelProvider(this)
            .get(TotalsViewModel::class.java)
        updateText(totalsViewModel.total)
    }
    ```

    The second line will help during device rotation. It will set the last total that was computed. If we remove this line and rebuild, then we will see **Total 0** every time we rotate, and after every click, we will see the previously computed total plus 1.

15. Update the View when the `fragment_split_one_button` button is clicked:

    ```
    private fun prepareViewModel() {
        val totalsViewModel =
            ViewModelProvider(this)
                .get(TotalsViewModel::class.java)
        updateText(totalsViewModel.total)
        view?.findViewById<Button>
        (R.id.fragment_split_one_button)
        ?.setOnClickListener {
            updateText(totalsViewModel.increaseTotal())
        }
    }
    ```

    The last few lines indicate that when a click is performed on the button, we tell `ViewModel` to recompute the total and set the new value.

16. Add the same `ViewModel` we used previously to our `SplitFragmentTwo`:

    ```
    class SplitFragmentTwo : Fragment() {
        override fun onCreateView(
            inflater: LayoutInflater,
            container: ViewGroup?,
            savedInstanceState: Bundle?
        ): View? {
    ```

```
            return inflater.inflate(
                R.layout.fragment_split_two, container, false)
        }
        override fun onViewCreated(view: View,
        savedInstanceState: Bundle?) {
            super.onViewCreated(view, savedInstanceState)
            val totalsViewModel = ViewModelProvider(this)
                .get(TotalsViewModel::class.java)
            updateText(totalsViewModel.total)
        }
        private fun updateText(total: Int) {
            view?.findViewById<TextView>(
            R.id.fragment_split_two_text_view)?.text =
            getString(R.string.total, total)
        }
    }
```

If we run the app now, we'll see that nothing has changed. The first fragment works as before, but the second fragment doesn't get any updates. This is because even though we defined one `ViewModel`, we have two instances of that `ViewModel` for each of our fragments.

We will need to limit the number of instances to one per fragment. We can achieve this by attaching our `ViewModel` to the `SplitActivity` lifecycle using a method called `requireActiviy`.

17. Let's modify our fragments. In both fragments, we need to find and change the following code:

    ```
    val totalsViewModel = ViewModelProvider(this).
    get(TotalsViewModel::class.java)
    ```

    We will change it to the following:

    ```
    val totalsViewModel = ViewModelProvider(requireActivity())
        .get(TotalsViewModel::class.java)
    ```

> **Note**
> Using ViewModels to communicate between fragments will only work when the fragments are placed in the same activity.

If we run the application, we should see the following:

Figure 11.3 – Output of Exercise 11.01

When the button is clicked, the total updates are on the top half of the screen but not on the bottom half. If we rotate the screen, the last value `ViewModel` had will be set on the second screen as well. This means that our application doesn't react properly to the changes in `ViewModel`.

This means that we will need a publisher-subscriber approach to monitor changes that occur in our data. In the next section, we will look at some common data streams that ViewModels can use to notify when the data is changed.

In this exercise, we implemented a ViewModel, which was responsible for incrementing an integer value that will be displayed on the screen. In the section that follows, we will connect data streams to react to changes when the number is incremented.

Data streams

When it comes to data observability, we have multiple approaches for implementation, whether manually built mechanisms, components from the Java language, third-party components, or finally to solutions developed particularly for Android. When it comes to Android, some of the most common solutions are `LiveData`, Flows from the Coroutines components, and RxJava.

The first one we will look at is `LiveData`, as it is part of the Android Architecture Components, which means that it is tailored specially to Android. We will then look at how we can use other types of data streams, which we will cover in more depth in future chapters.

LiveData

`LiveData` is a lifecycle-aware component that permits updates to your UI, but only if the UI is in an active state (for example, if the activity or fragment is in one of the `STARTED` or `RESUMED` states). To monitor changes on `LiveData`, you need an observer combined with a `LifecycleOwner`. When the activity is set to an active state, the observers will be notified when changes occur.

If the activity is recreated, then the observer will be destroyed, and a new one will be reattached. Once this happens, the last value of `LiveData` will be emitted to allow us to restore the state. Activities and fragments are `LifecycleOwners`, but fragments have a separate `LifecycleOwner` for the View states. Fragments have this particular `LifecycleOwner` due to their behavior in the `BackStack` fragment.

When fragments are replaced within the back stack, they are not fully destroyed; only their Views are. Some of the common callbacks that developers use to trigger processing logic are `onViewCreated()`, `onActivityResumed()`, and `onCreateView()`. If we were to register observers for `LiveData` in these methods, we might end up with scenarios where multiple observers will be created every time our fragment pops back onto the screen.

When updating a `LiveData` model, we are presented with two options: `setValue()` and `postValue()`. `setValue()` will deliver the result immediately and is meant to be called only on the UI thread. On the other hand, `postValue()` can be called on any thread. When `postValue()` is called, `LiveData` will schedule an update of the value on the UI thread and update the value when the UI thread becomes free.

In the `LiveData` class, these methods are protected, which means that there are subclasses that allow us to change the data. `MutableLiveData` makes the methods public, which gives us a simple solution for observing data in most cases. `MediatorLiveData` is a specialized implementation of `LiveData` that allows us to merge multiple `LiveData` objects into one (this is useful in situations where our data is kept in different repositories and we want to show a combined result).

`TransformLiveData` is another specialized implementation that allows us to convert one object into another (this helps us in situations where we grab data from one repository and we want to request data from another repository that depends on the previous data, as well as in situations where we want to apply extra logic to a result from a repository).

`Custom LiveData` allows us to create our own `LiveData` implementations (usually when we periodically receive updates, such as the odds in a sports betting app, stock market updates, and Facebook and Twitter feeds).

> **Note**
>
> It is a common practice to use LiveData in ViewModel. Holding LiveData in a fragment or activity will cause losses in data when configuration changes occur.

The following diagram shows how LiveData is connected to the lifecycle of LifecycleOwner:

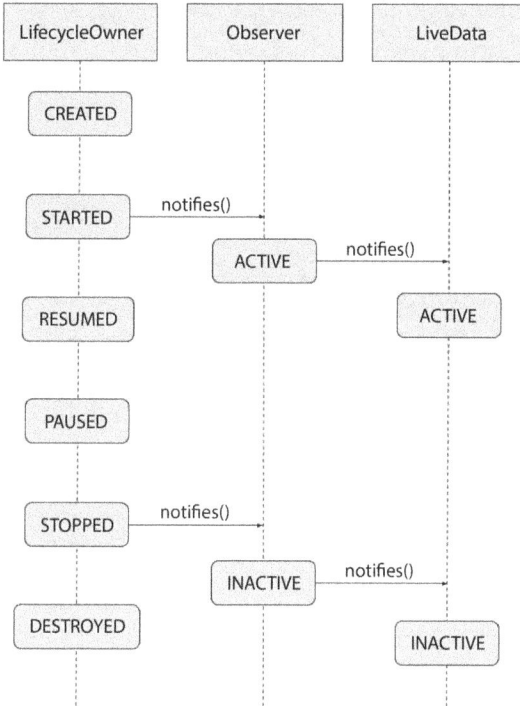

Figure 11.4 – The relationship between LiveData and lifecycle observers with LifecycleOwner

> **Note**
>
> We can register multiple observers for LiveData, and each observer can be registered for a different LifecycleOwner. In this situation, LiveData will become inactive, but only when all the observers are inactive.

In this section, we looked at how the LiveData component works and the benefits it provides for observing data from activities and fragments with regard to their lifecycles. In the following section, we will look at an exercise that uses LiveData.

Exercise 11.02 – observing with LiveData

Modify *Exercise 11.01 – shared ViewModel* so that when the button is clicked, both fragments will be updated with the total number of clicks.

Perform the following steps to achieve this:

1. Add the `LiveData` library to the `app/build.gradle` file:

    ```
    implementation "androidx.lifecycle:
    lifecycle-livedata-ktx:2.5.1"
    ```

2. `TotalsViewModel` should be modified so that it supports `LiveData`:

    ```
    class TotalsViewModel : ViewModel() {
        private val _total = MutableLiveData<Int>()
        val total: LiveData<Int> = _total
        init {
            _total.postValue(0)
        }
        fun increaseTotal() {
            _total.postValue((_total.value ?: 0) + 1)
        }
    }
    ```

    Here, we created `MutableLiveData`, a subclass of `LiveData` that allows us to change the value of the data. When `ViewModel` is created, we set the default value of 0, and then when we increase the total, we post the previous value plus 1.

    The reason we have duplicated representations for the total is that we wanted to keep the mutable component private to the class while exposing the non-mutable total to be observed by other objects.

3. Now, we need to modify our fragments so that they adjust to the new `ViewModel`. For `SplitFragmentOne`, we do the following:

    ```
    override fun onViewCreated(view: View,
    savedInstanceState: Bundle?) {
        super.onViewCreated(view, savedInstanceState)
            val totalsViewModel =
              ViewModelProvider(requireActivity())
                .get(TotalsViewModel::class.java)
        totalsViewModel.total.observe(
    ```

```
            viewLifecycleOwner, {
                updateText(it)
            })
            view.findViewById<Button>(
            R.id.fragment_split_one_button)
                .setOnClickListener {
                    totalsViewModel.increaseTotal()
                }
        }
```

4. For SplitFragmentTwo, we do the following:

```
        override fun onViewCreated(view: View,
            savedInstanceState: Bundle?) {
            super.onViewCreated(view, savedInstanceState)
            val totalsViewModel =
                ViewModelProvider(requireActivity())
                    .get(TotalsViewModel::class.java)
            totalsViewModel.total.observe(
            viewLifecycleOwner, {
                updateText(it)
            })
        }
```

If we look at the following line: `totalsViewModel.getTotal().observe(viewLifecycleOwner, { updateText(it) })`, the LifecycleOwner parameter for the observe method is called viewLifecycleOwner. This is inherited from the fragment class, and it helps when we observe data while the View that the fragment manages is being rendered. In our example, swapping viewLifecycleOwner with this would not have caused an impact.

But if our fragment had been part of a back stack feature, then there would have been the risk of creating multiple observers, which would have led to being notified multiple times for the same dataset.

5. Now, let's write a test for our new ViewModel. We will name it TotalsViewModelTest and place it in the test package, not androidTest. This is because we want this test to execute on our workstation, not the device:

```
    class TotalsViewModelTest {
        private lateinit var totalsViewModel:
```

```
            TotalsViewModel
    @Before
    fun setUp() {
        totalsViewModel = TotalsViewModel()
        assertEquals(0, totalsViewModel.total.value)
    }
    @Test
    fun increaseTotal() {
        val total = 5
        for (i in 0 until total) {
            totalsViewModel.increaseTotal()
        }
        assertEquals(4, totalsViewModel.total.value)
    }
}
```

6. In the preceding test, before testing begins, we assert that the initial value of LiveData is set to 0. Then, we write a small test in which we increase the total five times, and we assert that the final value is 5. Let's run the test and see what happens:

   ```
   java.lang.RuntimeException: Method getMainLooper in
   android.os.Looper not mocked.
   ```

7. A message similar to the preceding one will appear. This is because of how LiveData is implemented. Internally, it uses Handlers and Loopers, part of the Android framework, thus preventing us from executing the test. Luckily, there is a way around this. We will need the following configuration in our Gradle file for our test:

   ```
   testImplementation "androidx.arch.core:
   core-testing:2.1.0"
   ```

8. This adds a testing library to our testing code, not our application code. Now, let's add the following line to our code above the instantiation of the ViewModel class:

   ```
   class TotalsViewModelTest {
       @get:Rule
       val rule = InstantTaskExecutorRule()
       private val totalsViewModel = TotalsViewModel()
   ```

We have added a TestRule that says every time LiveData has its value changed, it will make the change instantly and will avoid using the Android Framework components.

Every test we write in this class will be impacted by this rule, thus giving us the freedom to play with the LiveData class for each new test method. If we run the test again, we will see the following:

```
java.lang.RuntimeException: Method getMainLooper in
android.os.Looper not mocked
```

9. Does this mean that our new rule didn't work? Not exactly. If you look in your TotalsViewModels class, you'll see this:

```
init {
        total.postValue(0)
}
```

10. This means that because we created the ViewModel class outside of the rule's scope, the rule will not apply. We can do two things to avoid this scenario: we can change our code to handle a null value that will be sent when we first subscribe to the LiveData class, or we can adjust our test so that we put the ViewModel class in the scope of the rule. Let's go with the second approach and change how we create our ViewModel class in the test. It should look something like this:

```
@get:Rule
val rule = InstantTaskExecutorRule()
private lateinit var totalsViewModel: TotalsViewModel
@Before
fun setUp() {
    totalsViewModel = TotalsViewModel()
    assertEquals(0, totalsViewModel.total.value)
}
```

11. Let's run the test again and see what happens:

```
java.lang.AssertionError:
Expected :4
Actual :5
```

See whether you can spot where the error in the test is, fix it, and then rerun it:

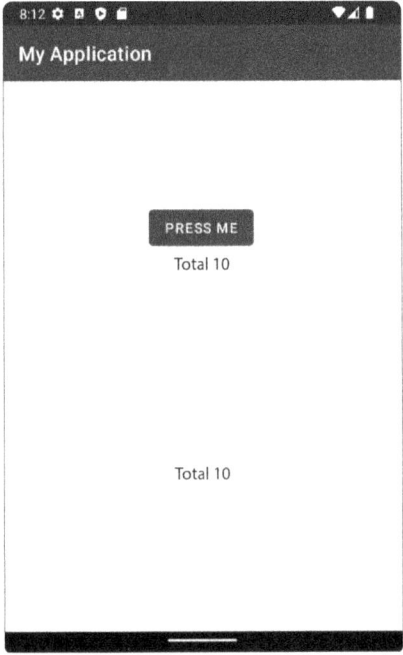

Figure 11.5 – Output of Exercise 11.02

The same output in landscape mode will look as follows:

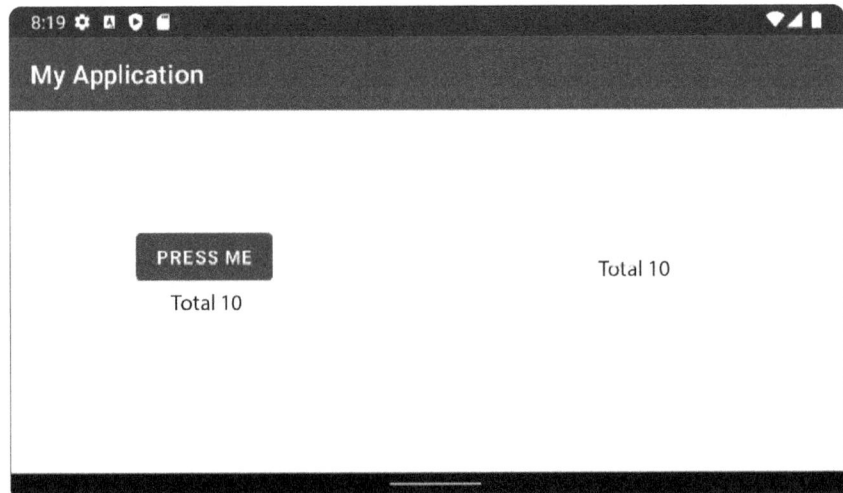

Figure 11.6 – Output of Exercise 11.02 in landscape mode

By looking at the preceding example, we can see how using a combination of the `LiveData` and `ViewModel` approaches helped us solve our problem while considering the particularities of the Android operating system:

- `ViewModel`: This helped us hold the data across device orientation changes, and it solved the issue of communicating between fragments
- `LiveData`: This helped us retrieve the most up-to-date information that we've processed while considering the fragment's lifecycle
- The combination of the two helped us efficiently delegate our processing logic, allowing us to unit test this processing logic

Additional data streams

One type of data stream that has gained popularity recently is the usage of Coroutines and Flows, mainly for their approach to asynchronous operations in Android. An example of Flows emitting data in a `ViewModel` would be as follows:

```
class TotalsViewModel : ViewModel() {
    private val _total = MutableStateFlow(0)
    val total: StateFlow<Int> = _total
    fun increaseTotal() {
        _total.value = _total.value + 1
    }
}
```

In the preceding snippet, we have the two total declarations for public and private usage. Instead of `LiveData`, we use `StateFlow`, which will emit the current value and all subsequent new values when we subscribe to it. Because it emits the last value, we must always set an initial value when we initialize it. If we want to subscribe to changes in the total value, we can use the following:

```
val totalsViewModel =
    ViewModelProvider(requireActivity())
        .get(TotalsViewModel::class.java)
viewLifecycleOwner.lifecycleScope.launch {
    repeatOnLifecycle(Lifecycle.State.CREATED) {
        totalsViewModel.total.collect {
            updateText(it)
        }
    }
}
```

The preceding snippet will subscribe to `StateFlow` every time `viewLifecycleOwner` enters the `CREATED` stage. This will connect `StateFlow` with the lifecycle of `Fragment` to prevent any possible leaks. We will explore the mechanics of Flows and Coroutines in future chapters.

Another example of a data stream is the RxJava library, which represents another mechanism for emitting data. The library is best used for performing asynchronous work and transformations, and because it's based on Java and not the Android operating system, it lacks any lifecycle awareness. For example, using RxJava in combination with ViewModels would look like the following:

```
class TotalsViewModel : ViewModel() {
    private val _total = BehaviorSubject.createDefault(0)
    val total: Observable<Int> = _total
    fun increaseTotal() {
        _total.onNext(_total.blockingLast())
    }
}
```

Here, we are using `BehaviorSubject` to replace `StateFlow`. `BehaviorSubject` has the same properties as state flow. It will keep the latest value and emit it when a component subscribes and all the new values after the subscription. Subscribing to the object looks like the following:

```
    private var disposable: Disposable? = null
    override fun onViewCreated(view: View,
    savedInstanceState: Bundle?) {
        super.onViewCreated(view, savedInstanceState)
        val totalsViewModel =
            ViewModelProvider(requireActivity())
                .get(TotalsViewModel::class.java)
        disposable = totalsViewModel.total.subscribe {
            updateText(it)
        }
    }
    override fun onDestroyView() {
        disposable?.dispose()
        super.onDestroyView()
    }
```

Here, we are using Disposable to hold the subscription in `onViewCreated`. In `onDestroyView`, we are disposing of the subscription to prevent any context leaks. This is an alternative to using lifecycle-aware components such as `LiveData` and `StateFlow`.

In this section, we looked at other types of streams of data that might be present in Android applications such as Kotlin Flows and RxJava, and analyzed their particularities. In the section that follows, we will look at how we can persist data using the Room library.

Room

The Room persistence library acts as a wrapper between your application code and the SQLite storage. You can think of SQLite as a database that runs without its own server and saves all the application data in an internal file that's only accessible to your application (if the device is not rooted).

Room sits between the application code and the SQLite Android Framework, and handles the necessary **create**, **read**, **update**, and **delete** (**CRUD**) operations while exposing an abstraction that your application can use to define the data and how you want the data to be handled. This abstraction comes in the form of the following objects:

- **Entities**: You can specify how you want your data to be stored and the relationships between your data
- **Data access object (DAO)**: The operations that can be done on your data
- **Database**: You can specify the configurations that your database should have (the name of the database and migration scenarios)

These can be seen in the following diagram:

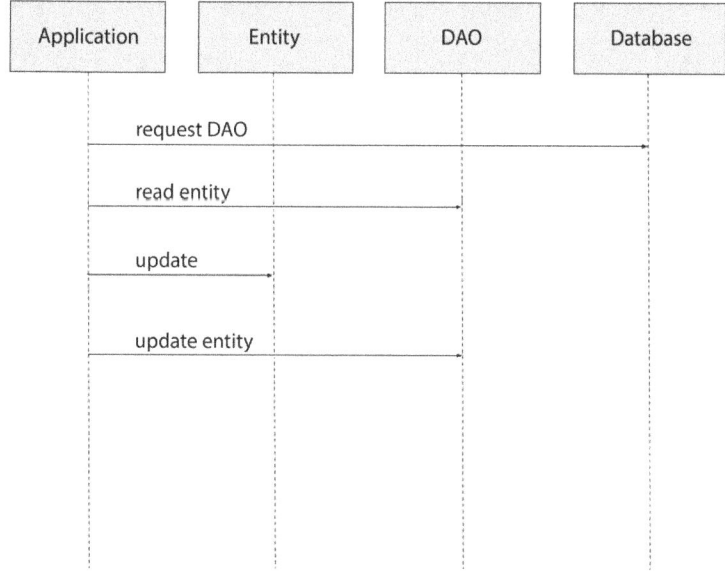

Figure 11.7 – The relationship between your application and the Room components

In the preceding diagram, we can see how the Room components interact with each other. It's easier to visualize this with an example. Let's assume you want to make a messaging app and store each message in your local storage. In this case, `Entity` would be a `Message` object, which will have an ID and will contain the contents of the message, the sender, the time, status, and so on.

In order to access messages from the local storage, you will need `MessageDao`, which will contain methods such as `insertMessage()`, `getMessagesFromUser()`, `deleteMessage()`, and `updateMessage()`. In addition, since it's a messaging application, you will need a `Contact` entity to hold information about the senders and receivers of a message.

The `Contact` entity will contain information such as a name, last time online, phone number, email, and so on. In order to access the contact information, you will need a `ContactDao` interface, which will contain `createUser()`, `updateUser()`, `deleteUser()`, and `getAllUsers()`. Both entities will create a matching table in SQLite, which contains the fields we defined inside the entity classes as columns. In order to achieve this, we'll have to create `MessagingDatabase` in which we will reference both entities.

In a world without Room or similar DAO libraries, we would need to use the Android Framework's SQLite components. This typically involves code when setting up our database, such as a query to create a table and applying similar queries for every table we would have. Every time we queried a table for data, we would need to convert the resulting object into a Java or Kotlin one.

By default, Room does not allow any operations on the UI thread to enforce the Android standards related to input-output operations. To make asynchronous calls to access data, Room is compatible with a number of libraries and frameworks, such as Kotlin coroutines, RxJava, and `LiveData`, on top of its default definitions.

We should now have an overview of how Room works and its main components. We will now look over each of these components and how we can use them for data persistence.

Entities

Entities serve two purposes: to define the structure of tables and to hold the data from a table row. Let's use our scenario of the messaging app and define two entities: one for the user and one for the message.

The `User` entity will contain information about who sent the messages, while the `Message` entity will contain information about the contents of a message, the time it was sent, and a reference to the sender of the message. The following code snippet provides an example of how entities are defined with Room:

```
@Entity(tableName = "messages")
data class Message(
    @PrimaryKey(autoGenerate = true) @ColumnInfo(name =
        "message_id") val id: Long,
```

```
    @ColumnInfo(name = "text", defaultValue = "") val text:
        String,
    @ColumnInfo(name = "time") val time: Long,
    @ColumnInfo(name = "user") val userId: Long,
)
@Entity(tableName = "users")
data class User(
    @PrimaryKey @ColumnInfo(name = "user_id") val id: Long,
    @ColumnInfo(name = "first_name") val firstName: String,
    @ColumnInfo(name = "last_name") val lastName: String,
    @ColumnInfo(name = "last_online") val lastOnline: Long
)
```

As you can see, entities are just *data classes* with annotations, which will tell Room how the tables should be built in SQLite. The annotations we used are as follows:

- The `@Entity` annotation defines the table. By default, the table name will be the name of the class. We can change the name of the table through the `tableName` method in the `Entity` annotation. This is useful in situations where we want our code obfuscated but wish to keep the consistency of the SQLite structure.

- `@ColumnInfo` defines configurations for a certain column. The most common one is the name of the column. We can also specify a default value, the SQLite type of the field, and whether the field should be indexed.

- `@PrimaryKey` indicates what in our entity will make it unique. Every entity should have at least one primary key. If your primary key is an integer or a long, then we can add the `autogenerate` field. This means that every entity that gets inserted into the Primary Key field is automatically generated by SQLite.

    Usually, this is done by incrementing the previous ID. If you wish to define multiple fields as primary keys, then you can adjust the `@Entity` annotation to accommodate this, such as the following:

    ```
    @Entity(tableName = "messages", primaryKeys = ["id",
    "time"])
    ```

Let's assume that our messaging application wants to send locations. Locations have a latitude, longitude, and name. We can add them to the `Message` class, but that would increase the complexity of the class. What we can do is create another entity and reference the ID in our class.

The problem with this approach is that we would then query the `Location` entity every time we queried the `Message` entity. Room has a third approach through the `@Embedded` annotation. Now, let's look at the updated `Message` entity:

```
@Entity(tableName = "messages")
data class Message(
    @PrimaryKey(autoGenerate = true) @ColumnInfo(name =
        "message_id") val id: Long,
    @ColumnInfo(name = "text", defaultValue = "") val text:
        String,
    @ColumnInfo(name = "time") val time: Long,
    @ColumnInfo(name = "user") val userId: Long,
    @Embedded val location: Location?
)
data class Location(
    @ColumnInfo(name = "lat") val lat: Double,
    @ColumnInfo(name = "long") val log: Double,
    @ColumnInfo(name = "location_name") val name: String
)
```

This code adds three columns (`lat`, `long`, and `location_name`) to the messages table. This allows us to avoid having objects with a large number of fields while keeping the consistency of our tables.

If we look at our entities, we'll see that they exist independently. The `Message` entity has a `userId` field, but nothing is preventing us from adding messages from invalid users. This may lead to situations where we collect data without any purpose. If we want to delete a particular user, along with their messages, we must do so manually. Room provides us with a way to define this relationship using a `ForeignKey`:

```
@Entity(
    tableName = "messages",
    foreignKeys = [ForeignKey(
        entity = User::class,
        parentColumns = ["user_id"],
        childColumns = ["user"],
        onDelete = ForeignKey.CASCADE
    )]
)
data class Message(
```

```
    @PrimaryKey(autoGenerate = true) @ColumnInfo(name =
        "message_id") val id: Long,
    @ColumnInfo(name = "text", defaultValue = "") val text:
        String,
    @ColumnInfo(name = "time") val time: Long,
    @ColumnInfo(name = "user") val userId: Long,
    @Embedded val location: Location?
)
```

In the preceding example, we added the `foreignKeys` field and created a new `ForeignKey` to the `User` entity, while for the parent column, we defined the `user_id` field in the `User` class, and for the child column, the `user` field in the `Message` class.

Every time we add a message to the table, there needs to be a `User` entry in the `users` table. If we try to delete a user and any messages from that user still exist, then, by default, this will not work because of the dependencies. However, we can tell Room to do a cascade delete, which will erase the user and the associated messages.

DAO

If entities specify how we define and hold our data, then DAOs specify what to do with that data. A DAO class is a place where we define our CRUD operations. Ideally, each entity should have a corresponding DAO, but there are situations where crossovers occur (usually, this happens when we have to deal with JOINs between two tables).

Continuing with our previous example, let's build some corresponding DAOs for our entity:

```
@Dao
interface MessageDao {
    @Insert(onConflict = OnConflictStrategy.REPLACE)
    fun insertMessages(vararg messages: Message)
    @Update
    fun updateMessages(vararg messages: Message)
    @Delete
    fun deleteMessages(vararg messages: Message)
    @Query("SELECT * FROM messages")
    fun loadAllMessages(): List<Message>
    @Query("SELECT * FROM messages WHERE user=:userId AND
        time>=:time")
    fun loadMessagesFromUserAfterTime(userId: String, time:
```

```kotlin
            Long): List<Message>
}
@Dao
interface UserDao {
    @Insert(onConflict = OnConflictStrategy.REPLACE)
    fun insertUser(user: User)
    @Update
    fun updateUser(user: User)
    @Delete
    fun deleteUser(user: User)
    @Query("SELECT * FROM users")
    fun loadAllUsers(): List<User>
}
```

In the case of our messages, we have defined the following functions: insert one or more messages, update one or more messages, delete one or more messages, and retrieve all the messages from a certain user that are older than a particular time. For our users, we can insert one user, update one user, delete one user, and retrieve all the users.

If you look at our `Insert` methods, you'll see we have defined that in the case of a conflict (when we try to insert something with an ID that already exists), it will replace the existing entry. The `Update` field has a similar configuration, but in our case, we have chosen the default. This means that nothing will happen if the update cannot occur.

The `@Query` annotation stands out from all the others. This is where we use SQLite code to define how our read operations work. `SELECT *` means we want to read all the data for every row in the table, which will populate all our entities' fields. The `WHERE` clause indicates a restriction that we want to apply to our query. We can also define a method like this:

```kotlin
@Query("SELECT * FROM messages WHERE user IN (:userIds) AND time>=:time")
fun loadMessagesFromUserAfterTime(userIds: List<String>, time:
Long): List<Message>
```

This allows us to filter messages from multiple users. We can define a new class like this:

```kotlin
data class TextWithTime(
    @ColumnInfo(name = "text") val text: String,
    @ColumnInfo(name = "time") val time: Long
)
```

Now, we can define the following query:

```
@Query("SELECT text,time FROM messages")
fun loadTextsAndTimes(): List<TextWithTime>
```

This will allow us to extract information from certain columns at a time, not the entire row.

Now, let's say that you want to add the user information of the sender to every message. Here, we'll need to use a similar approach to the one we used previously:

```
data class MessageWithUser(
    @Embedded val message: Message,
    @Embedded val user: User
)
```

By using the new data class, we can define this query:

```
@Query("SELECT * FROM messages INNER JOIN users on users.user_id=messages.user")
fun loadMessagesAndUsers(): List<MessageWithUser>
```

We now have the user information for every message we want to display. This will come in handy in scenarios such as group chats, where we should display the name of the sender of every message.

Setting up the database

What we have learned about so far is a bunch of DAOs and entities. Now, it's time to put them together. First, let's define our database:

```
@Database(entities = [User::class, Message::class],
version = 1)
abstract class ChatDatabase : RoomDatabase() {
    companion object {
        private lateinit var chatDatabase: ChatDatabase
        fun getDatabase(applicationContext: Context):
        ChatDatabase {
            if (!(::chatDatabase.isInitialized)) {
                chatDatabase =
                    Room.databaseBuilder(applicationContext
                    , chatDatabase::class.java, "chat-db")
                    .build()
```

```
                }
                return chatDatabase
            }
        }
        abstract fun userDao(): UserDao
        abstract fun messageDao(): MessageDao
    }
```

In the `@Database` annotation, we specify what entities go in our database and our version. Then, for every DAO, we define an abstract method in `RoomDatabase`. This allows the build system to build a subclass of our class in which it provides the implementations for these methods. The build system will also create the tables related to our entities.

The `getDatabase` method in the companion object illustrates how we create an instance of the `ChatDatabase` class. Ideally, there should be one instance of the database for our application due to the complexity involved in building a new database object. However, this can be better achieved through a dependency injection framework.

Let's assume you've released your chat application. Your database is currently version one, but your users are complaining that the message status feature is missing. You decide to add this feature in the next release. This involves changing the database structure, which can impact databases that have already built their structures.

Luckily, Room offers something called a migration. In the migration, we can define how our database changed between versions 1 and 2. So, let's look at our example:

```
    data class Message(
        @PrimaryKey(autoGenerate = true) @ColumnInfo(name =
            "message_id") val id: Long,
        @ColumnInfo(name = "text", defaultValue = "") val text:
            String,
        @ColumnInfo(name = "time") val time: Long,
        @ColumnInfo(name = "user") val userId: Long,
        @ColumnInfo(name = "status") val status: Int,
        @Embedded val location: Location?
    )
```

Here, we added the status flag to the `Message` entity. Now, let's look at `ChatDatabase`:

```
    Database(entities = [User::class, Message::class],
    version = 2)
```

```kotlin
abstract class ChatDatabase : RoomDatabase() {
    companion object {
        private lateinit var chatDatabase: ChatDatabase
        private val MIGRATION_1_2 = object : Migration(1, 2) {
            override fun migrate(database: SupportSQLiteDatabase) {
                database.execSQL("ALTER TABLE messages ADD
                    COLUMN status INTEGER")
            }
        }
        fun getDatabase(applicationContext: Context): ChatDatabase {
            if (!(::chatDatabase.isInitialized)) {
                chatDatabase =
                    Room.databaseBuilder(applicationContext,
                    chatDatabase::class.java, "chat-db")
                    .addMigrations(MIGRATION_1_2)
                    .build()
            }
            return chatDatabase
        }
    }
    abstract fun userDao(): UserDao
    abstract fun messageDao(): MessageDao
}
```

In our database, we've increased the version to 2 and added a migration between versions 1 and 2. Here, we added the status column to the table. We'll add this migration when we build the database.

Once we've released the new code, when the updated app is opened and the code to build the database is executed, it will compare the version of the stored data with the one specified in our class and notice a difference. Then, it will execute the specified migrations until it reaches the latest version. This allows us to maintain an application for years without impacting the user's experience.

If you look at our Message class, you may have noticed that we defined the time as Long. In Java and Kotlin, we have the Date object, which may be more useful than the timestamp of the message. Luckily, Room has a solution for this in the form of TypeConverter.

The following table shows what data types we can use in our code and the SQLite equivalent. Complex data types need to be brought down to these levels using TypeConverters:

Java/Kotlin	SQLite
String	TEXT
Byte, Short, Integer, Long, Boolean	INTEGER
Double, Float	REAL
Array<Byte>	BLOB

Figure 11.8 – The relationship between Kotlin/Java data types and the SQLite data types

Here, we've modified the `lastOnline` field so that it's of the `Date` type:

```
data class User(
    @PrimaryKey @ColumnInfo(name = "user_id") val id: Long,
    @ColumnInfo(name = "first_name") val firstName: String,
    @ColumnInfo(name = "last_name") val lastName: String,
    @ColumnInfo(name = "last_online") val lastOnline: Date
)
```

Here, we've defined a couple of methods that convert a `Date` object into `Long` and vice versa. The `@TypeConverter` annotation helps Room identify where the conversion takes place:

```
class DateConverter {
    @TypeConverter
    fun from(value: Long?): Date? {
        return value?.let { Date(it) }
    }
    @TypeConverter
    fun to(date: Date?): Long? {
        return date?.time
    }
}
```

Finally, we'll add our converter to Room using the `@TypeConverters` annotation:

```
@Database(entities = [User::class, Message::class],
version = 2)
@TypeConverters(DateConverter::class)
abstract class ChatDatabase : RoomDatabase() {
```

In the next section, we will look at some third-party frameworks.

Third-party frameworks

Room works well with third-party frameworks such as `LiveData`, RxJava, and coroutines. This solves two issues: multi-threading and observing data changes.

`LiveData` will make the `@Query` annotated methods in your DAOs reactive, which means that if new data is added, `LiveData` will notify the observers of this:

```
@Query("SELECT * FROM users")
fun loadAllUsers(): LiveData<List<User>>
```

Kotlin coroutines complement `LiveData` by making the `@Insert`, `@Delete`, and `@Update` methods asynchronous:

```
@Insert(onConflict = OnConflictStrategy.REPLACE)
suspend fun insertUser(user: User)
@Update
suspend fun updateUser(user: User)
@Delete
suspend fun deleteUser(user: User)
```

RxJava solves both issues: making the `@Query` methods reactive through components such as `Publisher`, `Observable`, or `Flowable` and making the rest of the methods asynchronous through `Completable`, `Single`, or `Maybe`:

```
@Insert(onConflict = OnConflictStrategy.REPLACE)
fun insertUser(user: User) : Completable
@Update
fun updateUser(user: User) : Completable
@Delete
fun deleteUser(user: User) : Completable
@Query("SELECT * FROM users")
fun loadAllUsers(): Flowable<List<User>>
```

Executors and threads come with the Java framework and can be a useful solution for solving threading issues with Room if none of the aforementioned third-party integrations are part of your project.

Your DAO classes will not suffer from any modifications; however, you will need the components that access your DAOs to adjust and use either an executor or a thread:

```
@Query("SELECT * FROM users")
fun loadAllUsers(): List<User>
@Insert(onConflict = OnConflictStrategy.REPLACE)
fun insertUser(user: User)
@Update
fun updateUser(user: User)
@Delete
fun deleteUser(user: User)
```

An example of accessing the DAO is as follows:

```
fun getUsers(usersCallback:()->List<User>){
    Thread(Runnable {
        usersCallback.invoke(userDao.loadUsers())
    }).start()
}
```

The preceding example will create a new thread and start it every time we want to retrieve the list of users. There are two major issues with this code:

- Thread creation is an expensive operation
- The code is hard to test

The solution to the first is to use `ThreadPools` and `Executors`. The Java framework offers a robust set of options when it comes to `ThreadPools`. A thread pool is a component responsible for thread creation and destruction and allows the developer to specify the number of threads in the pool. Multiple threads in a thread pool will ensure multiple tasks can be executed concurrently.

We can rewrite the preceding code as follows:

```
private val executor:Executor =
    Executors.newSingleThreadExecutor()
fun getUsers(usersCallback:(List<User>)->Unit){
    executor.execute {
        usersCallback.invoke(userDao.loadUsers())
    }
}
```

In the preceding example, we defined an executor that will use a pool of one thread. When we want to access the list of users, we move the query inside the executor, and when the data is loaded, our callback lambda will be invoked.

Exercise 11.03 – making a little room

You have been hired by a news agency to build a news application. The application will display a list of articles written by journalists. An article can be written by one or more journalists, and each journalist can write one or more articles. The data information for each article includes the article's title, content, and date.

The journalist's information includes their first name, last name, and job title. You will need to build a Room database that holds this information so it can be tested. Before we start, let's look at the relationship between the entities. In the chat application example, we defined the rule that one user can send one or multiple messages.

This relationship is known as a one-to-many relationship. That relationship is implemented as a reference between one entity to another (the user was defined in the message table in order to be connected to the sender).

In this case, we have a many-to-many relationship. To implement a many-to-many relationship, we need to create an entity that holds references that will link the other two entities. Let's get started:

1. Create a new Android Project with No Activity.

2. Let's start by adding the annotation processing plugin to `app/build.gradle`. This will read the annotations used by Room and generate the code necessary for interacting with the database:

   ```
   plugins {
       ...
       id 'kotlin-kapt'
   }
   ```

3. Next, let's add the Room libraries in `app/build.gradle`:

   ```
   def room_version = "2.2.5"
   implementation "androidx.room:
       room-runtime:$room_version"
   kapt "androidx.room:room-compiler:$room_version"
   ```

   The first line defines the library version, the second line brings in the Room library for Java and Kotlin, and the last line is for the Kotlin annotation processor. This allows the build system to generate boilerplate code from the Room annotations. After these changes to your Gradle files, you should get a prompt to sync your project, which you should click.

4. Let's define our entities in the main/java folder and the root package:

   ```
   @Entity(tableName = "article")
   data class Article(
       @PrimaryKey(autoGenerate = true) @ColumnInfo(
           name = "id") val id: Long = 0,
       @ColumnInfo(name = "title") val title: String,
       @ColumnInfo(name = "content") val content: String,
       @ColumnInfo(name = "time") val time: Long
   )
   @Entity(tableName = "journalist")
   data class Journalist(
       @PrimaryKey(autoGenerate = true) @ColumnInfo(name
           = "id") val id: Long = 0,
       @ColumnInfo(name = "first_name") val firstName:
           String,
       @ColumnInfo(name = "last_name") val lastName:
           String,
       @ColumnInfo(name = "job_title") val jobTitle:
           String
   )
   ```

5. Now, define the entity that connects the journalist to the article and the appropriate constraints in the main/java folder and the root package:

   ```
   @Entity(
       tableName = "joined_article_journalist",
       primaryKeys = ["article_id", "journalist_id"],
       foreignKeys = [ForeignKey(
           entity = Article::class,
           parentColumns = arrayOf("id"),
           childColumns = arrayOf("article_id"),
           onDelete = ForeignKey.CASCADE
       ), ForeignKey(
           entity = Journalist::class,
           parentColumns = arrayOf("id"),
           childColumns = arrayOf("journalist_id"),
           onDelete = ForeignKey.CASCADE
   ```

```
    )]
)
data class JoinedArticleJournalist(
    @ColumnInfo(name = "article_id") val articleId:
        Long,
    @ColumnInfo(name = "journalist_id") val
        journalistId: Long
)
```

In the preceding code, we defined our connecting entity. As you can see, we haven't defined an ID for uniqueness, but both the article and the journalist will be unique when used together. We also defined foreign keys for each of the other entities referred to by our entity.

6. Create the `ArticleDao` DAO in the `main/java` folder and the root package:

```
@Dao
interface ArticleDao {
    @Insert(onConflict = OnConflictStrategy.REPLACE)
    fun insertArticle(article: Article)
    @Update
    fun updateArticle(article: Article)
    @Delete
    fun deleteArticle(article: Article)
    @Query("SELECT * FROM article")
    fun loadAllArticles(): List<Article>
    @Query("SELECT * FROM article INNER JOIN
            joined_article_journalist ON
            article.id=joined_article_journalist
            .article_id WHERE
            joined_article_journalist.journalist_id=
            :journalistId")
    fun loadArticlesForAuthor(journalistId: Long):
        List<Article>
}
```

7. Now, create the `JournalistDao` data access object in the `main/java` folder and the root package:

```
@Dao
interface JournalistDao {
```

```
        @Insert(onConflict = OnConflictStrategy.REPLACE)
        fun insertJournalist(journalist: Journalist)
        @Update
        fun updateJournalist(journalist: Journalist)
        @Delete
        fun deleteJournalist(journalist: Journalist)
        @Query("SELECT * FROM journalist")
        fun loadAllJournalists(): List<Journalist>
        @Query("SELECT * FROM journalist INNER JOIN
            joined_article_journalist ON
            journalist.id=joined_article_journalist
            .journalist_id WHERE
            joined_article_journalist.article_id=
            :articleId")
        fun getAuthorsForArticle(articleId: Long):
            List<Journalist>
    }
```

8. Create the `JoinedArticleJournalistDao` DAO in the `main/java` folder and the root package:

```
    @Dao
    interface JoinedArticleJournalistDao {
        @Insert(onConflict = OnConflictStrategy.REPLACE)
        fun insertArticleJournalist(
          joinedArticleJournalist: JoinedArticleJournalist
        )
        @Delete
        Fun deleteArticleJournalist(
          joinedArticleJournalist: JoinedArticleJournalist
        )
    }
```

Let's analyze our code a little bit. For the articles and journalists, we can add, insert, delete, and update queries. For articles, we can extract all of the articles but also extract articles from a certain author.

We also have the option to extract all the journalists that wrote an article. This is done through a JOIN with our intermediary entity. For that entity, we define the options to insert (which will link an article to a journalist) and delete (which will remove that link).

9. Finally, let's define our `Database` class in the `main/java` folder and the root package:

    ```
    @Database(
        entities = [Article::class, Journalist::class,
            JoinedArticleJournalist::class],
        version = 1
    )
    abstract class NewsDatabase : RoomDatabase() {
        abstract fun articleDao(): ArticleDao
        abstract fun journalistDao(): JournalistDao
        abstract fun joinedArticleJournalistDao():
            JoinedArticleJournalistDao
    }
    ```

    We avoided defining the `getInstance` method here because we won't be calling the database anywhere. But if we don't do that, how will we know whether it works? The answer to this is that we'll test it. This won't be a test that will run on your machine but one that will run on the device. This means that we will create it in the `androidTest` folder.

10. Let's start by setting up the test data. Here, we will add some articles and journalists to the database then test retrieving, updating, and deleting the entries:

    ```
    @RunWith(AndroidJUnit4::class)
    class NewsDatabaseTest {
        @Test
        fun updateArticle() {
            val article = articleDao.loadAllArticles()[0]
            articleDao.updateArticle(article.copy(title =
                "new title"))
            assertEquals("new title", articleDao.
                loadAllArticles()[0].title)
        }
        @Test
        fun updateJournalist() {
            val journalist = journalistDao.
                loadAllJournalists()[0] journalistDao.
                updateJournalist(journalist.copy(jobTitle =
                "new job title"))
            assertEquals("new job title", journalistDao.
    ```

```
                loadAllJournalists()[0].jobTitle)
        }
    }
```

The complete code for this step can be found at `https://packt.link/6H8X2`.

Here, we have defined a few examples of how to test a Room database. What's interesting is how we build the database. Our database is an in-memory database. This means that all the data will be kept as long as the test is run and discarded afterward.

This allows us to start with a clean slate for each new state and avoids the consequences of each of our testing sessions affecting each other. In our test, we've set up 5 articles and 10 journalists. The first article was written by the top two journalists, while the second article was written by the first journalist.

The rest of the articles have no authors. By doing this, we can test our update and delete methods. For the delete method, we can test our foreign key relationship as well. In the test, we can see that if we delete article `1`, it will delete the relationship between the article and the journalists that wrote it.

When testing your database, you should add the scenarios that your app will use. Feel free to add other testing scenarios and improve the preceding tests in your own database. Note that if you are using the `androidTest` folder, then this will be an instrumented test, meaning that you will need an emulator or a device to test.

Activity 11.01 – a shopping notes app

You want to keep track of your shopping items, so you decide to build an app to save the items you wish to buy during your next trip to the store. The requirements for this are as follows:

- The UI will be split into two: top/bottom in portrait mode and left/right in landscape mode. The UI will look similar to what is shown in the following screenshot.
- The first half will display the number of notes, a text field, and a button. Every time the button is pressed, a note will be added with the text placed in the text field.
- The second half will display the list of notes.
- For each half, you will have a View model that will hold the relevant data.
- You should define a repository that will be used on top of the Room database to access your data.
- You should also define a Room database that will hold your notes.

- The note entity will have the following attributes: `id` and `text`:

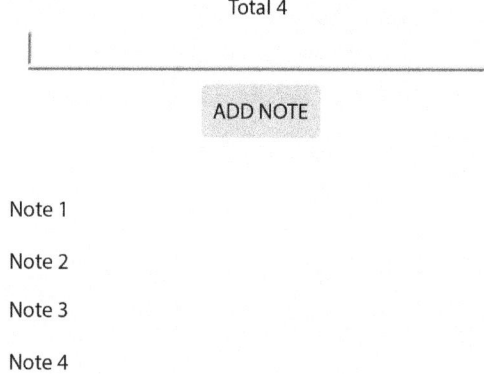

Figure 11.9 – Example of a possible output for activity 11.01

Perform the following steps to complete this activity:

1. Start with Room integration by creating the `Entity`, `Dao`, and `Database` methods. For `Dao`, the `@Query` annotated methods can directly return a `LiveData` object so that the observers can be directly notified if the data changes.
2. Define a template of our repository in the form of an interface.
3. Implement the repository. The repository will have one reference to the `Dao` object we defined previously. The code for inserting the data must be moved to a separate thread.
4. Create the `NotesApplication` class to provide one instance of the repository that will be used across the application. Make sure to update the `<application>` tag in the `AndroidManifest.xml` file to add your new application class.
5. Unit-test the repository and define `ViewModels`, as follows:

    I. Define `NoteListViewModel` and the associated test. This will have a reference to the repository and return the list of notes.

    II. Define `CountNotesViewModel` and the associated test. `CountViewModel` will have a reference to the repository and return the total number of notes as `LiveData`. It will also be responsible for inserting new notes.

    III. Define `CountNotesFragment` and the associated `fragment_count_notes.xml` layout. In the layout, define a `TextView`, which will display the total number, an `EditText` for the name of the new notes, and a button, which will insert the note that was introduced in `EditText`.

    IV. Define an adapter for the list of notes called `NoteListAdapter` and an associated layout file for the rows called `view_note_item.xml`.

V. Define the associated layout file, called `fragment_note_list.xml`, which will contain `RecyclerView`. The layout will be used by `NoteListFragment`, which will connect `NoteListAdapter` to `RecyclerView`. It will also observe the data from `NoteListViewModel` and update the adapter.

VI. Define `NotesActivity` with an associated layout for landscape mode and portrait mode.

6. Make sure you have all the necessary data in `strings.xml`.

> **Note**
> The solution to this activity can be found at `https://packt.link/ZhnDx`.

Summary

In this chapter, we analyzed the building blocks required to build a maintainable application. We also looked into one of the most common issues that developers come across when using the Android Framework, which is maintaining the states of objects during lifecycle changes.

We started by analyzing `ViewModels` and how they solve the issue of holding data during orientation changes. We added `LiveData` to `ViewModels` to show how the two complement each other and looked at how we can use other data streams with `ViewModels` and compare those with `LiveData`.

We then moved on to Room to show how we can persist data with minimal effort and without much SQLite boilerplate code. We also explored one-to-many and many-to-many relationships, as well as how to migrate data and break down complex objects into primitives for storage.

The activity we completed in this chapter serves as an example of what direction Android apps are heading in. However, this was not a complete example due to the numerous frameworks and libraries that you will discover, which give developers the flexibility to go in different directions.

The information you've learned in this chapter will serve you well for the next one, which will expand on the concept of repositories. This will allow you to save data that's been obtained from a server into a Room database.

The concept of persisting data will also be expanded as you will explore other ways to persist data, such as through `SharedPreferences`, `DataStore`, and files. Our focus will be on certain types of files: media files obtained from the camera of the device.

12
Persisting Data

This chapter goes in depth about data persistence in Android. By the end of the chapter, you will know multiple ways to store (persist) data directly on a device and the frameworks accessible to do this. When dealing with a filesystem, you will know how it's partitioned and how you can read and write files in different locations and use different frameworks.

In the previous chapter, you learned how to structure your code and save data. In the activity, you also had the opportunity to build a repository and use it to access and save data through Room. In this chapter, you will learn about alternative ways to persist data on a device through the Android filesystem and how it's structured into external and internal memory.

You'll also develop your understanding of read and write permissions, learn how to create the `FileProvider` class to offer other apps access to your files, and learn how you can save those files without requesting permissions on external drives. You'll also see how to download files from the internet and save them on a filesystem.

Another concept that will be explored in this chapter is using the **Camera** application to take photos and videos on your application's behalf and save them to external storage using FileProviders.

We will cover the following topics in the chapter:

- Preferences and DataStore
- Files
- Scoped storage

Technical requirements

The complete code for all the exercises and the activity in this chapter is available on GitHub at `https://packt.link/XlTwZ`

Preferences and DataStore

Imagine you are tasked with integrating a third-party API that uses something such as OAuth to implement logging in with Facebook, Google, and suchlike. The way these mechanisms work is as follows – they give you a token that you have to store locally and that can then be used to send other requests to access user data.

This raises several questions. How can you store that token? Do you use Room just for one token? Do you save the token in a separate file and implement methods for writing the file? What if that file has to be accessed in multiple places at the same time? `SharedPreferences` and `DataStore` are answers to these questions. `SharedPreferences` is a functionality that allows you to save Booleans, integers, floats, longs, strings, and sets of strings into an XML file.

When you want to save new values, you specify what values you want to save for the associated keys, and when you are done, you commit the change, which will trigger the save to the XML file in an asynchronous way. The `SharedPreferences` mappings are also kept in memory so that when you want to read these values, it's instantaneous, thereby removing the need for an asynchronous call to read the XML file.

We now have two ways to store data in key-value pairs in the form of `SharedPreferences` and `DataStore`. We will now look at how each of them works and the benefits each one provides.

SharedPreferences

The way to access the `SharedPreference` object is through the `Context` object:

```
val prefs = getSharedPreferences("my-prefs-file",
Context.MODE_PRIVATE)
```

The first parameter is where you specify the name of your preferences, and the second is how you want to expose a file to other apps. Currently, the best mode is the private one. All of the others present potential security risks.

If you want to write data into your preferences file, you first need to get access to the Preferences editor. The editor will give you access to write the data. You can then write your data in it. Once you finish writing, you will have to apply the changes that will trigger persistence to the XML file and change the in-memory values as well.

You have two choices to apply the changes on your preference file – `apply` or `commit`. Choosing `apply` will save your changes in memory instantly, but then writing to disk will be asynchronous, which is useful if you want to save data from from your app's main thread. `commit` does everything synchronously and gives you a Boolean result, informing you whether the operation was successful or not. In practice, `apply` tends to be favored over `commit`:

```
val editor = prefs.edit()
editor.putBoolean("my_key_1", true)
```

```
editor.putString("my_key_2", "my string")
editor.putLong("my_key_3", 1L)
editor.apply()
```

Now, you want to clear your entire data. The same principle will apply; you'll need `editor`, `clear`, and `apply`:

```
val editor = prefs.edit()
editor.clear()
editor.apply()
```

If you want to read the values you previously saved, you can use the `SharedPreferences` object to read the stored values. If there is no saved value, you can opt for a default value to be returned instead:

```
prefs.getBoolean("my_key_1", false)
prefs.getString("my_key_2", "")
prefs.getLong("my_key_3", 0L)
```

We should now have an idea about how we can persist data with `SharedPreferences`, and we can apply this in an exercise in the following section.

Exercise 12.01 – wrapping SharedPreferences

We're going to build an application that displays `TextView`, `EditText`, and a button. `TextView` will display the previously saved value in `SharedPreferences`. The user can type new text, and when the button is clicked, the text will be saved in `SharedPreferences` and `TextView` will display the updated text. We will need to use `ViewModel` and `LiveData` to make the code more testable.

In order to complete this exercise, we will need to create a `Wrapper` class, which will be responsible for saving the text. This class will return the value of the text as `LiveData`. This will be injected into our `ViewModel`, which will be bound to the activity:

1. Create a new Android Studio project with an empty activity.
2. Let's begin by adding the appropriate libraries to `app/build.gradle`:

    ```
    implementation "androidx.lifecycle:
        lifecycle-viewmodel-ktx:2.5.1"
    implementation "androidx.lifecycle:
        lifecycle-livedata-ktx:2.5.1"
    ```

3. Let's make our `Wrapper` class in the `main/java` folder in the `root` package, which will listen for changes in `SharedPreferences` and update the value of `LiveData` when the preferences change. The class will contain methods to save the new text and retrieve `LiveData`:

    ```
    const val KEY_TEXT = "keyText"
    ```

```kotlin
class PreferenceWrapper(private val sharedPreferences:
SharedPreferences) {
    private val textLiveData =
        MutableLiveData<String>()
    init {
        sharedPreferences
        .registerOnSharedPreferenceChangeListener {
        _, key ->
            when (key) {
                KEY_TEXT -> {
                    textLiveData.postValue(
                    sharedPreferences
                    .getString(KEY_TEXT, ""))
                }
            }
        }
    }
}
```

The complete code for this step can be found at https://packt.link/a2RuN.

Note the top of the file. We've added a listener so that when our `SharedPreferences` values change, we can look up the new value and update our `LiveData` model. This will allow us to observe `LiveData` for any changes and just update the UI.

The `saveText` method will open the editor, set the new value, and apply the changes. The `getText` method will read the last saved value, set it in `LiveData`, and return the `LiveData` object. This is helpful in scenarios where the app is opened and we want to access the last value prior to the app closing.

4. Now, let's set up the `Application` class with the instance of the preferences in the `main/java` folder in the root package:

```kotlin
class PreferenceApplication : Application() {
    lateinit var preferenceWrapper: PreferenceWrapper
    override fun onCreate() {
        super.onCreate()
        preferenceWrapper =
            PreferenceWrapper(getSharedPreferences(
```

```
            "prefs", Context.MODE_PRIVATE))
        }
    }
```

5. Now, let's add the appropriate attributes in the `application` tag to Android Manifest.xml:

   ```
   android:name=".PreferenceApplication"
   ```

6. Next, let's build the `ViewModel` component in the `main/java` folder in the root package:

   ```
   class PreferenceViewModel(private val
   preferenceWrapper: PreferenceWrapper) : ViewModel() {
       fun saveText(text: String) {
           preferenceWrapper.saveText(text)
       }
       fun getText(): LiveData<String> {
           return preferenceWrapper.getText()
       }
   }
   ```

7. Now, let's define our `activity_main.xml` layout file in the `res/layout` folder:

   ```
   <TextView
       android:id="@+id/activity_main_text_view"
       android:layout_width="wrap_content"
       android:layout_height="wrap_content"
       android:layout_marginTop="50dp"
       app:layout_constraintLeft_toLeftOf="parent"
       app:layout_constraintRight_toRightOf="parent"
       app:layout_constraintTop_toTopOf="parent" />

   <EditText
       android:id="@+id/activity_main_edit_text"
       android:layout_width="200dp"
       android:layout_height="wrap_content"
       android:inputType="none"
       app:layout_constraintLeft_toLeftOf="parent"
   ```

```
                app:layout_constraintRight_toRightOf="parent"
                app:layout_constraintTop_toBottomOf=
                    "@id/activity_main_text_view" />
```

The complete code for this step can be found at https://packt.link/2c5Ay.

8. Finally, in `MainActivity`, perform the following steps:

```
class MainActivity : AppCompatActivity() {
    override fun onCreate(savedInstanceState: Bundle?)
    {
        preferenceViewModel.getText().observe(this,
        Observer {
            findViewById<TextView>(
            R.id.activity_main_text_view) .text = it
        })
        findViewById<Button>(R.id.activity_main_button
        ) .setOnClickListener {
            preferenceViewModel.saveText(findViewById
            <EditText> (R.id.activity_main_edit_text)
            .text.toString())
        }
    }
}
```

The complete code for this step can be found at https://packt.link/ZRWNc.

The preceding code will produce the output presented in *Figure 12.1*:

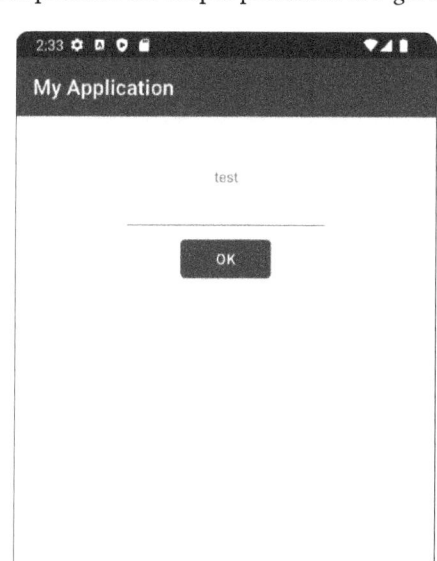

Figure 12.1 – Output of Exercise 12.01

Once you insert a value, try closing the application and reopening it. The app will display the last persisted value.

DataStore

The `DataStore` persistence library represents an alternative to `SharedPreferences` when we want to store data in key-value pairs through the Preference `DataStore`, or if we want to store entire objects through the Proto `DataStore`. Both libraries avoid dependencies with the Android framework (unlike `SharedPreferences`, which requires a `Context` object to be initialized) and are built using coroutines and flows, making them the ideal candidate when coroutines and flows are used in your project.

This integration allows the `DataStore` to notify subscribers of all changes, which means that developers no longer have to concern themselves with handling the changes:

```
val Context.dataStore: DataStore<Preferences> by
preferencesDataStore(name = "myDataStore")
val KEY_MY_INT = intPreferencesKey("my_int_key")
val KEY_MY_BOOLEAN =
```

```kotlin
        booleanPreferencesKey("my_boolean_key")
val KEY_MY_STRING = stringPreferencesKey("my_string_key")
class MyAppSettings(private val context: Context) {
    val myIntValue: Flow<Int> = context.dataStore.data
        .map { preferences ->
            preferences[KEY_MY_INT] ?: 0
        }
    val myBooleanValue: Flow<Boolean> =
        context.dataStore.data
        .map { preferences ->
            preferences[KEY_MY_BOOLEAN] ?: false
        }
    val myStringValue: Flow<String> =
        context.dataStore.data
        .map { preferences ->
            preferences[KEY_MY_STRING] ?: ""
        }
}
```

In the preceding snippet, we initialize `Context.dataStore` in the top-level Kotlin file. We then define three separate keys for separate types we want to read from. Inside `MyAppSettings`, we map the values from `context.dataStore.data` and extract the values from our keys.

If we want to store data in our `DataStore`, then we need to do the following:

```kotlin
class MyAppSettings(private val context: Context) {
    ...
    suspend fun saveMyIntValue(intValue: Int) {
        context.dataStore.edit { preferences ->
            preferences[KEY_MY_INT] = intValue
        }
    }
    suspend fun saveMyBooleanValue(booleanValue: Boolean) {
        context.dataStore.edit { preferences ->
            preferences[KEY_MY_BOOLEAN] = booleanValue
        }
    }
```

```
    suspend fun saveMyStringValue(stringValue: String) {
        context.dataStore.edit { preferences ->
            preferences[KEY_MY_STRING] = stringValue
        }
    }
}
```

The `suspend` keyword comes from coroutines, and it signals that we need to place the method invocation into an asynchronous call. `context.dataStore.edit` will make the preferences in the `DataStore` mutable and allow us to change the values.

Exercise 12.02 – Preference DataStore

We're going to build an application that displays `TextView`, `EditText`, and a button. `TextView` will display the value that was added to `DataStore`. The user can type new text, and when the button is clicked, the text will be saved in `DataStore` and `TextView` will display the updated text.

We will need to use `ViewModel` and `LiveData`. In the `ViewModel`, we will collect the data coming from `DataStore` and place it in a `LiveData` object:

1. Create a new Android Studio project with an empty activity.

2. Let's begin by adding the appropriate libraries to `app/build.gradle`:

    ```
    implementation "androidx.datastore:
        datastore-preferences:1.0.0"
    implementation "androidx.lifecycle:
        lifecycle-viewmodel-ktx:2.5.1"
    implementation "androidx.lifecycle:
        lifecycle-livedata-ktx:2.5.1"
    ```

3. Create a class called `SettingsStore` in the `main/java` folder in the root package, which will contain methods to load and save data from `DataStore`:

    ```
    val Context.dataStore: DataStore<Preferences> by
    preferencesDataStore(name = "settingsStore")
    val KEY_TEXT = stringPreferencesKey("key_text")
    class SettingsStore(private val context: Context) {
        val text: Flow<String> = context.dataStore.data
            .map { preferences ->
                preferences[KEY_TEXT] ?: ""
            }
    ```

```
           suspend fun saveText(text: String) {
               context.dataStore.edit { preferences ->
                   preferences[KEY_TEXT] = text
               }
           }
       }
```

In the preceding snippet, we have defined a key for storing the text, a field to retrieve the saved text, and a method to save it.

4. Create a new class called `SettingsViewModel` in the `main/java` folder in the root package, which will collect the data from `SettingsStore` in `LiveData` and invoke it to save the new text values:

```
       class SettingsViewModel(private val settingsStore:
       SettingsStore) : ViewModel() {
           private val _textLiveData =
               MutableLiveData<String>()
           val textLiveData: LiveData<String> = _textLiveData
           init {
               viewModelScope.launch {
                   settingsStore.text.collect {
                       _textLiveData.value = it
                   }
               }
           }
           fun saveText(text: String) {
               viewModelScope.launch {
                   settingsStore.saveText(text)
               }
           }
       }
```

In the preceding example, `viewModelScope` is an extension of `ViewModel` and represents `CoroutineScope`, which ensures that the background work is done while `ViewModel` is still active to avoid any possible leaks. Using this, we can collect the existing text into `LiveData` when `ViewModel` is initialized and then invoke the `saveText` method from `SettingsStore`.

5. Now, let's set up the `Application` class with the `SettingsStore` instance in the `main/java` folder in the root package:

   ```
   class SettingsApplication : Application() {
       lateinit var settingsStore: SettingsStore
       override fun onCreate() {
           super.onCreate()
           settingsStore = SettingsStore(this)
       }
   }
   ```

6. Next, let's add the appropriate attributes in the `application` tag to `AndroidManifest.xml`:

   ```
   android:name=".SettingsApplication"
   ```

7. Finally, let's define our `activity_main.xml` layout file in the `res/layout` folder:

   ```
   <TextView
       android:id="@+id/activity_main_text_view"
       android:layout_width="wrap_content"
       android:layout_height="wrap_content"
       android:layout_marginTop="50dp"
       app:layout_constraintLeft_toLeftOf="parent"
       app:layout_constraintRight_toRightOf="parent"
       app:layout_constraintTop_toTopOf="parent" />

   <EditText
       android:id="@+id/activity_main_edit_text"
       android:layout_width="200dp"
       android:layout_height="wrap_content"
       android:inputType="none"
       app:layout_constraintLeft_toLeftOf="parent"
       app:layout_constraintRight_toRightOf="parent"
       app:layout_constraintTop_toBottomOf=
           "@id/activity_main_text_view" />
   ```

The complete code for this step can be found at https://packt.link/8f854.

8. And finally, in MainActivity, perform the following steps:

```kotlin
class MainActivity : AppCompatActivity() {
    override fun onCreate(savedInstanceState: Bundle?)
    {
        super.onCreate(savedInstanceState)
        setContentView(R.layout.activity_main)
        val preferenceWrapper =
            (application as SettingsApplication)
            .settingsStore
        val preferenceViewModel =
            ViewModelProvider(this, object :
            ViewModelProvider.Factory {
                override fun <T : ViewModel>
                create(modelClass: Class<T>): T {
                    return SettingsViewModel(
                    preferenceWrapper) as T
                }
            }).get(SettingsViewModel::class.java)
    }
}
```

The complete code for this step can be found at https://packt.link/gydeC.

If we now run the application, we should see the following screen:

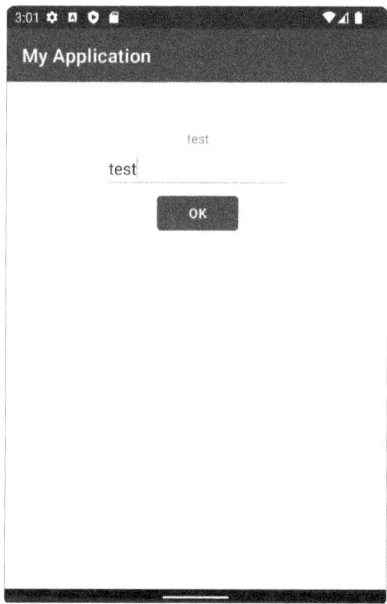

Figure 12.2 – Output of Exercise 12.02

If we enter a new text and click the **OK** button, we see that the text is instantly updated, unlike the previous exercise. This is because of the usage of flows and how DataStore will emit a new value for each change. We will look at flows and other reactive streams in future chapters.

In this exercise, we have looked at how the DataStore library works and its benefits, especially regarding streams of data. In the following chapters, we will continue to look at other ways of persisting data, using files.

Files

We've discussed Room, SharedPreferences, and DataStore and specified how the data they store is written to files. You may ask yourself, where are these files stored? These particular files are stored in internal storage. Internal storage is a dedicated space for every app that other apps are unable to access (unless a device is rooted). There is no limit to the amount of storage your app can use.

However, users have the ability to delete their app's files from the **Settings** menu. Internal storage occupies a smaller part of the total available space, which means that you should be careful when it comes to storing files there. There is also external storage. The files your app stores in external storage are accessible to other apps and the files from other apps are accessible to your one.

> **Note**
>
> In Android Studio, you can use the **Device File Explorer** tool to navigate through the files on a device or emulator. Internal storage is located in `/data/data/{packageName}`. If you have access to this folder, this means that the device is rooted. Using this, you can visualize the database files and the `SharedPreferences` files.

An example of how the **Device File Explorer** looks can be viewed in the following figure:

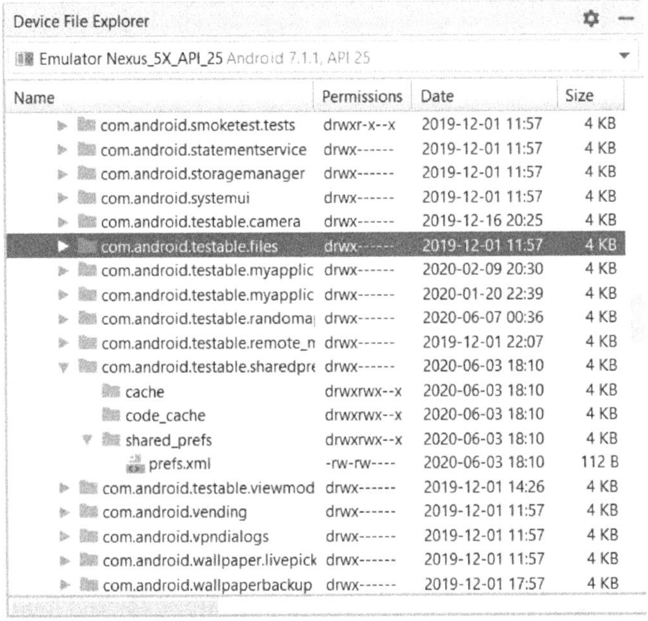

Figure 12.3 – Android's Device File Explorer for an emulated device

Internal storage

Internal storage requires no permissions from the user. To access the internal storage directories, you can use one of the following methods from the `Context` object:

- `getDataDir()`: Returns the root folder of your application sandbox.
- `getFilesDir()`: A dedicated folder for application files – recommended for usage.
- `getCacheDir()`: A dedicated folder where files can be cached. Storing files here does not guarantee that you can retrieve them later because the system may decide to delete this directory to free up memory. This folder is linked to the **Clear Cache** option in **Settings**.
- `getDir(name, mode)`: Returns a folder that will be created if it does not exist, based on the name specified.

When users use the **Clear Data** option from **Settings**, most of these folders will be deleted, bringing the app to a similar state as a fresh installation. When the app is uninstalled, these files will be deleted as well.

A typical example of reading a cache file is as follows:

```
val cacheDir = context.cacheDir
val fileToReadFrom = File(cacheDir, "my-file.txt")
val size = fileToReadFrom.length().toInt()
val bytes = ByteArray(size)
val tmpBuff = ByteArray(size)
val fis = FileInputStream(fileToReadFrom)
try {
    var read = fis.read(bytes, 0, size)
    if (read < size) {
        var remain = size - read
        while (remain > 0) {
            read = fis.read(tmpBuff, 0, remain)
            System.arraycopy(tmpBuff, 0, bytes,
                size - remain, read)
            remain -= read
        }
    }
} catch (e: IOException) {
    throw e
} finally {
    fis.close()
}
```

The preceding example will read from `my-file.txt`, located in the `Cache` directory, and will create `FileInputStream` for that file. Then, a buffer will be used that will collect the bytes from the file. The collected bytes will be placed in the `bytes` byte array, which will contain all of the data read from that file. Reading will stop when the entire length of the file has been read.

Writing to the `my-file.txt` file will look something like this:

```
val bytesToWrite = ByteArray(100)
val cacheDir = context.cacheDir
val fileToWriteIn = File(cacheDir, "my-file.txt")
```

```
        try {
            if (!fileToWriteIn.exists()) {
                fileToWriteIn.createNewFile()
            }
            val fos = FileOutputStream(fileToWriteIn)
            fos.write(bytesToWrite)
            fos.close()
        } catch (e: Exception) {
            e.printStackTrace()
        }
```

What the preceding example does is take the byte array you want to write, create a new `File` object, create the file if it doesn't exist, and write the bytes into the file through `FileOutputStream`.

> **Note**
>
> There are many alternatives to dealing with files. The readers (`StreamReader`, `StreamWriter`, and so on) are better equipped for character-based data. There are also third-party libraries that help with disk I/O operations. One of the most common third parties that help with I/O operations is called **Okio**. It started life as part of the `OkHttp` library, which is used in combination with Retrofit to make API calls. The methods provided by Okio are the same methods it uses to write and read data in HTTP communications.

External storage

Reading and writing in external storage requires user permission for reading and writing. If write permission is granted, then your app has the ability to read the external storage. Once these permissions are granted, then your app can do whatever it pleases on the external storage.

That may present a problem because users may not choose to grant these permissions. However, there are specialized methods that offer you the possibility to write to the external storage in folders dedicated to your application.

Some of the most common ways of accessing external storage are from the `Context` and `Environment` objects:

- `Context.getExternalFilesDir(mode)`: This method will return the path to the directory on the external storage dedicated to your application. Specifying different modes (pictures, movies, and so on) will create different subfolders, depending on how you want your files saved. This method *does not require permissions*.

- `Context.getExternalCacheDir()`: This will point toward an application's cache directory on the external storage. The same considerations should be applied to this `cache` folder as to the internal storage option. This method *does not require permissions*.
- The `Environment` class has access to paths of some of the most common folders on a device. However, on newer devices, apps may not have access to those files and folders.

> **Note**
> Avoid using hardcoded paths to files and folders. The Android operating system may shift the location of folders around, depending on the device or Android version.

FileProvider

This represents a specialized implementation of `ContentProviders` that is useful in organizing the file and folder structure of your application. It allows you to specify an XML file, in which you define how your files should be split between internal and external storage if you choose to do so. It also gives you the ability to grant access to other apps to your files by hiding the path and generating a unique URI to identify and query your file.

`FileProvider` lets you pick between six different folders, where you can set up your folder hierarchies:

- `Context.getFilesDir()` (files-path)
- `Context.getCacheDir()` (cache-path)
- `Environment.getExternalStorageDirectory()` (external-path)
- `Context.getExternalFilesDir(null)` (external-files-path)
- `Context.getExternalCacheDir()` (external-cache-path)
- The first result of `Context.getExternalMediaDirs()` (external-media-path)

The main benefits of `FileProvider` are the abstractions it provides in organizing your files while leaving a developer to define the paths in an XML file, and more importantly, if you choose to use it to store files in external storage, you do not have to ask for permissions from the user.

Another benefit is the fact that it makes sharing of internal files easier while giving a developer control of what files other apps can access without exposing their real location.

Let us understand better through the following example:

```
<paths
xmlns:android="http://schemas.android.com/apk/res/android">
    <files-path name="my-visible-name" path="/
    my-folder-name" />
</paths>
```

The preceding example will make `FileProvider` use the internal `files` directory and create a folder named `my-folder-name`. When the path is converted to a URI, then the URI will use `my-visible-name`.

The Storage Access Framework (SAF)

The SAF is a file picker introduced in Android KitKat that apps can use for their users to pick files, with a view to them being processed or uploaded. You can use it in your app in the following scenarios:

- Your app requires a user to process a file saved on a device by another app (photos and videos)
- You want to save a file on a device and give a user the choice of where to save the file and the name of the file
- You want to offer the files your application uses to other apps for scenarios similar to the first scenario in this list

This is again useful because your app will avoid read and write permissions and still write and access external storage. The way this works is based on intents. You can register for an activity result for `GetDocument` or `CreateDocument`. Then, in the activity result callback, the system will give you a URI that grants you temporary permissions to that file, allowing you to read and write.

Another benefit of the SAF is the fact that files don't have to be on a device. Apps such as Google Drive expose their content in the SAF, and when a Google Drive file is selected, it will be downloaded to the device and the URI will be sent as a result.

Another important thing to mention is the SAF's support for virtual files, meaning that it will expose Google docs, which have their own format, but when those docs are downloaded through the SAF, their formats will be converted to a common format such as PDF.

Asset files

Asset files are files you can package as part of your APK. If you've used an app that played certain videos or GIFs when the app is launched or as part of a tutorial, odds are that the videos were bundled with the APK. To add files to your assets, you need the `assets` folder inside your project. You can then group your files inside your assets using folders.

You can access these files at runtime through the `AssetManager` class, which itself can be accessed through the context object. `AssetManager` offers you the ability to look up the files and read them, but it does not permit any write operations:

```
val assetManager = context.assets
val root = ""
val files = assetManager.list(root)
files?.forEach {
```

```
            val inputStream = assetManager.open(root + it)
    }
```

The preceding example lists all files inside the root of the `assets` folder. The `open` function returns `inputStream`, which can be used to read the file information if necessary.

One common usage of the `assets` folder is for custom fonts. If your application uses custom fonts, then you can use the `assets` folder to store font files.

> **Note**
>
> For the following exercise, you will need an emulator. You can do so by selecting **Tools | AVD Manager** in Android Studio. Then, you can create one with the **Create Virtual Device** option, selecting the type of emulator, clicking **Next**, and then selecting an x86 image. Any image larger than Lollipop should be acceptable for this exercise. Next, you can give your image a name and click **Finish**.

Exercise 12.03 – copying files

Let's create an app that will keep a file named `my-app-file.txt` in the `assets` directory. The app will display two buttons called `FileProvider` and `SAF`. When the `FileProvider` button is clicked, the file will be saved on the external storage inside the app's external storage dedicated area (`Context.getExternalFilesDir(null)`). The `SAF` button will open the SAF and allow a user to indicate where the file should be saved.

In order to implement this exercise, follow these steps:

1. Define a file provider that will use the `Context.getExternalFilesDir(null)` location.
2. Copy `my-app-file.txt` to the preceding location when the `FileProvider` button is clicked.
3. Use `Intent.ACTION_CREATE_DOCUMENT` when the `SAF` button is clicked and copy the file to the location provided.
4. Use a separate thread for the file copy to comply with the Android guidelines.
5. Use the Apache IO library to help with the file copy functionality, by providing methods that allow us to copy data from `InputStream` to `OutputStream`.

The steps for completion are as follows:

1. Create a new Android Studio project with an empty activity.
2. Let's start with our Gradle configuration:

    ```
    implementation 'commons-io:commons-io:2.6'
    ```

3. Create the `my-app-file.txt` file in the `main/assets` folder. Feel free to fill it up with the text you want to be read. If the `main/assets` folder doesn't exist, then you can create it. To create the `assets` folder, you can right-click on the `main` folder, select **New**, then **Directory**, and name it `assets`.

   This folder will now be recognized by the build system, and any file inside it will also be installed on the device along with the app. You may need to switch **Project View** from **Android** to **Project** to be able to view this file structure.

4. We can also define a class that will wrap `AssetManager` in the `main/java` folder in the root package and define a method to access this particular file:

   ```
   class AssetFileManager(private val assetManager:
   AssetManager) {
       fun getMyAppFileInputStream() =
           assetManager.open("my-app-file.txt")
   }
   ```

5. Now, let's work on the `FileProvider` aspect. Create the `xml` folder in the `res` folder. Define `file_provider_paths.xml` inside the new folder. We will define `external-files-path`, name it `docs`, and place it in the `docs/` folder:

   ```
   <?xml version="1.0" encoding="utf-8"?>
   <paths>
       <external-files-path name="docs" path="docs/"/>
   </paths>
   ```

6. Next, we need to add `FileProvider` to the `AndroidManifest.xml` file and link it with the new path we defined inside the `<application` tag:

   ```
   <provider
       android:name=
           "androidx.core.content.FileProvider"
       android:authorities=
           "com.android.testable.files"
       android:exported="false"
       android:grantUriPermissions="true">
       <meta-data
           android:name="android.support
               .FILE_PROVIDER_PATHS"
   ```

```
                    android:resource="@xml/
                        file_provider_paths" />
          </provider>
```

The name will point to the `FileProvider` path that's part of the Android Support Library. The `authorities` field represents the domain your application has (usually the package name of the application).

The exported field indicates whether we wish to share our provider with other apps, and `grantUriPermissions` indicates whether we wish to grant other applications access to certain files through the URI. The metadata links the XML file we defined previously with `FileProvider`.

7. Define the `ProviderFileManager` class in the `main/java` folder in the root package, which is responsible for accessing the `docs` folder and writing data into the file:

```
class ProviderFileManager(
) {
    fun writeStream(name: String, inputStream:
    InputStream) {
        executor.execute {
            val fileToSave = File(getDocsFolder(),
            name)
            val outputStream =
                context.contentResolver
                .openOutputStream(fileToUriMapper
                .getUriFromFile(context, fileToSave),
                "rw")
            IOUtils.copy(inputStream, outputStream)
        }
    }
}
```

The complete code for this step can be found at https://packt.link/Gp0Ph.

`getDocsFolder` will return the path to the `docs` folder we defined in the XML. If the folder does not exist, then it will be created.

The `writeStream` method will extract the URI for the file we wish to save and, using the Android `ContentResolver` class, will give us access to the `OutputStream` class of the file we will be saving in. Note that `FileToUriMapper` doesn't exist yet. The code is moved into a separate class in order to make this class testable.

8. Create the `FileToUriMapper` class in the `main/java` folder in the root package:

```
class FileToUriMapper {
    fun getUriFromFile(context: Context, file: File):
    Uri {
        return FileProvider.getUriForFile(context,
        "com.android.testable.files", file)
    }
}
```

The `getUriForFile` method is part of the `FileProvider` class, and its role is to convert the path of a file into a URI that can be used by `ContentProviders` and `ContentResolvers` to access data. Because the method is static, it prevents us from testing properly.

9. Make sure that the following strings are added to `strings.xml`:

```
<string name="file_provider">FileProvider</string>
<string name="saf">SAF</string>
```

10. Let's now move on to defining our UI for the `activity_main.xml` file in the `res/layout` folder:

```
<Button
    android:id="@+id/activity_main_file_provider"
    android:layout_width="wrap_content"
    android:layout_height="wrap_content"
    android:layout_marginTop="200dp"
    android:text="@string/file_provider"
    app:layout_constraintEnd_toEndOf="parent"
    app:layout_constraintStart_toStartOf="parent"
    app:layout_constraintTop_toTopOf="parent" />

<Button
    android:id="@+id/activity_main_saf"
    android:layout_width="wrap_content"
    android:layout_height="wrap_content"
```

```
                    android:layout_marginTop="50dp"
                    android:text="@string/saf"
                    app:layout_constraintEnd_toEndOf="parent"
                    app:layout_constraintStart_toStartOf="parent"
                    app:layout_constraintTop_toBottomOf=
                        "@id/activity_main_file_provider" />
```

The complete code for this step can be found at https://packt.link/Pw37X.

11. Now, let's define our MainActivity class in the main/java folder in the root package:

```
    class MainActivity : AppCompatActivity() {

        override fun onCreate(savedInstanceState: Bundle?)
        {
            super.onCreate(savedInstanceState)
            setContentView(R.layout.activity_main)
            findViewById<Button>(
            R.id.activity_main_file_provider)
            .setOnClickListener {
                val newFileName = "Copied.txt"
                providerFileManager.writeStream(
                    newFileName, assetFileManager
                    .getMyAppFileInputStream())
            }
        }
    }
```

The complete code for this step can be found at https://packt.link/FBXgY.

For this example, we chose MainActivity to create our objects and inject data into the different classes we have. If we execute this code and click the FileProvider button, we don't see an output on the UI.

However, if we look at Android's **Device File Explorer**, we can locate where the file was saved. The path may be different on different devices and operating systems. The paths could be as follows:

- `mnt/sdcard/Android/data/<package_name>/files/docs`
- `sdcard/Android/data/<package_name>/files/docs`
- `storage/emulated/0/Android/data/<package_name>/files/docs`

The output will be as follows:

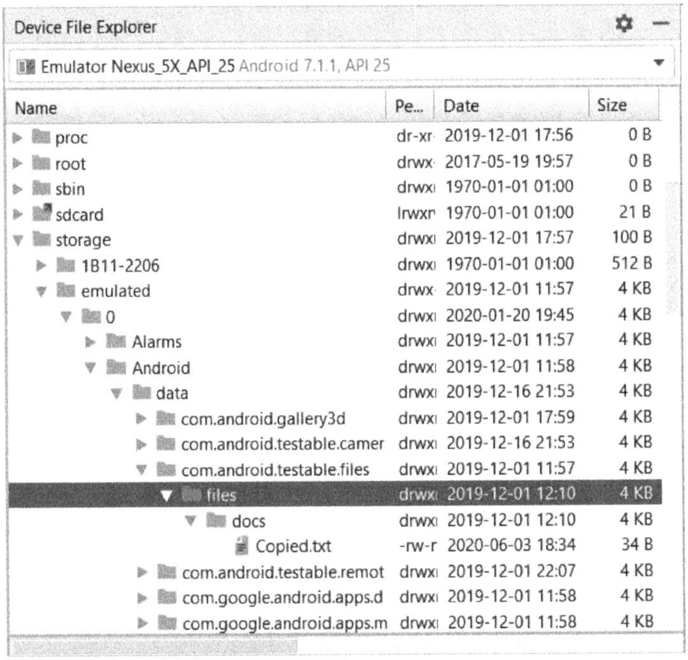

Figure 12.4 – Output of copy through FileProvider

12. Let's add the logic for the `SAF` button. We will need to start an activity pointing toward SAF with the `CREATE_DOCUMENT` intent, in which we specify that we want to create a text file.

   We will then need the result of SAF so that we can copy the file to the location selected by a user. In `MainActivity` in `onCreate`, we can add the following:

```
val createDocumentResult =
    registerForActivityResult(
    ActivityResultContracts.CreateDocument(
    "text/plain")) { uri ->
```

```
                uri?.let {
                    val newFileName = "Copied.txt"
                    providerFileManager
                        .writeStreamFromUri(newFileName,
                    assetFileManager
                        .getMyAppFileInputStream(), uri)
                }
            }
        findViewById<Button>(R.id.activity_main_saf)
            .setOnClickListener {
                createDocumentResult.launch("Copied.txt")
            }
```

What the preceding code will do is register for an `Activity` result when a user creates a new file. We will then invoke `writeStreamFromUri` from `ProviderFileManager` to save the contents from the file in the `assets` folder in the file created by the user. When the button is clicked, we will then launch the file creation screen from the SAF.

13. We now have the URI. We can add a method to `ProviderFileManager` that will copy our file to a location given by `uri`:

```
        fun writeStreamFromUri(name: String, inputStream:
        InputStream, uri:Uri){
            executor.execute {
                val outputStream =
                    context.contentResolver
                    .openOutputStream(uri, "rw")
                IOUtils.copy(inputStream, outputStream)
            }
        }
```

If we run the preceding code and click on the **SAF** button, we will see the output presented in *Figure 12.5*:

Figure 12.5 – Output of copy through the SAF

If you choose to save the file, the SAF will be closed and the callback from `registerForActivityResult` will be invoked, which will trigger the file copy. Afterward, you can navigate the Android Device File Manager tool to see whether the file was saved properly.

Scoped storage

Since Android 10 and with further updates in Android 11, the notion of scoped storage was introduced. The main idea behind this is to allow apps to gain more control of their files in external storage and prevent other apps from accessing these files.

The consequences of this mean that READ_EXTERNAL_STORAGE and WRITE_EXTERNAL_STORAGE will only apply to files a user interacts with (such as media files). This discourages apps from creating their own directories in external storage, instead sticking with the one already provided to them through `Context.getExternalFilesDir`.

FileProviders and the SAF are a good way of making your app comply with scoped storage practices, with one allowing the app to use `Context.getExternalFilesDir` and the other using the built-in File Explorer app, which will now avoid files from other applications in the `Android/data` and `Android/obb` folders in external storage.

Camera and media storage

Android offers a variety of ways to interact with media on its devices, from building your own camera application and controlling how users take photos and videos to using an existing camera application and instructing it on how to take photos and videos.

Android also comes with a `MediaStore` content provider, allowing applications to extract information about media files that are set on a device and shared between applications.

This is useful in situations where you want a custom display for media files that exist on a device (such as a photo or music player application) and in situations where you use the `MediaStore.ACTION_PICK` intent to select a photo from the device and want to extract the information about the selected media image (this is usually the case for older applications where the SAF cannot be used).

In order to use an existing camera application, you will need to use the `MediaStore.ACTION_IMAGE_CAPTURE` intent to start a camera application for a result and pass the URI of the image you wish to save. The user will then go to the camera activity and take the photo, and then you handle the result of the operation:

```
val imageCaptureLauncher =
    registerForActivityResult(
    ActivityResultContracts.TakePicture()){
    }
imageCaptureLauncher.launch(photoUri)
```

The `photoUri` parameter will represent the location of where you want your photo to be saved. It should point to an empty file with a JPEG extension. You can build this file in two ways:

- Create a file on the external storage using the `File` object (this requires the `WRITE_EXTERNAL_STORAGE` permission) and then use the `Uri.fromFile()` method to convert it into `URI` (this is no longer applicable on Android 10 and above)
- Create a file in a `FileProvider` location using the `File` object, and then use the `FileProvider.getUriForFile()` method to obtain the URI and grant it permissions if necessary (the recommended approach for when your app targets Android 10 and Android 11)

> **Note**
> The same mechanism can be applied to videos using `MediaStore.ACTION_VIDEO_CAPTURE`.

If your application relies heavily on camera features, then you can exclude the application from users whose devices don't have cameras by adding the `<uses-feature>` tag to the `AndroidManifest.xml` file. You can also specify the camera as non-required and query whether the camera is available using the `Context.hasSystemFeature(PackageManager.FEATURE_CAMERA_ANY)` method.

If you wish to have your file saved in `MediaStore`, there are multiple ways to achieve this:

1. Send an `ACTION_MEDIA_SCANNER_SCAN_FILE` broadcast with the URI of your media:

    ```
    val intent = Intent(Intent.ACTION_MEDIA_SCANNER_
    SCAN_FILE)
    intent.data = photoUri
    sendBroadcast(intent)
    ```

2. Use the media scanner to scan files directly:

    ```
    val paths = arrayOf("path1", "path2")
    val mimeTypes= arrayOf("type1", "type2")
    MediaScannerConnection.scanFile(context,paths,
        mimeTypes) { path, uri ->
    }
    ```

3. Insert the media into `ContentProvider` directly using `ContentResolver`:

    ```
    val contentValues = ContentValues()
    contentValues.put(MediaStore.Images
        .ImageColumns.TITLE, "my title")
    contentValues.put(MediaStore.Images
        .ImageColumns .DATE_ADDED, timeInMillis)
    contentValues.put(MediaStore.Images
        .ImageColumns .MIME_TYPE, "image/*")
    contentValues.put(MediaStore.Images
        .ImageColumns .DATA, "my-path")
    val newUri =
        contentResolver.insert(MediaStore.Video
        .Media.EXTERNAL_CONTENT_URI,
        contentValues)
    ```

```
newUri?.let {
    val outputStream = contentResolver
    .openOutputStream(newUri)
    // Copy content in outputstream
}
```

> **Note**
>
> The `MediaScanner` functionality no longer adds files from `Context.getExternal` `FilesDir` in Android 10 and above. Apps should rely on the `insert` method instead if they choose to share their media files with other apps.

Exercise 12.04 – taking photos

We're going to build an application that has two buttons; the first button will open a camera app to take a photo, and the second button will open the camera app to record a video. We will use `FileProvider` to save the photos to external storage (external-path) in two folders, `pictures` and `movies`.

The photos will be saved using `img_{timestamp}.jpg`, and the videos will be saved using `video_{timestamp}.mp4`. After a photo and video have been saved, you will copy the file from `FileProvider` into `MediaStore` so that they will be visible for other apps:

1. Create a new Android Studio project with an empty activity.

2. Let's add the libraries in `app/build.gradle`:

    ```
    implementation 'commons-io:commons-io:2.6'
    ```

3. We will need to request the `WRITE_EXTERNAL_STORAGE` permission for devices that predate Android 10, which means we need the following in `AndroidManifest.xml` outside of the `<application` tag:

    ```
    <uses-permission
        android:name="android.permission
        .WRITE_EXTERNAL_STORAGE"
        android:maxSdkVersion="28" />
    ```

4. Let's define a `FileHelper` class, which will contain methods that are harder to test in the `test` folder in the root package:

    ```
    class FileHelper(private val context: Context) {
        fun getUriFromFile(file: File): Uri {
            return FileProvider.getUriForFile(context,
    ```

```
            "com.android.testable.camera", file)
    }
    fun getPicturesFolder(): String =
        Environment.DIRECTORY_PICTURES
    fun getVideosFolder(): String =
        Environment.DIRECTORY_MOVIES
}
```

5. Let's define our `FileProvider` paths in `res/xml/file_provider_paths.xml`. Make sure to include the appropriate package name for your application in `FileProvider`:

```
<?xml version="1.0" encoding="utf-8"?>
<paths>
    <external-path
        name="photos"
        path="Android/data/com.android.testable
        .myapplication/files/Pictures" />
    <external-path
        name="videos"
        path="Android/data/com.android.testable
        .myapplication/files/Movies" />
</paths>
```

Let's add the file provider paths to the `AndroidManifest.xml` file inside the `<application` tag:

```
            <provider
                android:name=
                    "androidx.core.content.FileProvider"
                android:authorities=
                    "com.android.testable.camera"
                android:exported="false"
                android:grantUriPermissions="true">
                <meta-data
                    android:name="android.support
                    .FILE_PROVIDER_PATHS"
                    android:resource="@xml/
```

```
            file_provider_paths" />
    </provider>
```

6. Let's now define a model that will hold both Uri and the associated path for a file in the main/java folder in the root package:

    ```
    data class FileInfo(
        val uri: Uri,
        val file: File,
        val name: String,
        val relativePath:String,
        val mimeType:String
    )
    ```

7. Let's create a ContentHelper class in the main/java folder in the root package, which will provide us with the data required for ContentResolver. We will define two methods for accessing the photo and video content URI and two methods that will create ContentValues.

    We do this because of the static methods required to obtain URIs and create ContentValues, which makes this functionality hard to test. The following code is truncated for space. The full code you need to add can be found via the link that follows this code block:

    ```
    class MediaContentHelper {

        fun getImageContentUri(): Uri =
            if (android.os.Build.VERSION.SDK_INT >=
            android.os.Build.VERSION_CODES.Q) {
                MediaStore.Images.Media.getContentUri
                (MediaStore.VOLUME_EXTERNAL_PRIMARY)
            } else {
                MediaStore.Images.Media
                .EXTERNAL_CONTENT_URI
            }

        fun generateImageContentValues(fileInfo:
        FileInfo) = ContentValues().apply {
            this.put(MediaStore.Images.Media
            .DISPLAY_NAME, fileInfo.name)
            if (android.os.Build.VERSION.SDK_INT >=
    ```

```
                    android.os.Build.VERSION_CODES.Q) {
                this.put(MediaStore.Images.Media
                    .RELATIVE_PATH, fileInfo.relativePath)
            }
            this.put(MediaStore.Images.Media .MIME_TYPE,
                fileInfo.mimeType)
        }
```

The complete code for this step can be found at https://packt.link/DhOLR.

8. Now, let's create the `ProviderFileManager` class in the `main/java` folder in the root package, where we will define methods to generate files for photos and videos that will then be used by the camera and the methods that will save to the media store. Again, the code has been truncated for brevity. Please see the link that follows this code block for the full code that you need to use:

```
class ProviderFileManager(
) {
    fun generatePhotoUri(time: Long): FileInfo {
        val name = "img_$time.jpg"
        val file = File(
            context.getExternalFilesDir(
            fileHelper .getPicturesFolder()), name
        )
        return FileInfo(
            fileHelper.getUriFromFile(file),
            file, name,
            fileHelper.getPicturesFolder(),
            "image/jpeg"
        )
    }
```

The complete code for this step can be found at https://packt.link/ohv7a.

Note how we defined the root folders as `context.getExternalFilesDir(Environment.DIRECTORY_PICTURES)` and `context.getExternalFilesDir(Environment.DIRECTORY_MOVIES)`. This connects to `file_provider_paths.xml` and it will create a set of folders called `Movies` and `Pictures` in the application's dedicated folder in external storage. The `insertToStore` method is where the files will be then copied to the `MediaStore`.

First, we will create an entry into that store that will give us a URI for that entry. Next, we copy the contents of the files from the URI generated by `FileProvider` into `OutputStream`, pointing to the `MediaStore` entry.

9. Add the following strings to `strings.xml`:

   ```
   <string name="photo">Photo</string>
   <string name="video">Video</string>
   ```

10. Let's define the layout for our activity in `res/layout/activity_main.xml`:

    ```
    <Button
        android:id="@+id/photo_button"
        android:layout_width="wrap_content"
        android:layout_height="wrap_content"
        android:text="@string/photo" />

    <Button
        android:id="@+id/video_button"
        android:layout_width="wrap_content"
        android:layout_height="wrap_content"
        android:layout_marginTop="5dp"
        android:text="@string/video" />
    ```

    The complete code for this step can be found at `https://packt.link/6iSNp`.

11. Let's create the `MainActivity` class in the `main/java` folder in the root package, where we will check whether we need to request `WRITE_STORAGE_PERMISSION`, request it if we need to, and after it is granted, open the camera to take a photo or a video. As before, the code here has been truncated for brevity. You can access the full code using the link that follows the code block:

    ```
    class MainActivity : AppCompatActivity() {
        private lateinit var providerFileManager:
    ```

```
            ProviderFileManager
    private var photoInfo: FileInfo? = null
    private var videoInfo: FileInfo? = null
    private var isCapturingVideo = false

override fun onCreate(savedInstanceState:
Bundle?) {
    super.onCreate(savedInstanceState)
    setContentView(R.layout.activity_main)
    providerFileManager =
        ProviderFileManager(applicationContext,
        FileHelper(applicationContext),
        contentResolver, Executors
        .newSingleThreadExecutor(),
        MediaContentHelper()
        )
```

The complete code for this step can be found at https://packt.link/YeHWC. If we execute the preceding code, we will see the following:

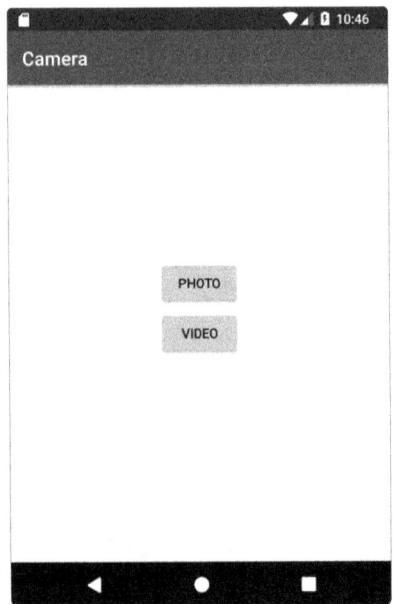

Figure 12.6 – Output of Exercise 12.04

12. By clicking on either of the buttons, you will be redirected to the camera application, where you can take a photo or a video if you are running the example on Android 10 and above. If you're running on lower Android versions, then the permissions will be asked first.

Once you have taken your photo and confirmed it, you will be taken back to the application. The photo will be saved in the location you defined in `FileProvider`:

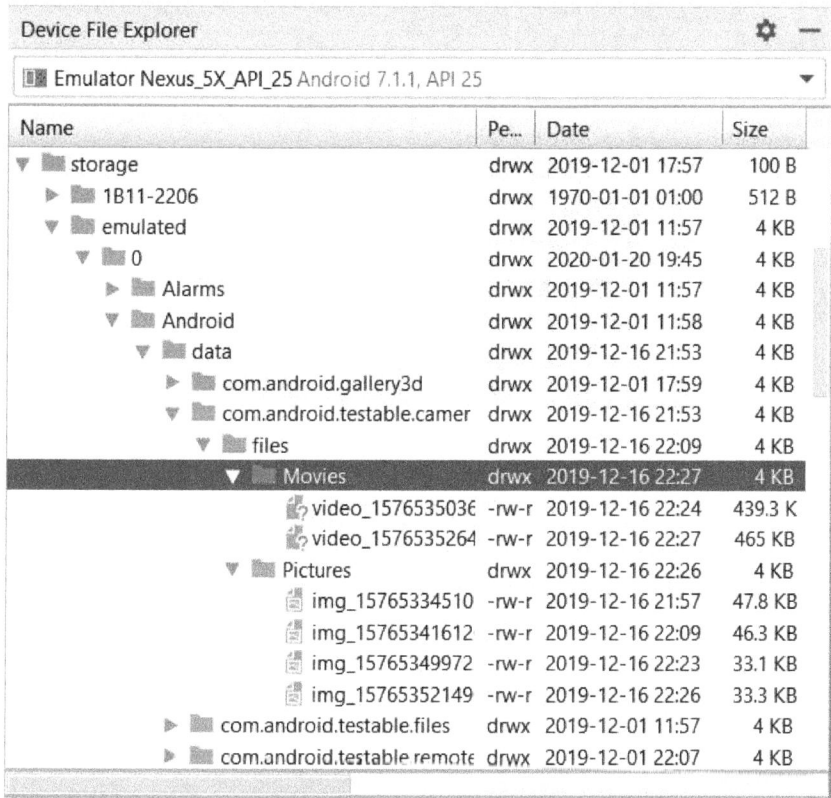

Figure 12.7 – The location of the captured files through the camera app

In the preceding screenshot, you can see where the files are located with the help of Android Studio's **Device File Explorer**. If you open any file-exploring app, such as the **Files**, **Gallery**, or **Google Photos** app, you will be able to see the videos and pictures taken.

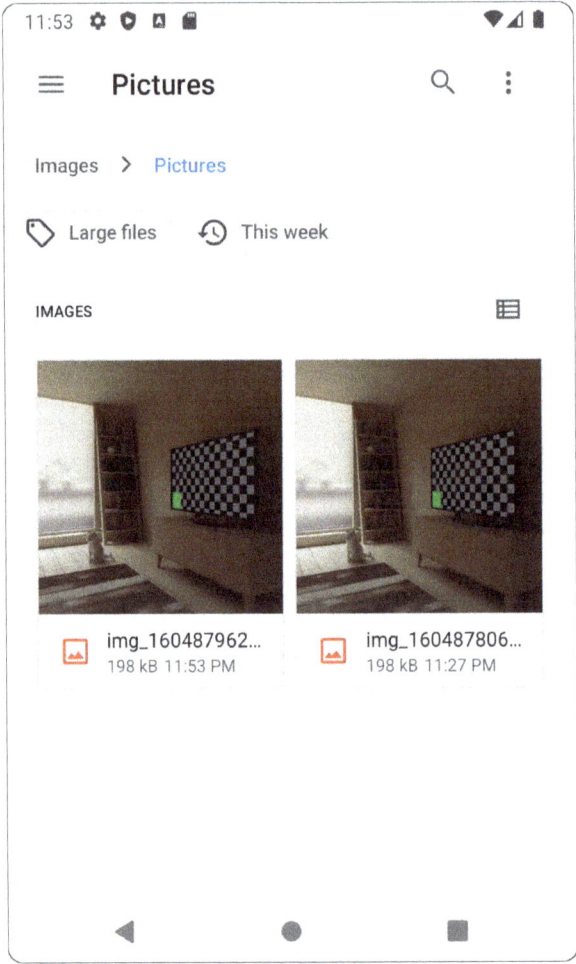

Figure 12.8 – The files from the app present in the File Explorer app

Activity 12.01 – dog downloader

You are tasked with building an application that will target Android versions above API 21 and display a list of URLs for dog photos. The URL you will connect to is `https://dog.ceo/api/breed/hound/images/random/{number}`, where `number` will be controlled through a **Settings** screen, where a user can choose the number of URLs they want to be displayed.

The **Settings** screen will be opened through an option presented on the home screen. When a user clicks on a URL, an image will be downloaded locally in the application's external cache path. While the image is being downloaded, the user will see an indeterminate progress bar. The list of URLs will be persisted locally using Room.

The technologies that we will use are the following:

- Retrofit to retrieve the list of URLs and download files
- Room to persist the list of URLs
- `SharedPreferences` to store the number of URLs to retrieve
- `FileProvider` to store the files in the cache
- Apache IO to write the files
- Repository to combine all the data sources
- `LiveData` and `ViewModel` to handle the logic from the user
- `RecyclerView` for the list of items

The response JSON will look similar to this:

```
{
    "message": [
        "https://images.dog.ceo/breeds/hound-
        afghan/n02088094_4837.jpg",
        "https://images.dog.ceo/breeds/hound-
        basset/n02088238_13908.jpg",
        "https://images.dog.ceo/breeds/hound-
        ibizan/n02091244_3939.jpg"
    ],
    "status": "success"
}
```

Perform the following steps to complete this activity:

1. Create an `api` package that will contain the network-related classes.
2. Create a data class that will model the response JSON.
3. Create a Retrofit `Service` class that will contain two methods. The first method will represent the API call to return a list of breeds, and the second method will represent the API call to download the file.
4. Create a `storage` package, and inside it, create a `room` package.
5. Create the `Dog` entity, which will contain an autogenerated ID and a URL.

6. Create the `DogDao` class, which will contain methods to insert a list of `Dog`s, delete all `Dog`s, and query all `Dog`s. The `delete` method is required because the API model does not have any unique identifiers.

7. Inside the `storage` package, create a `preference` package.

8. Inside the `preference` package, create a wrapper class around `SharedPreferences` that will return the number of URLs we need to use and set the number. The default will be `10`.

9. In `res/xml`, define your folder structure for `FileProvider`. The files should be saved in the root folder of the `external-cache-path` tag.

10. Inside the `storage` package, create a `filesystem` package.

11. Inside the `filesystem` package, define a class that will be responsible for writing `InputStream` into a file in `FileProvider`, using `Context.externalCacheDir`.

12. Create a `repository` package.

13. Inside the `repository` package, create a sealed class that will hold the result of an API call. The subclasses of the sealed class will be `Success`, `Error`, and `Loading`.

14. Define a `Repository` interface that will contain two methods, one to load the list of URLs and the other to download a file.

15. Define a `DogUi` model class that will be used in the UI layer of your application and that will be created in your repository.

16. Define a mapper class that will convert your API models into entities and entities into UI models.

17. Define an implementation for `Repository` that will implement the preceding two methods. The repository will hold references to `DogDao`, the Retrofit `Service` class, the `Preferences` wrapper class, the class managing the files, the `Dog` mapping class, and the `Executor` class for multithreading. When downloading the files, we will use the filename extracted from the URL.

18. Create a class that will extend `Application`, which will initialize the repository.

19. Define the `ViewModel` class used by your UI, which will have a reference to `Repository` and call it to load the URL list and download the images.

20. Define your UI, which will be composed of two activities:

    - `MainActivity` which displays the list of URLs and will have a click action to start the downloads. This activity will have a progress bar, which will be displayed when the download takes place. The screen will also have a **Settings** option, which will open the `SettingsActivity`.

    - The `SettingsActivity`, which will display `EditText` and `Button` and save the number of URLs.

> **Note**
> The solution to this activity can be found at `https://packt.link/z6g5j`.

Summary

In this chapter, we analyzed alternatives to Room when it comes to persisting data. We looked first at `SharedPreferences` and how it constitutes a handy solution for data persistence when it's in a key-value format and the amount of data is small. We also looked at `DataStore` and how we can use it like `SharedPreferences` but with built-in observability, which notifies us when values are changed.

Next, we looked over something that was continuously changing when it comes to the Android framework – the evolution of abstractions regarding a filesystem. We started with an overview of the types of storage that Android has and then took a more in-depth look at two of the abstractions – `FileProvider`, which your app can use to store files on a device and share them with others if necessary, and the SAF, which can be used to save files on the device in a location selected by a user.

We also used the benefits of `FileProvider` to generate URIs for files in order to use camera applications to take photos and record videos, saving them in the application's files while also adding them to `MediaStore`.

The activity performed in this chapter combines all the elements discussed previously to illustrate the point that even though you have to balance multiple sources inside an application, you can do it in a more readable way.

Note that for the activity and exercises in this chapter and the previous one, we kept having to use the `application` class to instantiate the data sources. In the next chapter, you will learn how to overcome this through dependency injection and see how it can benefit Android applications.

13
Dependency Injection with Dagger, Hilt, and Koin

This chapter covers the concept of dependency injection and the benefits it provides to an Android application. We will look at how we can perform dependency injection manually with the help of container classes. We will also cover some of the frameworks available for Android, Java, and Kotlin that can help developers when it comes to applying this concept. By the end of this chapter, you will be able to use Dagger 2 and Koin to manage your app's dependencies and know how to organize them efficiently.

In the previous chapter, we looked at how to structure code into different components, including ViewModels, API components, and persistence components. One of the difficulties that always emerged was the dependencies between all of these components, especially when it came to how we approached the unit tests for them.

We will cover the following topics in this chapter:

- Manual DI
- Dagger 2
- Hilt
- Koin

Technical requirements

The complete code for all the exercises and the activity in this chapter is available on GitHub at `https://packt.link/IIQmX`.

The necessity of dependency injection

We have constantly used the `Application` class to create instances of these components and pass them in the constructors of the components one layer above (we created the API and Room instances, then the Repository instances, and so on). What we were doing was a simplistic version of dependency injection.

Dependency injection (**DI**) is a software technique in which one object (the application) supplies the dependencies of another object (repositories, `ViewModels`). The reason for this is to increase the reusability and testability of the code and to shift the responsibility for creating instances from our components to the `Application` class.

One of the benefits of DI concerns how objects are created across the code base. DI separates the creation of an object from its usage. In other words, one object shouldn't care how another object is created; it should only be concerned with the interaction with the other object.

In this chapter, we will analyze three ways to inject dependencies in Android: Manual DI, Dagger, and Koin:

- **Manual DI**: This is a technique in which developers handle DI manually by creating container classes. In this chapter, we will examine how we can do this in Android. By studying how we manually manage dependencies, we will get some insight into how other DI frameworks operate and get a basis for how we can integrate these frameworks.

- **Dagger**: This is a DI framework developed for Java. It allows you to group your dependencies into different modules. You can also define components, where the modules are added to create the dependency graph, and which Dagger automatically implements to perform the injection. It relies on annotation processors to generate the necessary code to perform the injection. A specialized implementation of Dagger called **Hilt** is useful for Android applications because it removes a lot of boilerplate code and simplifies the process.

- **Koin**: This is a lightweight DI library developed for Kotlin. It doesn't rely on annotation processors; it relies on Kotlin's mechanisms to perform the injection. Here we can also split dependencies into modules.

In this chapter, we will explore how both these libraries work and the steps required to add them to a simple Android application.

Manual DI

In order to understand how DI works, we can first analyze how we can manually inject dependencies into different objects across an Android application. This can be achieved by creating container objects containing the dependencies required across the app.

You can also create multiple containers representing different scopes required across the application. Here, you can define dependencies that will only be required as long as a particular screen is displayed, and when the screen is destroyed, the instances can also be garbage collected.

A sample of a container that will hold instances as long as an application lives is shown here:

```kotlin
class AppContainer(applicationContext:Context) {
    val myRepository: MyRepository
    init {
        val retrofit =
            Retrofit.Builder().baseUrl(
            "https://google.com/").build()
        val myService=
            retrofit.create<MyService>(MyService::
            class.java)
        val database =
            Room.databaseBuilder(applicationContext,
            MyDatabase::class.java, "db").build()
        myRepository = MyRepositoryImpl(myService,
            database.myDao())
    }
}
```

An `Application` class using that container looks something like the following:

```kotlin
class MyApplication : Application() {
    lateinit var appContainer: AppContainer
    override fun onCreate() {
        super.onCreate()
        appContainer = AppContainer(this)
    }
}
```

As you can see in the preceding example, the responsibility for creating the dependencies shifted from the `Application` class to the `Container` class. Activities across the code base can still access the dependencies using the following command:

```kotlin
        override fun onCreate(savedInstanceState: Bundle?) {
            ....
            val myRepository = (application as
            MyApplication).appContainer. myRepository
            ...
        }
```

Modules with a limited scope could be used for something such as creating the `ViewModel` factories, which, in turn, are used by the framework to create `ViewModel`:

```kotlin
class MyContainer(private val myRepository: MyRepository) {
    fun geMyViewModelFactory(): ViewModelProvider.Factory {
        return object : ViewModelProvider.Factory {
            override fun <T : ViewModel?>
            create(modelClass: Class<T>): T {
                return MyViewModel(myRepository) as T
            }
        }
    }
}
```

An activity or fragment can use this particular container to initialize `ViewModel`:

```kotlin
class MyActivity : AppCompatActivity() {
    private lateinit var myViewModel: MyViewModel
    private lateinit var myContainer: MyContainer
    override fun onCreate(savedInstanceState: Bundle?) {
        super.onCreate(savedInstanceState)
        ....
        val myRepository = (application as
            MyApplication).appContainer.myRepository
        myContainer = MyContainer (myRepository)
        myViewModel = ViewModelProvider(this,
            myContainer.geMyViewModelFactory())
            .get(MyViewModel::class.java)
    }
}
```

Again, we can see here that the responsibility of creating the `Factory` class was shifted from the `Activity` class to the `Container` class. `MyContainer` could be expanded to provide instances required by `MyActivity` in situations where the lifecycle of those instances should be the same as the activity, or the constructor could be expanded to provide instances with a different lifecycle.

Now, let's apply some of these examples to an exercise.

Exercise 13.01 – manual injection

In this exercise, we will write an Android application that will apply the concept of manual DI. The application will have a Repository, which will generate a random number, and a `ViewModel` object with a `LiveData` object responsible for retrieving the number generated by the Repository and publishing it in the `LiveData` object.

In order to do so, we will need to create two containers that will manage the following dependencies:

- Repository
- A `ViewModel` factory responsible for creating `ViewModel`

The app itself will display the randomly generated number each time a button is clicked:

1. Create a new Android Studio Project with an empty activity.
2. Let's start by adding the `ViewModel` and `LiveData` libraries to the `app/build.gradle` file:

   ```
   implementation "androidx.lifecycle:
       lifecycle-viewmodel-ktx:2.5.1"
   implementation "androidx.lifecycle:
       lifecycle-livedata-ktx:2.5.1"
   ```

3. Next, let's write a `NumberRepository` interface in the `main/java` folder in the root package, which will contain a method to retrieve an integer:

   ```
   interface NumberRepository {
       fun generateNextNumber(): Int
   }
   ```

4. Now, we will provide the implementation for this in the `main/java` folder in the root package. We can use the `java.util.Random` class to generate a random number:

   ```
   class NumberRepositoryImpl(private val random: Random)
   : NumberRepository {
       override fun generateNextNumber(): Int {
           return random.nextInt()
       }
   }
   ```

5. We will now move on to the `MainViewModel` class in the `main/java` folder in the root package, which will contain a `LiveData` object containing each generated number from the repository:

```
class MainViewModel(private val numberRepository:
NumberRepository) : ViewModel() {
    private val _numberLiveData =
        MutableLiveData<Int>()
    val numberLiveData: LiveData<Int> =
        _numberLiveData
    fun generateNextNumber() {
        _numberLiveData.postValue(numberRepository
            .generateNextNumber())
    }
}
```

6. Next, let's move on to create our **user interface** (**UI**) containing `TextView` for displaying the number and `Button` for generating the next random number. This will be part of the `res/layout/activity_main.xml` file:

```
<TextView
    android:id="@+id/activity_main_text_view"
    android:layout_width="wrap_content"
    android:layout_height="wrap_content" />
<Button
    android:id="@+id/activity_main_button"
    android:layout_width="wrap_content"
    android:layout_height="wrap_content"
    android:text="@string/randomize" />
```

The complete code for this step can be found at https://packt.link/lr5Fx.

7. Make sure to add the string for the button to the `res/values/strings.xml` file:

```
<string name="randomize">Randomize</string>
```

8. Now, let's create our `Application` class in the `main/java` folder in the root package:

```
class RandomApplication : Application() {
    override fun onCreate() {
```

                    super.onCreate()
            }
        }

9. Let's also add the `Application` class to the `AndroidManifest.xml` file in the `application` tag:

    ```
    <application
        ...
        android:name=".RandomApplication"
    .../>
    ```

10. Now, let's create our first container responsible for managing the `NumberRepository` dependency in the `main/java` folder in the root package:

    ```
    class ApplicationContainer {
        val numberRepository: NumberRepository =
        NumberRepositoryImpl(Random())
    }
    ```

11. Next, let's add this container to the `RandomApplication` class:

    ```
    class RandomApplication : Application() {
        val applicationContainer = ApplicationContainer()
        override fun onCreate() {
            super.onCreate()
        }
    }
    ```

12. We now move on to creating `MainContainer` in the `main/java` folder in the root package, which will need a reference to the `NumberRepository` dependency and will provide a dependency to the `ViewModel` factory required to create `MainViewModel`:

    ```
    class MainContainer(private val numberRepository:
    NumberRepository) {
        fun getMainViewModelFactory():
        ViewModelProvider.Factory {
            return object : ViewModelProvider.Factory {
                override fun <T : ViewModel>
                create(modelClass: Class<T>): T {
    ```

```kotlin
            return MainViewModel(numberRepository)
                as T
        }
    }
}
```

13. Finally, we can modify `MainActivity` to inject our dependencies from our containers and connect the UI elements to display the output:

```kotlin
class MainActivity : AppCompatActivity() {
    override fun onCreate(savedInstanceState: Bundle?) {
        super.onCreate(savedInstanceState)
        setContentView(R.layout.activity_main)
        val mainContainer =
            MainContainer((application as
            RandomApplication).applicationContainer
            .numberRepository)
        val viewModel = ViewModelProvider(this,
            mainContainer.getMainViewModelFactory()
        ).get(MainViewModel::class.java)
        viewModel.numberLiveData.observe(this,
        Observer {
            findViewById<TextView>(
            R.id.activity_main_text_view).text =
            it.toString()
        }
        )
        findViewById<TextView>(
        R.id.activity_main_button).setOnClickListener
        {
            viewModel.generateNextNumber()
        }
    }
}
```

In the highlighted code, we can see that we are using the repository defined in `ApplicationContainer` and injecting it into `MainContainer`, which will then inject it into `ViewModel` through `ViewModelProvider.Factory`. The preceding example should render the output presented in *Figure 13.1*:

Figure 13.1 – Emulator output of exercise 13.01 displaying a randomly generated number

Manual DI is an easy way to set up your dependencies in situations where the app is small, but it can become extremely difficult as the app grows. Imagine if, in *Exercise 13.01, Manual injection*, we had two classes that extended from `NumberRepository`. How would we handle such a scenario? How would developers know which one went in what activity? These types of questions have become very common in most of the well-known apps on Google Play, which is why manual DI is rarely used. When used, it normally takes the form of a DI framework similar to the ones we will look over next.

Dagger 2

Dagger 2 offers a comprehensive way to organize your application's dependencies. It has the advantage of being adopted first on Android by the developer community before Kotlin was introduced. This is one of the reasons that many Android applications use Dagger as their DI framework.

Another advantage the framework holds is for Android projects written in Java because the library is developed in the same language. The framework was initially developed by Square (Dagger 1) and later transitioned to Google (Dagger 2). We will cover Dagger 2 in this chapter and describe its benefits.

Some of the key functionality that Dagger 2 provides is listed here:

- Injection
- Dependencies grouped in modules
- Components used to generate dependency graphs
- Qualifiers
- Scopes
- Subcomponents

Annotations are the key elements when dealing with Dagger because it generates the code required to perform the DI through an annotation processor. The main annotations can be grouped as follows:

- **Provider**: Classes that are annotated with `@Module` are responsible for providing an object (dependent object) that can be injected
- **Consumer**: The `@Inject` annotation is used to define a dependency
- **Connector**: An `@Component`-annotated interface defines the connection between the provider and the consumer

You will need to add the following dependencies in the `app/build.gradle` file to add Dagger to your project:

```
implementation 'com.google.dagger:dagger:2.44.2'
kapt 'com.google.dagger:dagger-compiler:2.44.2'
```

Since we are dealing with annotation processors, in the same `build.gradle` file, you will need to add the plugin for them:

```
apply plugin: 'com.android.application'
apply plugin: 'kotlin-android'
apply plugin: 'kotlin-kapt'
```

We should now have an idea of how Dagger 2 goes about performing DI. Next, we will look at each group of annotations Dagger 2 offers.

Consumers

Dagger uses `javax.inject.Inject` to identify objects that require injection. There are multiple ways to inject dependencies, but the recommended ways are through constructor injection and field injection. Constructor injection looks similar to the following code:

```
import javax.inject.Inject
class ClassA @Inject constructor()
```

```
class ClassB @Inject constructor(private val classA:
ClassA)
```

When constructors are annotated with @Inject, Dagger will generate the Factory classes responsible for instantiating the objects. In the example of ClassB, Dagger will try to find the appropriate dependencies that fit the signature of the constructor, which, in this example, is ClassA, which Dagger already created an instance for.

If you do not want Dagger to manage the instantiation of ClassB but still have the dependency on ClassA injected, you can use field injection, which will look something like this:

```
import javax.inject.Inject
class ClassA @Inject constructor()
class ClassB {
    @Inject
    lateinit var classA: ClassA
}
```

In this case, Dagger will generate the necessary code just to inject the dependency between ClassB and ClassA.

Providers

You will find yourself in situations where your application uses external dependencies. That means that you cannot provide instances through constructor injections. Another situation where constructor injection is not possible is when interfaces or abstract classes are used.

In this situation, Dagger can provide the instance using the @Provides annotation. You will then need to group the methods where instances are provided into modules annotated with @Module:

```
import dagger.Module
import dagger.Provides
class ClassA
class ClassB(private val classA: ClassA)
@Module
object MyModule {
    @Provides
    fun provideClassA(): ClassA = ClassA()
    @Provides
```

```
        fun provideClassB(classA: ClassA): ClassB =
            ClassB(classA)
}
```

As you can see in the preceding example, `ClassA` and `ClassB` don't have any Dagger annotations. A module was created that will provide the instance for `ClassA`, which will then be used to provide the instance for `ClassB`. In this case, Dagger will generate a `Factory` class for each of the `@Provides` annotated methods.

Connectors

Assuming we will have multiple modules, we must combine them in a graph of dependencies that can be used across the application. Dagger offers the `@Component` annotation. This is usually used for an interface or an abstract class that will be implemented by Dagger.

Along with assembling the dependency graph, components also offer the functionality to add methods to inject dependencies into a certain object's members. In components, you can specify provision methods that return dependencies provided in the modules:

```
import dagger.Component
@Component(modules = [MyModule::class])
interface MyComponent {
    fun inject(myApplication: MyApplication)
}
```

For the preceding `Component`, Dagger will generate a `DaggerMyComponent` class, and we can build it as described in the following code:

```
import android.app.Application
import javax.inject.Inject
class MyApplication : Application() {
    @Inject
    lateinit var classB: ClassB
    override fun onCreate() {
        super.onCreate()
        val component = DaggerMyComponent.create()
        //needs to build the project once to generate
        //DaggerMyComponent.class
        component.inject(this)
    }
}
```

The `Application` class will create the Dagger dependency graph and component. The `inject` method in `Component` allows us to perform DI on the variables in the `Application` class annotated with `@Inject`, giving us access to the `ClassB` object defined in the module.

Qualifiers

You can use qualifiers if you want to provide multiple instances of the same class (such as injecting different strings or integers across an application). These are annotations that can help you identify instances. One of the most common ones is the `@Named` qualifier, as described in the following code:

```
@Module
object MyModule {
    @Named("classA1")
    @Provides
    fun provideClassA1(): ClassA = ClassA()
    @Named("classA2")
    @Provides
    fun provideClassA2(): ClassA = ClassA()
    @Provides
    fun provideClassB(@Named("classA1") classA: ClassA):
    ClassB = ClassB(classA)
}
```

In this example, we create two instances of `ClassA` and give them different names. We then use the first instance whenever possible to create `ClassB`. We can also create custom qualifiers instead of the `@Named` annotation, as described in the following code:

```
import javax.inject.Qualifier
@Qualifier
@MustBeDocumented
@kotlin.annotation.Retention(AnnotationRetention.RUNTIME)
annotation class ClassA1Qualifier
@Qualifier
@MustBeDocumented
@kotlin.annotation.Retention(AnnotationRetention.RUNTIME)
annotation class ClassA2Qualifier
```

The module can be updated like this:

```
@Module
object MyModule {
    @ClassA1Qualifier
    @Provides
    fun provideClassA1(): ClassA = ClassA()
    @ClassA2Qualifier
    @Provides
    fun provideClassA2(): ClassA = ClassA()
    @Provides
    fun provideClassB(@ClassA1Qualifier classA: ClassA):
    ClassB = ClassB(classA)
}
```

Scopes

If you want to keep track of the lifecycle of your components and your dependencies, you can use scopes. Dagger offers a `@Singleton` scope. This usually indicates that your component will live as long as your application.

Scoping has no impact on the lifecycle of the objects; they are built to help developers identify the lifecycles of objects. Giving your components one scope and grouping your code to reflect that scope is recommended.

Some common Dagger scopes on Android are related to the activity or fragment:

```
import javax.inject.Scope
@Scope
@MustBeDocumented
@kotlin.annotation.Retention(AnnotationRetention.RUNTIME)
annotation class ActivityScope
@Scope
@MustBeDocumented
@kotlin.annotation.Retention(AnnotationRetention.RUNTIME)
annotation class FragmentScope
```

The annotation can be used in the module where the dependency is provided:

```
@ActivityScope
@Provides
fun provideClassA(): ClassA = ClassA()
```

The code for Component will be as follows:

```
@ActivityScope
@Component(modules = [MyModule::class])
interface MyComponent {
}
```

The preceding example indicates that Component can only use objects with the same scope. If any of the modules that are part of Component contain dependencies with different scopes, Dagger will throw an error indicating that there is something wrong with the scopes.

Subcomponents

Something that goes hand-in-hand with scopes is subcomponents. They allow you to organize your dependencies for smaller scopes. One common use case on Android is to create subcomponents for activities and fragments. Subcomponents inherit dependencies from the parent, and they generate a new dependency graph for the scope of the subcomponent.

Let's assume we have a separate module, as shown here:

```
class ClassC
@Module
object MySubcomponentModule {
    @Provides
    fun provideClassC(): ClassC = ClassC()
}
```

A Subcomponent that will generate a dependency graph for that module would look something like the following:

```
import dagger.Subcomponent
@ActivityScope
@Subcomponent(modules = [MySubcomponentModule::class])
interface MySubcomponent {
```

```
        fun inject(mainActivity: MainActivity)
}
```

The parent component would need to declare the new component, as shown in the following code snippet:

```
import dagger.Component
@Component(modules = [MyModule::class])
interface MyComponent {
    fun inject(myApplication: MyApplication)
    fun createSubcomponent(mySubcomponentModule:
    MySubcomponentModule): MySubcomponent
}
```

And you can inject `ClassC` into your activity as follows:

```
@Inject
    lateinit var classC: ClassC
    override fun onCreate(savedInstanceState: Bundle?) {
        super.onCreate(savedInstanceState)
        (application as MyApplication).component
        .createSubcomponent(MySubcomponentModule)
        .inject(this)
}
```

With this knowledge, let's move on to the exercise.

Exercise 13.02 – Dagger injection

In this exercise, we will write an Android application that will apply the concept of DI with Dagger. The application will have the same `Repository` and `ViewModel` defined in *Exercise 13.01, Manual injection*.

We will need to use Dagger to expose the same two dependencies:

- `Repository`: This will have the `@Singleton` scope and will be provided by `ApplicationModule`. Now, `ApplicationModule` will be exposed as part of `ApplicationComponent`.
- `ViewModelProvider.Factory`: This will have the custom-defined scope named `MainScope` and will be provided by `MainModule`. Now, `MainModule` will be exposed by `MainSubComponent`. Also, `MainSubComponent` will be generated by `ApplicationComponent`.

The app itself will display a randomly generated number each time a button is clicked. To achieve this, take the following steps:

1. Create a new Android Studio Project with Empty Activity.

2. Let's start by adding Dagger and the `ViewModel` libraries to the `app/build.gradle` file:

   ```
   implementation 'com.google.dagger:dagger:2.44.2'
   kapt 'com.google.dagger:dagger-compiler:2.44.2'
   implementation "androidx.lifecycle:
       lifecycle-viewmodel-ktx:2.5.1"
   implementation "androidx.lifecycle:
       lifecycle-livedata-ktx:2.5.1"
   ```

3. We also need the `kapt` plugin in the `app/build.gradle` module. Attach the plugin as shown here:

   ```
   apply plugin: 'kotlin-kapt'
   ```

4. We now need to add the `NumberRepository`, `NumberRepositoryImpl`, `Main ViewModel`, and `RandomApplication` classes and build our UI with `Main Activity`. This can be done by following *steps 2-9* from *Exercise 13.01, Manual injection*.

5. Now, let's move on to `ApplicationModule` in the `main/java` folder in the root package, which will provide the `NumberRepository` dependency:

   ```
   @Module
   class ApplicationModule {
       @Provides
       fun provideRandom(): Random = Random()
       @Provides
       fun provideNumberRepository(random: Random):
       NumberRepository =
       NumberRepositoryImpl(random)
   }
   ```

6. Now, let's create `MainModule` in the `main/java` folder in the root package, which will provide the instance of `ViewModel.Factory`:

   ```
   @Module
   class MainModule {
       @Provides
       fun provideMainViewModelFactory(numberRepository:
   ```

```
            NumberRepository): ViewModelProvider.Factory {
        return object : ViewModelProvider.Factory {
            override fun <T : ViewModel>
            create(modelClass: Class<T>): T {
                return MainViewModel(numberRepository)
                    as T
            }
        }
    }
}
```

7. Now, let's create `MainScope` in the `main/java` folder in the root package:

   ```
   @Scope
   @MustBeDocumented
   @kotlin.annotation.Retention(AnnotationRetention
   .RUNTIME)
   annotation class MainScope
   ```

8. We will need `MainSubcomponent` in the `main/java` folder in the root package, which will use the preceding scope:

   ```
   @MainScope
   @Subcomponent(modules = [MainModule::class])
   interface MainSubcomponent {
       fun inject(mainActivity: MainActivity)
   }
   ```

9. Next, we will require `ApplicationComponent` in the `main/java` folder in the root package:

   ```
   @Singleton
   @Component(modules = [ApplicationModule::class])
   interface ApplicationComponent {
       fun createMainSubcomponent(): MainSubcomponent
   }
   ```

10. Next, we modify the `RandomApplication` class to add the code required to initialize the Dagger dependency graph:

    ```
    class RandomApplication : Application() {
        lateinit var applicationComponent:
        ApplicationComponent
        override fun onCreate() {
            super.onCreate()
            applicationComponent =
            DaggerApplicationComponent.create()
        }
    }
    ```

11. We now modify the `MainActivity` class to inject `ViewModelProvider.Factory` and initialize `ViewModel` so that we can display the random number:

    ```
    class MainActivity : AppCompatActivity() {
        @Inject
        lateinit var factory: ViewModelProvider.Factory
        override fun onCreate(savedInstanceState: Bundle?)
        {
            (application as RandomApplication)
                .applicationComponent
                .createMainSubcomponent()
                .inject(this)
            super.onCreate(savedInstanceState)
        }
    }
    ```

    The complete code for this step can be found at https://packt.link/A7ozE.

12. We will need to navigate to `Build` and click on `Rebuild project` in Android Studio so that Dagger will generate the code for performing the DI.

    If you run the preceding code, it will build an application that will display a different random output when you click the button:

542 Dependency Injection with Dagger, Hilt, and Koin

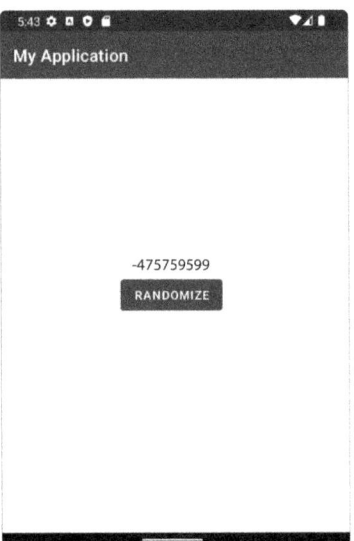

Figure 13.2 – Emulator output of exercise 13.02 displaying a randomly generated number

13. *Figure 13.3* shows what the application looks like. You can view the generated Dagger code in the `app/build` folder:

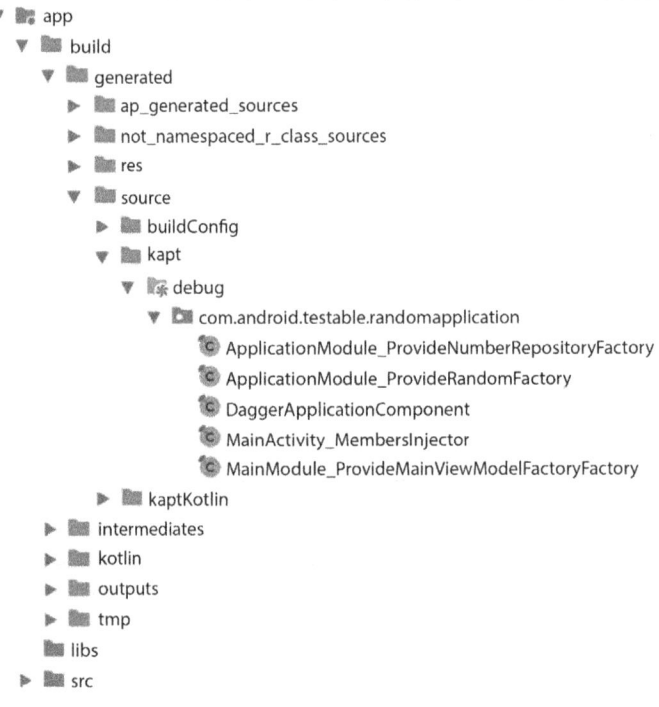

Figure 13.3 – Generated Dagger code for Exercise 13.02

In *Figure 13.3*, we can see the code that Dagger generated to satisfy the relationship between dependencies. For every dependency that needs to be injected, Dagger will generate an appropriate `Factory` class (based on the `Factory` design pattern), which will be responsible for creating the dependency.

Dagger also looks at the places where dependencies will need to be injected and generates an `Injector` class, which will have the responsibility of assigning the value to the dependency (in this case, it will assign the value to the members annotated with `@Inject` in the `MainActivity` class).

Finally, Dagger creates implementations for the interfaces that have the `@Component` annotation. In the implementation, Dagger will handle how the modules are created and also provide a builder in which developers can specify how modules can be built.

A common setup you will find for Android applications when it comes to organizing their dependencies is as follows:

- `ApplicationModule`: This is where dependencies common for the entire project are defined. Objects such as context, resources, and other Android framework objects can be provided here.
- `NetworkModule`: This is where dependencies related to API calls are stored.
- `StorageModule`: This is where dependencies related to persistence are stored. It can be split into `DatabaseModule`, `FilesModule`, `SharedPreferencesModule`, and so on.
- `ViewModelsModule`: This is where dependencies to `ViewModels` or the `ViewModel` factories are stored.
- `FeatureModule`: This is where dependencies are organized for a particular activity or fragment with their own `ViewModel`. Here, either subcomponents or Android injectors are used for this purpose.

We've raised some questions about how manual DI can go wrong. Now we have seen how Dagger can address these issues. Although it does the job, and it does it quickly when it comes to performance, it is also a complex framework with a very steep learning curve.

Hilt

When we use Dagger in an Android application, there is a bit of boilerplate code we are forced to write. Some of it is around dealing with the lifecycles of objects linked with Activities and Fragments, which leads us to create subcomponents; other parts are around the usage of ViewModels.

An attempt to simplify Dagger for Android was made with the Dagger-Android library, but later on, a new library was developed on top of Dagger called **Hilt**. This library simplifies much of the Dagger usage through the usage of new annotations, which leads to more boilerplate code that can be generated.

To use Hilt in a project, we will need the following:

```
apply plugin: 'com.android.application'
apply plugin: 'kotlin-android'
apply plugin: 'kotlin-kapt'
apply plugin: 'com.google.dagger.hilt.android'
```

Or depending on how your project uses Gradle, you might need to use the following:

```
plugins {
  ...
  id 'kotlin-kapt'
  id 'com.google.dagger.hilt.android'
}
```

In both cases, you need a plugin to process annotations and a separate plugin to process Hilt in your project.

To add Hilt to your project you need the following:

```
dependencies {
    implementation "com.google.dagger:hilt-android:2.44.2"
    kapt "com.google.dagger:hilt-compiler:2.44.2"
}
```

The first change Hilt makes is in the `Application` class. Instead of needing to invoke a particular Dagger component to be initialized, with Hilt, you can just use the `@HiltAndroidApp` annotation:

```
@HiltAndroidApp
class MyApplication : Application() {
}
```

The preceding snippet will let Hilt know the entry point into your application and it will start generating the dependency graph.

Another benefit of Hilt comes when interacting with Android components such as `Activities`, `Fragments`, `Views`, `Services`, and `BroadcastReceivers`. For these we can use the `@AndroidEntryPoint` annotation to inject dependencies into each of these classes, which looks like the following:

```
@AndroidEntryPoint
class MyActivity : AppCompatActivity() {
```

```
    @Inject
    lateinit var myObject: MyObject
}
```

In the above snippet, the usage of `@AndroidEntryPoint` allows Hilt to inject `myObject` into `MyActivity`. A similar approach can be used for injecting dependencies into `ViewModels`, through the `@HiltViewModel` annotation:

```
@HiltViewModel
class MyViewModel @Inject constructor(private val myObject:
MyObject) : ViewModel()
```

In the above snippet, the `@HiltViewModel` annotation allows Hilt to inject `myObject` into `MyViewModel`. We can also observe the `@Inject` annotation, carried over from Dagger, not requiring the usage of modules.

When it comes to modules, Hilt continues the approach from Dagger with one minor addition: the usage of the `@InstallIn` annotation. This associates the annotated module with a particular component. Hilt provides a set of prebuilt components such as `SingletonComponent`, `ViewModelComponent`, `ActivityComponent`, `FragmentComponent`, `ViewComponent`, and `ServiceComponent`.

Each of these components links the lifecycle of the dependencies inside the annotated module to the lifecycles of the application, `ViewModel`, `Activity`, `Fragment`, `View`, and `Service`:

```
@Module
@InstallIn(SingletonComponent::class)
class MyModule {
    @Provides
    fun provideMyObject(): MyObject = MyObject()
}
```

In the preceding snippet, we can see what a `@Module` looks like in Hilt and how we can use the `@InstallIn` annotation to specify that `MyObject` lives as long as our application lives.

When it comes to instrumented tests, Hilt provides useful annotations for changing the dependencies for the tests. If we want to take advantage of these features, then we need the following dependencies for tests:

```
    androidTestImplementation 'com.google.dagger:
        hilt-android-testing:2.44.2'
    kaptAndroidTest 'com.google.dagger:
        hilt-android-compiler:2.44.2'
```

We can then go to our test and introduce Hilt into it as follows:

```
@HiltAndroidTest
class MyInstrumentedTest {
    @get:Rule
    var hiltRule = HiltAndroidRule(this)

    @Inject
    lateinit var myObject: MyObject
    @Before
    fun init() {
        hiltRule.inject()
    }
}
```

In the preceding snippet, the `@HiltAndroid` test and `hiltRule` are used to swap the dependencies used in the application with the test dependencies. The call to inject is what allows us to inject the `MyObject` dependency into the test class. To provide the test dependencies, we can write a new module in the `androidTest` folder as follows:

```
@Module
@TestInstallIn(
    components = [SingletonComponent::class],
    replaces = [MyModule::class]
)
class MyTestModule {
    @Provides
    fun provideMyObject(): MyObject = MyTestObject()
}
```

Here, we are using the `@TestInstallIn` annotation, which will replace the existing `MyModule` from the dependency graph with `MyTestModule`, which can provide a different sub-class of the dependency we want to swap.

For Hilt to be initialized for the instrumented tests, we will need to define a custom test runner to provide a test application from the Hilt library. The runner might look like the following:

```
class HiltTestRunner : AndroidJUnitRunner() {

    override fun newApplication(cl: ClassLoader?, name:
```

```
        String?, context: Context?): Application {
        return super.newApplication(cl,
        HiltTestApplication::class.java.name, context)
    }
}
```

This runner will need to be registered in `build.gradle` of the module running the test:

```
android {
    ...
    defaultConfig {
        ...
        testInstrumentationRunner "{app_package_name}
            .HiltTestRunner"
    }
}
```

In this section, we studied the Hilt library and its benefits when it comes to removing boilerplate code that was required using Dagger.

Exercise 13.03 – Hilt injection

Modify *Exercise 13.02, Dagger injection*, such that the @Component and @Subcomponent classes are removed and Hilt is used instead:

1. Add the Hilt plugin in the top-level `build.gradle` file:

    ```
    plugins {
        ...
        id 'com.google.dagger.hilt.android' version
        '2.44.2' apply false
    }
    ```

2. Add the Hilt plugin in `app/build.gradle`:

    ```
    plugins {
        id 'com.android.application'
        id 'org.jetbrains.kotlin.android'
        id 'kotlin-kapt'
        id 'com.google.dagger.hilt.android'
    }
    ```

3. In the same file, replace the Dagger dependencies with Hilt dependencies and add the fragments extension library used for generating `ViewModel`:

   ```
   implementation "com.google.dagger:
       hilt-android:2.44.2"
   kapt "com.google.dagger:hilt-compiler:2.44.2"
   implementation 'androidx.fragment:
       fragment-ktx:1.5.5'
   ```

4. Delete `ApplicationComponent`, `MainModule`, `MainScope`, and `MainSubcomponent` from the project.

5. Add the `@InstallIn` annotation to `ApplicationModule`:

   ```
   @Module
   @InstallIn(SingletonComponent::class)
   class ApplicationModule {
   }
   ```

6. Remove all the code from inside `RandomApplication` and add the `@HiltAndroidApp` annotation:

   ```
   @HiltAndroidApp
   class RandomApplication : Application()
   ```

7. Modify `MainViewModel` to add the `@HiltViewModel` and `@Inject` annotations:

   ```
   @HiltViewModel
   class MainViewModel @Inject constructor(private val
   numberRepository: NumberRepository) :
       ViewModel() {
       …
   }
   ```

8. Modify `MainActivity` to instead inject `MainViewModel`, remove all the component dependencies that were deleted previously, and add the `@AndroidEntryPoint` annotation:

   ```
   @AndroidEntryPoint
   class MainActivity : AppCompatActivity() {
       private val mainViewModel: MainViewModel by
       viewModels()
   }
   ```

The complete code for this step can be found at https://packt.link/k7hs7.

In the preceding snippet, we use the `viewModels` method to obtain the `MainViewModel` dependency. This is a mechanism built into the extension functions from `androidx.fragment:fragment-ktx:1.5.5`, which will look for factories that will obtain the instance of our `ViewModel`.

If we run the code, we should see the following output:

Figure 13.4 – Output of exercise 13.03

We can see how much we can simplify an application's code using Hilt instead of Dagger. For example, we no longer have to deal with the `@Component` and `@Subcomponent` annotated classes and managing subcomponents in the application component, and also, we don't need to manually initialize the dependency graph from the `Application` class because Hilt handles this for us. These are some of the main reasons why Hilt became the most adopted library for dependency injection in Android applications.

Koin

Koin is a lighter framework that is suitable for smaller apps. It requires no code generation and is built based on Kotlin's functional extensions. It is also a **domain-specific language** (**DSL**). You may have noticed that when using Dagger, a lot of code must be written to set up the DI. Koin's approach to DI solves most of those issues, allowing faster integration.

Koin can be added to your project by adding the following dependency to your `build.gradle` file:

```
implementation "io.insert-koin:koin-core:3.2.2"
```

To set up Koin in your application, you need the `startKoin` call with the DSL syntax:

```
class MyApplication : Application() {
    override fun onCreate() {
        super.onCreate()
        startKoin {
            androidLogger(Level.INFO)
            androidContext(this@MyApplication)
            androidFileProperties()
            modules(myModules)
        }
    }
}
```

Here, you can configure what your application context is (in the `androidContext` method), specify property files to define Koin configurations (in `androidFileProperties`), state the Logger Level for Koin, which will output in `LogCat` results of Koin operations depending on the Level (in the `androidLogger` method), and list the modules your application uses. A similar syntax is used to create the modules:

```
class ClassA
class ClassB(private val classB: ClassA)
    val moduleForClassA = module {
        single { ClassA() }
    }
    val moduleForClassB = module {
        factory { ClassB(get()) }
    }
    override fun onCreate() {
```

```
        super.onCreate()
        startKoin {
            androidLogger(Level.INFO)
            androidContext(this@MyApplication)
            androidFileProperties()
            modules(listOf(moduleForClassA,
            moduleForClassB))
        }
    }
}
```

In the preceding example, the two objects will have two different lifecycles. When a dependency is provided using the **single** notation, only one instance will be used across the entire application lifecycle. This is useful for repositories, databases, and API components, where multiple instances will be costly for the application.

The **factory** notation will create a new object every time an injection is performed. This may be useful in situations when an object needs to live as long as an activity or fragment.

The dependency can be injected using the by inject() method or the get() method, as shown in the following code:

```
class MainActivity : AppCompatActivity() {
  val classB: ClassB by inject()
}
override fun onCreate(savedInstanceState: Bundle?) {
    super.onCreate(savedInstanceState)
    val classB: ClassB = get()
}
```

Koin also offers the possibility of using qualifiers with the help of the named() method when the module is created. This allows you to provide multiple implementations of the same type (for example, providing two or more list objects with different content):

```
val moduleForClassA = module {
    single(named("name")) { ClassA() }
}
```

One of Koin's main features for Android applications is scopes for activities and fragments, and can be defined as shown in the following code snippet:

```
val moduleForClassB = module {
    scope(named<MainActivity>()) {
        scoped { ClassB(get()) }
    }
}
```

The preceding example connects the lifecycle of the `ClassB` dependency to the lifecycle of `MainActivity`. In order for you to inject your instance into your activity, you will need to extend the `ScopeActivity` class. This class is responsible for holding a reference as long as the activity lives. Similar classes exist for other Android components such as Fragments (`ScopeFragment`) and services (`ScopeService`):

```
class MainActivity : ScopeActivity() {
    val classB: ClassB by inject()
}
```

You can inject the instance into your activity using the `inject()` method. This is useful when you wish to limit who gets to access the dependency. In the preceding example, if another activity had wanted to access the reference to `ClassB`, then it wouldn't be able to find it in the scope.

Another feature that comes in handy for Android is the `ViewModel` injections. To set this up, you will need to add the library to `build.gradle`:

```
implementation "io.insert-koin:koin-android:3.2.2"
```

If you recall, `ViewModels` require `ViewModelProvider.Factories` in order to be instantiated. Koin automatically solves this, allowing `ViewModels` to be injected directly and to handle the factory work:

```
val moduleForClassB = module {
    factory {
        ClassB(get())
    }
    viewModel { MyViewModel(get()) }
}
```

To inject the dependency of ViewModel into your activity, you can use the viewModel() method:

```
class MainActivity : AppCompatActivity() {
    val model: MyViewModel by viewModel()
}
```

Alternatively, you can use the method directly:

```
override fun onCreate(savedInstanceState: Bundle?) {
    super.onCreate(savedInstanceState)
    val model : MyViewModel = getViewModel()
}
```

As we can see in the preceding setup, Koin takes full advantage of Kotlin's language features and reduces the amount of boilerplate required to define your modules and their scopes.

Exercise 13.04 – Koin injection

Here, we will write an Android application that will perform DI using Koin. The application will be based on *Exercise 13.01, Manual injection*, by keeping NumberRepository, NumberRepositoryImpl, MainViewModel, and MainActivity. The following dependencies will be injected:

- Repository: As part of a module named appModule.
- MainViewModel: This will rely on Koin's specialized implementation for ViewModels. This will be provided as part of a module named mainModule and will have the MainActivity scope.

Perform the following steps to complete the exercise:

1. The app will display a randomly generated number each time a button is clicked. Let's start by adding the Koin libraries:

```
implementation "androidx.lifecycle:
    lifecycle-viewmodel-ktx:2.5.1"
implementation "androidx.lifecycle:
    lifecycle-livedata-ktx:2.5.1"
implementation "io.insert-koin:koin-android:3.2.2"
implementation "io.insert-koin:koin-core:3.2.2"
testImplementation 'junit:junit:4.13.2'
```

2. Next, define the `appModule` variable inside the `RandomApplication` class. This will have a similar structure to `AppModule` with the Dagger setup:

```
class RandomApplication : Application() {
    val appModule = module {
        single {
            Random()
        }
        single<NumberRepository> {
            NumberRepositoryImpl(get())
        }
    }
}
```

3. Now, let's add the activity module variable after `appModule`:

```
val mainModule = module {
    scope(named<MainActivity>()) {
        scoped {
            MainViewModel(get())
        }
    }
}
```

4. Next, let's initialize `Koin` in the `onCreate()` method of `RandomApplication`:

```
super.onCreate()
startKoin {
    androidLogger()
    androidContext(this@RandomApplication)
    modules(listOf(appModule, mainModule))
}
```

5. Finally, let's inject the dependencies into the activity:

```
class MainActivity : ScopeActivity() {
    private val mainViewModel: MainViewModel by inject()
}
```

The complete code for this step can be found at https://packt.link/0Njdv.

6. If you run the preceding code, the app should work as per the previous examples. However, if you check `LogCat`, you will see a similar output to this:

```
[Koin]: [init] declare Android Context
[Koin]: bind type:'android.content.Context' ~
[type:Single,primary_type:'android.content.Context']
[Koin]: bind type:'android.app.Application' ~
[type:Single,primary_type:'android.app.Application']
[Koin]: bind type:'java.util.Random' ~
[type:Single,primary_type:'java.util.Random']
[Koin]: bind type:'com.android.testable.randomapplication
.NumberRepository' ~ [type:Single,primary_type:'com.
android .testable.randomapplication.NumberRepository']
[Koin]: total 5 registered definitions
[Koin]: load modules in 0.4638 ms
```

In *Figure 13.5*, we can see the same output as in previous exercises:

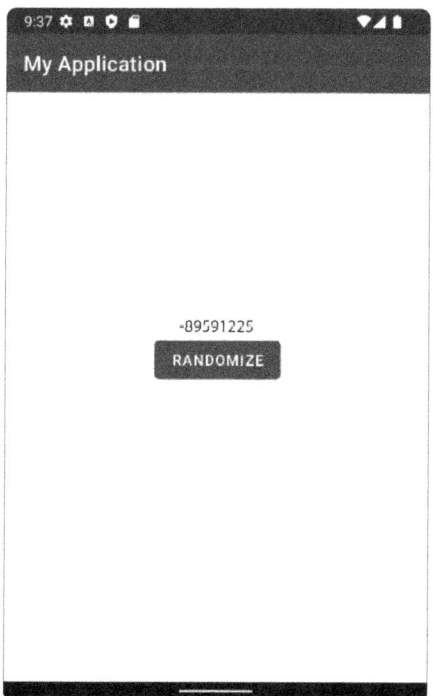

Figure 13.5 – Emulator output of exercise 13.04 displaying a randomly generated number

As we can see from this exercise, Koin is much faster and easier to integrate, especially with its `ViewModel` library. This comes in handy for small projects, but its performance will be impacted once projects grow.

Activity 13.01 – injected repositories

In this activity, you are going to create an app in Android Studio that connects to a sample API, `https://jsonplaceholder.typicode.com/posts`, using the Retrofit library and retrieves a list of posts from the web page, which will then be displayed on the screen.

You will then need to set up a UI test in which you will check whether the data is asserted correctly on the screen, but instead of connecting to the actual endpoint, you will provide dummy data for the test to display on the screen. You will use the DI concept to swap the dependencies using Hilt when the app is executed as opposed to when the app is being tested.

In order to achieve this, you will need to build the following:

- A network component that is responsible for downloading and parsing the JSON file
- A repository that accesses the data from the API layer
- A `ViewModel` instance that accesses the Repository
- An activity with `RecycleView` that displays the data
- One UI test that will assert the rows and use a dummy object to generate the API data

> **Note**
> Error handling can be avoided for this activity.

Perform the following steps to complete this activity:

1. In Android Studio, create an application with `Empty Activity` (`MainActivity`) and add an `api` package where your API calls are stored.
2. Define a class responsible for the API calls.
3. Create a `repository` package.
4. Define a `repository` interface with one method, returning `LiveData` with the list of posts.
5. Create the implementation for the `repository` class.
6. Create a `ViewModel` instance to call the `repository` to retrieve the data.
7. Create an adapter for the rows of the UI.
8. Create the activity that will render the UI.
9. Set up a Hilt module that will initialize the network-related dependencies.

10. Create a Hilt module that will be responsible for defining the dependencies required for the activity.
11. Set up the UI tests and a test application and provide a separate `RepositoryModule` class, which will return a dependency holding dummy data.
12. Implement the UI test.

> **Note**
> The solution to this activity can be found at `https://packt.link/3xfkt`.

Summary

In this chapter, we analyzed the concept of DI and how it should be applied to separate concerns and prevent objects from having the responsibility of creating other objects and how this is of great benefit for testing. We started the chapter by analyzing the concept of manual DI. This served as a good example of how DI works and how it can be applied to an Android application; it served as the baseline when comparing the DI frameworks.

We also analyzed two of the most popular frameworks that help developers inject dependencies. We started with a powerful and fast framework called Dagger 2, which relies on annotation processors to generate code to perform an injection. We then looked at how Hilt reduced the complexity of Dagger for Android applications. We also investigated Koin, a lightweight framework written in Kotlin with slower performance but simpler integration and a lot of focus on Android components.

The exercises in this chapter were intended to explore how the same problem can be solved using multiple solutions and compare the degrees of difficulty between the solutions. In the activities for this chapter, we leveraged Dagger's, Hilt's, and Koin's modules to inject certain dependencies when running the app and other dependencies when running the tests on an application that uses `ViewModels`, repositories, and APIs to load data.

This is designed to show the seamless integration of multiple frameworks that achieve different goals. In the chapter's activity, we looked at how we can use Hilt to swap dependencies for testing purposes and inject dummy data that we can then assert whether it is displayed on the screen.

In the following chapters, you will have the opportunity to build upon the knowledge acquired thus far by adding concepts related to threading and how to handle background operations. In addition, you will get the opportunity to explore libraries such as RxJava and its reactive approach to threading, and you will also learn about coroutines, which take a different approach to threading.

You will also observe how coroutines and RxJava can combine very effectively with libraries such as Room and Retrofit. Finally, you will be able to combine all of these concepts in a robust application that will have a high degree of scalability for the future.

Part 4: Polishing and Publishing an App

In this part, we will look at how we can load data asynchronously with coroutines and flows and how we can integrate them into different architecture patterns, which further helps with how we can structure an application's code.

Next, we will look at how we can render animations in the user interface with `CoordinatorLayout` and `MotionLayout`. Finally, we will learn about the process involved in publishing an application on Google Play.

We will cover the following chapters in this section:

- *Chapter 14, Coroutines and Flow*
- *Chapter 15, Architecture Patterns*
- *Chapter 16, Animations and Transitions with CoordinatorLayout and MotionLayout*
- *Chapter 17, Launching Your App on Google Play*

14
Coroutines and Flow

This chapter introduces you to background operations and data manipulations with Coroutines and Flow. You'll also learn how to manipulate and display the data using `LiveData` transformations and Kotlin Flow operators.

By the end of this chapter, you will be able to use Coroutines and Flow to manage network calls in the background. You will also be able to manipulate data with `LiveData` transformations and Flow operators.

You learned the basics of Android app development and implemented features such as RecyclerViews, notifications, fetching data from web services, and services. You also gained skills in the best practices for testing and persisting data. In the previous chapter, you learned about dependency injection. Now, you will learn about background operations and data manipulation.

Some Android applications work on their own. However, most apps would need a backend server to retrieve or process data. These operations may take a while, depending on the internet connection, device settings, and server specifications. If long-running operations are run in the main **user interface (UI)** thread, the application will be blocked until the tasks are completed. The application might become unresponsive and prompt the user to close and stop using it.

To avoid this, tasks that can take an indefinite amount of time must be run asynchronously. An asynchronous task means it can run in parallel to another task or in the background. For example, while fetching data from a data source asynchronously, your UI can still be displayed and user interaction can occur.

You can use libraries such as Coroutines and Flow for asynchronous operations. We'll discuss both in this chapter.

We will cover the following key topics in this chapter:

- Using Coroutines on Android
- Transforming `LiveData`
- Using Flow on Android

Technical requirements

The complete code for all the exercises and the activity in this chapter is available on GitHub at `https://packt.link/puLUO`

Let's get started with Coroutines.

Using Coroutines on Android

Coroutines were added in Kotlin 1.3 to manage background tasks such as making network calls and accessing files or databases. Kotlin coroutines are Google's official recommendation for asynchronous programming on Android. Their Jetpack libraries, such as LifeCycle, WorkManager, and Room, now include support for coroutines.

With coroutines, you can write your code in a sequential way. The long-running task can be made into a suspending function, which, when called, can pause the thread without blocking it. When the suspending function is done, the current thread will resume execution. This will make your code easier to read and debug.

To mark a function as a suspending function, you can add the `suspend` keyword to it; for example, if you have a function that calls the `getMovies` function, which fetches `movies` from your endpoint and then displays it:

```
val movies = getMovies()
displayMovies(movies)
```

You can make the `getMovies()` function a suspending function by adding the `suspend` keyword:

```
suspend fun getMovies(): List<Movies> { ... }
```

Here, the calling function will invoke `getMovies` and pause. After `getMovies` returns a list of movies, it will resume its task and display the movies.

Suspending functions can only be called in other suspending functions or from a coroutine. Coroutines have a context, which includes the coroutine dispatcher. Dispatchers specify what thread the coroutine will use. There are three dispatchers you can use:

- `Dispatchers.Main`: Used to run on Android's main thread
- `Dispatchers.IO`: Used for network, file, or database operations
- `Dispatchers.Default`: Used for CPU-intensive work

To change the context for your coroutine, you can use the `withContext` function for the code that you want to use a different thread with. For example, in your suspending function, `getMovies`, which gets movies from your endpoint, you can use `Dispatchers.IO`:

```
suspend fun getMovies(): List<Movies> {
    withContext(Dispatchers.IO) { ... }
}
```

In the next section, we will cover how to create coroutines.

Creating coroutines

You can create a coroutine with the `async` and `launch` keywords. The `launch` keyword creates a coroutine and doesn't return anything. On the other hand, the `async` keyword returns a value that you can get later with the `await` function.

The `async` and `launch` keywords must be created from `CoroutineScope`, which defines the lifecycle of the coroutine. For example, the coroutine scope for the main thread is `MainScope`. You can then create coroutines with the following:

```
MainScope().async { ... }
MainScope().launch { ... }
```

You can also create your own `CoroutineScope` instead of using `MainScope` by creating one with `CoroutineScope` and passing in the context for the coroutine. For example, to create `CoroutineScope` for use on a network call, you can define the following:

```
val scope = CoroutineScope(Dispatchers.IO)
```

The coroutine can be canceled when the function is no longer needed, such as when you close the activity. You can do that by calling the `cancel` function from `CoroutineScope`:

```
scope.cancel()
```

A ViewModel also has a default `CoroutineScope` for creating coroutines: `viewModelScope`. Jetpack's LifeCycle also has the `lifecycleScope` that you can use. `viewModelScope` is canceled when the ViewModel has been destroyed; `lifecycleScope` is also canceled when the lifecycle is destroyed. Thus, you no longer need to cancel them.

In the next section, you will learn how to add coroutines to your project.

Adding coroutines to your project

You can add coroutines to your project by adding the following code to your `app/build.gradle` file dependencies:

```
implementation 'org.jetbrains.kotlinx:
    kotlinx-coroutines-core:1.6.4'
implementation 'org.jetbrains.kotlinx:
    kotlinx-coroutines-android:1.6.4'
```

`kotlinx-coroutines-core` is the main library for coroutines, while `kotlinx-coroutines-android` adds support for the main Android thread.

You can add coroutines in Android when making a network call or fetching data from a local database.

If you're using Retrofit 2.6.0 or above, you can mark the endpoint function as a suspending function with suspend:

```
@GET("movie/latest")
suspend fun getMovies() : List<Movies>
```

Then, you can create a coroutine to call the suspending `getMovies` function and display the list:

```
CoroutineScope(Dispatchers.IO).launch {
    val movies = movieService.getMovies()
    withContext(Dispatchers.Main) {
        displayMovies(movies)
    }
}
```

You can also use LiveData for the response of your coroutines. `LiveData` is a Jetpack class that can hold observable data. You can add `LiveData` to your Android project by adding the following dependency:

```
implementation 'androidx.lifecycle:
lifecycle-livedata-ktx:2.5.1'
```

Let's try to use coroutines in an Android project.

Exercise 14.01 – using coroutines in an Android app

For this chapter, you will work with an application that displays popular movies using The Movie Database API. Go to `https://developers.themoviedb.org` and register for an API key. In this exercise, you will be using coroutines to fetch a list of popular movies:

1. Open the `Popular Movies` project in Android Studio in the `Chapter14` directory from this book's code repository.

2. Open the `AndroidManifest.xml` file and add the `INTERNET` permission inside the manifest tag but outside the application tag:

   ```
   <uses-permission android:name="android.permission
   .INTERNET" />
   ```

3. Open the `app/build.gradle` file and add the dependencies for the Kotlin Coroutines:

   ```
   implementation 'org.jetbrains.kotlinx:
       kotlinx-coroutines-core:1.6.4'
   implementation 'org.jetbrains.kotlinx:
       kotlinx-coroutines-android:1.6.4'
   ```

   These will allow you to use coroutines in your project.

4. Also, add the dependencies for the ViewModel and `LiveData` extension libraries:

   ```
   implementation 'androidx.lifecycle:
       lifecycle-livedata-ktx:2.5.1'
   implementation 'androidx.lifecycle:
       lifecycle-viewmodel-ktx:2.5.1'
   ```

5. Open the `MovieService` interface and replace it with the following code:

   ```
   interface MovieService {
       @GET("movie/popular")
       suspend fun getPopularMovies(@Query("api_key")
       apiKey: String): PopularMoviesResponse
   }
   ```

   This will mark `getPopularMovies` as a suspending function.

6. Open `MovieRepository` and add `apiKey` (with the value from the Movie Database API):

   ```
   private val apiKey = "your_api_key_here"
   ```

7. In the `MovieRepository` file, add the movies and error `LiveData` for the list of movies:

   ```
   private val movieLiveData =
       MutableLiveData<List<Movie>>()
   private val errorLiveData =
       MutableLiveData<String>()
   val movies: LiveData<List<Movie>>
       get() = movieLiveData
   val error: LiveData<String>
       get() = errorLiveData
   ```

8. Add the suspending `fetchMovies` function to retrieve the list from the endpoint:

   ```
   suspend fun fetchMovies() {
       try {
           val popularMovies =
               movieService.getPopularMovies(apiKey)
           movieLiveData.postValue(popularMovies
               .results)
       } catch (exception: Exception) {
           errorLiveData.postValue(
           "An error occurred: ${exception.message}")
       }
   }
   ```

9. Open `MovieApplication` and add a property for `movieRepository`:

   ```
   class MovieApplication: Application() {
       lateinit var movieRepository: MovieRepository
   }
   ```

10. Override the `onCreate` function of the `MovieApplication` class and initialize `movieRepository`:

    ```
    override fun onCreate() {
        super.onCreate()

        val retrofit = Retrofit.Builder()
            .baseUrl("https://api.themoviedb.org/3/")
            .addConverterFactory(
    ```

```
            MoshiConverterFactory.create())
        .build()
    val movieService = retrofit.create(
        MovieService::class.java)

    movieRepository = MovieRepository(movieService)
}
```

11. Update the contents of `MovieViewModel` with the following code:

    ```
    init {
        fetchPopularMovies()
    }
    val popularMovies: LiveData<List<Movie>>
    get() = movieRepository.movies
    val error: LiveData<String> =
        movieRepository.error
    private fun fetchPopularMovies() {
        viewModelScope.launch(Dispatchers.IO) {
            movieRepository.fetchMovies()
        }
    }
    ```

    The `fetchPopularMovies` function has a coroutine, using `viewModelScope`, that will fetch the movies from `movieRepository`.

12. Open the `MainActivity` class. At the end of the `onCreate` function, create movie Repository and `movieViewModel`:

    ```
    val movieRepository =
        (application as MovieApplication).movieRepository
    val movieViewModel =
        ViewModelProvider(
        this, object: ViewModelProvider.Factory {
        override fun <T : ViewModel> create(modelClass:
        Class<T>): T {
            return MovieViewModel(movieRepository) as T
        }
    }) [MovieViewModel::class.java]
    ```

13. After that, add an observer to popularMovies and error LiveData from movie ViewModel:

    ```
    movieViewModel.popularMovies.observe(this) {
    popularMovies ->
        movieAdapter.addMovies(popularMovies
            .filter {
                it.releaseDate.startsWith(
                    Calendar.getInstance()
                    .get(Calendar.YEAR)
                    .toString()
                )
            }
            .sortedByDescending { it.popularity }
        )
    }
    movieViewModel.error.observe(this) { error ->
        if (error.isNotEmpty()) Snackbar.make(
        recyclerView, error, Snackbar
        .LENGTH_LONG).show()
    }
    ```

    This will update the activity's RecyclerView with the movies fetched. The list of movies is filtered using Kotlin's `filter` function to only include movies released this year. They are then sorted by popularity using Kotlin's `sortedByDescending` function.

14. Run the application. You will see that the app will display a list of popular movie titles from the current year, sorted by popularity:

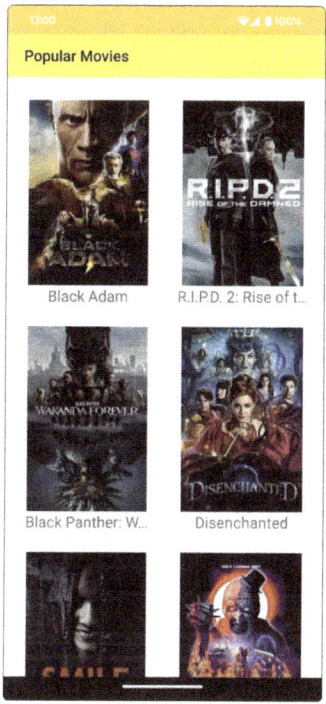

Figure 14.1 – The app displaying popular movies released this year, sorted by popularity

15. Click on a movie, and you will see its details, such as its release date and an overview:

Figure 14.2 – The movie details screen

You have used coroutines and `LiveData` to retrieve and display a list of popular movies from a remote data source without blocking the main thread.

Before passing `LiveData` into the UI for display, you can also transform the data first. You will learn about that in the next section.

Transforming LiveData

Sometimes, the `LiveData` you pass from the ViewModel to the UI layer needs to be processed first before displaying. For example, you can only select a part of the data or do some processing on it first. In the previous exercise, you filtered the data to only select popular movies from the current year.

To modify `LiveData`, you can use the `Transformations` class. It has two functions, `Transformations.map` and `Transformations.switchMap`, that you can use.

`Transformations.map` modifies the value of `LiveData` into another value. This can be used for tasks such as filtering, sorting, or formatting the data. For example, you can transform `movieLiveData` into string `LiveData` from the movie's title:

```
private val movieLiveData: LiveData<Movie>
val movieTitleLiveData : LiveData<String> =
    Transformations.map(movieLiveData) { it.title }
```

When `movieLiveData` changes value, `movieTitleLiveData` will also change based on the movie's title.

With `Transformations.switchMap`, you can transform the value of a `LiveData` into another `LiveData`. This is used when you want to do a specific task involving a database or network operation with the original `LiveData`. For example, if you have a `LiveData` representing a movie `id` object, you can transform that into movie `LiveData` by applying the `getMovieDetails` function, which returns `LiveData` of movie details from the `id` object (such as from another network or database call):

```
private val idLiveData: LiveData<Int> = MutableLiveData()
val movieLiveData : LiveData<Movie> =
    Transformations.switchMap(idLiveData) {
        getMovieDetails(it) }
fun getMovieDetails(id: Int) : LiveData<Movie> = { ... }
```

Let's use `LiveData` transformations on the list of movies fetched using coroutines.

Exercise 14.02 – LiveData transformations

In this exercise, you will transform the `LiveData` list of movies before passing them to the observers in the `MainActivity` file:

1. Open the `Popular Movies` project you worked on in the previous exercise in Android Studio.
2. Open the `MainActivity` file. In the `movieViewModel.popularMovies` observer in the `onCreate` function, remove the filter and the `sortedByDescending` function calls. The code should look like the following:

   ```
   movieViewModel.getPopularMovies().observe(this,
   Observer { popularMovies ->
       movieAdapter.addMovies(popularMovies)
   })
   ```

   This will now display all movies in the list without them being sorted by popularity.

3. Run the application. You should see all movies (even those from the past year), not sorted by popularity:

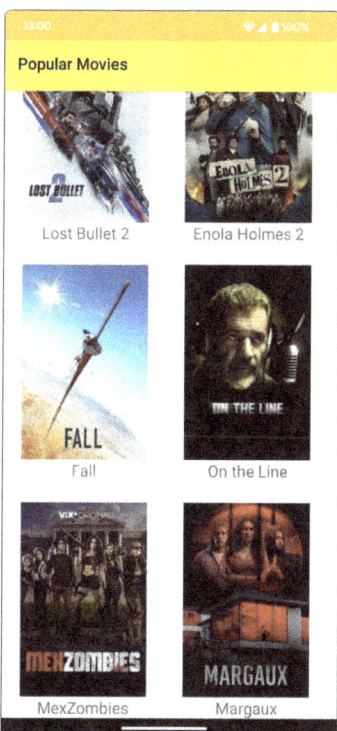

Figure 14.3 – The app with unsorted popular movies

4. Open the `MovieViewModel` class and update `popularMovies` with `LiveData` transformations to filter and sort the movies:

   ```
   val popularMovies: LiveData<List<Movie>>
   get() = movieRepository.movies.map { list ->
   list.filter {
       val cal = Calendar.getInstance()
       it.releaseDate.startsWith(
           "${cal.get(Calendar.YEAR)}"
       )
   }.sortedByDescending { it.popularity }
   }
   ```

   This will select the movies released this year and sort them by title before passing them to the UI observer in `MainActivity`.

5. Run the application. You will see that the app shows a list of popular movies from the current year, sorted by popularity:

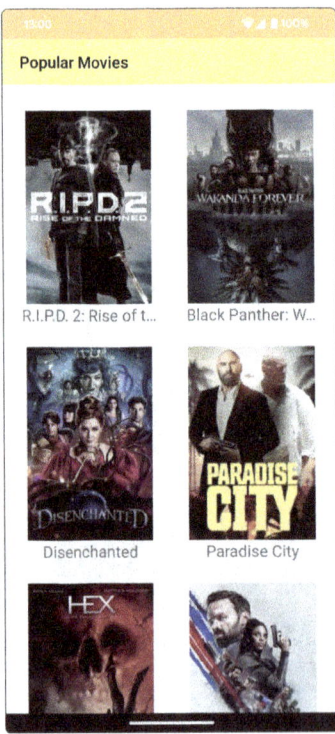

Figure 14.4 – The app with the movies released last month sorted by popularity

You have used `LiveData` transformations to modify the list of movies to select only the ones released in the previous month. They were also sorted by popularity before passing them to the observers in the UI layer.

In the next section, you will learn about Kotlin Flows.

Using Flow on Android

In this section, you will look into using Flows for asynchronous programming in Android. Flow, an asynchronous stream library built on top of Kotlin Coroutines, is ideal for live data updates in your application. Android Jetpack libraries include Room, WorkManager, and Jetpack Compose, and third-party libraries support Flow.

A Flow of data is represented by the `kotlinx.coroutines.flow.Flow` interface. Flows emit multiple values of the same type one at a time. For example, `Flow<String>` is a Flow that emits string values.

A flow starts to emit values when you call the suspending `collect` function from a coroutine or another suspending function. In the following example, the `collect` function was called from the coroutine created using the `launch` builder of `lifecycleScope`:

```
class MainActivity : AppCompatActivity() {
    ...
    override fun onCreate(savedInstanceState: Bundle?) {
        ...
        lifecycleScope.launch {
            viewModel.fetchMovies().collect { movie ->
                Log.d("movies", "${movie.title}")
            }
        }
    }
}
class MovieViewModel : ViewModel() {
    ...
    fun fetchMovies(): Flow<Movie> { ... }
}
```

Here, the `collect{}` function was called on `viewModel.fetchMovies()`. This will start the Flow's emission of movies; each movie title is then logged.

To change the CoroutineContext where the Flow runs, you can use the flowOn() function to change the Dispatcher. The previous example can be updated with a different Dispatcher, as shown in the following code:

```
override fun onCreate(savedInstanceState: Bundle?) {
    ...
    lifecycleScope.launch {
        viewModel.fetchMovies()
            .flowOn(Dispatchers.IO)
            .collect { movie ->
                Log.d("movies", "${movie.title}")
            }
    }
}
```

In this example, the Dispatcher for the Flow will be changed to Dispatchers.IO. Calling flowOn will only change the functions before it, not the functions and operators after.

In the next section, you will learn about collecting Flows on Android.

Collecting Flows on Android

In Android, Flows are usually collected in the Activity or Fragment for display in the UI. Moving the app in the background will not stop the data collection. The app must not do so and continue updating the screen to avoid memory leaks and prevent wasting resources.

You can safely collect flows in the UI layer by manually handling lifecycle changes or by using Lifecycle.repeatOnLifecycle and Flow.flowWithLifecycle, available in the lifecycle-runtime-ktx library, starting with version 2.4.0.

To use it in your project, add the following to your app/build.gradle dependencies:

```
implementation 'androidx.lifecycle:
    lifecycle-runtime-ktx:2.4.1'
```

This adds the lifecycle-runtime-ktx library to your project, so you can use both Lifecycle.repeatOnLifecycle and Flow.flowWithLifecycle.

Lifecycle.repeatOnLifecycle(state, block) will suspend the parent coroutine until the lifecycle is destroyed and executes the suspending block code when the Lifecycle is at least in the state provided. The flow will stop when the Lifecycle moves out of the state and restart when the lifecycle moves back to the state. Lifecycle.repeatOnLifecycle must be called on Activity's onCreate or on Fragment's onViewCreated.

When using `Lifecycle.State.STARTED` for state, the `repeatOnLifecycle` will start Flow collection when the Lifecycle is started and stop when the Lifecycle is stopped (`onStop()` is called).

If you use `Lifecycle.State.RESUMED`, the start will be when the Lifecycle is resumed, and the stop will be when `onPause` is called or when the Lifecycle is paused.

The following example shows how you can use `Lifecycle.repeatOnLifecycle`:

```
class MainActivity : AppCompatActivity() {
    ...

    override fun onCreate(savedInstanceState: Bundle?) {
        ...
        lifecycleScope.launch {
            repeatOnLifecycle(Lifecycle.State.STARTED) {
                viewModel.fetchMovies()
                    .collect { movie ->
                        Log.d("movies", "${movie.title}")
                    }
            }
        }
    }
}
```

In this class, `repeatOnLifecycle` with `Lifecycle.State.STARTED` starts collecting the Flow of movies when the lifecycle is started and stops when the lifecycle is stopped.

`Flow.flowWithLifecycle` is another way to safely collect Flows in Android. It emits values from the Flow and operators preceding the call (the upstream Flow) when the lifecycle is at least in the state you set or the default, `Lifecycle.State.STARTED`. Internally, it uses `Lifecycle.repeatOnLifecycle`. The following example shows how you can use `Flow.flowWithLifecycle`:

```
class MainActivity : AppCompatActivity() {
    ...

    override fun onCreate(savedInstanceState: Bundle?) {
        ...
        lifecycleScope.launch {
            viewModel.fetchMovies()
                .flowWithLifecycle(lifecycle,
```

```
                    Lifecycle.State.STARTED)
                    .collect { movie ->
                        Log.d("movies", "${movie.title}")
                    }
            }
        }
    }
```

Here, we used `flowWithLifecycle` with `Lifecycle.State.STARTED` to collect the Flow of movies when the lifecycle is started and stop when the lifecycle is stopped.

In the following section, you will learn how to create Flows.

Creating Flows with Flow Builders

You can create Flows using the Flow Builders from the Kotlin Flow API. The following are the Flow Builders you can use:

- `flow{}`: This creates a new Flow from a suspendable lambda block. You can send values using the `emit` function.
- `flowOf()`: This creates a Flow from the specified value or the `vararg` values.
- `asFlow()`: This is an extension function used to convert a type (sequence, array, range, or collection) into a Flow.

The following example shows how to use the Flow Builders in an application:

```
class MovieViewModel : ViewModel() {
    ...

    fun fetchMovies: Flow<List<Movie>> = flow {
        fetchMovieList().forEach { movie -> emit(movie) }
    }

    fun fetchTop3Titles: Flow<List<String>> {
        val movies = fetchTopMovies()
        return flowOf(movies[0].title,
            movies[1].title, movies[2].title)
    }
```

```
    fun fetchMovieIds: Flow<Int> {
        return fetchMovies().map { it.id }.asFlow()
    }
}
```

In this example, `fetchMovies` created a Flow using `flow{}` and emitted each movie from the list. The `fetchTop3Titles` function uses `flowOf` to create a Flow with the titles of the first three movies. Finally, `fetchMovieIds` converted the list of IDs into a Flow of movie IDs using the `asFlow` function.

In the next section, you will learn about the Kotlin Flow operators you can use with Flows.

Using operators with Flows

There are built-in Flow operators you can use with Flows. You can collect Flows with terminal operators and transform Flows with intermediate operators.

Terminal operators, such as the `collect` function used in the previous examples, are used to collect Flows. The following are the other terminal operators you can use:

- `count`
- `first` and `firstOrNull`
- `last` and `lastOrNull`
- `fold`
- `reduce`
- `single` and `singleOrNull`
- `toCollection`, `toList`, and `toSet`

These operators work similarly to the Kotlin `Collection` function with the same name.

You can use Intermediate operators to modify a Flow and return a new one. They can also be chained. The following Intermediate operators work the same as the Kotlin collection functions with the same name:

- `filter`, `filterNot`, `filterNotNull`, and `filterIsInstance`
- `map` and `mapNotNull`
- `onEach`
- `runningReduce` and `runningFold`
- `withIndex`

Additionally, there is a `transform` operator you can use to apply your own operation. For example, this class has a Flow that uses the `transform` operator:

```
class MovieViewModel : ViewModel() {
    ...
    fun fetchTopRatedMovie(): Flow<Movie> {
        return fetchMoviesFlow()
            .transform {
                if(it.voteAverage > 0.6f) emit(it)
            }
    }
}
```

Here, the `transform` operator was used in the Flow of movies to only emit the ones whose `voteAverage` is higher than `0.6` (60%).

There are also size-limiting Kotlin Flow operators such as `drop`, `dropWhile`, `take`, and `takeWhile`, which function similarly to the Kotlin collection functions of the same name.

Let's add Kotlin Flow into an Android project.

Exercise 14.03 – using Flow in an Android application

In this exercise, you will update the Popular Movies app to use Kotlin Flow in fetching the list of movies:

1. Open the Popular Movies project from the previous exercise in Android Studio.
2. Go to the `MovieRepository` class and remove the `movies` and `error` LiveData. Then, replace the `fetchMovies` function with the following:

```
fun fetchMovies(): Flow<List<Movie>> {
    return flow {
        emit(movieService
        .getPopularMovies(apiKey).results)
    }.flowOn(Dispatchers.IO)
}
```

This changes the `fetchMovies` function to use Kotlin Flow. The Flow will emit the list of movies from `movieService.getPopularMovies`, and it will flow on the `Dispatchers.IO` dispatcher.

3. Open the `MovieViewModel` class. In the class declaration, add a dispatcher parameter with a default value of `Dispatchers.IO`:

   ```
   class MovieViewModel(
       private val movieRepository: MovieRepository,
       private val dispatcher: CoroutineDispatcher =
           Dispatchers.IO
   ) : ViewModel() {
       ...
   }
   ```

   This will be the dispatcher that will be used later for the Flow.

4. Replace the `popularMovies` LiveData with the following:

   ```
   private val _popularMovies = MutableStateFlow(
       emptyList<Movie>())
   val popularMovies: StateFlow<List<Movie>> =
       _popularMovies
   ```

   You will use these for the value of the list of movies from `MovieRepository`. `StateFlow` is an observable Flow that emits state updates to the collectors, while `MutableStateFlow` is a `StateFlow` that you can change the value. In Android, `StateFlow` can be an alternative to LiveData.

5. Remove the `error` LiveData and replace it with the following:

   ```
   private val _error = MutableStateFlow("")
   val error: StateFlow<List<String>> =_error
   ```

   You will use these for handling when the Flow encounters an exception.

6. Change the content of the `fetchPopularMovies` function with the following:

   ```
   private fun fetchPopularMovies() {
       viewModelScope.launch(dispatcher) {
           movieRepository.fetchMovies().catch {
               _error.value =
                   "An exception occurred:
                   ${it.message}"
           }.collect {
               _popularMovies.value = it
           }
   ```

            }
        }

   This will collect the list of movies from `movieRepository` and set it to `MutableStateFlow` in `_popularMovies` (and `StateFlow` in `popularMovies`).

7. Open the `app/build.gradle` file and add the following in the dependencies:

   ```
   implementation 'androidx.lifecycle:
   lifecycle-runtime-ktx:2.5.1'
   ```

   This allows you to use `lifecycleScope` for collecting the Flows in `MainActivity`.

8. Go to the `MainActivity` file and remove the lines of code for observing `popularMovies` and `error` from `MovieViewModel`. Add the following to collect the Flow from `MovieViewModel`:

   ```
   lifecycleScope.launch {
       repeatOnLifecycle(Lifecycle.State.STARTED) {
           launch {
               movieViewModel.popularMovies.collect {
                   movies ->movieAdapter.addMovies(
                   movies)
               }
           }
           launch {
               movieViewModel.error.collect { error ->
                   if (error.isNotEmpty()) Snackbar
                   .make(recyclerView, error, Snackbar
                   .LENGTH_LONG).show()
               }
           }
       }
   }
   ```

9. Run the application. The app will display the list of movies, as shown in the following screenshot:

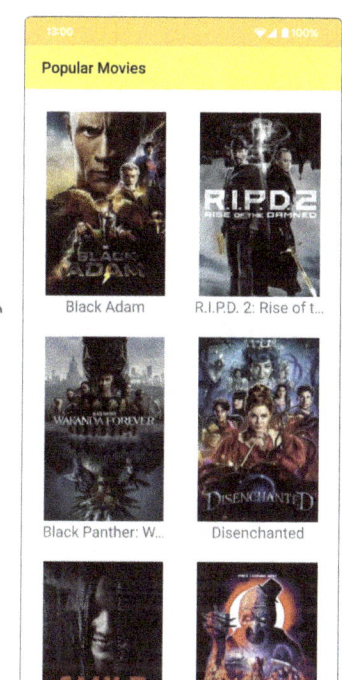

Figure 14.5 – The app displaying popular movies

In this exercise, you added Kotlin Flow to an Android project. `MovieRepository` returns the list of movies as a Flow, which was collected in `MovieViewModel`. `MovieViewModel` uses `StateFlow`, which was then collected in `MainActivity` for displaying in RecyclerView.

Let's move on to the next activity.

Activity 14.01 – creating a TV Guide app

A lot of people watch television. Most of the time, though, they are not sure what TV shows are currently airing. Suppose you wanted to develop an app that can display a list of these shows from the Movie Database API's `tv/on_the_air` endpoint using Kotlin Flow.

The app will have two screens: the main screen and the details screen. On the main screen, you will display a list of the TV shows that are on the air. The TV shows will be sorted by name. Clicking on a TV show will open the details screen, which displays more information about the selected TV show.

The following steps are for the completion of the activity:

1. Create a new project in Android Studio and name it TV Guide. Set its package name.
2. Add the INTERNET permission in the AndroidManifest.xml file.
3. Add the dependencies for Retrofit, Coroutines, Moshi, Lifecycle, and other libraries in your app/build.gradle file.
4. Add a layout_margin dimension value.
5. Create a view_tv_show_item.xml layout file with ImageView for the poster and TextView for the name of the TV show.
6. In the activity_main.xml file, remove the Hello World TextView and add a RecyclerView to the list of TV shows.
7. Create a TVShow model class.
8. Create another class named TVResponse for the response you get from the API endpoint for the TV shows on air.
9. Create a new activity named DetailsActivity with activity_details.xml as the layout file.
10. Open the AndroidManifest.xml file and add the parentActivityName attribute in the DetailsActivity declaration.
11. In activity_details.xml, add the views for the details of the TV show.
12. Open DetailsActivity and add the code to display the details of the TV show selected.
13. Create a TVShowAdapter adapter class for the list of TV shows.
14. Create a TelevisionService class for adding the Retrofit method.
15. Create a TVShowRepository class with a constructor for tvService, and properties for apiKey and tvShows.
16. Create a function to retrieve the list of TV shows from the endpoint.
17. Create a TVShowViewModel class with a constructor for TVShowRepository. Add the tvShows and error StateFlow and a fetchTVShows function that collects the Flow from the repository.
18. Create an application class named TVApplication with a property for TVShowRepository.
19. Set TVApplication as the value for the application in the AndroidManifest.xml file.

20. Open `MainActivity` and add the code to update the RecyclerView when the Flow from `ViewModel` updates its value. Add a function that will open the details screen when clicking on a TV show from the list.

21. Run your application. The app will display a list of TV shows. Clicking on a TV show will open the details activity, which displays the show details. The main screen and details screen will be similar to the following screenshot:

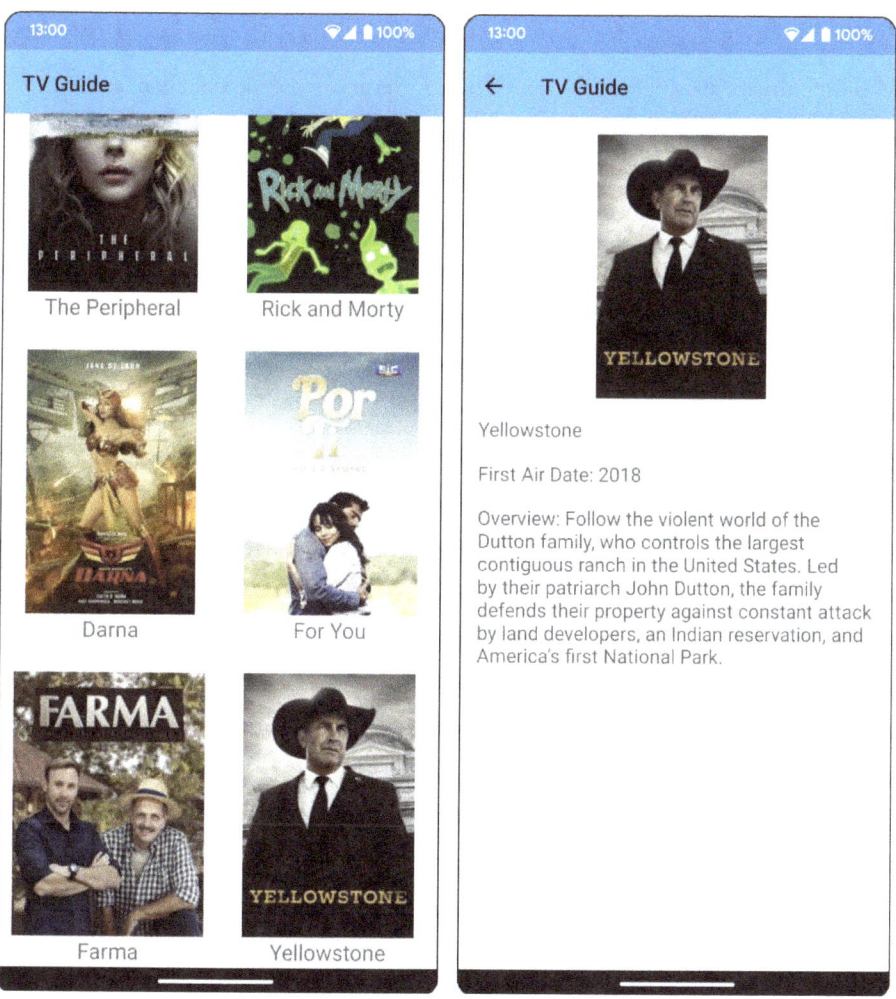

Figure 14.6 – The main screen and details screen of the TV Guide app

> **Note**
> The solution to this activity can be found at `https://packt.link/By7eE`.

Summary

This chapter focused on doing background operations with Coroutines and Flow. Background operations are used for long-running tasks such as accessing data from the local database or a remote server.

You started with the basics of using Kotlin coroutines, Google's recommended solution for asynchronous programming. You learned that you can make a background task into a suspending function with the `suspend` keyword. Coroutines can be started with the `async` or `launch` keywords.

You learned how to create suspending functions and how to start coroutines. You also used dispatchers to change the thread where a coroutine runs. Then, you used coroutines for performing network calls and modified the data retrieved with the `map` and `switchMap` `LiveData` transformation functions.

You then moved on to using Kotlin Flow in an Android app to load the data in the background. To safely collect flows in the UI layer, prevent memory leaks, and avoid wasting resources, you can use `Lifecycle.repeatOnLifecycle` and `Flow.flowWithLifecycle`.

You learned about using Flow Builders to create Flows. The `flow` builder function creates a new Flow from a suspending lambda block and then you can send values with `emit()`. The `flowOf` function creates a Flow that emits the value or the `vararg` values. You can use the `asFlow()` extension function to convert collections and functional types into a Flow.

Finally, you explored Flow operators and learned how to use them with Kotlin Flows. Terminal operators are used to start the collection of the Flow. With Intermediate operators, you can transform a Flow into another Flow.

In the next chapter, you will learn about architecture patterns. You will learn about patterns such as **Model-View-ViewModel** (**MVVM**) and how you can improve the architecture of your app.

15
Architecture Patterns

This chapter will introduce you to architectural patterns you can use for your Android projects. It covers using the **Model-View-ViewModel** (**MVVM**) pattern, adding ViewModels, and using data binding. You will also learn about using the Repository pattern for caching data and WorkManager for scheduling data retrieval and storage.

By the end of the chapter, you will be able to structure your Android project using MVVM and data binding. You will also be able to use the Repository pattern with the Room library to cache data and WorkManager to fetch and save data at a scheduled interval.

In the previous chapter, you learned about using Coroutines and Flow for background operations and data manipulation. Now, you will learn about architectural patterns so you can improve your application.

When developing an Android application, you may tend to write most of the code (including business logic) in activities or fragments. This will make your project hard to test and maintain later. As your project grows and becomes more complex, the difficulty also increases. You can improve your projects with architectural patterns.

Architectural patterns are general solutions for designing and developing parts of applications, especially for large apps. There are architectural patterns you can use to structure your project into different layers (the presentation layer, the **user interface** (**UI**) layer, and the data layer) or functions (observer/observable). With architectural patterns, you can organize your code in a way that makes it easier for you to develop, test, and maintain.

For Android development, commonly used patterns include **Model-View-Controller** (**MVC**), **Model-View-Presenter** (**MVP**), and MVVM. The recommended architectural pattern is MVVM, which will be discussed in this chapter. You will also learn about data binding, the Repository pattern using the Room library, and WorkManager.

We will cover the following topics in the chapter:

- Getting started with MVVM
- Binding data on Android with data binding
- Using Retrofit and Moshi

- Implementing the Repository pattern
- Using WorkManager

Technical requirements

The complete code for all the exercises and the activity in this chapter is available on GitHub at https://packt.link/PZNNT

Getting started with MVVM

MVVM allows you to separate the UI and business logic. When you need to redesign the UI or update the Model/business logic, you only need to touch the relevant component without affecting the other components of your app. This will make it easier for you to add new features and test your existing code. MVVM is also useful in creating huge applications that use a lot of data and views.

With the MVVM architectural pattern, your application will be grouped into three components:

- **Model**: This represents the data layer
- **View**: This is the UI that displays the data
- **ViewModel**: This fetches data from Model and provides it to View

The MVVM architectural pattern can be understood better through the following diagram:

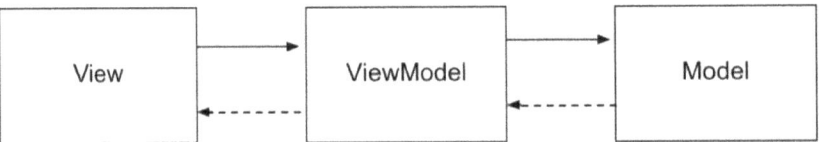

Figure 15.1 – The MVVM architectural pattern

The Model contains the data of the application. The activities, fragments, and layouts that your users see and interact with are the Views in MVVM. Views only deal with how the app looks. They let ViewModel know about user actions (such as opening an activity or clicking on a button).

ViewModel links View and Model. ViewModels also perform the business logic processing and transform them for display in the View. Views subscribe to the ViewModel and update the UI when a value changes.

You can use Jetpack's ViewModel to create the ViewModel classes for your app. Jetpack's ViewModel manages its own lifecycle so you don't need to handle it yourself.

You can add ViewModel to your project by adding the following code in your `app/build.gradle` file dependencies:

```
implementation 'androidx.lifecycle:
lifecycle-viewmodel-ktx:2.5.1'
```

For example, if you're working on an app that displays movies, you could have `MovieViewModel`. This ViewModel will have a function that fetches a list of movies:

```
class MovieViewModel : ViewModel() {
    private val _movies: MutableStateFlow<List<Movie>>
    fun movies: StateFlow<List<Movie>> { ... }
    ...
}
```

In your activity, you can create ViewModel using `ViewModelProvider`:

```
class MainActivity : AppCompatActivity() {
    private val movieViewModel by lazy {
        ViewModelProvider(this).get(MovieViewModel::
        class.java)
    }
    ...
}
```

Then, you can connect to the `movies` Flow from `ViewModel` and automatically update the list on the UI when the list of movies changes:

```
override fun onCreate(savedInstanceState: Bundle?) {
    ...

    lifecycleScope.launch {
        repeatOnLifecycle(Lifecycle.State.STARTED) {
            launch {
                movieViewModel.popularMovies.collect {
                movies ->
                    movieAdapter.addMovies(movies)
                }
            }
        }
    }
```

```
        }
        ...
}
```

Views are notified when values in `ViewModel` have changed. You can also use data binding to connect `View` with the data from `ViewModel`. You will learn more about data binding in the next section.

Binding data on Android with data binding

View binding and data binding are two ways to bind data to Android Views. View binding is a simpler and faster binding, which you can use to replace `findViewById` in your code. Data binding is more powerful and can be customized to connect your data with layout variables and expressions.

With data binding, you can link the views in your layout to data from a source such as a ViewModel. Instead of adding code to find the views in the layout file and updating them when the value from the ViewModel changes, data binding can handle that for you automatically.

To use data binding in your Android project, you should add the following in the `android` block of the `app/build.gradle` file:

```
buildFeatures {
    dataBinding true
}
```

In the `layout` file, you must wrap the root element with a `layout` tag. Inside the `layout` tag, you need to define the `data` element for the data to be bound to this `layout` file:

```
<layout xmlns:android=
    "http://schemas.android.com/apk/res/android">
    <data>
        <variable name="movie" type=
        "com.example.model.Movie"/>
    </data>
    <ConstraintLayout ... />
</layout>
```

The `movie` layout variable represents the `com.example.model.Movie` class that will be displayed in the layout. To set the attribute to fields in the data model, you need to use the `@{}` syntax. For example, to use the movie's title as the text value of `TextView`, you can use the following:

```
<TextView
```

```
    ...
    android:text="@{movie.title}"/>
```

You also need to change your activity file. If your layout file is named `activity_movies.xml`, the data binding library will generate a binding class named `ActivityMoviesBinding` in your project's build files. In the activity, you can replace the `setContentView(R.layout.activity_movies)` line with the following:

```
val binding: ActivityMoviesBinding = DataBindingUtil
    .setContentView(this, R.layout.activity_movies)
```

You can also use the `inflate` method of the `binding` class or the `DataBindingUtil` class:

```
val binding: ActivityMoviesBinding = ActivityMoviesBinding
    .inflate(getLayoutInflater())
```

Then, you can set the `movie` instance to bind in the layout with the `layout` variable named `movie`:

```
val movieToDisplay = ...
binding.movie = movieToDisplay
```

If you are using `LiveData` or `Flow` as the item to bind to the layout, you need to set `lifeCycleOwner` for the `binding` variable. `lifeCycleOwner` specifies the scope of the object. You can use the activity as `lifeCycleOwner` of the `binding` class:

```
binding.lifeCycleOwner = this
```

With this, when the values in `ViewModel` change their value, `View` will automatically update with the new values.

You set the movie title in `TextView` with `android:text="@{movie.title}"`. The data binding library has default binding adapters that handle the binding to the `android:text` attribute. Sometimes, there are no default attributes that you can use. You can create your own binding adapter. For example, if you want to bind the list of movies for `RecyclerView`, you can create a custom `BindingAdapter` call:

```
@BindingAdapter("list")
fun bindMovies(view: RecyclerView, movies: List<Movie>?) {
    val adapter = view.adapter as MovieAdapter
    adapter.addMovies(movies ?: emptyList())
}
```

This will allow you to add an `app:list` attribute to `RecyclerView` that accepts a list of movies:

```
app:list="@{movies}"
```

Let's try implementing data binding on an Android project.

Exercise 15.01– using data binding in an Android project

In the previous chapter, you worked on an application that displays popular movies using the Movie Database API. For this chapter, you will be improving the app using MVVM. You can use the `Popular Movies` project from the previous chapter or make a copy of it. In this exercise, you will add data binding to bind the list of movies from `ViewModel` to the UI:

1. Open the `Popular Movies` project in Android Studio.
2. Open the `app/build.gradle` file and add the following in the `android` block:

    ```
    buildFeatures {
        dataBinding true
    }
    ```

    This enables data binding for your application.

3. Add the `kotlin-kapt` plugin at the end of the plugins block in your `app/build.gradle` file:

    ```
    plugins {
        ...
        id 'kotlin-kapt'
    }
    ```

    The `kotlin-kapt` plugin is the Kotlin annotation processing tool, which is needed for using data binding.

4. Create a new file called `RecyclerViewBinding` that contains the binding adapter for the `RecyclerView` list:

    ```
    @BindingAdapter("list")
    fun bindMovies(view: RecyclerView, movies:
    List<Movie>?) {
        val adapter = view.adapter as MovieAdapter
        adapter.addMovies(movies ?: emptyList())
    }
    ```

This will allow you to add an `app:list` attribute for `RecyclerView` where you can pass the list of movies to be displayed. The list of movies will be set to the adapter, updating `RecyclerView` in the UI.

5. Open the `activity_main.xml` file and wrap everything inside a `layout` tag:

   ```
   <layout xmlns:android=
       "http://schemas.android.com/apk/res/android"
       xmlns:app="http://schemas.android.com/apk/res-auto"
       xmlns:tools="http://schemas.android.com/tools">
       <androidx.constraintlayout.widget.ConstraintLayout
           ... >
       </androidx.constraintlayout.widget
           .ConstraintLayout>
   </layout>
   ```

   With this, the data binding library will be able to generate a binding class for this layout.

6. Inside the `layout` tag and before the `ConstraintLayout` tag, add a data element with a variable for `viewModel`:

   ```
   <data>
       <variable
           name="viewModel"
           type="com.example.popularmovies
           .MovieViewModel" />
   </data>
   ```

   This creates a `viewModel` layout variable that corresponds to your `MovieViewModel` class.

7. In `RecyclerView`, add the list to be displayed with `app:list`:

   ```
   app:list="@{viewModel.popularMovies}"
   ```

   `popularMovies` from `MovieViewModel.getPopularMovies` will be passed as the list of movies for `RecyclerView`.

8. Open `MainActivity`. In the `onCreate` function, replace the `setContentView` line with the following:

   ```
   val binding: ActivityMainBinding = DataBindingUtil
       .setContentView(this, R.layout.activity_main)
   ```

This sets the layout file to be used and creates a binding object.

9. Remove the collection of popularMoviesView from movieViewModel.

10. Add the following after the initialization of movieViewModel:

    ```
    binding.viewModel = movieViewModel
    binding.lifecycleOwner = this
    ```

    This binds movieViewModel to the viewModel layout variable in the activity_main.xml file.

11. Run the application. It should work as usual, displaying the list of popular movies, where clicking on one will open the details of the movie selected:

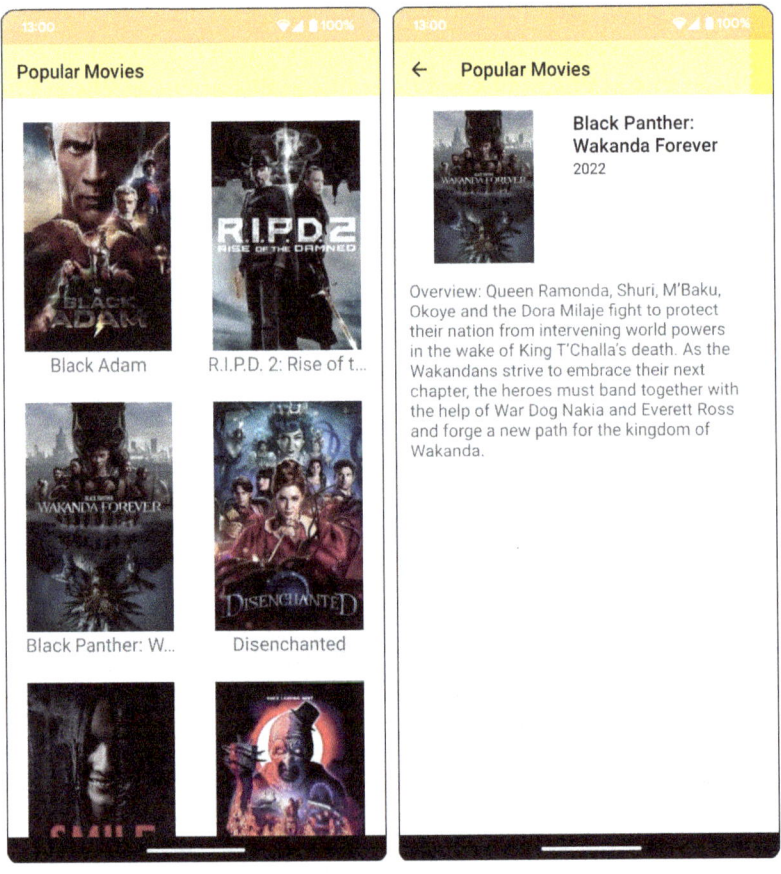

Figure 15.2 – The main screen (left) with the year's popular movies and the details screen (right) with more information about the selected movie

In this exercise, you have used data binding on an Android project.

Data binding links the Views to the ViewModel. The ViewModel retrieves the data from the Model. Some of the libraries you can use to fetch data are Retrofit and Moshi, which you will learn more about in the next section.

Using Retrofit and Moshi

When connecting to your remote network, you can use Retrofit. Retrofit is an HTTP client that makes it easy to implement creating requests and retrieving responses from your backend server.

You can add Retrofit to your project by adding the following code to your app/build.gradle file dependencies:

```
implementation 'com.squareup.retrofit2:retrofit:2.9.0'
```

You can then convert the JSON response from Retrofit by using Moshi, a library for parsing JSON into Java objects. For example, you can convert the JSON string response from getting the list of movies into a ListofMovie object for display and storage in your app.

You can add the Moshi Converter to your project by adding the following code to your app/build.gradle file dependencies:

```
implementation 'com.squareup.retrofit2:converter-moshi:2.9.0'
```

In your Retrofit builder code, you can call addConverterFactory and pass Moshi ConverterFactory:

```
Retrofit.Builder()
    ...
    .addConverterFactory(MoshiConverterFactory.create())
    ...
```

You can call the data layer from the ViewModel. To reduce its complexity, you can use the Repository pattern for loading and caching data. You will learn about this in the next section.

Implementing the Repository pattern

Instead of ViewModel directly calling the services for getting and storing data, it should delegate that task to another component, such as a repository.

With the Repository pattern, you can move the code in the `ViewModel` that handles the data layer into a separate class. This reduces the complexity of `ViewModel`, making it easier to maintain and test. The repository will manage where the data is fetched and stored, just as if the local database or the network service were used to get or store data:

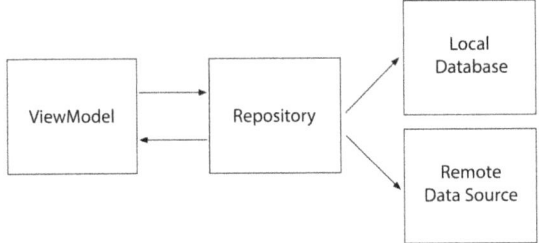

Figure 15.3 – ViewModel with the Repository pattern

In `ViewModel`, you can add a property for the repository:

```
class MovieViewModel(val repository: MovieRepository):
ViewModel() { ...}
```

`ViewModel` will get the movies from the repository, or it can listen to them. It will not know where you actually got the list from.

You can create a repository interface that connects to a data source, such as in the following example:

```
interface MovieRepository {
    fun getMovies(): List<Movie>
}
```

The `MovieRepository` interface has a `getMovies` function that your repository implementation class will override to fetch movies from the data source. You can also have a single repository class that handles the fetching of data from either the local database or from your remote endpoint.

When using the local database as the data source for your repository, you can use the Room library, which makes it easier for you to work with the SQLite database by writing less code and having compile-time checks on queries.

You can add Room to your project by adding the following code to your `app/build.gradle` file dependencies:

```
implementation 'androidx.room:room-runtime:2.4.3'
implementation 'androidx.room:room-ktx:2.4.3'
kapt 'androidx.room:room-compiler:2.4.3'
```

Let's try adding the Repository pattern with Room to an Android project.

Exercise 15.02 – using Repository with Room in an Android project

You have added data binding in the `Popular Movies` project in the previous exercise. In this exercise, you will update the app with the Repository pattern.

When opening the app, it fetches the list of movies from the network. This takes a while. You will cache this data into the local database every time you fetch them. When the user opens the app next time, the app will immediately display the list of movies from the database on the screen. You will be using Room for data caching:

1. Open the `Popular Movies` project that you used in the previous exercise.

2. Open the `app/build.gradle` file and add the dependencies for the Room library:

   ```
   implementation 'androidx.room:room-runtime:2.4.3'
   implementation 'androidx.room:room-ktx:2.4.3'
   kapt 'androidx.room:room-compiler:2.4.3'
   ```

3. Open the `Movie` class and add an `Entity` annotation for it:

   ```
   @Entity(tableName = "movies", primaryKeys = [("id")])
   data class Movie( ... )
   ```

   The `Entity` annotation will create a table named `movies` for the list of movies. It also sets `id` as the primary key of the table.

4. Make a new package called `com.example.popularmovies.database`. Create a `MovieDao` data access object for accessing the `movies` table:

   ```
   @Dao
   interface MovieDao {
   @Insert(onConflict = OnConflictStrategy.REPLACE)
   fun addMovies(movies: List<Movie>)
   @Query("SELECT * FROM movies")
   fun getMovies(): List<Movie>
   }
   ```

   This class contains a function for adding a list of movies in the database and another for getting all the movies from the database.

5. Create a `MovieDatabase` class in the `com.example.popularmovies.database` package:

   ```
   @Database(entities = [Movie::class], version = 1)
   abstract class MovieDatabase : RoomDatabase() {
       abstract fun movieDao(): MovieDao
   ```

```kotlin
        companion object {
            @Volatile
            private var instance: MovieDatabase? = null
            fun getInstance(context: Context):
            MovieDatabase {
                return instance ?: synchronized(this) {
                    instance ?: buildDatabase(
                    context).also { instance = it }
                }
            }
            private fun buildDatabase(context: Context):
            MovieDatabase {
                return Room.databaseBuilder(context,
                MovieDatabase::class.java, "movie-db")
                .build()
            }
        }
    }
```

This database has a version of 1, a single entity for `Movie`, and the data access object for the movies. It also has a `getInstance` function to generate an instance of the database.

6. Update the `MovieRepository` class with constructors for `movieDatabase`:

   ```kotlin
   class MovieRepository(private val movieService:
   MovieService, private val movieDatabase:
   MovieDatabase) { ... }
   ```

7. Update the `fetchMovies` function:

   ```kotlin
   fun fetchMovies(): Flow<List<Movie>> {
       return flow {
           val movieDao: MovieDao =
               movieDatabase.movieDao()
           val savedMovies = movieDao.getMovies()
           if(savedMovies.isEmpty()) {
               val movies = movieService
                   .getPopularMovies(apiKey).results
               movieDao.addMovies(movies)
   ```

```
            emit(movies)
        } else {
            emit(savedMovies)
        }
    }.flowOn(Dispatchers.IO)
}
```

It will fetch the movies from the database. If there's nothing saved yet, it will retrieve the list from the network endpoint and then save it.

8. Open `MovieApplication` and in the `onCreate` function, replace the `movieRepository` initialization with the following:

```
val movieDatabase =
    MovieDatabase.getInstance(applicationContext)
movieRepository =
    MovieRepository(movieService, movieDatabase)
```

9. Run the application. It will display the list of popular movies, and clicking on one will open the details of the movie selected. If you turn off mobile data or disconnect from the wireless network, it will still display the list of movies, which is now cached in the database:

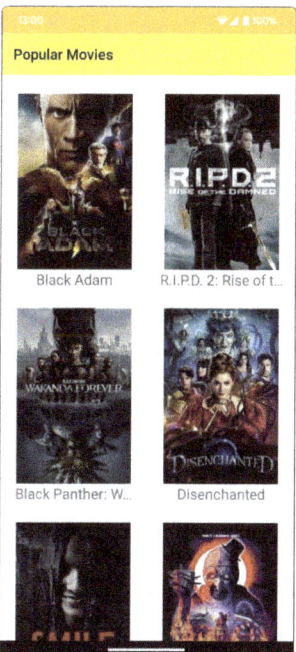

Figure 15.4 – The Popular Movies app using Repository with Room

In this exercise, you have improved the app by moving the loading and storing of data into a repository. You have also used Room to cache the data.

The repository fetches the data from the data source. If there's no data stored in the database yet, the app will call the network to request the data. This can take a while. You can improve the user experience by pre-fetching data at a scheduled time so the next time the user opens the app, they will already see the updated contents. You can do this with WorkManager, which we will discuss in the next section.

Using WorkManager

WorkManager is a Jetpack library for background operations that can be delayed and can run based on the constraints you set. It is ideal for doing something that must be run but can be done later or at regular intervals, regardless of whether the app is running or not.

You can use WorkManager to run tasks such as fetching the data from the network and storing it in your database at scheduled intervals. WorkManager will run the task even if the app has been closed or if the device restarts. This will keep your database up to date with your backend.

You can add WorkManager to your project by adding the following code to your `app/build.gradle` file dependencies:

```
implementation 'androidx.work:work-runtime:2.7.1'
```

WorkManager can call the repository to fetch and store data from either the local database or the network server.

Let's try adding WorkManager to an Android project.

Exercise 15.03 – adding WorkManager to an Android Project

In the previous exercise, you added the Repository pattern with Room to cache data in the local database. The app can now fetch the data from the database instead of the network. Now, you will be adding WorkManager to schedule a task for fetching data from the server and saving it to the database at scheduled intervals:

1. Open the Popular Movies project you used in the previous exercise.
2. Open the `app/build.gradle` file and add the dependency for the WorkManager library:

   ```
   implementation 'androidx.work:work-runtime:2.7.1'
   ```

   This will allow you to add the WorkManager workers to your app.

3. Open `MovieRepository` and add a suspending function for fetching movies from the network using `apiKey` from `movieDatabase` and saving them to the database:

   ```
   suspend fun fetchMoviesFromNetwork() {
       val movieDao: MovieDao = movieDatabase.movieDao()
       try {
           val popularMovies = movieService
               .getPopularMovies(apiKey)
           val moviesFetched = popularMovies.results
           movieDao.addMovies(moviesFetched)
       } catch (exception: Exception) {
           Log.d("MovieRepository", "An error occurred:
               ${exception.message}")
       }
   }
   ```

   This will be the function that will be called by the `Worker` class that will be running to fetch and save the movies.

4. Create the `MovieWorker` class in the `com.example.popularmovies` package:

   ```
   class MovieWorker(private val context: Context,
   params: WorkerParameters) : Worker(context, params) {
       override fun doWork(): Result {
           val movieRepository = (context as
           MovieApplication).movieRepository
           CoroutineScope(Dispatchers.IO).launch {
               movieRepository.fetchMoviesFromNetwork()
           }
           return Result.success()
       }
   }
   ```

5. Open `MovieApplication` and at the end of the `onCreate` function, schedule `MovieWorker` to retrieve and save the movies:

   ```
   override fun onCreate() {
       ...
       val constraints =
   ```

```
            Constraints.Builder().setRequiredNetworkType(
            NetworkType.CONNECTED).build()
        val workRequest = PeriodicWorkRequest
            .Builder(MovieWorker::class.java, 1,
            TimeUnit.HOURS).setConstraints(constraints)
            .addTag("movie-work").build()
        WorkManager.getInstance(
            applicationContext).enqueue(workRequest)
    }
```

This schedules `MovieWorker` to run every hour when the device is connected to the network. `MovieWorker` will fetch the list of movies from the network and save it to the local database.

6. Run the application. Close it and make sure the device is connected to the internet. After more than an hour, open the application again and check whether the list of movies displayed has been updated. If not, try again in a few hours. The list of movies displayed will be updated regularly, around every hour, even if the app has been closed.

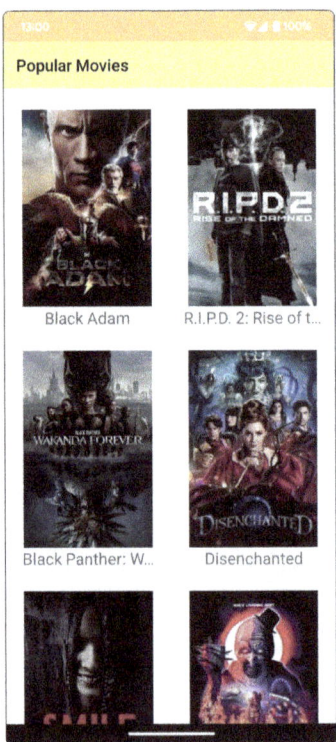

Figure 15.5 – The Popular Movies app updates its list with WorkManager

In this exercise, you added WorkManager to your application to automatically update the database with the list of the movies retrieved from the network.

Activity 15.01 – revisiting the TV Guide app

In the previous chapter, you developed an app that can display a list of TV shows that are on the air. The app had two screens: the main screen and the details screen. On the main screen, there's a list of TV shows. When clicking on a TV show, the details screen will be displayed with the details of the selected show.

When running the app, it takes a while to display the list of shows. Update the app to cache the list so it will be immediately displayed when opening the app. Also, improve the app by using MVVM with data binding and adding WorkManager.

You can use the TV Guide app you worked on in the previous chapter or download it from the GitHub repository (https://packt.link/Eti8M). The following steps will help guide you through this activity:

1. Open the TV Guide app in Android Studio. Open the app/build.gradle file and add the kotlin-kapt plugin, the data binding dependency, and the dependencies for Room and WorkManager.
2. Create a binding adapter class for RecyclerView.
3. In activity_main.xml, wrap everything inside a layout tag.
4. Inside the layout tag and before the ConstraintLayout tag, add a data element with a variable for ViewModel.
5. In RecyclerView, add the list to be displayed with app:list.
6. In MainActivity, replace the line for setContentView with the DataBindingUtil.setContentView function.
7. Replace the observer from TVShowViewModel with the data binding code.
8. Add an Entity annotation in the TVShow class.
9. Create a TVDao data access object for accessing the TV shows table.
10. Create a TVDatabase class.
11. Update TVShowRepository with a constructor for tvDatabase.
12. Update the fetchTVShows function to get the TV shows from the local database. If there's nothing there yet, retrieve the list from the endpoint and save it in the database.
13. Add a suspending fetchTVShowsFromNetwork function to get the TV shows from the network and save them to the database.
14. Create the TVShowWorker class.

15. Open the `TVApplication` file. In `onCreate`, schedule `TVShowWorker` to retrieve and save the shows.
16. Run your application. The app will display a list of TV shows. Clicking on a TV show will open the details activity, which displays the movie details. The main screen and details screen will be similar to the following:

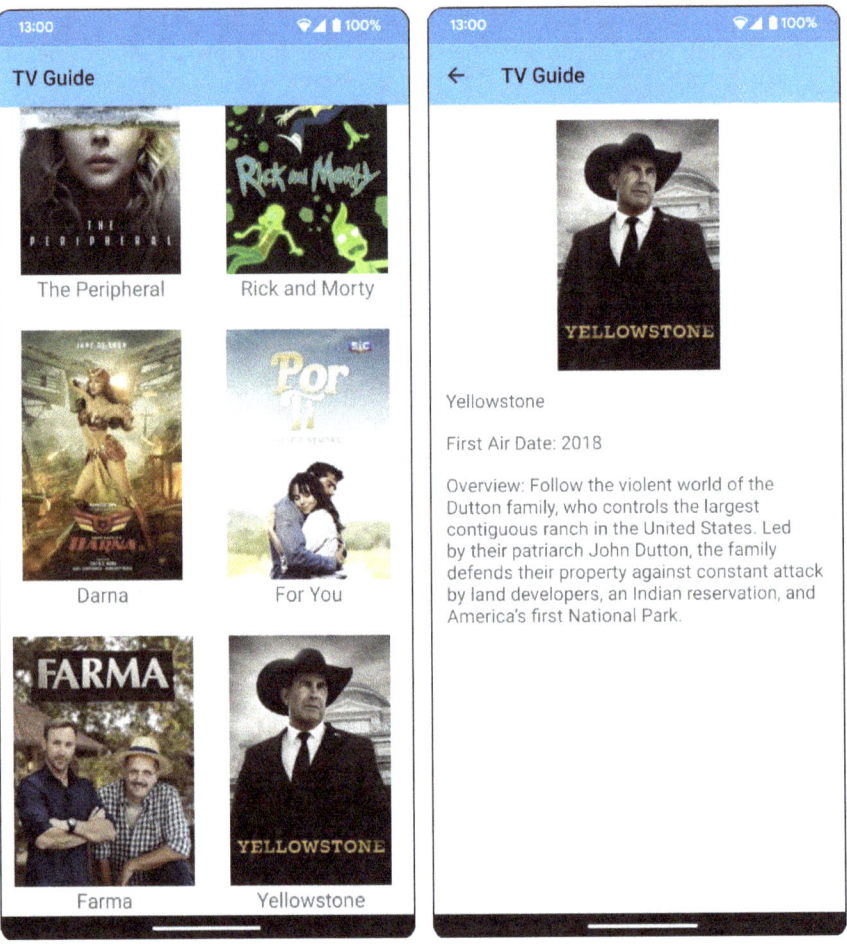

Figure 15.6 – The main screen and details screen of the TV Guide app

> **Note**
> The solution to this activity can be found at `https://packt.link/By7eE`.

Summary

This chapter focused on architectural patterns for Android. You started with the MVVM architectural pattern. You learned about its three components: the Model, the View, and the ViewModel. You also used data binding to link the View with the ViewModel.

Next, you learned about how the Repository pattern can be used to cache data. Then, you learned about WorkManager and how you can schedule tasks such as retrieving data from the network and saving that data to the database to update your local data.

In the next chapter, you will learn how to improve the look and design of your apps with animations. You will add animations and transitions to your apps with `CoordinatorLayout` and `MotionLayout`.

16
Animations and Transitions with CoordinatorLayout and MotionLayout

This chapter will introduce you to animations and how to handle changing between layouts. It offers a description of moving objects using `MotionLayout` and the Motion Editor in Android, along with a detailed explanation of constraint sets. The chapter also covers modifying paths and adding keyframes to a frame's motion.

By the end of this chapter, you will be able to create animations using `CoordinatorLayout` and `MotionLayout` and use the Motion Editor in Android Studio to create `MotionLayout` animations.

In the previous chapter, you learned about architecture patterns such as MVVM. You now know how to improve the architecture of an app. Next, we will learn how to use animations to enhance our app's look and feel and make it different and better than other apps.

Sometimes, the apps we develop can look a little plain, so we can include some moving parts and delightful animations in our apps to make them livelier and the UI and user experience better. For example, we can add visual cues so that a user will not be confused about what to do next and can be guided through what steps they can take.

Animations while loading can entertain a user while content is being fetched or processed. Pretty animations when the app encounters an error can help prevent users from getting angry about what has happened and can inform them of what options they have.

In this chapter, we'll start by looking at some of the traditional ways of doing animations with Android. We'll end the chapter by looking at the newer `MotionLayout` option. Let's get started with activity transitions, which are one of the easiest and most used animations.

We will cover the following topics in this chapter:

- Activity transitions
- Animations with `CoordinatorLayout`
- Animations with `MotionLayout`

Technical requirements

The complete code for all the exercises and the activity in this chapter is available on GitHub at `https://packt.link/G8RoL`.

Activity transitions

When opening and closing an activity, Android will play a default transition. We can customize the activity transition to reflect the brand and/or differentiate our app. Activity transitions are available, starting with Android 5.0 Lollipop (API level 21).

Activity transitions have two parts – the enter transition and the exit transition. The enter transition defines how the activity and its views will be animated when the activity is opened. The exit transition, meanwhile, describes how the activity and views are animated when the activity is closed, or a new activity is opened. Android supports the following built-in transitions:

- **Explode**: This moves views in or out from the center
- **Fade**: This view slowly appears or disappears
- **Slide**: This moves views in or out from the edges

Now, let's see how we can add activity transitions in the following section. There are two ways to add activity transitions – through XML and through code. First, we will learn how to add transitions via XML, and then via code.

Adding activity transitions through XML

You can add activity transitions through XML. The first step is to enable window content transitions. This is done by adding the following in the activity's theme in `themes.xml`:

```
<item name="android:windowActivityTransitions">true</item>
```

After that, you can then add the enter and exit transitions with the `android:windowEnterTransition` and `android:windowExitTransition` style attributes. For example, if you want to use the default transitions from `@android:transition/`, the attributes you will need to add are as follows:

```
<item name="android:windowEnterTransition"> @android:
```

```
            transition/slide_left</item>
    <item name="android:windowExitTransition"> @android:
        transition/explode</item>
```

Your `themes.xml` file would then look as follows:

```
    <style name="AppTheme" parent=
        "Theme.AppCompat.Light.DarkActionBar">
        ...
        <item name="android:
            windowActivityTransitions">true</item>
        <item name="android:windowEnterTransition">
            @android:transition/slide_left</item>
        <item name="android:windowExitTransition">
            @android:transition/explode</item>
    </style>
```

Activity transitions are enabled with `<item name="android:windowActivityTransitions">true</item>`. The `<item name="android:windowEnterTransition">@android:transition/slide_left</item>` attribute sets the enter transition, while `@android:transition/explode` is the exit transition file, as set by the `<item name="android:windowExitTransition">@android:transition/explode</item>` attribute.

In the next section, you will learn how to add activity transitions through coding.

Adding activity transitions through code

Activity transitions can also be added programmatically. The first step is to enable window content transitions. You can do that by calling the following function in your activity before the call to `setContentView()`:

```
  window.requestFeature(Window.FEATURE_CONTENT_TRANSITIONS)
```

You can add the enter and exit transactions afterward with `window.enterTransition` and `window.exitTransition` respectively. We can use the built-in `Explode()`, `Slide()`, and `Fade()` transitions from the `android.transition` package. For example, if we want to use `Explode()` as an enter transition and `Slide()` as an exit transition, we can add the following code:

```
  window.enterTransition = Explode()
  window.exitTransition = Slide()
```

Remember to wrap these calls with a check for `Build.VERSION.SDK_INT >= Build.VERSION_CODES.LOLLIPOP` if your app's minimum supported SDK is lower than 21.

Now that you know how to add entry and exit activity transitions through code or XML, you need to learn how to activate the transition when opening the activity. We will do that in the next section.

Starting an activity with an activity transition

Once you have added activity transitions to an activity (either through XML or by coding), you can activate the transition when opening the activity. Instead of the `startActivity(intent)` call, you should pass in a bundle with the transition animation. To do that, start your activity with the following code:

```
startActivity(intent, ActivityOptions
.makeSceneTransitionAnimation(this).toBundle())
```

The `ActivityOptions.makeSceneTransitionAnimation(this).toBundle()` argument will create a bundle with the enter and exit transition we specified for the activity (via XML or with code).

Let's try out what we have learned so far by adding activity transitions to an app.

Exercise 16.01 – creating activity transitions in an app

In many establishments, it is common to leave a tip (often called a **gratuity**). This is a sum of money given to show appreciation for a service – for example, to the waiting staff in a restaurant. The tip is provided in addition to the basic charge denoted on the final bill.

Throughout this chapter, we will be working with an application that calculates the amount that should be given as a tip. This value will be based on the amount of the bill (the basic charge) and the extra percentage that the user wants to give. The user will input both values, and the app will calculate the tip value.

In this exercise, we will be customizing the activity transition between the input and the output screen:

1. In Android Studio, open the Tip Calculator project in the `Chapter16` directory from the book code repository.
2. Run the application. Tap on the **Compute** button and note what happens when opening `OutputActivity` and going back. There is a default animation while `MainActivity` is being closed and `OutputActivity` is being opened and closed.
3. Now, let's start adding transition animations. Open `themes.xml` and update the activity theme with the `windowActivityTransitions`, `windowEnterTransition`, and `windowExitTransition` style attributes:

   ```
   <item name="android:windowActivityTransitions">
   ```

```
      true</item>
    <item name="android:windowEnterTransition">
      @android:transition/explode</item>
    <item name="android:windowExitTransition">
      @android:transition/slide_left</item>
```

This will enable the activity transition, add an explode enter transition, and add a slide left exit transition to the activity.

4. Go back to the `MainActivity` file and replace `startActivity(intent)` with the following:

```
startActivity(intent, ActivityOptions
    .makeSceneTransitionAnimation(this).toBundle())
```

This will open `OutputActivity` with the transition animation we specified in the XML file (which we set in the previous step).

5. Run the application. You will see that the animation when opening and closing `MainActivity` and `OutputActivity` has changed. When the Android UI opens `OutputActivity`, note that the text is moving toward the center. While closing, the views slide to the left:

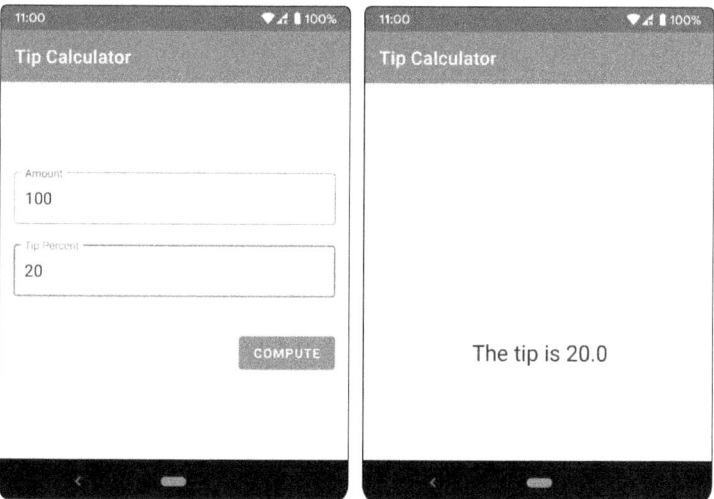

Figure 16.1 – The app screens – the input screen (on the left) and the output screen (on the right)

We have added an activity transition to an app. When we open a new activity, the new activity's enter transition will be played. Its exit transition will play when the activity is closed.

Sometimes, when we open another activity from one activity, there is a common element that is present in both activities. In the next section, we will learn about adding this shared element transition.

Adding a shared element transition

There are times when an application moves from one activity to another and there is a common element that is present in both activities. We can add an animation to this shared element to highlight to users the link between the two activities.

In a movie application, for example, an activity with a list of movies (with a thumbnail image) can open a new activity with details of the selected movie, along with a full-sized image at the top. Adding a shared element transition for the image will link the thumbnail on the list activity to the image on the details activity.

The shared element transition has two parts – the enter transition and the exit transition. These transitions can be done through XML or code.

The first step is to enable a window content transition. You can do this by adding the activity's theme to `themes.xml` with the following:

```
<item name="android:windowContentTransitions">true</item>
```

You can also do this programmatically by calling the following function in your activity before the call to `setContentView()`:

```
window.requestFeature(Window.FEATURE_CONTENT_TRANSITIONS)
```

The `android:windowContentTransitions` attribute with a `true` value and `window.requestFeature(Window.FEATURE_CONTENT_TRANSITIONS)` will enable the window content transition.

Afterward, you can add the shared element enter transition and the shared element exit transition. If you have `enter_transition.xml` and `exit_transition.xml` in your `res/transitions` directory, you can add the shared element enter transition by adding the following style attribute:

```
<item name="android:windowSharedElementEnterTransition"> @transition/enter_transition</item>
```

You can also do this through code with the following lines:

```
val enterTransition = TransitionInflater.from(this)
    .inflateTransition(R.transition.enter_transition)
window.sharedElementEnterTransition = enterTransition
```

The `windowSharedElementEnterTransition` attribute and `window.sharedElementEnterTransition` will set our enter transition to the `enter_transition.xml` file.

To add the shared element exit transition, you can add the following style attributes:

```xml
<item name="android:windowSharedElementExitTransition">
@transition/exit_transition</item>
```

This can be done programmatically with the following lines of code:

```kotlin
val exitTransition = TransitionInflater.from(this)
    .inflateTransition(R.transition.exit_transition)
window.sharedElementExitTransition = exitTransition
```

The `windowSharedElementExitTransition` attribute and `window.sharedElementExitTransition` will set our exit transition to the `exit_transition.xml` file.

You have learned how to add shared element transitions. In the next section, we'll learn how to start an activity with the shared element transition.

Starting an activity with the shared element transition

Once you have added the shared element transition to an activity (either through XML or programmatically), you can activate the transition when opening the activity. Before you do that, add a `transitionName` attribute. Set its value as the same text for the shared element in both activities.

For example, in `ImageView`, we can add a `transition_name` value for the `transitionName` attribute:

```xml
<ImageView
    ...
    android:transitionName="transition_name"
    android:id="@+id/sharedImage"
    ... />
```

To start the activity with shared elements, we will be passing in a bundle with the transition animation. To do that, start your activity with the following code:

```kotlin
startActivity(intent, ActivityOptions
.makeSceneTransitionAnimation(this, sharedImage,
"transition_name").toBundle());
```

The `ActivityOptions.makeSceneTransitionAnimation(this, sharedImage, "transition_name").toBundle()` argument will create a bundle with the shared element (`sharedImage`) and the transition name (`transition_name`).

If you have more than one shared element, you can pass the variable arguments of `Pair<View, String>` of `View`, and the transition name `String` instead. For example, if we have the view's button and image as shared elements, we can do the following:

```
val buttonPair: Pair<View, String> = Pair(button, "button")
val imagePair: Pair<View, String> = Pair(image, "image")
val activityOptions = ActivityOptions
    .makeSceneTransitionAnimation(this, buttonPair,
    imagePair)
startActivity(intent, activityOptions.toBundle())
```

This will start the activity with two shared elements (the button and image).

> **Note**
> Remember to import `android.util.Pair` instead of `kotlin.Pair`, as `makeSceneTransitionAnimation` is expecting the pair from the Android SDK.

Let's try out what we have learned so far by adding shared element transitions to the *Tip Calculator* app.

Exercise 16.02 – creating the shared element transition

In the first exercise, we customized the activity transitions for `MainActivity` and `OutputActivity`. In this exercise, we will be adding an image to both activities. This shared element will be animated when moving from the input screen to the output screen. We'll be using the app launcher icon (`res/mipmap/ic_launcher`) for `ImageView`. You can change yours instead of using the default one:

1. Open the `Tip Calculator` project we developed in the previous exercise.
2. Go to the `activity_main.xml` file and add `ImageView` at the top of the amount text field:

```
<ImageView
    android:id="@+id/image"
    ...
    android:transitionName="transition_name"
    ... />
```

The complete code for this step can be found at https://packt.link/NvDO2.

The transitionName value of transition_name will be used to identify this as a shared element.

3. Change the top constraint of the TextInputLayout with ID amount_text_layout by changing app:layout_constraintTop_toTopOf="parent" to the following:

    ```
    app:layout_constraintTop_toBottomOf="@id/image"
    ```

    This will move the amount_text_layout below the image.

4. Now, open the activity_output.xml file and add an image above the tip TextView, with a height and width of 200 dp and a scaleType of fitXY to fit the image to the dimensions of ImageView:

    ```
    <ImageView
        android:id="@+id/image"
        ...
        android:transitionName="transition_name"
        ... />
    ```

    The complete code for this step can be found at https://packt.link/jpgVe.

    The transitionName value of transition_name is the same as the value for ImageView from MainActivity.

5. Open MainActivity and change the startActivity code to the following:

    ```
    val image: ImageView = findViewById(R.id.image)
    startActivity(intent, ActivityOptions
        .makeSceneTransitionAnimation(this, image,
        "transition_name").toBundle())
    ```

    This will start a transition from ImageView in MainActivity, with the ID image transitioning to another ImageView in OutputActivity, whose transitionName value is also transition_name.

6. Run the application. Provide an amount and percentage, and then tap on the **Compute** button. You will see that the image in the input activity appears to enlarge and position itself in OutputActivity:

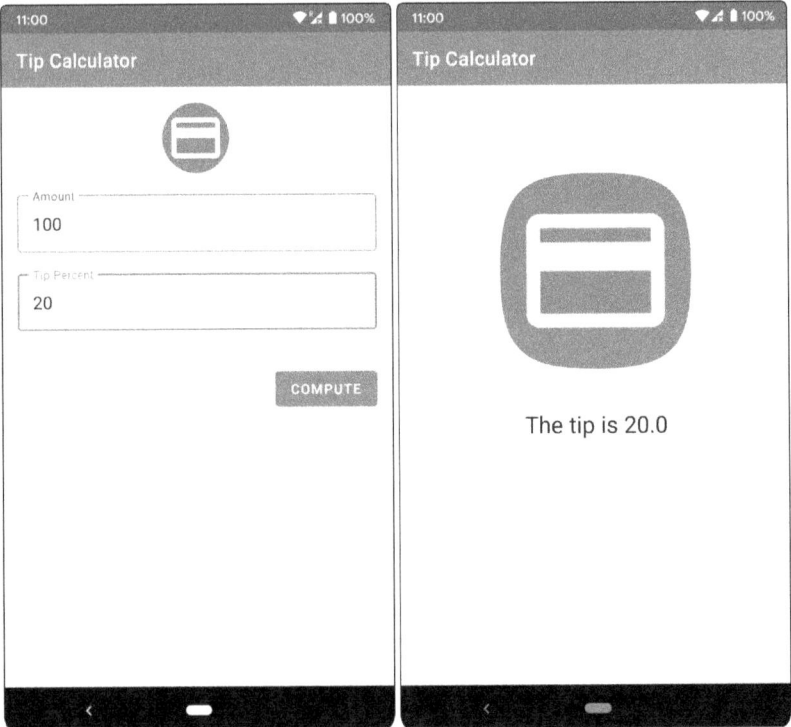

Figure 16.2 – The app screens – the input screen (on the left) and the output screen (on the right)

We have learned about adding activity transitions and shared element transitions. Now, let's look into animating views inside a layout. If we have more than one element inside, it might be difficult to animate each element. `CoordinatorLayout` can be used to simplify this animation. We will discuss this in the next section.

Animations with CoordinatorLayout

`CoordinatorLayout` is a layout that handles the motions between its child views. When you use `CoordinatorLayout` as the parent view group, you can animate the views inside it with little effort. You can add `CoordinatorLayout` to your project by adding your `app/build.gradle` file dependencies with the following:

```
implementation 'androidx.coordinatorlayout:
coordinatorlayout:1.2.0'
```

This will allow us to use `CoordinatorLayout` in our layout files.

Let's say we have a layout file with a floating action button inside `CoordinatorLayout`. When tapping on the floating action button, the UI displays a `Snackbar` message.

> **Note**
> `Snackbar` is an Android widget that provides a brief message to a user at the bottom of the screen.

If you use any layout other than `CoordinatorLayout`, `Snackbar` with its message will be rendered on top of the floating action button. If we use `CoordinatorLayout` as the parent view group, the layout will push the floating action button upwards, display `Snackbar` below it, and move it back when `Snackbar` disappears. *Figure 16.3* shows how the layout adjusts to prevent `Snackbar` from being on top of the floating action button:

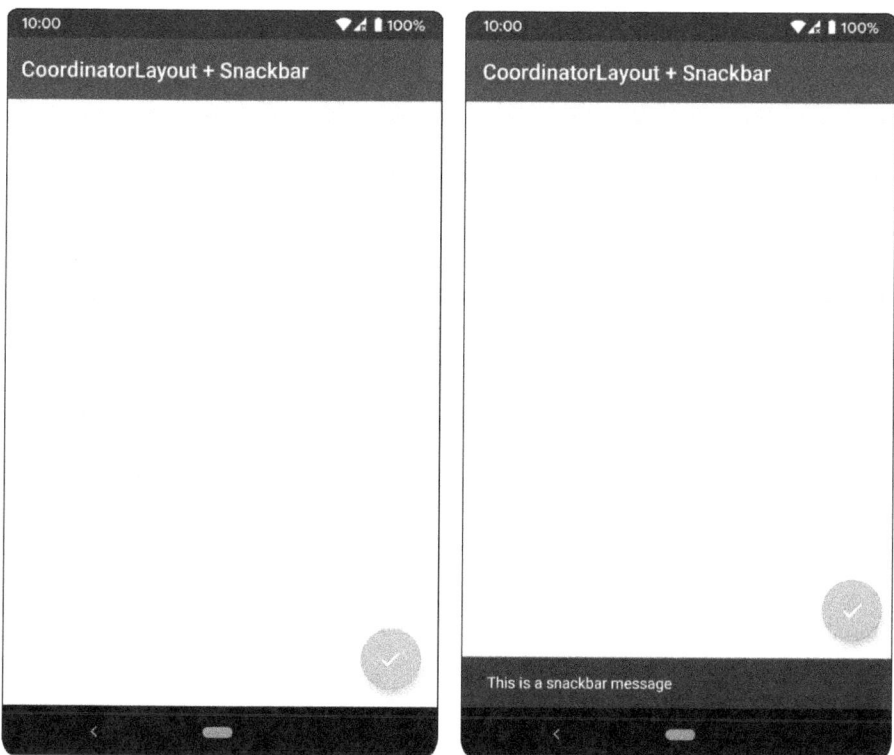

Figure 16.3 – The left screenshot displays the UI before and after the Snackbar message is shown. The screenshot on the right shows the UI while the Snackbar is visible

The floating action button moves and gives space to the `Snackbar` message because it has a default behavior called `FloatingActionButton.Behavior`, a subclass of `CoordinatorLayout.Behavior`. The `FloatingActionButton.Behavior` subclass moves the floating action button while `Snackbar` is being displayed so that `Snackbar` won't cover the floating action button.

Not all views have the `CoordinatorLayout` behavior. To implement custom behavior, you can start by extending `CoordinatorLayout.Behavior`. You can then attach it to the view with the `layout_behavior` attribute. For example, if we made `CustomBehavior` in the `com.example.behavior` package for a button, we can update the button in the layout with the following:

```
...
<Button
    ...
    app:layout_behavior="com.example.behavior.CustomBehavior">
    .../>
```

We have learned how to create animations and transitions with `CoordinatorLayout`. In the next section, we will look into another layout, `MotionLayout`, which allows developers more control over motion.

Animations with MotionLayout

Creating animations in Android is sometimes time-consuming. You need to work on XML and code files even to create simple animations. More complicated animations and transitions take even more time to make.

To help developers easily make animations, Google created `MotionLayout`. This is a new way to create motion and animations through XML. It is available starting at API level 14 (Android 4.0).

With `MotionLayout`, we can animate the position, width/height, visibility, alpha, color, rotation, elevation, and other attributes of one or more views. Normally, some of these are hard to do with code, but `MotionLayout` allows us to easily adjust them using declarative XML so that we can focus more on our application.

Let's get started by adding `MotionLayout` to our application.

Adding MotionLayout

To add `MotionLayout` to your project, you just need to add the dependency for `ConstraintLayout` 2.0. `ConstraintLayout` 2.0 is the new version of `ConstraintLayout`, with new features added, including `MotionLayout`. Add to your `app/build.gradle` file dependencies with the following:

```
implementation 'androidx.constraintlayout:constraintlayout:2.1.4'
```

This will add the latest version of ConstraintLayout (2.1.4 at the time of writing) to your app.

After adding the dependency, we can now use `MotionLayout` to create animations. We'll be doing that in the next section.

Creating animations with MotionLayout

`MotionLayout` is a subclass of our good old friend ConstraintLayout. To create animations with `MotionLayout`, open the layout file where we will add the animations. Replace the root ConstraintLayout container with `androidx.constraintlayout.motion.widget.MotionLayout`.

The animation itself won't be in the layout file but in another XML file, called `motion_scene`. This will specify how `MotionLayout` will animate the views inside it. `motion_scene` files should be placed in the `res/xml` directory. The layout file will link to this `motion_scene` file with the `app:layoutDescription` attribute in the root view group. Your layout file should look similar to the following:

```
<?xml version="1.0" encoding="utf-8"?>
<androidx.constraintlayout.motion.widget.MotionLayout
    ...
    app:layoutDescription="@xml/motion_scene">
    ...
</androidx.constraintlayout.motion.widget.MotionLayout>
```

To create animations with `MotionLayout`, we must have the initial state and final state of our views. `MotionLayout` will automatically animate the transition between the two. You can specify these two states in the same `motion_scene` file. If you have a lot of views inside the layout, you can also use two different layouts for the beginning and ending states of the animation.

The root container of the `motion_scene` file is `motion scene`. This is where we add the constraints and animation for `MotionLayout`. It contains the following:

- **ConstraintSet**: Specifies the beginning and ending position and style for the view/layout to animate
- **Transition**: Specifies the start, end, duration, and other details of the animation to be done on the views

Let's try adding animations with `MotionLayout` by adding it to our *Tip Calculator* app.

Exercise 16.03 – adding animations with MotionLayout

In this exercise, we will be updating our *Tip Calculator* app with a `MotionLayout` animation. In the output screen, the image above the tip text will move down when tapped and will go back to its original position when tapped again:

1. Open the *Tip Calculator* project in Android Studio 4.0 or higher.

2. Open the `activity_output.xml` file and change the root `ConstraintLayout` tag to `MotionLayout`. Change `androidx.constraintlayout.widget.ConstraintLayout` to the following:

    `androidx.constraintlayout.motion.widget.MotionLayout`

3. Add `app:layoutDescription="@xml/motion_scene"` to the `MotionLayout` tag. The IDE will warn you that this file does not yet exist. Ignore that for now, as we will be adding it in the next step. Your file should look similar to this:

    ```
    <?xml version="1.0" encoding="utf-8"?>
    <androidx.constraintlayout.motion.widget.MotionLayout
        ...
        app:layoutDescription="@xml/motion_scene">
        ...
    </androidx.constraintlayout.motion.widget
    .MotionLayout>
    ```

4. Create a `motion_scene.xml` file in the `res/xml` directory. This will be our motion_scene file where the animation configuration will be defined. Use `motion_scene` as the root element for the file.

5. Add the starting `Constraint` element by adding the following to the `motion_scene` file:

    ```
    <ConstraintSet android:id="@+id/start_constraint">
        <Constraint
            android:id="@id/image"
            .../>
    </ConstraintSet>
    ```

The complete code for this step can be found at https://packt.link/jdJrD.

This is how the image looks at the current position (constrained to the top of the screen).

6. Next, add the ending `Constraint` element by adding the following to the `motion_scene` file:

```
<ConstraintSet android:id="@+id/end_constraint">
    <Constraint
        android:id="@id/image"
        ... />
</ConstraintSet>
```

The complete code for this step can be found at https://packt.link/jdJrD.

In the ending animation, `ImageView` will be at the bottom of the screen.

7. Let's now add in the transition for our `ImageView` after the `ConstraintSet`:

```
<Transition
    app:constraintSetEnd="@id/end_constraint"
    app:constraintSetStart="@id/start_constraint"
    app:duration="2000">
    <OnClick
        app:clickAction="toggle"
        app:targetId="@id/image" />
</Transition>
```

Here, we're specifying the start and end constraints, which will animate for 2,000 milliseconds (2 seconds). We also added an `OnClick` event to `ImageView`. The toggle will animate the view from start to end, and if the view is already in the end state, it will animate back to the start state.

8. Run the application and tap on `ImageView`. It will move straight downward in around 2 seconds. Tap it again, and it will move back up in 2 seconds. *Figure 16.4* shows the start and the end of this animation:

Figure 16.4 – The starting animation (left) and the ending animation (right)

In this exercise, we have animated `ImageView` in `MotionLayout` by specifying the start constraint, end constraint, and transition with a duration and `OnClick` event. `MotionLayout` automatically plays the animation from the start to the end position (which to us looks like it's moving up or down in a straight line automatically when tapped).

We have created animations with `MotionLayout`. In the next section, we will be using Android Studio's Motion Editor to create `MotionLayout` animations.

The Motion Editor

Android Studio, starting with version 4.0, includes the Motion Editor. The Motion Editor can help developers create animations with `MotionLayout`. This makes it easier for developers to create and preview transitions and other motions, instead of doing it by hand and running the app to see the changes. The editor will also generate the corresponding files automatically.

You can convert your ConstraintLayout to `MotionLayout` in the Layout Editor by right-clicking the preview and clicking on the **Convert to MotionLayout** item. Android Studio will do the conversion and also create the motion scene file for you.

Animations with MotionLayout   621

When viewing a layout file that has `MotionLayout` as the root in the **Design** view, the Motion Editor UI will be included in the **Design** view, as shown in *Figure 16.5*:

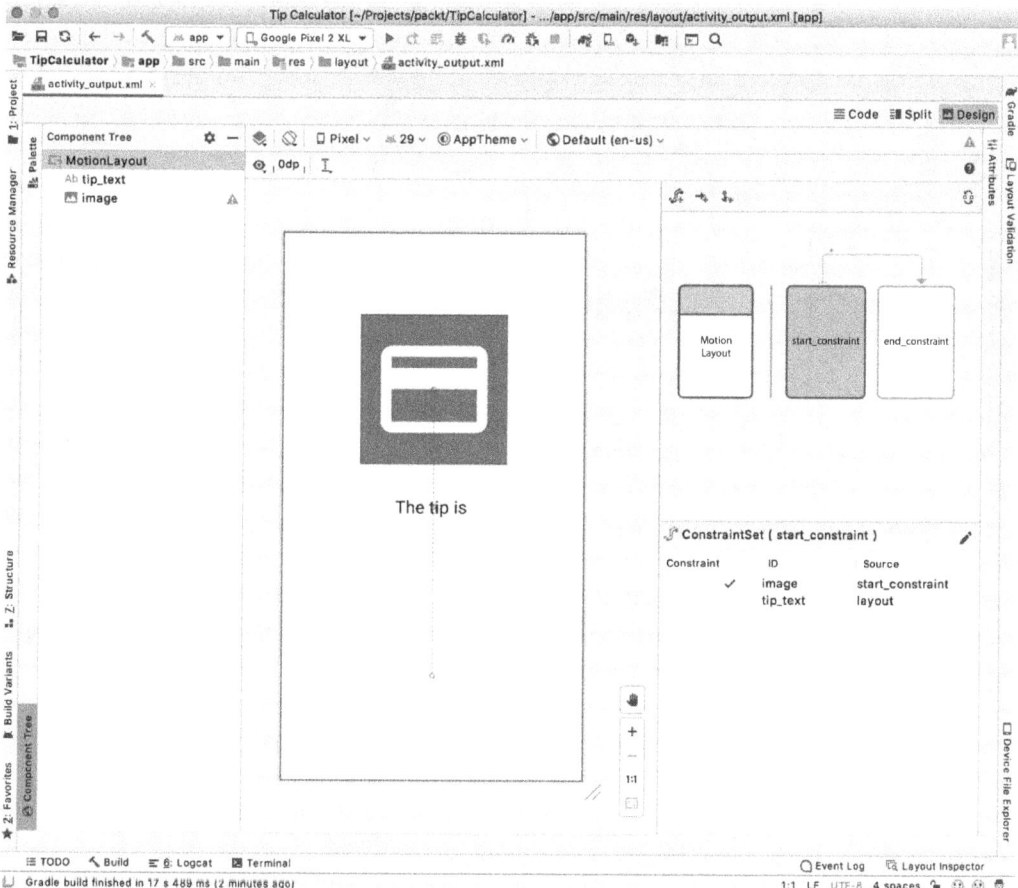

Figure 16.5 – The Motion Editor in Android Studio 4.0

In the upper-right window (the **Overview** panel), you can see a visualization of `MotionLayout` and the start and end constraint. The transition is displayed as an arrow coming from the start constraint. The dot near the start constraint shows the click action for the transition. *Figure 16.6* shows the **Overview** panel with `start_constraint` selected:

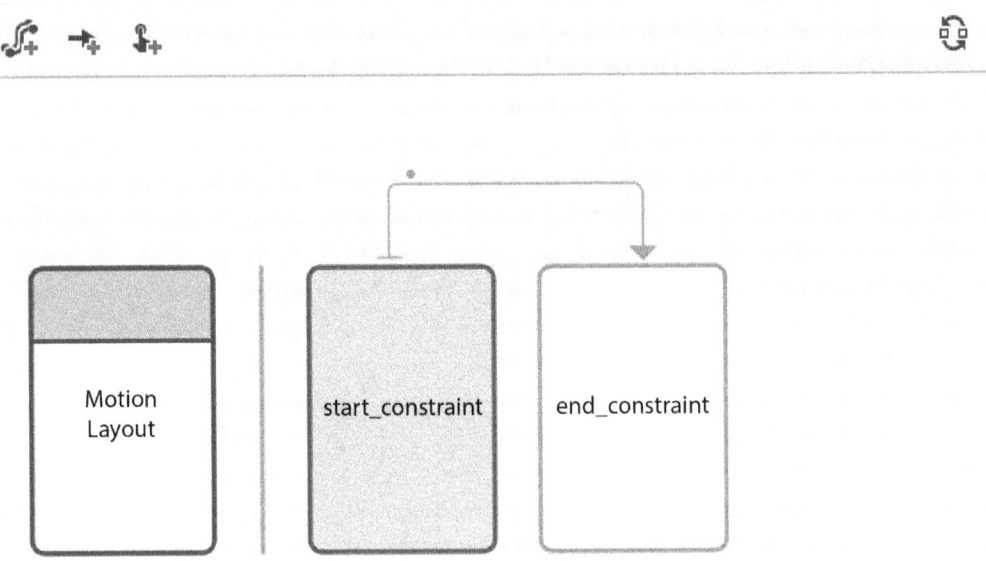

Figure 16.6 – The Motion Editor's Overview panel with start_constraint selected

The bottom-right window is the **Selection** panel, which shows the views in the constraint set or `MotionLayout` selected in the **Overview** panel. It can also show the transitions when the transition arrow is selected. *Figure 16.7* shows the **Selection** panel when `start_constraint` is selected:

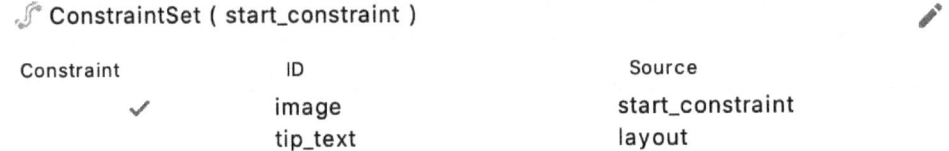

Figure 16.7 – The Motion Editor's Selection panel showing ConstraintSet for start_constraint

When you click on `MotionLayout` on the left of the **Overview** panel, the **Selection** panel below will display the views and their constraints, as shown in *Figure 16.8*:

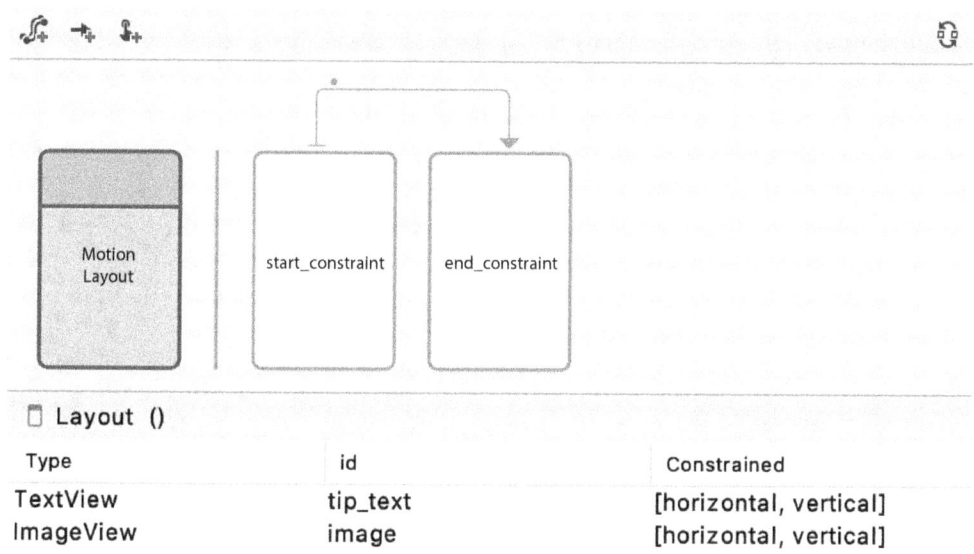

Figure 16.8 – The Overview and Selection panels when MotionLayout is selected

When you click on start_constraint or end_constraint, the preview window on the left will display how the start or end state looks. The **Selection** panel will also show the view and its constraints. Take a look at *Figure 16.9* to see how it looks when start_constraint is selected:

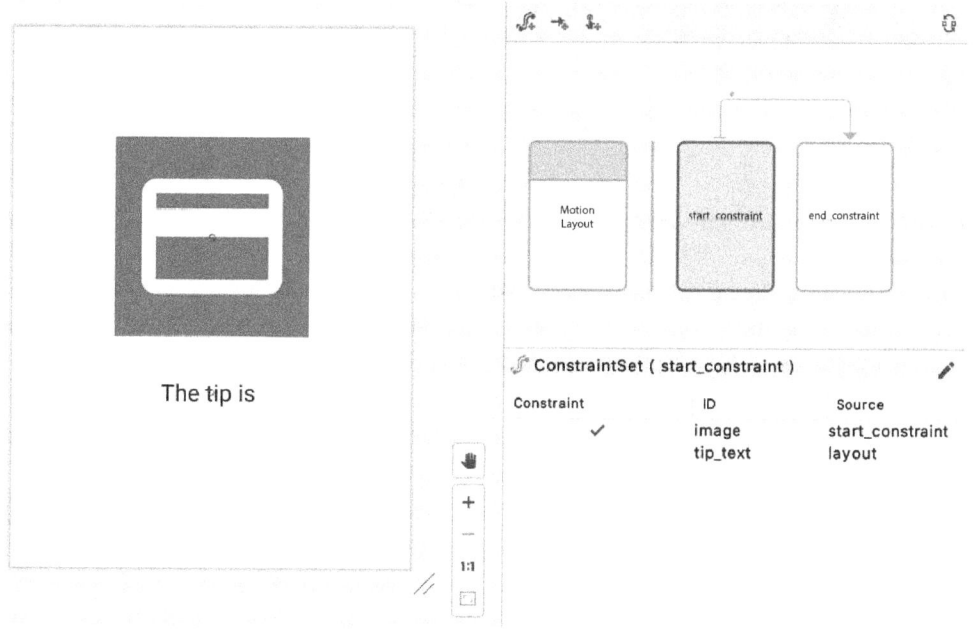

Figure 16.9 – How the Motion Editor looks when start_constraint is selected

Figure 16.10 shows how the Motion Editor will look if you select `end_constraint`:

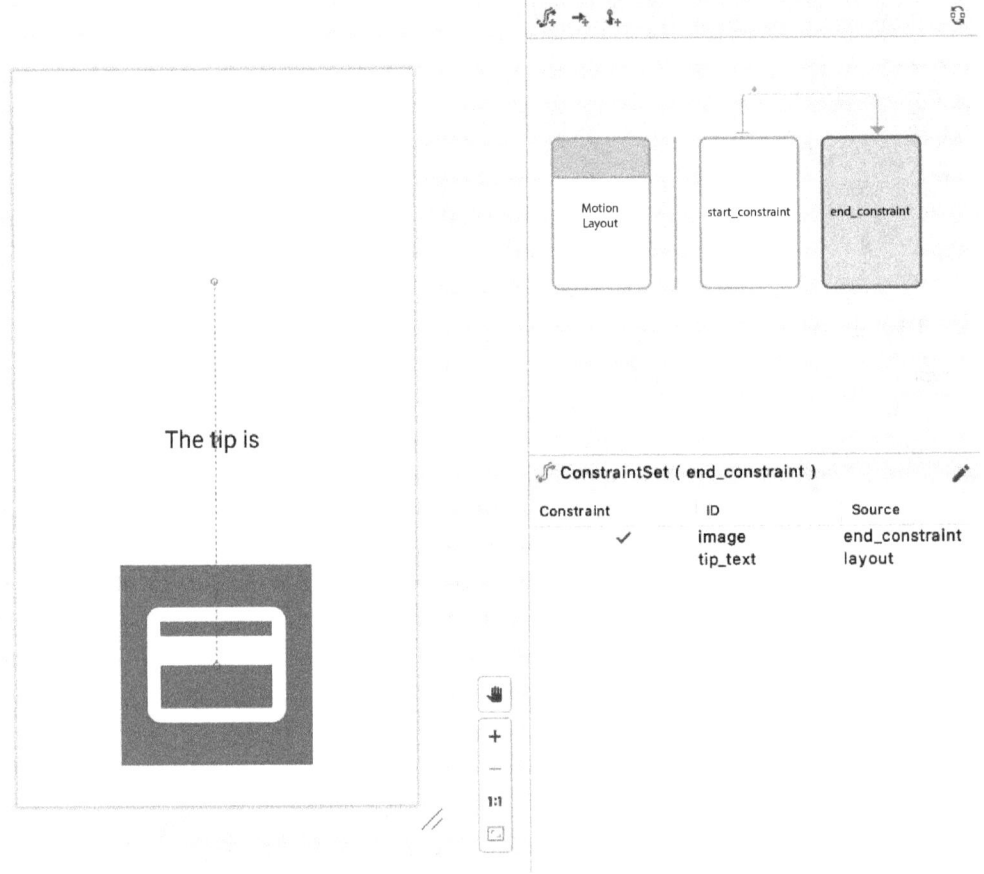

Figure 16.10 – How the Motion Editor looks with end_constraint selected

The arrow connecting `start_constraint` and `end_constraint` represents the transition of `MotionLayout`. On the **Selection** panel, there are controls to play or go to the first or last state. You can also drag the arrow to a specific position. *Figure 16.11* shows how it looks in the middle (50% of the animation):

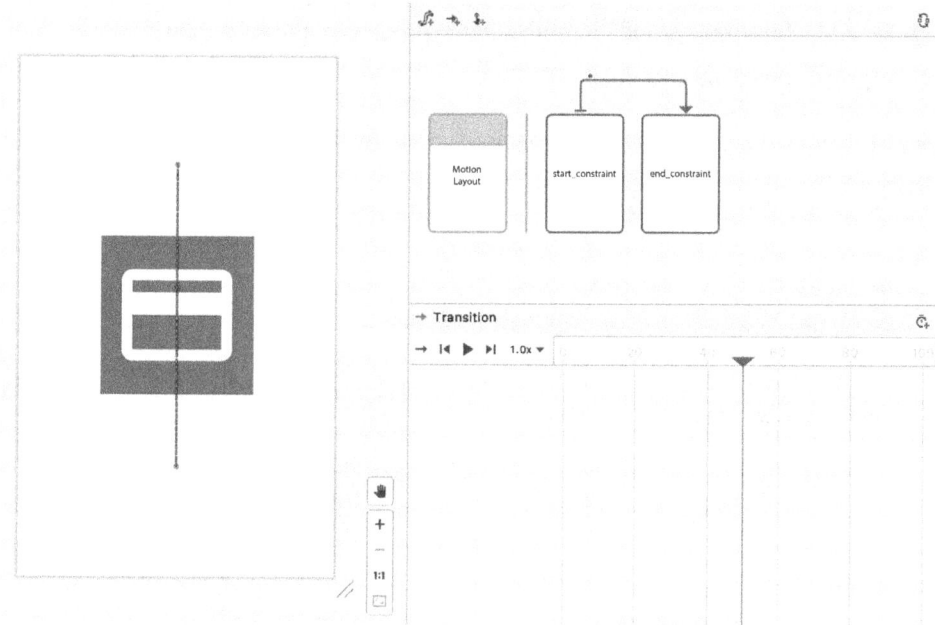

Figure 16.11 – The transition in the middle of the animation

During the development of animations with `MotionLayout`, it would be better if we could debug the animations to make sure we're doing the animations correctly. We'll discuss how we can do this in the next section.

Debugging MotionLayout

To help you visualize the `MotionLayout` animation before running the app, you can show the motion path and the animation's progress in the Motion Editor. The motion path is the straight route that the object to animate will take from the start to the end state.

To show the path and/or progress animation, we can add a `motionDebug` attribute to the `MotionLayout` container.

We can use the following values for `motionDebug`:

- `SHOW_PATH`: This displays the path of the motion only
- `SHOW_PROGRESS`: This displays the animation progress only

- SHOW_ALL: This displays both the path and the progress of the animation
- NO_DEBUG: This hides all animations

To display the MotionLayout path and progress, we can use the following:

```
<androidx.constraintlayout.motion.widget.MotionLayout
    ...
    app:motionDebug="SHOW_ALL"
    ...>
```

The SHOW_ALL value will display the path and the progress of the animation. *Figure 16.12* shows how it will look when we use SHOW_PATH and SHOW_PROGRESS:

Figure 16.12 – Using SHOW_PATH (left) shows the animation path, while SHOW_PROGRESS (right) shows the animation progress

While motionDebug sounds like something that only appears in debug mode, it will also appear in release builds, so it should be removed when you're preparing your app for publishing.

During the `MotionLayout` animation, the start constraint will transition to the end constraint, even if there's an element or elements that can block the objects in motion. We'll discuss how we can avoid this from happening in the next section.

Modifying the MotionLayout path

In an animation with `MotionLayout`, the UI will play the motion from the start constraint to the end constraint, even if there are elements in the middle that can block our moving views. For example, if `MotionLayout` involves text that moves from the top to the bottom of the screen and vice versa, and we add a button to the middle, the button will cover the moving text.

Figure 16.13 shows how the **OK** button blocks the moving text in the middle of the animation:

Figure 16.13 – The OK button is blocking the middle of the text animation

`MotionLayout` plays the animation from the start to the end constraint in a straight path and adjusts the views, based on the specified attributes. We can add keyframes between the start and end constraints to adjust the animation path and/or the view attributes. For example, during the animation, as well as changing the position of the moving text to avoid the button, we can also change the attributes of the text or other views.

Keyframes can be added in `KeyFrameSet` as a child of the transition attribute of `motion_scene`. We can use the following keyframes:

- `KeyPosition`: This specifies the view's position at a specific point during the animation to adjust the path
- `KeyAttribute`: This specifies the view's attributes at a specific point during the animation
- `KeyCycle`: This adds oscillations during animations
- `KeyTimeCycle`: This allows cycles to be driven by time instead of animation progress
- `KeyTrigger`: This adds an element that can trigger an event based on the animation progress

We will focus on `KeyPosition` and `KeyAttribute`, as `KeyCycle`, `KeyTimeCycle`, and `KeyTrigger` are more advanced keyframes.

`KeyPosition` allows us to change the location of views in the middle of the `MotionLayout` animation. It has the following attributes:

- `motionTarget`: This specifies the object controlled by the keyframe.
- `framePosition`: Numbered from 1 to 99, this specifies the percentage of the motion when the position is changed. For example, 25 means it is at one-quarter of the animation, and 50 is the halfway point.
- `percentX`: This specifies how much the *x* value of the path will be modified.
- `percentY`: This specifies how much the *y* value of the path will be modified.
- `keyPositionType`: This specifies how `KeyPosition` modifies the path.

The `keyPositionType` attribute can have the following values:

- `parentRelative`: `percentX` and `percentY` are specified based on the parent of the view
- `pathRelative`: `percentX` and `percentY` are specified based on the straight path from the start constraint to the end constraint
- `deltaRelative`: `percentX` and `percentY` are specified based on the position of the view

For example, if we want to modify the path of `TextView` with the `text_view` ID at the exact middle of the animation (50%), by moving it 10% by *x* and 10% by *y*, relative to the parent container of `TextView`, we will have the following key position in `motion_scene`:

```
<KeyPosition
    app:motionTarget="@+id/text_view"
    app:framePosition="50"
    app:keyPositionType="parentRelative"
    app:percentY="0.1"
    app:percentX="0.1"
/>
```

Meanwhile, `KeyAttribute` allows us to change the attributes of views while the `MotionLayout` animation is ongoing. Some of the view attributes we can change are `visibility`, `alpha`, `elevation`, `rotation`, `scale`, and `translation`. `KeyAttribute` has the following attributes:

- `motionTarget`: This specifies the object controlled by the keyframe.
- `framePosition`: Numbered from 1 to 99, this specifies the percentage of the motion when the view attributes are applied. For example, 20 means it is one-fifth of the animation, and 75 is the three-quarter point of the animation.

Let's try adding keyframes to the *Tip Calculator* app. When the `ImageView` animates, it goes on top of the text displaying the tip. We'll fix that with keyframes.

Exercise 16.04 – modifying the animation path with keyframes

In the previous exercise, we animated the image to move straight down when tapped (or upward when it's already at the bottom). When it is in the middle, the image is covering the tip `TextView`. We'll be solving this issue in this exercise by adding `KeyFrame` to `motion_scene` using Android Studio's Motion Editor:

1. Open the *Tip Calculator* app with Android Studio 4.0 or higher.
2. Open the `activity_output.xml` file in the `res/layout` directory.
3. Add `app:motionDebug="SHOW_ALL"` to the `MotionLayout` container. This will allow us to see the path and progress information in Android Studio and on our device/emulator. Your `MotionLayout` container will look like the following:

    ```
    <androidx.constraintlayout.motion.widget.MotionLayout
        ...
        app:motionDebug="SHOW_ALL">
    ```

4. Run the app and make a computation. On the output screen, tap on the image. Look at the tip text while the animation is in progress. Note that the text is covered by the image in the middle of the animation, as shown in *Figure 16.14*:

Figure 16.14 – The image hides the TextView displaying the tip

5. Go back to the `activity_output.xml` file in Android Studio. Make sure it's opened in the **Design** view.

6. In the **Overview** panel at the top right, click the arrow connecting `start_constraint` and `end_constraint`. Drag the down arrow in the **Selection** panel to the middle (50%), as shown in *Figure 16.15*:

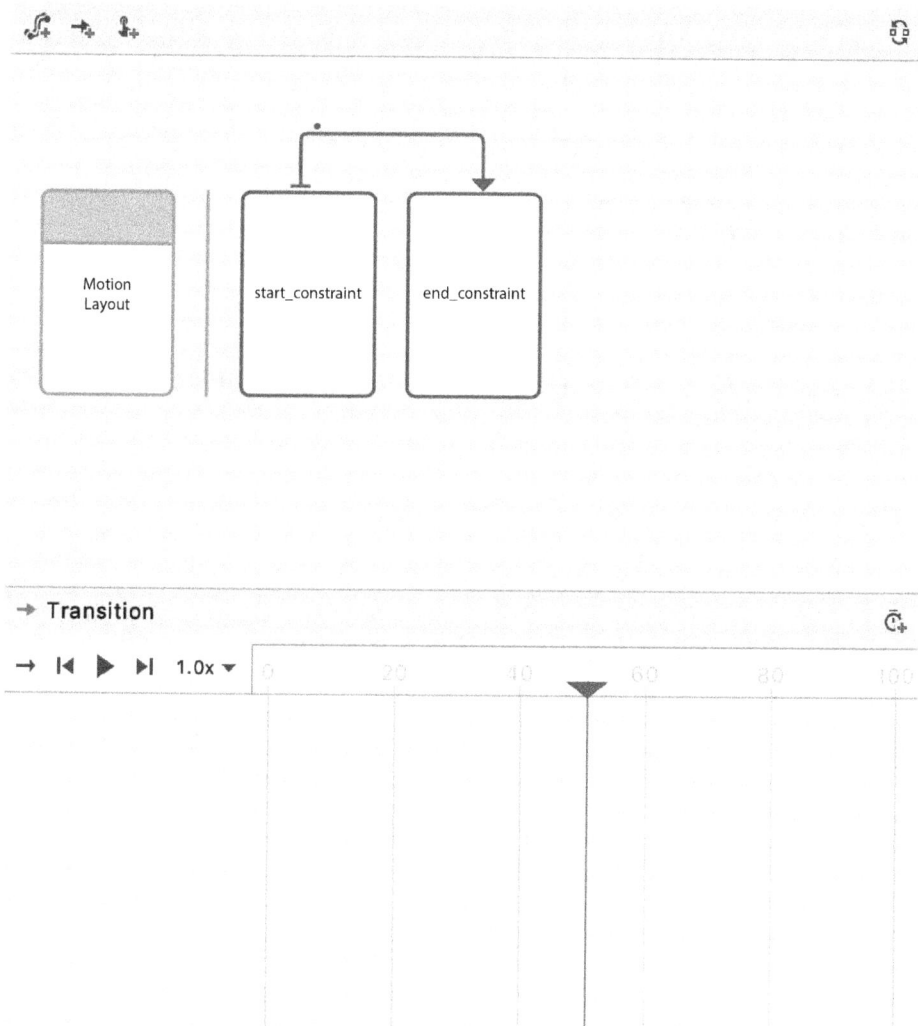

Figure 16.15 – Selecting the arrow representing transition between start and end constraints

7. Click the **Create KeyFrames** icon to the right of **Transition** in the **Selection** panel (the one with a green + symbol), as shown in *Figure 16.16*:

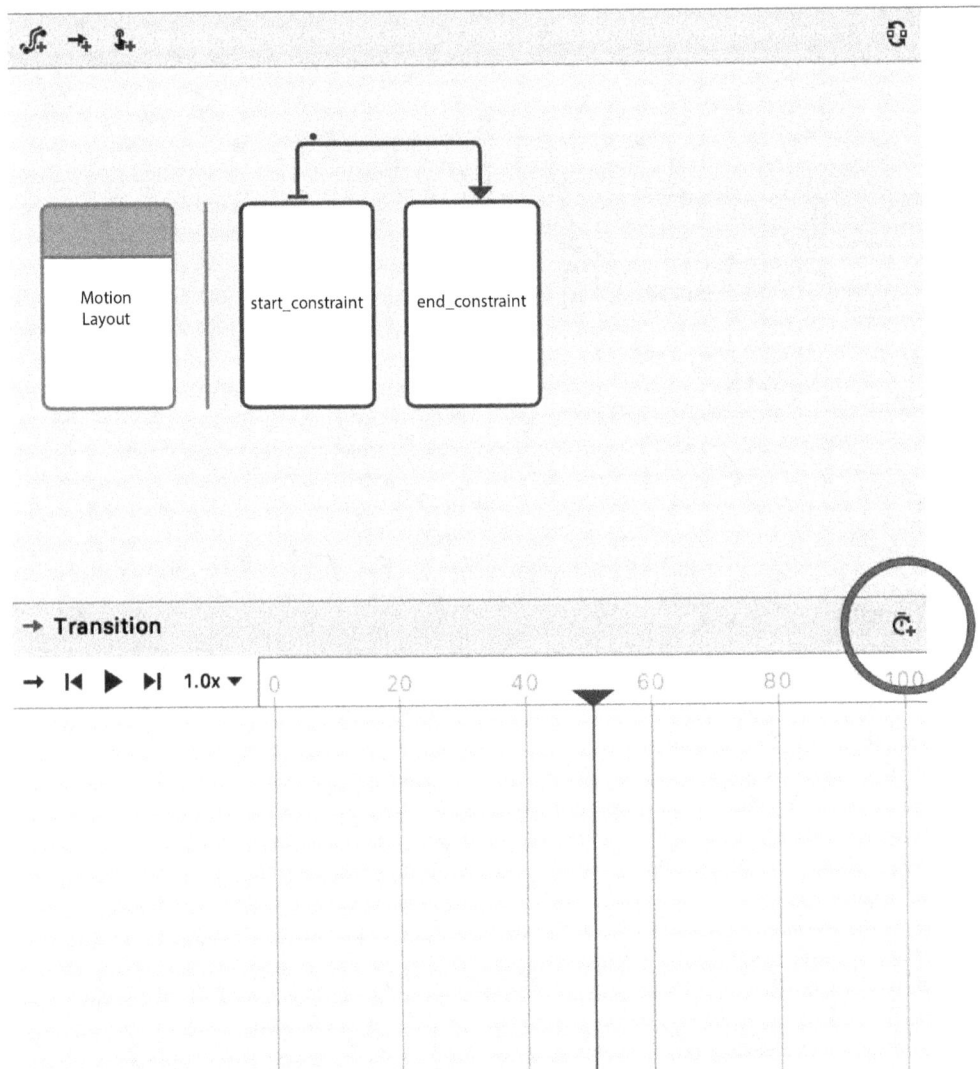

Figure 16.16 – The Create KeyFrames icon

8. Select **KeyPosition**. We will be using **KeyPosition** to adjust the image in the output screen so that it will not overlap with the text containing the tip.

9. Select **ID**, choose **image**, and set the input position to 50. The type is parentRelative, and the **PercentX** value is 1.5, as shown in *Figure 16.17*. This will add a **KeyPosition** attribute for the image in the middle (50%) of the transition, which is 1.5 times relative to the *x* axis of the parent:

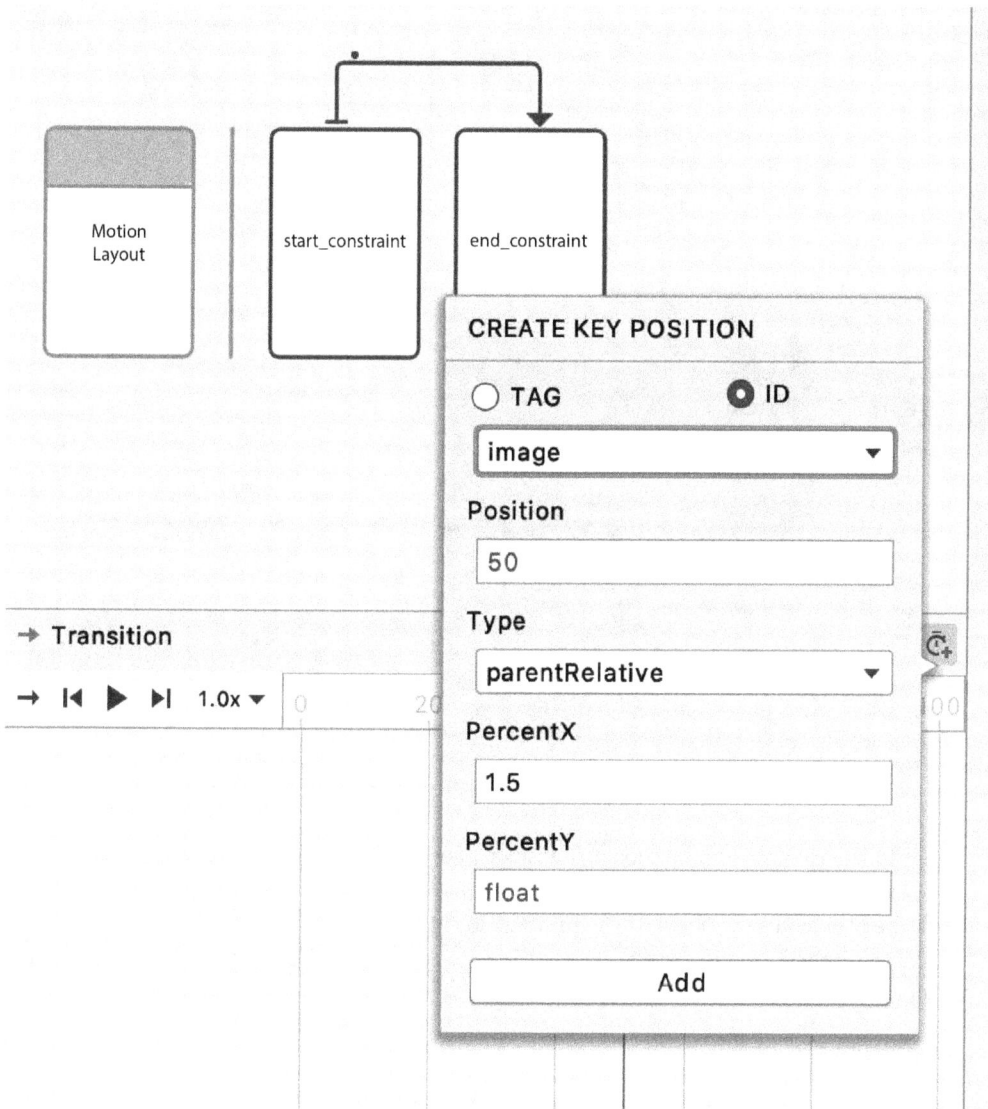

Figure 16.17 – Provide the input to the key position to be made

10. Click the **Add** button. You will see in the **Design** preview, as shown in the following figure, that the motion path is no longer a straight line. At position 50 (the middle of the animation), the text will no longer be covered by `ImageView`. That will now be to the right of `TextView`:

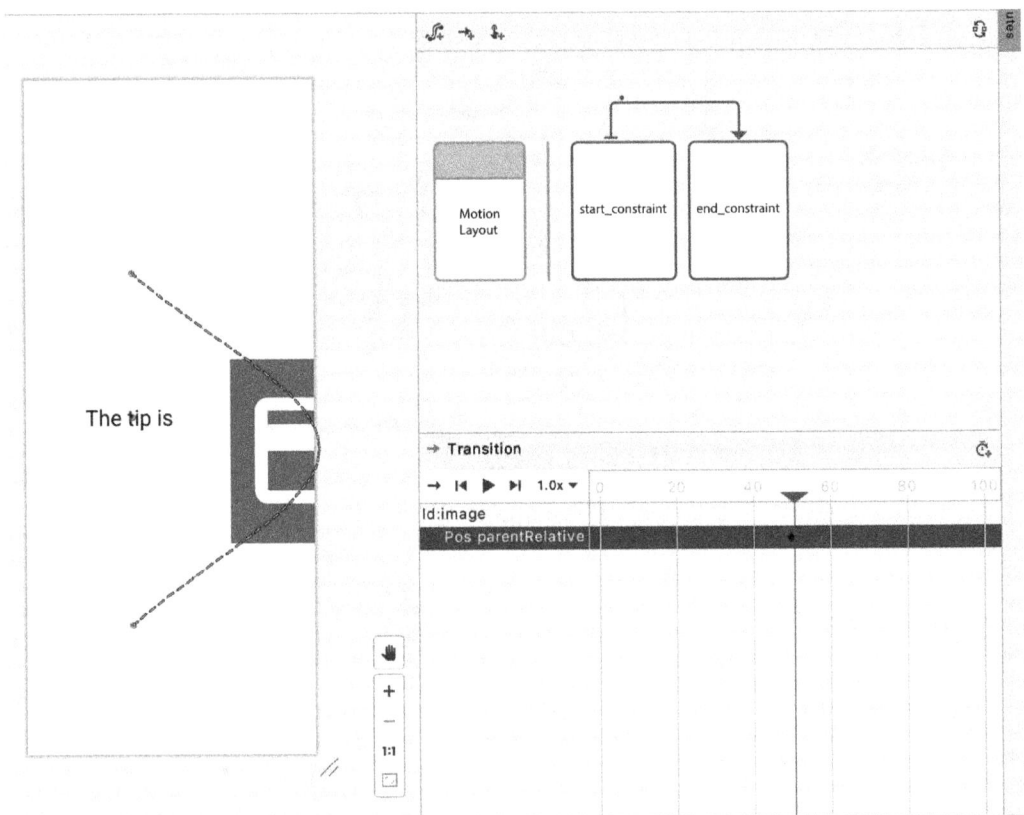

Figure 16.18 – Path will now be curved; Transition panel will have new item for KeyPosition

11. Click the play icon to see how it will animate. Run the application to verify it on a device or emulator. You will see that the animation is now curving to the right instead of taking its previous straight path, as shown in *Figure 16.19*:

Figure 16.19 – The animation now avoids TextView with the tip

12. The Motion Editor will automatically generate the code for KeyPosition. If you go to the motion_scene.xml file, you will see that the Motion Editor added the following code in the transition attribute:

```xml
<KeyFrameSet>
    <KeyPosition
        app:framePosition="50"
        app:keyPositionType="parentRelative"
        app:motionTarget="@+id/image"
        app:percentX="1.5" />
</KeyFrameSet>
```

A KeyPosition attribute was added to the keyframes during the transition. At 50% of the animation, the image's *x* position is moved 1.5 times relative to its parent view. This allows the image to avoid other elements during the animation process.

In this exercise, you have added a key position that will adjust the MotionLayout animation, ensuring that it will not block or be blocked by another view in its path.

Let's test everything you've learned by doing another activity.

Activity 16.01 – Password Generator

Using a strong password is important to secure our online accounts. It must be unique and must include uppercase and lowercase letters, numbers, and special characters. In this activity, you will develop an app that can generate a strong password.

The app will have two screens – the input screen and the output screen. In the input screen, a user can provide the length of a password and specify whether it must have uppercase or lowercase letters, numbers, or special characters.

The output screen will display three possible passwords, and when the user selects one, the other passwords will move away, and a button will display to copy the password to the clipboard. You should customize the transition from the input screen to the output screen.

The steps to complete are as follows:

1. Create a new project in Android Studio 4.0 or higher and name it `Password Generator`. Set its package name and **Minimum SDK**.
2. Add the `MaterialComponents` dependency to your `app/build.gradle` file.
3. Update the dependency for `ConstraintLayout`.
4. Make sure that the activity's theme uses one from `MaterialComponents` in the `themes.xml` file.
5. In the `activity_main.xml` file, remove the `Hello World` `TextView` and add the input text field for the length.
6. Add the code for checkboxes for uppercase, numbers, and special characters.
7. Add a **Generate** button at the bottom of the checkboxes.
8. Create another activity and name it `OutputActivity`.
9. Customize the activity transition from the input screen (`MainActivity`) as `OutputActivity`. Open `themes.xml` and update the activity theme with the `windowActivityTransitions`, `windowEnterTransition`, and `windowExitTransition` style attributes.
10. Update the end of the `onCreate` function in `MainActivity`.
11. Update the code for `androidx.constraintlayout.widget.ConstraintLayout` in the `activity_output.xml` file.
12. Add `app:layoutDescription="@xml/motion_scene"` and `app:motionDebug="SHOW_ALL"` to the `MotionLayout` tag.
13. Add three instances of `TextView` to the output activity for the three passwords generated.
14. Add a **Copy** button at the bottom of the screen.
15. Add the `generatePassword` function in `OutputActivity`.

16. Add the code to generate the three passwords based on user input, and add a `ClickListener` component to the **Copy** button for a user to copy the selected password to the clipboard.
17. In `OutputActivity`, create an animation per the password `TextView`.
18. Create `ConstraintSet` for the default view.
19. Add `ConstraintSet` when the first, second, and third passwords are selected.
20. Next, add `Transition` when each password is selected.
21. Run the application by going to the **Run** menu and clicking the **Run app** menu item.
22. Input a length, select the checkboxes for uppercase, numbers, and special characters, and tap on the **Generate** button. Three passwords will be displayed.
23. Select one and the rest will move out of view. A **Copy** button will also be displayed. Click on it and check whether the password you selected is now on the clipboard. The initial and final state of the output screen will be similar to *Figure 16.20*:

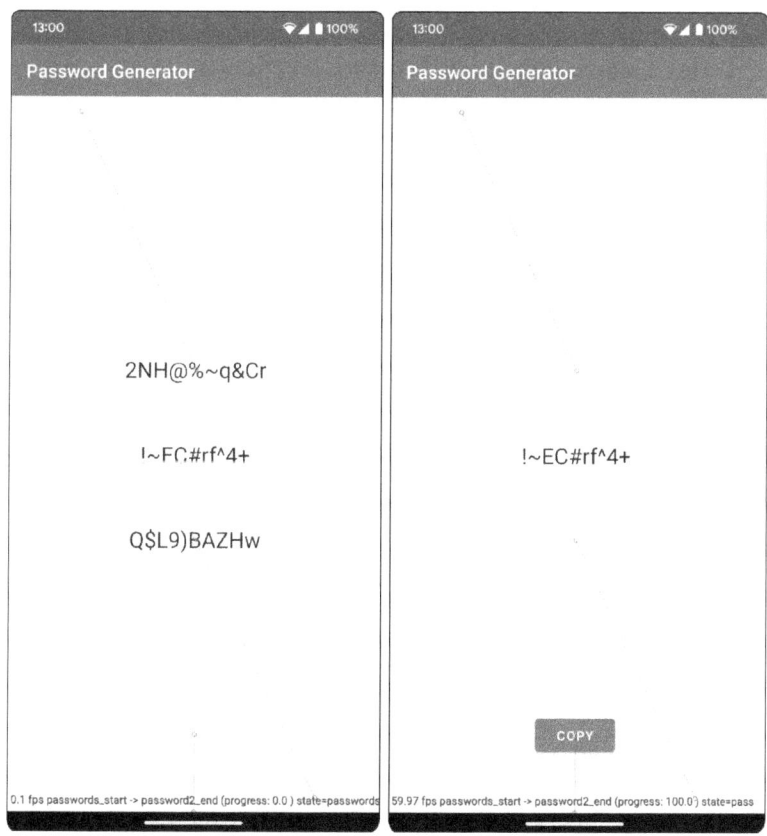

Figure 16.20 – The start and end state of MotionLayout in the Password Generator app

> **Note**
> The solution to this activity can be found at `https://packt.link/By7eE`.

Summary

This chapter covered creating animations and transitions with `CoordinatorLayout` and `MotionLayout`. Animations can improve the usability of our app and make it stand out compared to other apps.

We started by customizing the transition when opening and closing an activity with activity transitions. We also learned about adding shared element transitions when an activity and the activity that it opens both contain the same elements, enabling us to highlight this link between the shared elements to users.

We learned how we can use `CoordinatorLayout` to handle the motion of its child views. Some views have built-in behaviors that handle how they work inside `CoordinatorLayout`. You can add custom behaviors to other views too. Then, we moved on to using `MotionLayout` to create animations by specifying the start constraint, end constraint, and the transition between them. We also looked into modifying the motion path by adding keyframes in the middle of an animation. We learned about keyframes such as `KeyPosition`, which can change a view's position, and `KeyAttribute`, which can change the view's style. We also looked into using the Motion Editor in Android Studio to simplify the creation and previewing of animations and modifying the path.

In the next chapter, we'll learn about the Google Play store. We'll discuss how you can create an account and prepare your apps for release, as well as how you can publish them for users to download and use.

17
Launching Your App on Google Play

This chapter will introduce you to the Google Play console, release channels, and the entire release process. It covers creating a Google Play Developer account, setting up the store entry for our developed app, and creating a key store (including coverage of the importance of passwords and where to store files). We'll also learn about app bundles and APK, looking at how to generate the app's APK or AAB file. Later in the chapter, we'll set up release paths, open beta, and closed alpha, and finally, we'll upload our app to the store and download it on a device.

By the end of this chapter, you will be able to create your own Google Play Developer account, prepare your signed APK or app bundle for publishing, and publish your first application on Google Play.

You learned how to add animations and transitions with `CoordinatorLayout` and `MotionLayout` in *Chapter 16, Animations and Transitions with CoordinatorLayout and MotionLayout*. Now, you are ready to develop and launch Android applications.

After developing Android apps, they will only be available on your devices and emulators. You must make them available to everyone so they can download them. In turn, you will acquire users, and you can earn from them. The official marketplace for Android apps is Google Play. With Google Play, the apps and games you release will be available to over two billion active Android devices globally. There are also other marketplaces where you can publish your apps, but they are beyond the scope of this book.

In this chapter, we're going to learn about launching your apps on Google Play. We'll start with preparing the apps for release and creating a Google Play Developer account. Then, we'll move on to uploading your app and managing app releases.

We will cover the following topics in this chapter:

- Preparing your apps for release
- Creating a developer account
- Uploading the app to Google Play
- Managing app releases

Preparing your apps for release

Android Studio normally signs your build using a debug key. This debug build allows you to build and test your app quickly. To publish your app on Google Play, you must create a release build signed with your own key. This release build will not be debuggable and can be optimized for size.

The release build must also have the correct version information. Otherwise, you won't be able to publish a new app or update an already published app.

Let's start with adding versions to your app.

Versioning apps

The version of your app is important for the following reasons:

- Users can see the version they have downloaded. They can use this when checking whether there's an update or whether there are known issues when reporting bugs/problems with the app.
- The device and Google Play use the version value to determine whether an app can or should be updated.
- Developers can also use this value to add feature support to specific versions. They can also warn or force users to upgrade to the latest version to get important fixes on bugs or security issues.

An Android app has two versions: `versionCode` and `versionName`. Now, `versionCode` is an integer that is used by developers, Google Play, and the Android system while `versionName` is the string that the users see on the Google Play page for your app.

The initial release of an app should have a `versionCode` value of 1, and you should increase it for each new release.

`versionName` can be in *x.y* format (where *x* is the major version, and *y* is the minor version). You can also use semantic versioning, as in *x.y.z*, by adding the patch version with *z*.

To learn more about semantic versioning, refer to https://semver.org.

In the module's `build.gradle` files, `versionCode` and `versionName` are automatically generated when you create a new project in Android Studio. They are in the `defaultConfig` block under the `android` block. An example `build.gradle` file shows these values:

```
android {
    compileSdk 33
    defaultConfig {
        applicationId "com.example.app"
        minSdk 21
        targetSdk 33
        versionCode 1
        versionName "1.0"
        ...
    }
    ...
}
```

When publishing updates, the new package being released must have a higher `versionCode` value because users cannot downgrade their apps and can only download new versions.

After ensuring that the app version is correct, the next step in the release process is to get a keystore to sign the app. This will be discussed in the next section.

Creating a keystore

Android apps, when run, are automatically signed by a debug key. However, before it can be published on Google Play Store, an app must be signed with a release key. To do so, you must have a keystore. If you don't have one yet, you can create one in Android Studio.

Exercise 17.01 – creating a keystore in Android Studio

In this exercise, we'll use Android Studio to make a keystore that can be used to sign Android apps. Follow these steps to complete this exercise:

1. Open a project in Android Studio.
2. Go to the **Build** menu and then click on **Generate Signed Bundle or APK**:

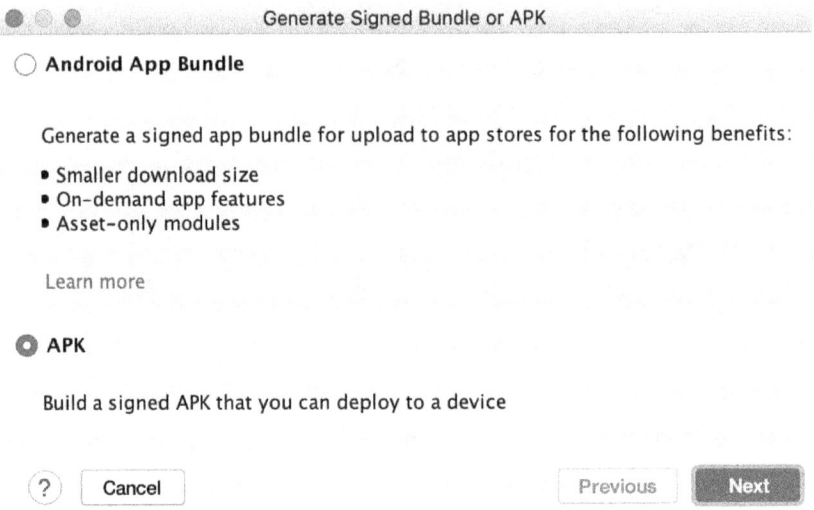

Figure 17.1 – The Generate Signed Bundle or APK dialog

An **APK** file is the file format by which users can install your app. The Android App Bundle is a new file publishing format that allows Google Play to distribute specific and smaller APKs to devices, so developers don't need to release and manage multiple APKs to support different devices.

3. Make sure either **APK** or **Android App Bundle** is selected, and then click the **Next** button. Here, you can choose an existing keystore or create a new one:

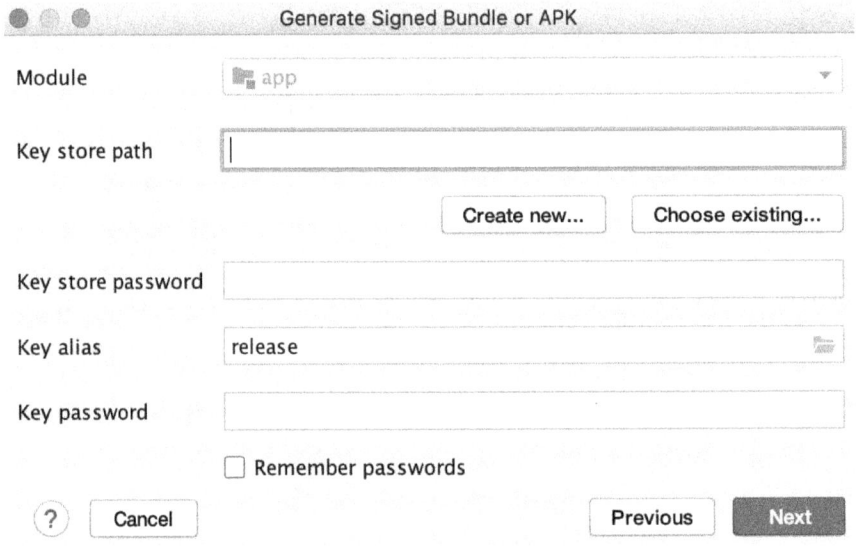

Figure 17.2 – The Generate Signed Bundle or APK dialog after selecting APK and pressing the Next button

4. Click the **Create new...** button. The **New Key Store** dialog will then appear:

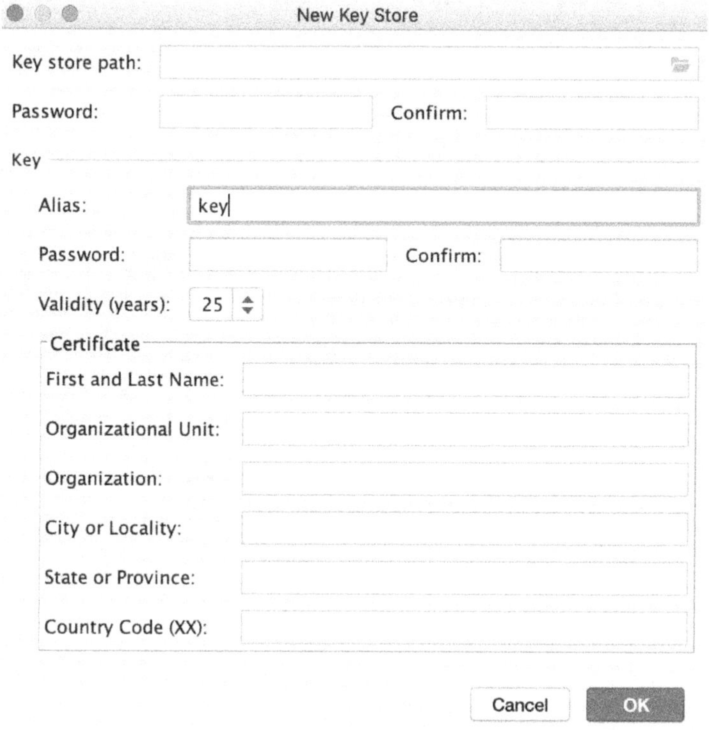

Figure 17.3 – The New Key Store dialog

5. In the **Key store path** field, choose the location where the keystore file will be saved. You can click on the folder icon on the right to select your folder and type the filename. The value will be similar to `users/packt/downloads/keystore.keystore`.

6. Provide the password in both the **Password** and **Confirm** fields.

7. In the **Certificate** section under **Key**, input values into the **First and Last Name**, **Organizational Unit**, **Organization**, **City or Locality**, **State or Province**, and **Country Code** fields. Only one of these is required, but it's good to provide all the information.

8. Click the **OK** button. If there is no error, the keystore will be created in the path you provided, and you will be back in the **Generate Signed Bundle or APK** dialog with the keystore values so you can continue generating the APK or app bundle. You could close the dialog if you only wanted to create a keystore.

In this exercise, you have created your own keystore, which you can use to sign applications to be published to Google Play.

You can also use the command line to generate a keystore if you prefer to use that. The `keytool` command is available in the **Java Development Kit (JDK)**. The command is as follows:

```
keytool -genkey -v -keystore my-key.jks -keyalg RSA -keysize
  2048 -validity 9125 -alias key-alias
```

This command creates a 2,048-bit RSA keystore in the current working directory, valid for 9,125 days (25 years), with a `my-key.jks` filename and an `key-alias` alias. You can change the validity, filename, and alias to your preferred values. The command line will prompt you to input the keystore password, then prompt you to enter it again to confirm.

It will then ask you for the first and last name, organizational unit, organization name, city or locality, state or province, and country code, one at a time. Only one of these is required; you can press the *Enter* key if you want to leave something blank. It is good practice, though, to provide all the information.

After the country code prompt, you will be asked to verify the input provided. You can type `yes` to confirm. You will then be asked to provide the password for the key alias. If you want it to be the same as the keystore password, you can press *Enter*. The keystore will then be generated.

Now that you have a keystore for signing your apps, you need to know how you can keep it safe. You'll learn about that in the next section.

Storing the keystore and passwords

You need to keep the keystore and passwords in a safe and secure place because if you lose the keystore and/or its credentials, you will no longer be able to release updates for your apps. If a hacker also gains access to these, they may be able to update your apps without your consent.

You can store the keystore in your CI/build server or on a secure server.

Keeping the credentials is a bit tricky, as you will need them later when signing releases for app updates. One way you can do this is by including this information in your project's `app/build.gradle` file.

In the `android` block, you can have `signingConfigs`, which references the keystore file, its password, and the key's alias and password:

```
android {
    ...
    signingConfigs {
        release {
            storeFile file("keystore-file")
            storePassword "keystore-password"
            keyAlias "key-alias"
```

```
            keyPassword "key-password"
        }
    }
    ...
}
```

Under the buildTypes release block in the project's build.gradle file, you can specify the release config in the signingConfigs block:

```
buildTypes {
    release {
        ...
        signingConfig signingConfigs.release
    }
    ...
}
```

Storing the signing configs in the build.gradle file is not that secure, as someone who has access to the project or the repository can compromise the app.

You can store these credentials in environment variables to make them more secure. With this approach, even if malicious people get access to your code, the app updates will still be safe as the signing configurations are not stored in your code but on the system. An environment variable is a key-value pair that is set outside your **integrated development environment** (IDE) or project, for example, on your own computer or on a build server.

To use environment variables for keystore configurations in Gradle, you can create environment variables for the store file path, store password, key alias, and key password. For example, you can use the KEYSTORE_FILE, KEYSTORE_PASSWORD, KEY_ALIAS, and KEY_PASSWORD environment variables.

On macOS and Linux, you can set an environment variable by using the following command:

```
export KEYSTORE_PASSWORD=securepassword
```

If you're using Windows, it can be done with this command:

```
set KEYSTORE_PASSWORD=securepassword
```

This command will create a KEYSTORE_PASSWORD environment variable with securepassword as the value. In the app/build.gradle file, you can then use the values from the environment variables:

```
storeFile System.getenv("KEYSTORE_FILE")
storePassword System.getenv("KEYSTORE_PASSWORD")
keyAlias System.getenv("KEY_ALIAS")
keyPassword System.getenv("KEY_PASSWORD")
```

Your keystore will be used to sign your app for release so you can publish it on Google Play. We'll discuss that in the next section.

Signing your apps for release

When you run an application on an emulator or an actual device, Android Studio automatically signs it with the debug keystore. To publish it on Google Play, you must sign the APK or app bundle with your own key, using a keystore you made in Android Studio or from the command line.

Suppose you have added the signing config for the release build in your build.gradle file; you can automatically build a signed APK or app bundle by selecting the release build in the **Build Variants** window. You then need to go to the **Build** menu, click on the **Build Bundle(s)** item, and then select either **Build APK(s)** or **Build Bundle(s)**. The APK or app bundle will be generated in the app/build/output directory of your project.

Exercise 17.02 – creating a signed APK

In this exercise, we will create a signed APK for an Android project using Android Studio:

1. Open a project in Android Studio.
2. Go to the **Build** menu and then click on the **Generate Signed Bundle or APK** menu item:

Preparing your apps for release   647

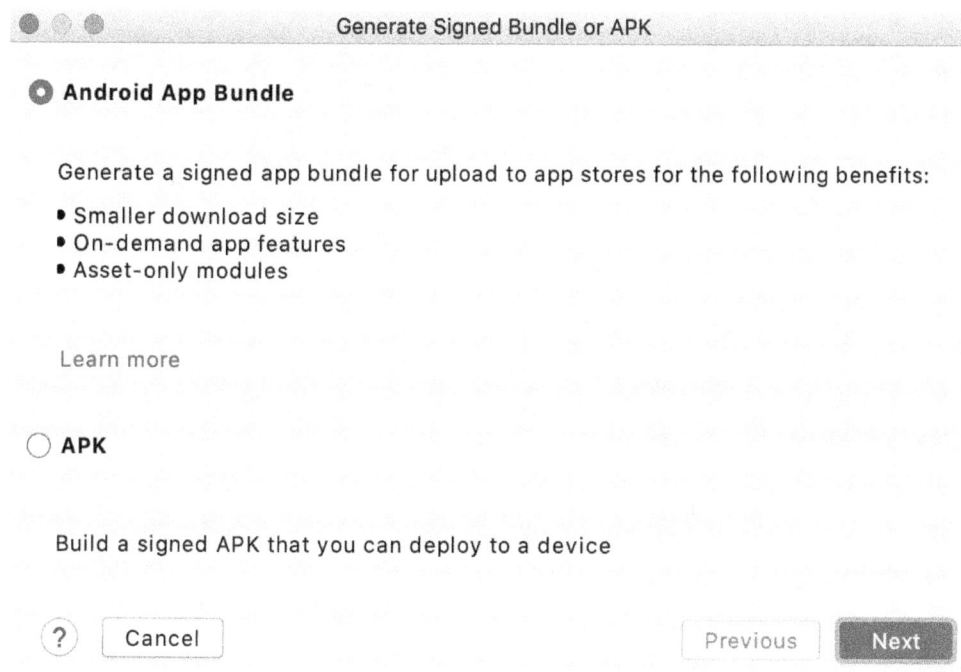

Figure 17.4 – The Generate Signed Bundle or APK dialog

3. Select **APK** and then click the **Next** button:

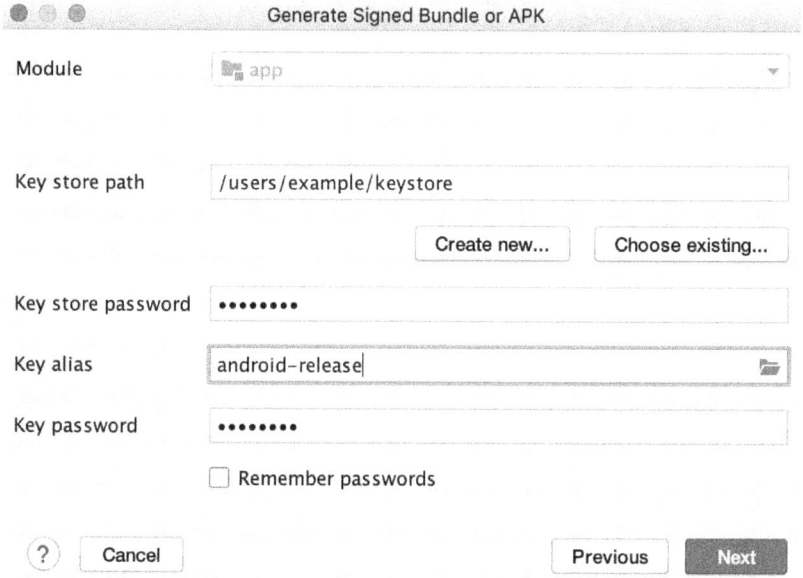

Figure 17.5 – The Generate Signed Bundle or APK dialog after clicking the Next button

4. Choose the keystore you made in *Exercise 17.01 – creating a keystore in Android Studio*.
5. Provide the password that you set for the keystore you created in the **Key store password** field.
6. In the **Key alias** field, click the icon on the right side and select the key alias.
7. Provide the alias password that you set for the keystore in the **Key password** field.
8. Click the **Next** button.
9. Choose the destination folder where the signed APK will be generated.
10. In the **Build Variants** field, make sure the **release** variant is selected:

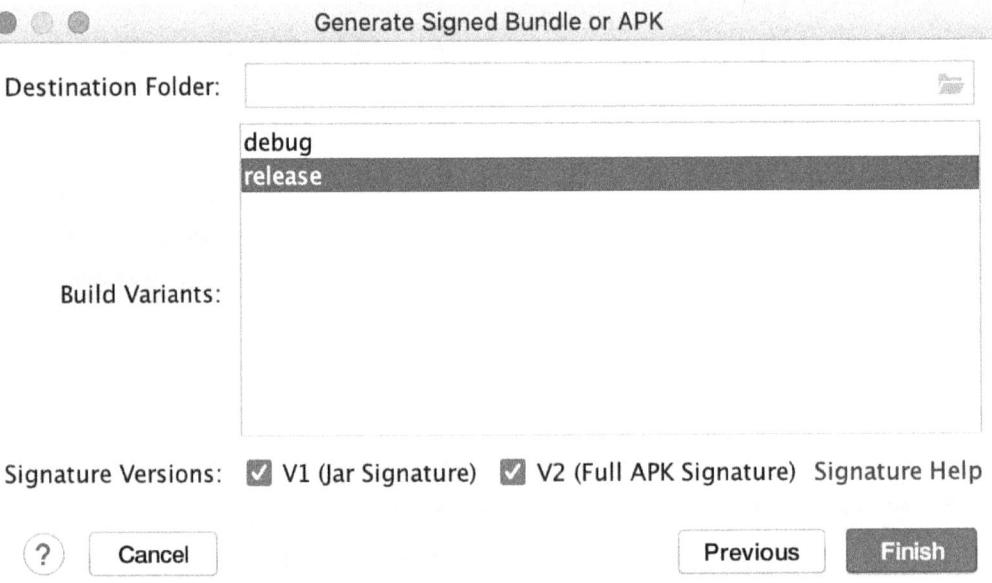

Figure 17.6 – Choose the release build in the Generate Signed Bundle or APK dialog

For the signature version, select both **V1** and **V2**. **V2 (Full APK Signature)** is a whole-file scheme that increases your app security and makes it faster to install. This is only available for Android 7.0 Nougat and above. If you are targeting a version lower than that, you should also use **V1 (Jar Signature)**, which is the old way of signing APKs but is less secure than **V2**.

11. Click the **Finish** button. Android Studio will build the signed APK. An IDE notification will inform you that the signed APK was generated. You can click on **locate** to go to the directory where the signed APK file is:

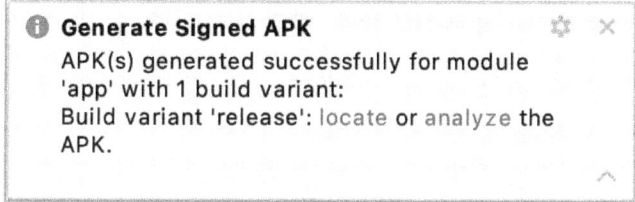

Figure 17.7 – A pop-up notification for successfully signed APK generation

In this exercise, you have made a signed APK, which you can now publish on Google Play. In the next section, you will learn about Android App Bundle, which is a new way of packaging apps for release.

Android app bundle

The traditional way of releasing Android apps is through an APK or an application package. This APK file is the one downloaded to users' devices when they install your app. This one big file contains all the strings, images, and other resources for all device configurations.

As you support more device types and more countries, this APK file will grow in size. The APK that users download will contain things that are not really needed for their devices. This will be an issue for you as users with low storage might not have enough space to install your app. Users with expensive data plans or slow internet connections might also avoid downloading the app if it's too big. They might also uninstall your app to save storage space.

Some developers have built an published multiple APKs to avoid these issues. However, it's a complicated and inefficient solution, especially when you target different screen densities, CPU architectures, and languages. In addition, that would be too many APK files to maintain per release.

Android App Bundle is a new way of packaging apps for publishing. You just generate a single app bundle file (using Android Studio 3.2 and upward) and upload it on Google Play. Google Play will automatically generate the base APK file and the APK files for each device configuration, CPU architecture, and language. Users who install your app will only download the necessary APKs for their device. This will be smaller in size compared to a universal APK.

This will work for devices running on Android 5.0 Lollipop and upward; for those below it, the APK files that will be generated are only for device configuration and CPU architecture. All the languages and other resources will be included in each APK file.

Exercise 17.03 – creating a signed app bundle

In this exercise, we will create a signed app bundle for an Android project using Android Studio:

1. Open a project in Android Studio.
2. Go to the **Build** menu, then click on the **Generate Signed Bundle or APK** menu item:

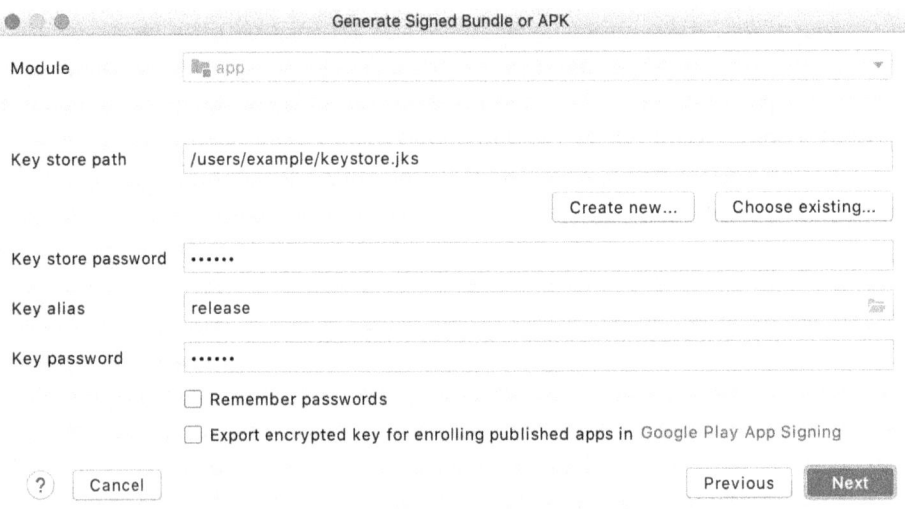

Figure 17.8 – The Generate Signed Bundle or APK dialog

3. Select **Android App Bundle**, then click the **Next** button:

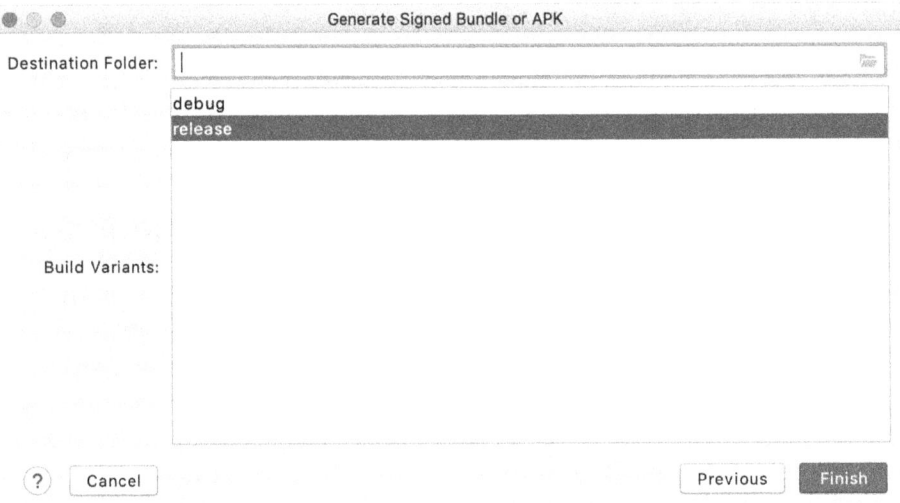

Figure 17.9 – The Generate Signed Bundle or APK dialog after clicking the Next button

4. Choose the keystore you made in *Exercise 17.01 – creating a keystore in Android Studio*.
5. Provide the password that you set for the keystore you created, in the **Key store password** field.
6. In the **Key alias** field, click the icon on the right side and select the key alias.
7. Provide the alias password that you set for the keystore you created in the **Key password** field.
8. Click the **Next** button.
9. Choose the destination folder to generate the signed app bundle.
10. In the **Build Variants** field, make sure the **release** variant is selected:

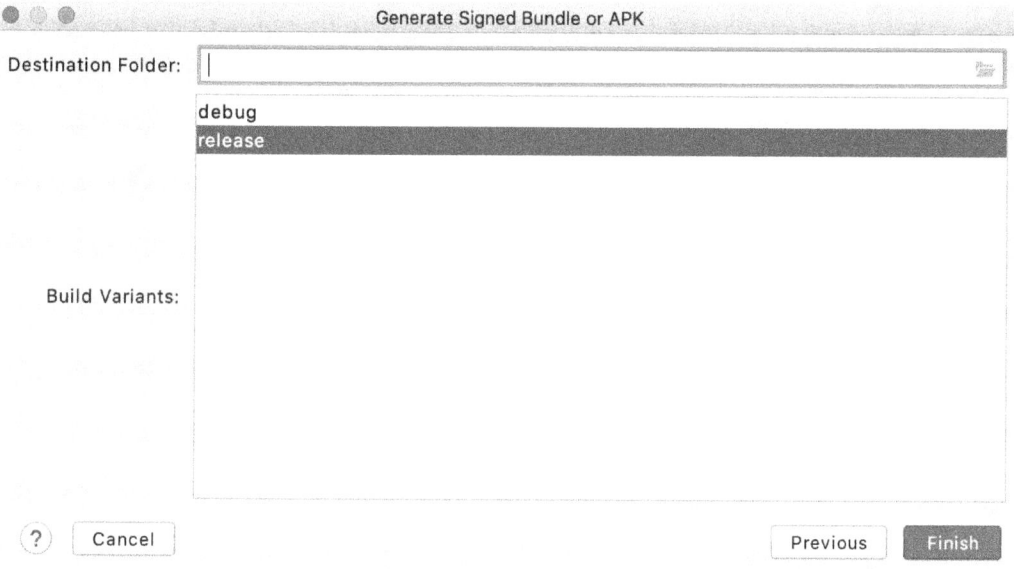

Figure 17.10 – Choose the release build in the Generate Signed Bundle or APK dialog

11. Click the **Finish** button. Android Studio will build the signed app bundle. An IDE notification will inform you that the signed app bundle was generated. You can click on **locate** to go to the directory where the signed app bundle file is:

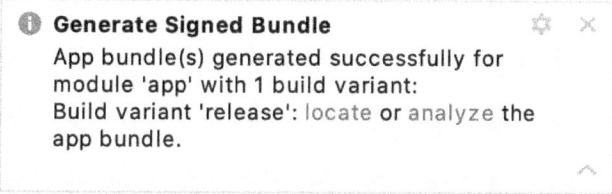

Figure 17.11 – Pop-up notification that the signed app bundle was generated

In this exercise, you have made a signed app bundle that you can now publish on Google Play.

To be able to publish your app to Google Play Store with the Android app bundle format, you will first need to opt-in to Google Play's app signing feature. We will discuss Google Play app signing in the next section.

App signing by Google Play

Google Play provides a service called app signing that allows Google to manage and protect your app signing keys and automatically re-sign your app for users.

With the Google Play app signing service, you can let Google generate the signing key or upload your own. You can also create a different upload key for additional security. You can sign the app with the upload key and publish the app on the Play Console.

Google will check the upload key, remove it, and use the app signing key to re-sign the app for distribution to users. When app signing is enabled for the app, the upload key can be reset. If you lose the upload key or feel that it is already compromised, you can simply contact Google Play developer support, verify your identity, and get a new upload key.

It is easy to opt into app signing when publishing a new app. In the Google Play Console (`https://play.google.com/console`), you can go to the **Release Management** | **App Releases** section and select **Continue** in the **Let Google manage and protect your app signing key** section. The key you originally used to sign the app will become the upload key, and Google Play will generate a new app signing key.

You can also configure existing apps to use app signing. This is available in the **Release** | **Setup** | **App Signing** section of the app in the Google Play Console. You need to upload your existing app signing key and generate a new upload key.

Once you enroll in Google Play app signing, you can no longer opt out. Also, if you use third-party services, you must use the app signing key's certificate. This is available in **Release Management** | **App Signing**.

App signing also enables you to upload an app bundle, and Google Play will automatically sign and generate APK files, which users will download when they install your app.

In the next section, you will create a Google Play developer account, so you can publish an app's signed APK or app bundle to Google Play.

Creating a developer account

To publish applications on Google Play, the first step that you need to take is to create a Google Play developer account. Head over to `https://play.google.com/console/signup` and log in with your Google account. If you don't have one, you should create one first.

We recommend using a Google account that you plan to use in the long term instead of a throwaway one. Read the developer distribution agreement and agree to the terms of service.

> **Note**
> If your goal is to sell paid apps or add in-app products to your apps/games, you must also create a merchant account. This is not available in all countries, unfortunately. We won't cover this here, but you can read more about it on the registration page or at `https://packt.link/LDncA`.

You must pay a $25 registration fee to create your Google Play Developer account. (This is a one-time payment). The fee must be paid using a valid debit/credit card, but some prepaid/virtual credit cards also work. What you can use varies by location/country.

The final step is to complete the account details, such as the developer's name, email address, website, and phone number. These, which can be updated later, will form the developer information displayed on your app's store listing.

After completing the registration, you will receive a confirmation email. It may take a few hours (up to 48 hours) for your payment to be processed and your account registered, so be patient. Ideally, you should do this in advance, even if your app is not yet ready, so you can easily publish the app once it's ready for release.

When you have received the confirmation email from Google, you can start publishing apps and games to Google Play.

In the next section, we will discuss uploading apps to Google Play.

Uploading an app to Google Play

Once you have an app ready for release and a Google Play Developer account, you can go to the Google Play Console (`https://play.google.com/console`) to publish the app.

To upload an app, go to the Play Console, click **All Apps**, and then click **Create app**. Provide the name of the application and the default language. In the **App or game** section, set whether it's an app or game. Likewise, in the **Free or paid** section, set whether it's free or paid. Create your store listing, prepare the app release, and roll out the release. We'll have a look at the detailed steps in this section.

Creating a store listing

The store listing is what users first see when they open your app's page on Google Play. If the app is already published, you can go to **Grow**, **Store presence**, and select **Main store listing**.

App Details

You will navigate to the **App details** page. On the **App details** page, you need to fill in the following fields:

- **App name**: Here, you provide your app's name (the maximum number of characters is 50).
- **Short description**: Here, you provide a short text summarizing your app (the maximum number of characters is 80).
- **Full description**: This is the long description for your app. The limit is 4,000 characters, so you can add a lot of relevant information here, such as what its features are and things users need to know.

> **Note**
>
> For the product details, you can add localized versions depending on the languages/countries where your app will be released.
>
> Your app title and description must not contain copyrighted materials and spam, as this might result in your app being rejected.

Graphic Assets

In this section, provide the following details:

- An icon (a high-resolution icon that is 512 x 512).
- A feature graphic (1,024 x 500).
- 2–8 screenshots of the app. If your app supports other form factors (tablet, TV, or Wear OS), you should also add screenshots for each form factor.

You can also add promo graphics and videos if you have any. Your app can be rejected if you use graphics that violate Google Play policy, so ensure that the images you use are your own and don't include copyrighted or inappropriate content.

Preparing the release

Before preparing your release, ensure that your build is signed with a signature key. If you're publishing an app update, make sure that it is of the same package name, signed with the same key, and with a version code higher than the current one on the Play Store.

You must also make sure you follow the developer policy (to avoid any violations) and make sure that your app follows the app quality guidelines. More of these are listed on the launch checklist, which you can see at `https://support.google.com/googleplay/android-developer/`.

APK/app bundle

You can upload an APK or the newer format: Android App Bundle. Go to **Release** and then **App Releases**. This will display a summary of active and draft releases in each track.

There are different tracks where you can release the app:

- Production
- Open testing
- Closed testing
- Internal testing

We'll discuss the release tracks in detail in this chapter's *Managing app releases* section.

Select the track where you will create the release. For the production track, you can select **Manage** on the left. For the other tracks, click **Testing** first, and then select the track. To release on a closed testing track, you must also select **Manage track** and then create a new track by clicking on **Create track**.

Once done, you can click **Create new release** at the top right of the page. In the **Android App Bundles and APKs to add** section, you can upload your APK or app bundle.

Make sure that the app bundle or APK file is signed by your release signing key. The Google Play Console will not accept it if it's not properly signed. If you're publishing an update, the version code for the app bundle or APK must be higher than the existing version.

You can also add a release name and release notes. The release name is for the developer's use to track the release and won't be visible to users. By default, the version name of the APK or app bundle uploaded is set as the release name. The release notes form the text that will be shown on the Play page and will inform users of what the updates to the app are.

The text for the release notes must be added inside the tags for the language. For example, the opening and closing tags for the default US English language are `<en-US>` and `</en-US>`. If your app supports multiple languages, each language tag will be displayed in the field for the release notes by default. You can then add the release notes for each language.

If you have already released the app, you can copy the release notes from previous releases and reuse or modify them by clicking the **Copy from a previous release** button and selecting from the list.

When you click the **Save** button, the release will be saved, and you can go back to it later. The **Review release** button will take you to the screen where you can review and roll out the release.

Rolling out a release

If you're ready to roll out your release, go to the Play Console and select your app. Go to **Release** and select your release track. Click the **Releases** tab and then click on the **Edit** button next to the release:

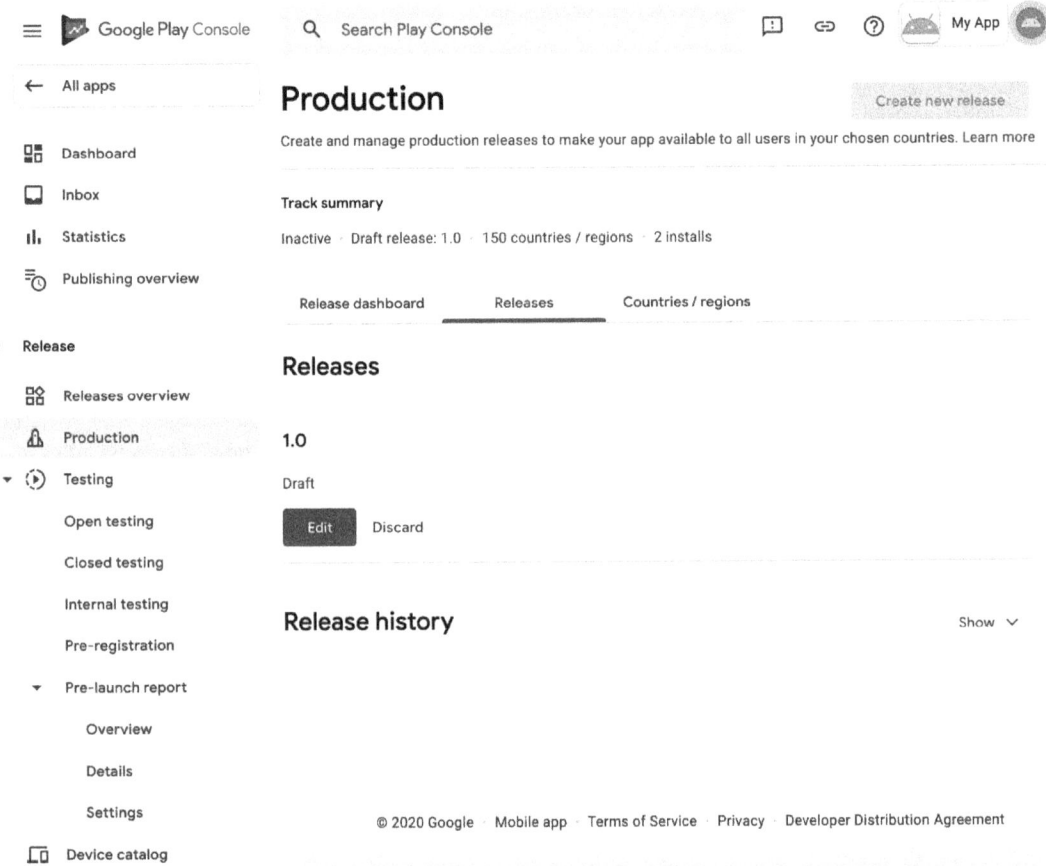

Figure 17.12 – A draft release on the Production track

You can review the APK or app bundle, release names, and release notes. Click the **Review release** button to start the rollout for the release. The Play Console will open the **Review and release** screen. Here, you can review the release information and check whether there are warnings or errors.

If you are updating an app, you can also select the rollout percentage when creating another release. Setting it to 100% means it will be available for all your users to download. When you set it to a lower percentage, for example, 50%, the release will be available to half of your existing users.

If you're confident with the release, select the **START ROLLOUT TO PRODUCTION** button at the bottom of the page. After publishing your app, it will take a while (7 days or longer for new apps) before it is reviewed. You can see the status in the top-right corner of the Google Play Console. These statuses include the following:

- **Pending publication** (your new app is being reviewed)
- **Published** (your app is now available on Google Play)
- **Rejected** (your app wasn't published because of a policy violation)
- **Suspended** (your app violated Google Play policy and was suspended)

If there are issues with your app, you can resolve them and resubmit the app. Your app can be rejected for reasons such as copyright infringement, impersonation, and spam.

Once the app has been published, users can now download it. It can take some time before the new app or the app update becomes live on Google Play. If you're trying to search for your app on Google Play, it might not be searchable. Make sure you publish it on the production or open track.

Managing app releases

You can slowly release your apps on different tracks to test them before publicly rolling them out to users. You can also do timed publishing to make the app available on a certain date instead of automatically publishing it once approved by Google.

Release tracks

When creating a release for an app, you can choose between four different tracks:

- Production is where everyone can see the app.
- Open testing is targeted at wider public testing. The release will be available on Google Play, and anyone can join the beta program and test.
- Closed testing is intended for small groups of users testing pre-release versions.
- Internal testing is for the developer/tester builds while developing/testing an app.

The internal, closed, and open tracks allow developers to create a special release and allow real users to download it while the rest are on the production version. This allows you to know whether the release has bugs and quickly fix them before rolling it out to everyone. User feedback on these tracks will also not affect the public reviews/ratings of your app.

The ideal way is to release it first on internal tracks during development and internal testing. When a pre-release version is ready, you can create a closed test for a small group of trusted people/users/testers. Then, you can create an open test to allow other users to try your app before the full launch to production.

To go to each track and manage releases, you can go to the **Release** section of the Google Play Console and select **Production** or **Testing** and then the **Open testing**, **Closed testing**, or **Internal testing** tracks.

The feedback channel and opt-in link

In the internal, closed, and open tracks, there is a section for **Feedback URL or email address** and **How testers join your test**. You can provide an email address or a website under **Feedback URL or email address** to which testers can send their feedback. This is displayed when they opt into your testing program.

In the **How testers join your test** section, you can copy the link to share with your testers. They can then join the testing program using this link.

Internal testing

This track is for builds while developing/testing the app. Releases here will be quickly available on Google Play for internal testers. In the **Testers** tab, there's a **Testers** section. You can choose an existing list or create a new one. There is a maximum of 100 testers for an internal test.

Closed testing

On the **Testers** tab, you can choose **Email list** or **Google Groups** for the testers. If you choose email lists, choose a list of testers or create a new list. There is a maximum of 2,000 testers for a closed test.

If you select **Google Groups**, you can provide the Google Group email address (for example, `the-alpha-group@googlegroups.com`), and all the members of that group will become testers.

Open testing

In the **Testers** tab, you can set **Unlimited** or **Limited number** for the testers. The minimum number of testers for the limited testing that you can set is 1,000.

In the open, closed, and internal tracks, you can add users to be your testers for your applications. You will learn how to add testers in the next section.

Staged rollouts

When rolling out app updates, you can release them to a small group of users first. Then, when the release has issues, you can stop the rollout or publish another update to fix the issues. If there are none, you can slowly increase the rollout percentage. This is called **staged rollout**.

If you have published an update to less than 100% of your users, you can go to the Play Console, select **Release**, click the track, then select the **Releases** tab. Below the release you want to update, you can see the **Manage rollout** drop-down menu. It will have options to update or halt the rollout.

You can select **Manage rollout**, then **Update rollout** to increase the percentage of the rollout of the release. A dialog will appear where you can input the rollout percentage. You can also click the **Update** button to update the percentage.

A 100% rollout will make the release available to all of your users. Any percentage below that means the release will only be available to that percentage of your users.

If a major bug or crash is found during a staged rollout, you can go to the Play Console, select **Release**, click the track, then select the **Releases** tab. Under the release you want to update, select **Manage rollout**, then **Halt rollout**. A dialog will appear with additional information. Add an optional note, then click the **Halt rollout** button to confirm:

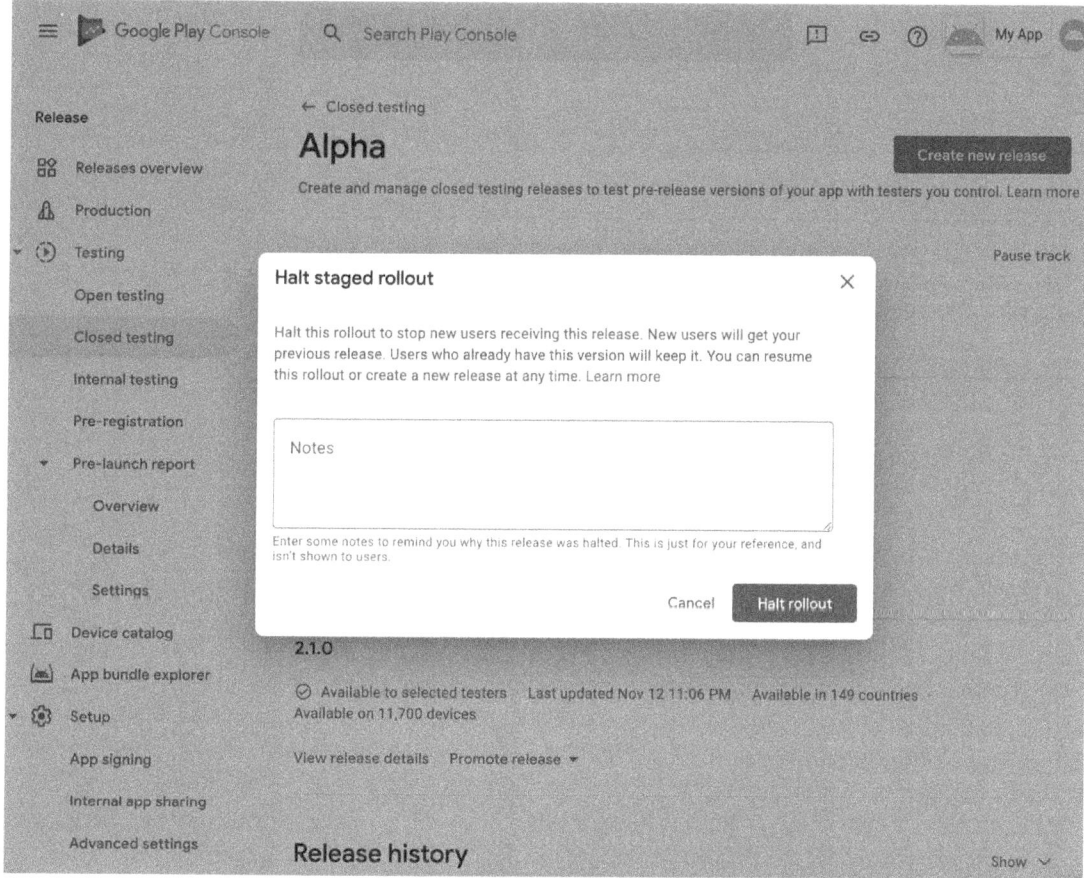

Figure 17.13 – The dialog for halting a staged rollout

When a staged rollout is halted, the release page in your track page will be updated with **Rollout halted** and a **Resume rollout** button:

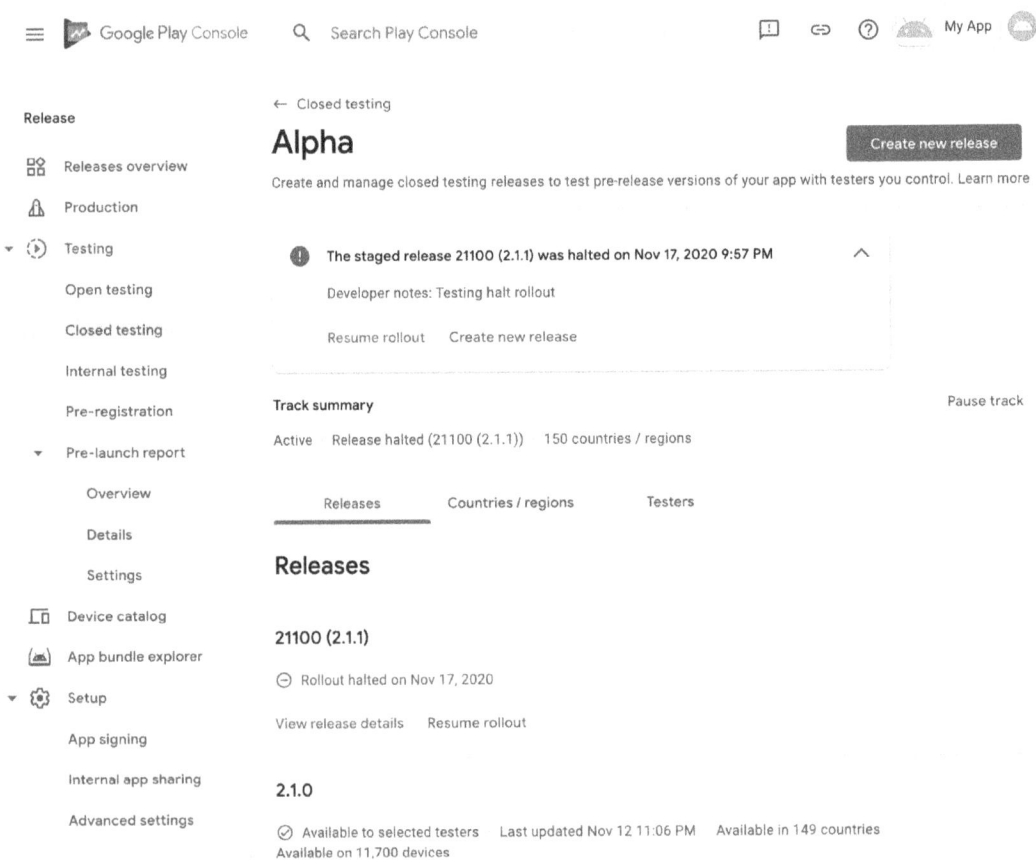

Figure 17.14 – The release page for a halted staged rollout

If you have fixed the issue, for example, in the backend, and there's no need to release a new update, you can resume your staged rollout. To do that, go to the Play Console, select **Release**, click the track, then select the **Releases** tab. Choose the release and click the **Resume rollout** button. In the **Resume staged rollout** dialog, you can update the percentage and click **Resume rollout** to continue the rollout.

Managed publishing

When you roll out a new release on Google Play, it will be published in a few minutes. You can change it to be published at a later time. This is useful when targeting a specific day, for example, the same day as an iOS/web release or after a launch date.

Managed publishing must be set up before creating and releasing the update for which you want to control the publishing. When you select your app on the Google Play Console, you can select **Publishing Overview** on the left side. In the **Managed publishing** section, click on the **Turn on managed publishing** button:

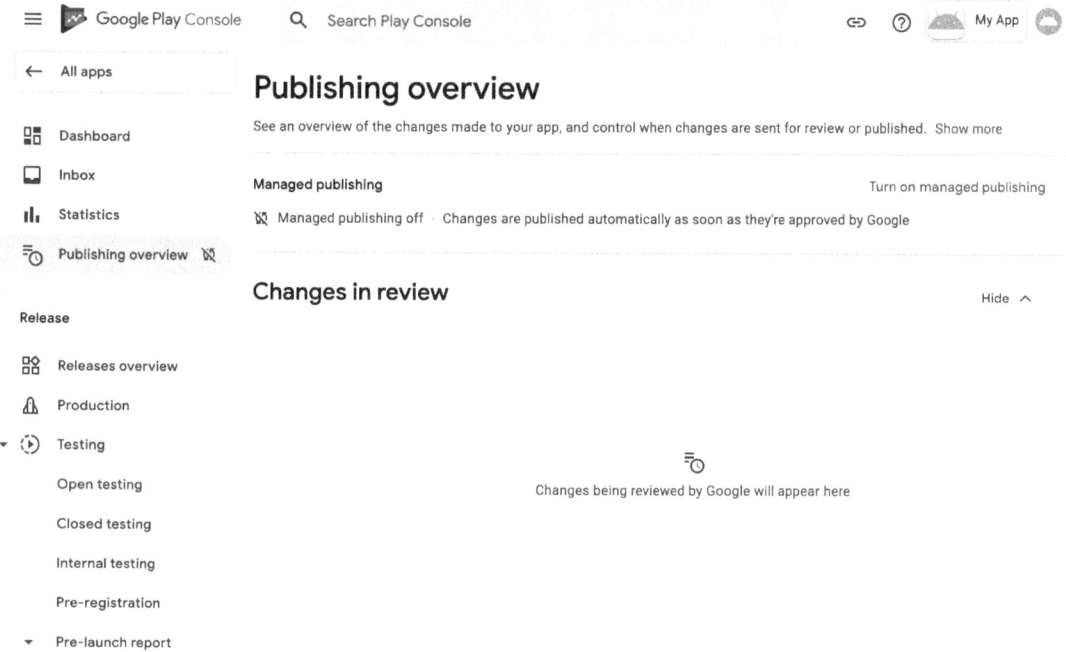

Figure 17.15 – Managed publishing in Publishing overview

Managed publishing dialog will be displayed. Here, you can turn managed publishing on or off, then click the **Save** button.

When you turn on **Managed publishing**, you can continue adding and submitting updates to the app. You can see these changes in **Publishing overview** under the **Changes in review** section.

Once the changes have been approved, **Changes in review** will be empty and moved to the **Changes ready to publish** section. There, you can click on the **Publish changes** button. In the dialog that appears, you can click on the **Publish changes** button to confirm. Your update will then be published instantly:

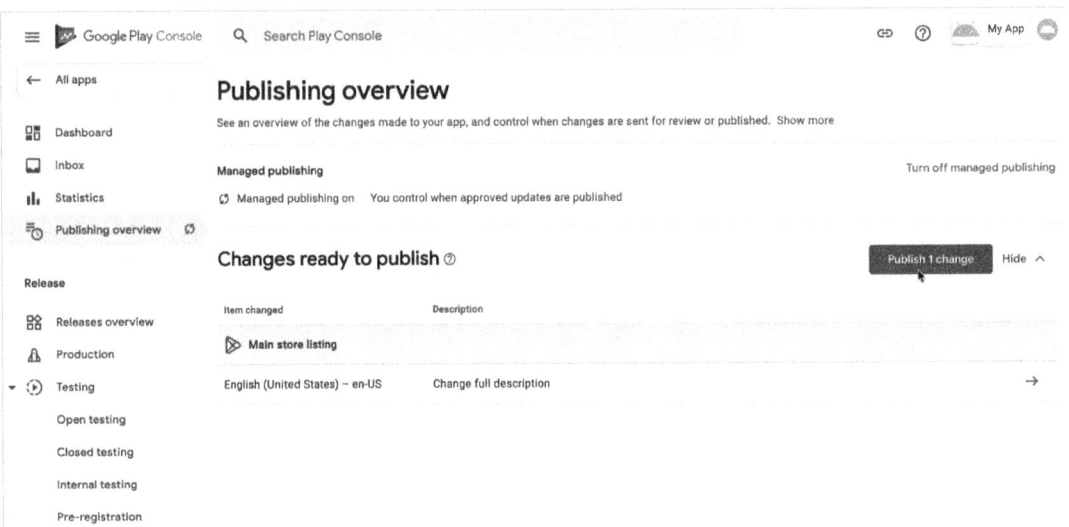

Figure 17.16 – Changes ready to publish

In this section, you learned about the release tracks where you can test your releases, perform staged rollouts for your releases, and manage your publishing time.

Let's test everything you've learned by doing an activity.

Activity 17.01 – publishing an app

As the final activity of this book, you are tasked with creating a Google Play Developer account and publishing a newly developed Android app that you have built. You could publish one of the apps you've built as part of this book or another project that you've developed. You can use the following steps as guidelines:

1. Go to the Google Play Developer Console (`https://play.google.com/console`) and create an account.
2. Create a keystore that you can use for signing the release build.
3. Generate an Android app bundle for release.
4. Publish the app on an open beta track before releasing it to the production track.

> **Note**
> The detailed steps for publishing an app have been explained throughout this chapter, so no separate solution is available for this activity. You can follow the exercises of this chapter to successfully complete the preceding steps. The exact steps required will be unique to your app and will depend on the settings you want to use.

Summary

This chapter covered the Google Play Store: from preparing a release to creating a Google Play Developer account and finally publishing your app. We started with versioning your apps, generating a keystore, creating an APK file or Android app bundle and signing it with a release keystore, and storing the keystore and its credentials. We then moved on to registering an account on the Google Play Console, uploading your APK file or app bundle, and managing releases.

This is the culmination of the work done throughout this book—publishing your app and opening it up to the world is a great achievement and demonstrates your progress throughout this course.

Throughout this book, you have gained many skills, from the basics of Android app development and building up to implementing features such as RecyclerViews, fetching data from web services, notifications, and testing. You have seen how to improve your apps with best practices, architecture patterns, and animations, and finally, you have learned how to publish them to Google Play.

This is still just the start of your journey as an Android developer. There are many more advanced skills for you to develop as you continue to build more complex apps of your own and expand upon what you have learned here.

Remember that Android is continuously evolving, so keep yourself up to date with the latest Android releases. You can go to `https://developer.android.com/` to find the latest resources and further immerse yourself in the Android world.

Index

A

activity
interacting, with intents 77
starting, with shared element transition 611, 612

Activity Callbacks
logging 55-62

Activity lifecycle 32, 52, 53
callback 55

Activity state
restoring 63
restoring, in layouts 63-71
restoring, with Callbacks 71-77
saving 63
saving, in layouts 63-71
saving, with Callbacks 71-77

activity transitions 606
activity, starting with 608
adding, through code 607
adding, through XML 606, 607
creating, in app 608, 609
enter transition 606
exit transition 606

Add A Cat button
implementing 274-277

Alpha, Red, Green, and Blue (ARGB) 302

Android
coroutines, using 562, 563
data binding 588-590
data binding, implementing 590-593
Kotlin Flow, collecting 574-576
Kotlin Flow, using 573, 574
WorkManager, adding to 598-600

Android App Bundle 642, 649

Android application
coroutines, using 565-570
input validation 46, 47
interactive UI elements, adding to display bespoke greeting to user 38-45
Kotlin Flow, using 578-581
producing, to create RGB colors 47
structure 30-38
Views, accessing in layout files 46

Android Emulator 9

Android Jetpack
components 442, 443
Data streams 453
Room 463
ViewModel 443-445

Android Manifest 16, 17
internet permission, configuring 18-22

Android Package (APK) file 642
Android project
  creating, with Android Studio
Android SDK Build-Tools 9
Android SDK Platform 9
Android SDK Platform-Tools 9
Android SDK Tools 9
Android software development kit (SDK) components 8
Android Studio
  keystore, creating 641-644
  project, creating for app 4-8
  testing, tips 382-386
animations
  adding, with MotionLayout 618-620
  creating, with MotionLayout 617
  path, modifying with keyframes 629-635
  with CoordinatorLayout 614-616
  with MotionLayout 616
API response
  image URL, extracting from 215-218
APK/App bundle 655
app
  bottom navigation, adding to 181-191
  creating, to find location of parked car 308, 309
  creating, with navigation drawer 164-181
  releases, managing 657
  signing, by Google Play 652
  tests, types 373
app-level build.gradle file 23-27
Application Not Responding (ANR) time window 321
application programming interface (API) 203-205
  data, reading from 208-213

app navigation
  tabbed navigation, using 191-198
apps
  preparing, for release 640
  publishing 662
  signing, for release 646
  uploading, to Google Play 653
  versions, adding to 640, 641
ARGB color space 36
Arrange-Act-Assert (AAA) 375
asset files 500

B

background task
  starting, with WorkManager 312-315
background work
  executing, with WorkManager class 315-321
bottom navigation 181
  adding, to app 181-191
built-in transitions
  explode 606
  fade 606
  slide 606

C

camera storage 509, 510
click events
  responding to 253-256
  responding to, in RecyclerView 251, 252
code
  activity transitions, adding through 607
connectors 534, 535
consumers 532
context leak 441
continuous integration 384

converters 207
CoordinatorLayout
    for animations 614-616
coroutines
    adding, to project 564
    creating 563
    using, in Android app 565-570
    using, on Android 562, 563
create, read, update, and delete
        (CRUD) operation 463
current weather
    displaying 228, 229
custom markers 299-303
    adding, where map was clicked 303-308

D

Dagger 524
Dagger 2 531
    applying 538-543
    connector 532-535
    consumer 532
    functionalities 532
    provider 532, 533
    qualifiers 535, 536
    scopes 536, 537
    subcomponents 537, 538
data
    fetching, from network endpoint 205-208
    reading, from API 208-213
data access object (DAO) 463, 467-469
data binding
    implementing, on Android 590-593
    on Android 588-590
DataStore 484-491

preferences, creating 491-495
data streams 461-463
    LiveData 454, 455
debug build 640
declarative approach 338
dependency injection (DI) 373, 524
    benefits 524
dependency inversion 373
developer account
    creating 652, 653
Device File Explorer 496
dog downloader 518-520
domain-specific language (DSL) 550
double integration 403-411
dual-pane layouts
    with static fragments 128-145
dynamic fragments 145
    adding, to activity 146-150

E

end-to-end tests 372
enter transition 606
entities 464-467
Espresso 401, 402
executors and threads 473, 474
exit transition 606
explicit intent 77
Extensible Markup Language
        (XML) 203-205
external storage 498, 499

F

factory notation 551
feature modules 27
FileProvider 499, 500
files 495
  asset files 500
  copying 501-508
  external storage 498, 499
  internal storage 496-498
  Storage Access Framework (SAF) 500
Flow Builders
  asflow() 576
  flow() 576
  flowOf() 576
  used, for creating Kotlin Flow 576, 577
Foreground Service
  SCA's work, tracking with 326-334
  using 321-326
fragment lifecycle 104, 105, 161
  adding 108-116
  adding, statically to activity 117-127
  onActivityCreated 106
  onAttach 105
  onCreate 106
  onCreateView 106
  onDestroy 107
  onDestroyView 107
  onDetach 107
  onPause 107
  onResume 106
  onStart 106
  onStop 107
  onViewCreated 106
Fresco
  URL 218

G

Glide
  URL 218
Google Play
  app, signing 652
  apps, uploading to 653
Gradle 22
  using, to manage app dependencies 22
gratuity 608
GSON 205

H

Hilt 524, 543
  using 544-549
Hypertext Transfer Protocol (HTTP) 204

I

image
  loading, from obtained URL 221-228
images
  loading, from remote URL 218-221
image URL
  extracting, from API response 215-218
imperative approach 337
implicit intents 77
instrumentation framework 412
integrated development environment (IDE) 4, 645
integration tests 372, 396
  double integration 403-411
  libraries 396
integration tests, libraries
  Espresso 401, 402
  Robolectric 396-400

IntelliJ IDEA IDE 4
Intel x86 Emulator Accelerator
    (HAXM installer) 9
intents 77
    for retrieving result, from Activity 84-95
    using 78-83
internal storage 496-498
items
    swiping, to remove from
        RecyclerView 266-269

J

Jackson 205
Java Development Kit (JDK) 644
JavaScript Object Notation (JSON) 203-205
    response, parsing 213-215
Java virtual machine (JVM) 22
Jetpack 7
Jetpack Compose 337-341
    adding, to existing projects 363-366
    screen 342-345
    testing 417-420
    theming 350-357
    used, for creating app 366, 367
Jetpack navigation 150
    graph, adding 151-155
    quiz, creating on planets 156-159
JUnit 374-381
    features 374
JUnit library 372

K

keyframes
    animation path, modifying with 629-635
keystore
    creating 641

creating, in Android Studio 641-644
saving 644-646
keytool command 644
Koin 524, 550
    injected repositories 556, 557
    using 553-556
    working 550-552
Kotlin 4
Kotlin Flow
    collecting, on Android 574-576
    creating, with Flow Builders 576, 577
    operators, using 577, 578
    TV Guide app, creating 581-584
    using, in Android application 578-581
    using, on Android 573, 574

L

launch mode 95
    reference link 96
    setting 96-100
layout
    RecyclerView, adding to 232-235
LiveData 454, 455
    observing with 456-461
LiveData transformation 570-572
location permission
    requesting, from user 286-292
login form
    creating 101

M

managed publishing 660
Manual DI 524
    applying 527-531
    working 524-526

map clicks
   responding 299-303
Material Design 27
   app, theming 27-30
Material Design Components (MDC) 28
mdpi 35
media storage 509, 510
migration 470
mocking 386
Mockito 386-392
   sum of numbers, testing 392-395
Model-View-Controller (MVC) 585
Model-View-Presenter (MVP) 585
Model-View-ViewModel
   (MVVM) 208, 243, 585-588
   data binding, on Android 588-590
Moshi 205
   using 593
Motion Editor 620-625
MotionLayout
   adding 616, 617
   animations, adding with 618-620
   animations, creating with 617
   debugging 625-627
   for animations 616
   path, modifying 627-629

N

navigation
   overview 162
navigation drawer 162, 163
   used, for creating App 164-181
network endpoint
   data, fetching from 205-208

O

Okio 498
onActivityCreated 106
onAttach 105
onCreate 106
onCreateView 106
onDestroy 107
onDestroyView 107
onDetach 107
one-to-many relationship 475
onPause 107
onResume 106
onStart 106
onStop 107
onViewCreated 106
operators
   using, with Kotlin Flow 577, 578
orchestrator framework 412

P

Password Generator 636, 637
passwords
   saving 644-646
permission
   requesting, from user 282-286
permissions, categories
   dangerous 17
   normal 17
   signature 17
photos
   saving, to external storage with
      FileProvider 511-517
   taking 511-517
Picasso
   URL 218

primary and secondary app navigation
  building 198
project-level build.gradle file 22, 23
Project Object Model (POM) 26
providers 533

Q

qualifiers 535, 536

R

recomposition 338
RecyclerView
  Add A Cat button, implementing 274-277
  adding, to layout 232-235
  items, adding interactively 272-274
  items, swiping to remove from 266-269
  item types, supporting 256-260
  list of items, managing 277-279
  populating 235-251
  responding, to click events in 251-256
  swipe, adding to delete
    functionality 269-272
  titles, adding to 260-266
release
  apps, preparing for 640
  apps, signing for 646
  preparing 654
  rolling out 656, 657
release build 640
release tracks 657, 658
  closed testing 658
  feedback channel 658
  internal testing 658

open testing 658
opt-in link 658
reminder to drink water activity 334, 335
remote URL
  images, loading from 218-221
Repository pattern
  implementing 593, 594
Repository pattern, with Room
  using, in Android 595-598
representational state transfer
    (REST) 203-205
  reference link 204
Retrofit 204
  reference link 206
  using 593
Robolectric 396-400
  features 397
Robot 414
Room 463
  creating 475-480
  DAO 467-469
  database, setting up 469-473
  entities 464-467
  relationship, with application 464
  third-party frameworks 473, 474
RxJava 473

S

scoped storage 508, 509
  camera storage 509, 510
  dog downloader 518-520
  media storage 509, 510
  photos, taking 511-517
scopes 536
SDK Patch Applier v4 9

Secret Cat Agents (SCAs) 311
  work, tracking with Foreground Service 326-334
semantic versioning
  reference link 640
services 312
shared element transition
  activity, starting with 611, 612
  adding 610, 611
  creating 612-614
SharedPreferences 484, 485
  wrapping 485-489
shopping notes app
  building 480-482
signed APK
  creating 646-648
signed app bundle
  creating 650-652
single notation
  using 551
staged rollouts 658, 659
state hoisting 346
static fragments 127
  dual-pane layouts with 128-145
Storage Access Framework (SAF) 500
store listing
  App Details 654
  creating 653
  Graphic Assets 654
subcomponents 537, 538

T

tabbed navigation 191
  primary and secondary app navigation, building 198
  using, for app navigation 191-198

test-driven development (TDD) 372
  developing with 436, 437
  testing 431, 432
  using, to calculate sum of numbers 433-436
testing
  types 372-374
testing pyramid 373
testing, types
  Espresso 372
  Mockito 372
  Robolectric 372
themes
  applying 357-363
titles
  adding, to RecyclerView 260-266

U

UI tests 412-417
  Jetpack Compose, testing 417-420
  random waiting times 420-431
Uniform Resource Locators (URLs) 205
unit tests 372
user
  actions, handling 345-350
  location permission, requesting from 286-292
  permission, requesting from 282-286
user interface (UI) 443, 528, 561, 585
user's current location
  obtaining 292-298

V

versions
  adding, to apps 640, 641
ViewBinding 46

ViewModel 443, 445
  analyzing 445-453
virtual device
  setting up 8-15

W

WorkManager 598
  adding, to Android 598-600
  TV Guide app, revisiting 601, 602
  used, for starting background task 312-315
  using 598
WorkManager class
  used, for executing background
    work 315-321

X

XML
  activity transitions, adding through 606, 607

www.packtpub.com

Subscribe to our online digital library for full access to over 7,000 books and videos, as well as industry leading tools to help you plan your personal development and advance your career. For more information, please visit our website.

Why subscribe?

- Spend less time learning and more time coding with practical eBooks and Videos from over 4,000 industry professionals
- Improve your learning with Skill Plans built especially for you
- Get a free eBook or video every month
- Fully searchable for easy access to vital information
- Copy and paste, print, and bookmark content

Did you know that Packt offers eBook versions of every book published, with PDF and ePub files available? You can upgrade to the eBook version at www.packtpub.com and as a print book customer, you are entitled to a discount on the eBook copy. Get in touch with us at customercare@packtpub.com for more details.

At www.packtpub.com, you can also read a collection of free technical articles, sign up for a range of free newsletters, and receive exclusive discounts and offers on Packt books and eBooks.

Other Books You May Enjoy

If you enjoyed this book, you may be interested in these other books by Packt:

Kotlin Design Patterns and Best Practices

Alexey Soshin

ISBN: 978-1-80181-572-7

- Implement all the classical design patterns using the Kotlin programming language
- Apply reactive and concurrent design patterns to make your application more scalable
- Discover best practices in Kotlin and explore its new features
- Understand the key principles of functional programming and learn how they apply to Kotlin
- Find out how to write idiomatic Kotlin code and learn which patterns to avoid
- Harness the power of Kotlin to design concurrent and reliable systems with ease
- Create an effective microservice with Kotlin and the Ktor framework

Simplifying Android Development with Coroutines and Flows

Jomar Tigcal

ISBN: 978-1-80181-624-3

- Understand how coroutines and flows differ from existing ways
- Apply asynchronous programming in Android with coroutines and flows
- Find out how to build your own coroutines and flows in Android
- Handle, manipulate, and combine data in coroutines and flows
- Handle cancellations and exceptions from coroutines and flows
- Discover how to add tests for your coroutines and flows
- Integrate coroutines and flows into your Android projects

Packt is searching for authors like you

If you're interested in becoming an author for Packt, please visit `authors.packtpub.com` and apply today. We have worked with thousands of developers and tech professionals, just like you, to help them share their insight with the global tech community. You can make a general application, apply for a specific hot topic that we are recruiting an author for, or submit your own idea.

Share your thoughts

Now you've finished *How to Build Android Apps with Kotlin, Second Edition*, we'd love to hear your thoughts! Scan the QR code below to go straight to the Amazon review page for this book and share your feedback or leave a review on the site that you purchased it from.

`https://www.amazon.in/review/create-review/error?asin=1837634939`

Your review is important to us and the tech community and will help us make sure we're delivering excellent quality content.

Download a free PDF copy of this book

Thanks for purchasing this book!

Do you like to read on the go but are unable to carry your print books everywhere?

Is your eBook purchase not compatible with the device of your choice?

Don't worry, now with every Packt book you get a DRM-free PDF version of that book at no cost.

Read anywhere, any place, on any device. Search, copy, and paste code from your favorite technical books directly into your application.

The perks don't stop there, you can get exclusive access to discounts, newsletters, and great free content in your inbox daily

Follow these simple steps to get the benefits:

1. Scan the QR code or visit the link below

https://packt.link/free-ebook/9781837634934

2. Submit your proof of purchase
3. That's it! We'll send your free PDF and other benefits to your email directly